# Gauge Theories of the Strong, Weak, and Electromagnetic Interactions

||||||||||||||||||||||||||||||||||||||||||||||||||||||||||||||||||||||||||||

## SECOND EDITION

# Gauge Theories of the Strong, Weak, and Electromagnetic Interactions

## SECOND EDITION

# Gauge Theories of the Strong, Weak, and Electromagnetic Interactions

### SECOND EDITION

## Chris Quigg

PRINCETON UNIVERSITY PRESS • PRINCETON AND OXFORD

«Où habitez-vous ? — Au voisinage de l'inconnu.»

«Que faites-vous ? — Je cherche à deviner la présence des secrets.»

— Jean-Bertrand Pontalis, *En marge des jours*

# Contents ||||||||||||||||||||||||||||||||||||||||||||||||||||||||||||||||||||||||||||||||

# Contents ||||||||||||||||||||||||||||||||||||||||||||||||||||||||||||||||||||||||||||||||||||||

# Preface ||||||||||||||||||||||||||||||||||||||||||||||||||||||||||||||||||||||||||||||||||||||||||||||||||||||||||||||||||||||||||||||||||||

$\mathbf{M}$y purpose in writing this book has been to present a straightforward introduction to the idea of gauge theories and to two new laws of nature—quantum
chromodynamics and the electroweak theory. By straightforward, I mean that little
effort will be expended on field-theoretic technicalities and that correspondingly
little formal sophistication will be demanded of the reader. The physical situations
dealt with are, however, "advanced" topics of current interest in theoretical and
experimental research. An assiduous reader will, I hope, come to understand
the logic of local gauge symmetries and gauge theories, acquire the ability to
compute the consequences of these theories, and gain a perspective on the body of
experimental evidence and today's open questions. In particular, the reader should
be fully prepared to take part in the great wave of exploration led by experiments at
the Large Hadron Collider.

The idea of gauge theories is rooted in the classic investigations of Hermann
Weyl in the 1920s, and the notion that symmetries generate interactions was given
full expression in the work of Yang and Mills a quarter-century later. More recently,
the principle of local gauge invariance has blossomed into a unifying theme that
seems capable of embracing and even synthesizing all the elementary interactions.
The emergence of gauge theories has been coupled with the recognition that a
fundamental description of the subnuclear particles must be based upon the idea
that the strongly interacting particles, or hadrons, are composed of quarks. Together
with leptons, such as the electron and neutrino, quarks seem to be the elementary
particles—structureless and indivisible—at least at the present limits of resolution.

Thus we possess today a coherent point of view and a single language
appropriate for the description of all subnuclear phenomena. This development,
which is the work of many hands, has not only made of particle physics a much
more unified subject, it has also helped us to perceive common interests and to
make common cause with other specialties, notably astrophysics and cosmology,
condensed-matter physics, atomic physics, and nuclear physics.

Experimental support for the new paradigm of quarks and leptons with interactions prescribed by gauge symmetries is impressive in its consistency, diversity,
and strength. The case for quarks consists in the spectroscopy of hadrons, the
evidence for pointlike constituents within hadrons, the $\psi/J$ and $\Upsilon$ families of
heavy mesons with their quasi-atomic spectra, and more. In support of gauge
theories and the unification of elementary interactions, we may cite the triumphs
of the unified theory of weak and electromagnetic interactions with its implication
of neutral weak currents and corollary prediction of charm, its prediction of the
electroweak gauge bosons $W^{\pm}$ and $Z^0$, and a host of quantitative successes. The
theory of strong interactions, quantum chromodynamics, has been validated in
many particulars: the scale dependence of the strong coupling (known as asymptotic

freedom), precise perturbative calculations, and a deepening understanding of the low-energy regime in which the strong interactions are, indeed, strong. The similarity among quarks and leptons and the mathematical resemblance among the gauge theories of the fundamental interactions motivate an audacious program of "grand unification" in which the strong, weak, and electromagnetic interactions are different manifestations of a single, underlying symmetry.

This book has grown out of graduate courses given at the University of Chicago, l'Université de Paris, Cornell University, and Princeton University; lecture series presented at Fermilab and CERN; and courses taught in summer schools around the world. Experience with diverse audiences has persuaded me both of the need for a nontechnical treatment for specialists in elementary particle physics and of the desirability that the gauge theory perspective be part of the education of every physics graduate student. In preparing this new edition, I have kept in mind three sorts of readers:

1. Graduate students in physics who have completed (at least) a graduate course in quantum mechanics, including a study of relativistic quantum mechanics and the rudiments of Feynman diagrams.

2. Experimental physicists working in particle physics and neighboring fields.

3. Physicists who have not specialized in high-energy physics but wish to gain an appreciation of the essential ideas of gauge theories and the "standard model" of particle physics.

The book is intended to serve as a text for a special topics course for advanced graduate students and as a monograph for reference and self-study. It may also be used, selectively, to supplement a course on relativistic quantum mechanics.

I have developed the subject as a logical whole, while bearing in mind that it should be possible to open the book at random and make sense of what is written. There is both a coherence and a progression, in that issues raised in early chapters are recalled, amplified, refined, and—in some cases—resolved later in the text. There is an accompanying evolution of the physical concepts and mathematical techniques.

I have stressed the essential interplay between theory and experiment, with respect to qualitative phenomena, symmetries, and specific numbers. Not least because of the mutual stimulation of observation and abstraction, I have included in the text detailed calculations of experimental observables and have posed a number of similarly explicit problems. Among the fondest memories of my physics infancy are those of a summer vacation during which I came upon Heitler's classic treatise on the quantum theory of radiation and Feynman's two slender volumes on quantum electrodynamics and the theory of fundamental processes and began to understand how to compute and how to learn from experiment. Not every beginning reader will experience a similar epiphany, but I have sought to provide the opportunity. There is another reason for undertaking explicit calculations, which is to be struck by some of the consequences of gauge invariance. This can be done only by witnessing at first hand the miraculous cooperation among Feynman diagrams that leads, among other things, to an acceptable high-energy behavior of the weak interactions.

At the end of each chapter, I have provided an extensive reading list. These annotated bibliographies offer amplification of points of conceptual or technical interest, introduce further applications, or guide the reader to alternative presentations of the text material. It should not be necessary to consult the documents cited in order to master the contents of the chapter. Rather, they are intended as a selective guide to optional further study.

Although a detailed description of the topics covered in this book can be found in the table of contents, it is appropriate to present a brief overview here. The opening chapter contains a capsule review of elementary-particle phenomenology and a preview of issues to be elaborated later, including the importance of the 1-TeV scale. Chapters 2–5 develop the basic theoretical concepts, proceeding from the elementary implications of symmetry principles through electrodynamics, non-Abelian gauge theories, and the notion of spontaneously broken symmetries. The emphasis in this part of the book is on conceptual matters, but applications are not entirely neglected. The last four chapters are devoted to the gauge theories that describe the strong, weak, and electromagnetic interactions. Two chapters are concerned with the electroweak theory—its successes, the search for the agent that hides the electroweak symmetry, and indications that the electroweak theory cannot be the last word. Quantum chromodynamics, the theory of the strong interactions, is the subject of chapter 8. The final chapter deals with the program of unifying the fundamental interactions, the principal ideas and consequences of which are studied in the context of the minimal SU(5) model. A brief epilogue highlights some open questions for experiment and theory. Matters of convention and other technical issues are relegated to three appendices. Additional material relating to this book can be found at http://press.princeton.edu/titles/10156.html.

It is a pleasure to express my appreciation to many people and institutions for their contributions to this work. In the course of preparing this volume, I have enjoyed the warm hospitality of a number of institutions, including CERN (the European Laboratory for Particle Physics), the Institute for Theoretical Particle Physics in Karlsruhe, Ludwig-Maximilians University and the Technical University of Munich, NIKHEF (the Dutch National Institute for Subatomic Physics), and the Laboratoire de Physique Théorique de l'École Normale Supérieure, Paris. My visits to German institutions were generously underwritten by an Alexander von Humboldt Foundation Senior Scientist Award. I thank Luis Álvarez-Gaumé, Gerhard Buchalla, Andrzej Buras, Johann Kühn, Eric Laenen, and Ulrich Nierste for welcoming me to their institutions. I am immensely grateful to generations of leaders of the Fermi National Accelerator Laboratory and to my Fermilab colleagues for the stimulating environment I have enjoyed there for many years. I thank the Office of Science of the U.S. Department of Energy for a long season of research support.

Many physicists have contributed to my understanding of gauge theories, and the reactions and questions of students have influenced my presentation of the material. I owe special thanks to Carl Albright, Bill Bardeen, Chris Bouchard, Mu-Chun Chen, Estia Eichten, Keith Ellis, Chris Hill, Ian Hinchliffe, Andreas Kronfeld, Ken Lane, Joe Lykken, Olga Mena, Jon Rosner, and Hank Thacker. Many of the figures were executed using Jos Vermaseren's AXODRAW. I am grateful to Ingrid Gnerlich, Samantha Hasey, Karen Carter, Linda Thompson, and the Princeton University Press team for making this book a reality.

Finally, I thank my wife Elizabeth and our children David and Katherine for their encouragement, support, and kindness.

CHRIS QUIGG
Batavia, 2013

# One

## Introduction

Over the past three decades, an animated conversation between experiment and theory has brought us to a new and radically simple conception of matter. Fundamental particles called quarks and leptons make up the everyday world, and new laws of nature—in the form of theories of the strong, weak, and electromagnetic forces—govern their interactions. Quantum chromodynamics, the theory of the strong interaction among quarks, and the electroweak theory have both been abstracted from experiment, refined within the framework of local gauge symmetries, and validated to an extraordinary degree through confrontation with experiment. What we have learned suggests paths to a more complete picture of nature—perhaps a unified theory of the fundamental particles and interactions.

But the triumph of this new picture is incomplete. We are still searching for a missing piece, the agent of electroweak symmetry breaking. We also need to discover what accounts for the masses of the electron and the other leptons and quarks, without which there would be no atoms, no chemistry, no liquids or solids—no stable structures. In the standard electroweak theory, both tasks are the work of the Higgs mechanism. Moreover, we have reason to believe that the electroweak theory is imperfect and that new symmetries or new dynamical principles are required to make it fully robust.

To extend our understanding, particle physicists from around the world have launched remarkable experiments using the Large Hadron Collider in Geneva, Switzerland, a superconducting synchrotron 27 km in circumference, in which counterrotating proton beams collide at c.m. energies planned to reach 14 TeV. We do not know what the new wave of exploration will find, but the discoveries we make and the new puzzles we encounter are certain to change the face of particle physics and echo through neighboring sciences. Decisive results on the agent of electroweak symmetry breaking appear imminent.

This book is devoted to an exposition of the logic, structure, and phenomenology of our "standard model" of particle physics, from the experimental systematics and theoretical constructs that underlie it to the highly successful form that joins quantum chromodynamics to the electroweak theory. We shall see how the idea of gauge theories—interactions derived from symmetries observed in nature—makes it possible to capture the regularities embodied in earlier theoretical descriptions in an economical and richly predictive framework that gives new understanding and suggests new consequences. The standard model displays in full measure the attributes of an exemplary theory expressed in Heinrich Hertz's celebration of classical electrodynamics:

> One cannot study Maxwell's marvelous electromagnetic theory of light without sometimes having the feeling that these mathematical formulae have an independent existence and an intelligence of their own, that they are wiser than we are, wiser even than their inventor, that they give back to us more than was originally put into them [1].

The utility of the quark model as a classification tool that provides a systematic basis for hadron spectroscopy has long been appreciated. The quark language also provides an apt description of the dynamics of hadronic interactions. The quark-parton model, refined by quantum chromodynamics, underlies a quantitative phenomenology of deeply inelastic lepton–hadron scattering, electron–positron annihilation into hadrons, hard scattering of hadrons, and decays of hadrons, especially those containing heavy quarks.

An elementary particle, in the time-honored sense of the term, is structureless and indivisible. Although history cautions that the physicist's list of elementary particles is dependent upon experimental resolution—and thus subject to revision with the passage of time—it has also rewarded the hope that interactions among the elementary particles of the moment would be simpler and more fundamental than those among composite systems. Neither quarks nor leptons exhibit any structure on a scale of about $10^{-16}$ cm, the currently attained resolution. We thus have no experimental reason but tradition to suspect that they are not the ultimate elementary particles. Accordingly, we idealize the quarks and leptons as pointlike particles, remembering that elementarity is subject to experimental test.

Analyses of collision phenomena suggest that quarks behave as free particles within hadrons, and yet the nonobservation of isolated free quarks encourages the idealization that quarks must be permanently confined within the hadrons. This apparently paradoxical state of affairs requires that the strong interaction among quarks be of a rather particular sort. No rigorous theoretical demonstration of the confinement hypothesis has yet been given, but it is widely held that quantum chromodynamics (QCD) contains the necessary elements. In common with other non-Abelian gauge theories, QCD exhibits an effective interaction strength that decreases at short distances and grows at large distances. This property—asymptotic (ultraviolet) freedom vs. infrared slavery—suggests a resolution of the parton-model conundrum. Monte Carlo simulations of the gauge-theory vacuum provide strong numerical evidence for quark confinement.

The appeal of a unified theory of the weak and electromagnetic interactions is at once aesthetic and practical. The effective weak-interaction Lagrangian that evolved from Fermi's description of nuclear $\beta$-decay and provided a serviceable low-energy phenomenology is now seen to be the limiting form of a renormalizable field theory. At the same time, neutral-current interactions predicted by the new electroweak theory have been found to occur at approximately the strength of the long-studied charged-current interactions. The observed neutral currents are neutral not only with respect to electric charge, but with respect to all other additive quantum numbers as well. To accommodate this property in the theory requires the introduction of a new quark species, bearing a new additive quantum number known as charm. This, too, has subsequently been observed in experiments. The price for this neat picture includes the prediction of several hypothetical particles: the intermediate vector bosons, $W^+$, $W^-$, and $Z^0$, that carry the weak interactions. Definite predictions for the masses and properties of the intermediate bosons have been confirmed by experiment.

Electromagnetism is a force of infinite range, whereas the influence of the charged-current weak interaction responsible for radioactive beta decay spans only distances shorter than about $10^{-15}$ cm, less than 1% of the proton radius. If these two interactions, so different in their range and apparent strength, originate in a common gauge symmetry, that symmetry must be spontaneously broken. That is to say, the vacuum state of the universe must not respect the full symmetry. How the electroweak gauge symmetry is spontaneously broken is one of the most urgent and challenging questions before particle physics. The standard-model answer is an elementary scalar field whose self-interactions select a vacuum state in which the full electroweak symmetry is hidden. Experiments in 2012 have discovered a new particle that—at first look—fits the profile of the Higgs boson, as the elementary scalar is known. We do not yet know whether this observation means that a fundamental Higgs field exists or a different agent breaks electroweak symmetry. General arguments imply that the Higgs boson or other new physics is required on the TeV energy scale. Indirect constraints from global analyses of electroweak measurements suggest that the mass of the standard-model Higgs boson is less than 200 GeV. Once its mass is assumed, the properties of the Higgs boson follow from the electroweak theory. Finding the Higgs boson or its replacement is one of the great campaigns now under way in both experimental and theoretical particle physics. The answer—expected soon—will steer the future development of the electroweak theory.

One measure of the electroweak theory's sweep is that its predictions hold over a prodigious range of distances, from about $10^{-18}$ m to more than $10^8$ m. The origins of the theory lie in the discovery of Coulomb's law in tabletop experiments by Cavendish and Coulomb. It was stretched to longer and shorter distances by the progress of experiment. In the long-distance limit, the classical electrodynamics of a massless photon suffices. At shorter distances than the human scale, classical electrodynamics was superseded by quantum electrodynamics (QED), which is now subsumed in the electroweak theory, tested at energies up to a few hundred GeV.

Because the charged-current weak interactions are purely left-handed, it is not possible to construct a self-consistent theory of the weak and electromagnetic interactions based solely on leptons or solely on quarks. They must come in matched sets. This fact suggests a deep connection between the quarks (which experience the strong interactions) and the leptons (which do not). That observation, in turn, motivates a description that gathers both quarks and leptons into extended families—a unified theory of the strong, weak, and electromagnetic interactions. Such a theory can give an understanding of the low-energy strengths of the individual interactions. Another consequence of the unification of forces is the implication of new forces that can transform quarks into leptons.

For all its triumphs, the standard model is not entirely satisfying. The electroweak theory does not make specific predictions for the masses of the quarks and leptons or for the mixing among different flavors. It leaves unexplained how an elementary Higgs-boson mass could remain below 1 TeV in the face of quantum corrections that tend to lift it toward the Planck scale or a unification scale. The Higgs field thought to pervade all of space to hide the electroweak symmetry contributes a vacuum energy density far in excess of what is observed. And the standard model, even when extended to a unified theory of the strong, weak, and electromagnetic interactions, responds inadequately to challenges raised by astronomical observations, including the dark-matter problem and the predominance of matter over antimatter in the universe. These shortcomings argue for physics beyond the standard model.

The remainder of this chapter is a concise review of some of the primitive concepts of particle phenomenology that serve as a basis for our development of the standard model. In succeeding chapters, we shall assemble a detailed description of the electroweak theory and quantum chromodynamics, the two pillars of the standard model, with close attention to the experimental foundations. Our goal is not only to exhibit the successes of the two theories and to establish their utility for reliable calculations, but also to highlight unfinished business and point to the need for future developments. We shall also see how QCD and the electroweak theory might be joined into a unified theory of the strong, weak, and electromagnetic interactions. The text closes by posing essential questions for theory and experiment.

## 1.1 ELEMENTS OF THE STANDARD MODEL OF PARTICLE PHYSICS

Our picture of matter is based on the identification of a set of pointlike spin-$\frac{1}{2}$ constituents: the (up, down, charm, strange, top, and bottom) quarks,

$$\begin{pmatrix} u \\ d \end{pmatrix}_{\mathrm{L}}, \quad \begin{pmatrix} c \\ s \end{pmatrix}_{\mathrm{L}}, \quad \begin{pmatrix} t \\ b \end{pmatrix}_{\mathrm{L}}, \tag{1.1.1}$$

and the leptons (electron, muon, and tau, plus three neutrinos),

$$\begin{pmatrix} \nu_e \\ e^- \end{pmatrix}_{\mathrm{L}}, \quad \begin{pmatrix} \nu_\mu \\ \mu^- \end{pmatrix}_{\mathrm{L}}, \quad \begin{pmatrix} \nu_\tau \\ \tau^- \end{pmatrix}_{\mathrm{L}}, \tag{1.1.2}$$

where the subscript L denotes the left-handed components, plus a few fundamental forces derived from gauge symmetries. The quarks are influenced by the strong interaction and so carry *color*, the strong-interaction charge, whereas the leptons do not feel the strong interaction and are colorless. Each of the six quark flavors comes in three distinct colors: red, green, and blue. The right-handed fermions are weak-interaction singlets. By pointlike, we understand that the quarks and leptons show no evidence of internal structure at the current limit of our resolution, ($r \lesssim 10^{-18}$ m) [2].

The notion that the quarks and leptons are elementary—structureless and indivisible—is necessarily provisional. *Elementarity* is one of the aspects of our picture of matter that we test ever more stringently as we improve the resolution

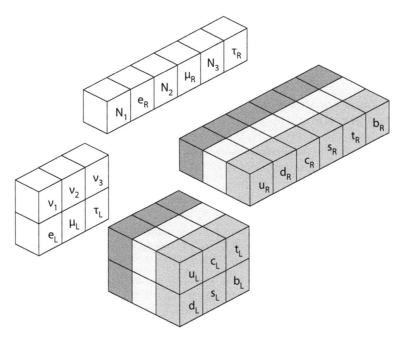

**Figure 1.1.** The left-handed $SU(2)_L$ doublets and right-handed $SU(2)_L$ singlets of color-triplet quarks and color-singlet leptons from which the standard model of particle physics is constructed.

with which we can examine the quarks and leptons. For the moment, the world's most powerful microscope is the Large Hadron Collider at CERN, where the ATLAS and CMS Collaborations have studied $pp$ collisions at c.m. energy $\sqrt{s} = 8$ TeV. For the production of hadron jets at transverse energy $E_\perp$, we may roughly estimate the resolution as $r \approx (\hbar c)/E_\perp \approx 2 \times 10^{-19}$ TeV m/$E_\perp$ [3].

The left-handed and right-handed fermions behave very differently under the influence of the charged-current weak interactions. In 1956, Wu and collaborators [4] studied the $\beta$-decay $^{60}\text{Co} \rightarrow {}^{60}\text{Ni}\ e^- \bar{\nu}_e$ and observed a correlation between the direction $\hat{p}_e$ of the outgoing electron and the spin vector $\vec{J}$ of the polarized $^{60}\text{Co}$ nucleus. Spatial reflection, or parity, leaves the (axial vector) spin unchanged, $P : \vec{J} \rightarrow \vec{J}$, but reverses the electron direction, $P : \hat{p}_e \rightarrow -\hat{p}_e$. Accordingly, the correlation $\vec{J} \cdot \hat{p}_e$ is manifestly *parity violating*. Experiments in the late 1950s (cf. §6.3) established that (charged-current) weak interactions are left-handed, and motivated the construction of a manifestly parity-violating theory of the weak interactions with only a left-handed neutrino $\nu_L$.

Perhaps our familiarity with parity violation in the weak interactions has dulled our senses a bit. It seems to me that nature's broken mirror—the distinction between left-handed and right-handed fermions—qualifies as one of the great mysteries. Even if we will not get to the bottom of this mystery next week or next year, it should be prominent in our consciousness—and among the goals we present to others as the aspirations of our science.

The family relationships among the quarks and the leptons are depicted in figure 1.1. To excellent approximation, the observed charged-current weak interactions connect up quarks with down, charm with strange, and top with bottom. At the current limits of experimental sensitivity, the charged-current interactions conserve electron number, muon number, and tau number. Although the right-handed quarks and charged leptons do not participate in charged-current weak interactions, their existence is implied by charged-fermion masses and the characteristics of the strong and electromagnetic interactions. We take the weak-isospin doublets as evidence for $SU(2)_L$ gauge symmetry and infer $SU(3)_c$ gauge symmetry from the three quark colors, interpreted as a continuous symmetry. As we shall see, the successful electroweak theory incorporates a $U(1)_Y$ weak-hypercharge phase symmetry, along with $SU(2)_L$. We have already commented that the electroweak gauge symmetry must be hidden, $SU(2)_L \otimes U(1)_Y \to U(1)_{EM}$, with the phase symmetry of electromagnetism the residual symmetry.

The discovery of neutrino oscillations (cf. §6.7), which implies that neutrinos have mass, motivates the inclusion of the right-handed neutrinos, $N_i$, in figure 1.1. The right-handed neutrinos would be sterile—inert with respect to the known $SU(3)_c \otimes SU(2)_L \otimes U(1)_Y$ interactions.

The representation of the quarks and leptons in figure 1.1 invites not only speculations about the symmetries that lead to the strong, weak, and electromagnetic interactions, but also questions about possible relations between quarks and leptons or between the left-handed and right-handed particles.

Let us now look in slightly greater detail at each of the ingredients of the standard model of particle physics.

## 1.2 LEPTONS

The leptons can exist as free particles and, therefore, can be studied directly. Three charged leptons—the electron, muon, and tau—are firmly established, the electron and muon by direct observation and the short-lived tau ($c\tau = 87\,\mu$m) through its decay products. The gyromagnetic ratios of the electron and muon have been measured with remarkable precision (cf. problem 1.6). They differ from the value of 2 expected for Dirac particles only by tiny fractions, which have been calculated in QED. This strongly supports their identification as point particles. The electron neutrino and muon neutrino are, likewise, well known (cf. problem 1.9). Decay characteristics of the $\tau$-lepton imply the existence of a distinct tau neutrino, and the first examples of $\nu_\tau$ interactions have been observed. On short baselines, neutrino flavor is strictly correlated with the flavor of the charged lepton to which it couples. Some properties of the known leptons are summarized in table 6.4.

The traditional names of the neutrinos follow from the structure of the charged weak current, which is represented by the weak-isospin doublets

$$\psi_1 = \begin{pmatrix} \nu_e \\ e \end{pmatrix}_L, \quad \psi_2 = \begin{pmatrix} \nu_\mu \\ \mu \end{pmatrix}_L, \quad \psi_3 = \begin{pmatrix} \nu_\tau \\ \tau \end{pmatrix}_L, \tag{1.2.1}$$

The leptonic charged current thus has the (vector minus axial vector) form [5]

$$J_\lambda^{(\pm)} = \sum_i \bar{\psi}_i \tau_\pm \gamma_\lambda (1 - \gamma_5) \psi_i, \tag{1.2.2}$$

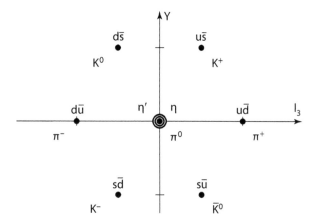

**Figure 1.2.** SU(3) weight diagram for the pseudoscalar nonet = **1** ⊕ **8**.

where $\tau_\pm \equiv \frac{1}{2}(\tau_1 \pm i\tau_2)$ are the Pauli isospin matrices

$$\tau_+ = \begin{pmatrix} 0 & 1 \\ 0 & 0 \end{pmatrix}, \ \tau_- = \begin{pmatrix} 0 & 0 \\ 1 & 0 \end{pmatrix}. \tag{1.2.3}$$

The implied universality of weak-interaction transition matrix elements holds to high accuracy for the electronic and muonic transitions and has been verified within 1.5% for the tau family. We shall see that the identification of weak-isospin doublets leads, in addition, to an understanding of the structure of the neutral weak current.

We have not penetrated the origin of the simple and orderly family pattern exhibited by the leptons. Indeed, many apparent facts and regularities are inadequately comprehended; therefore, many questions present themselves. Why are there three doublets of leptons? Might more be found? Is the difference between the number of leptons and antileptons in the universe a conserved quantity? What is the pattern of lepton masses? Is the separate conservation of an additive electron number, muon number, and tau number—apart from the effect of neutrino mixing—an exact, or only approximate, statement?

## 1.3 QUARKS

Quarks are the fundamental constituents of the strongly interacting particles, the hadrons. They experience all the known interactions: strong, weak and electromagnetic, and gravitational. Quarks have much in common with the leptons, but there is a crucial distinction. Free quarks have never been observed, so it has been necessary to adduce indirect evidence for their existence and their properties.

Gell-Mann and Zweig proposed quarks in 1964 by as a means for understanding the SU(3) classification of the hadrons known to them [6]. The light mesons occur only in SU(3) singlets and octets. For example, the nine pseudoscalars are shown in the familiar (strangeness versus third component of isospin) hexagonal array in figure 1.2. Similarly, the light baryons are restricted to singlets, octets, and decimets of SU(3). The lowest-lying baryons are displayed in figures 1.3 and 1.4. The observation that no higher representations are indicated is far more restrictive

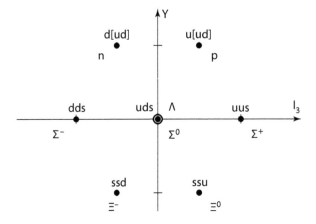

**Figure 1.3.** The $J^P = \frac{1}{2}^+$ baryon octet.

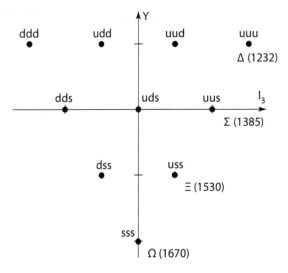

**Figure 1.4.** The $J^P = \frac{3}{2}^+$ baryon decimet.

than the mere fact that SU(3) is a good classification symmetry and requires explanation.

The observed patterns can be understood in terms of the hypothesis that hadrons are composite structures built up from an elementary triplet of spin-$\frac{1}{2}$ quarks, corresponding to the fundamental representation of SU(3). The three "flavors" of quarks, commonly named *up*, *down*, and *strange*, have the properties summarized in table 1.1, where $I$ is (strong) isospin, $S$ is strangeness, $B$ is baryon number, $Y$ is (strong) hypercharge, and $Q$ is electric charge. The quark masses are indicated in figure 9.6. A meson composed of a quark and antiquark ($q\bar{q}$) then lies in the SU(3) representations

$$3 \otimes 3^* = 1 \oplus 8, \tag{1.3.1}$$

TABLE 1.1
Properties of the Light Quarks

| Quark | $I$ | $I_3$ | $S$ | $B$ | $Y = B + S$ | $Q$ |
|-------|-----|-------|-----|-----|-------------|-----|
| $u$ | $\frac{1}{2}$ | $\frac{1}{2}$ | $0$ | $\frac{1}{3}$ | $\frac{1}{3}$ | $\frac{2}{3}$ |
| $d$ | $\frac{1}{2}$ | $-\frac{1}{2}$ | $0$ | $\frac{1}{3}$ | $\frac{1}{3}$ | $-\frac{1}{3}$ |
| $s$ | $0$ | $0$ | $-1$ | $\frac{1}{3}$ | $-\frac{2}{3}$ | $-\frac{1}{3}$ |

TABLE 1.2
Some Properties of the Heavy Quarks

| Quark | $I$ | $Q$ | Charm | Beauty | Truth | Mass (GeV) |
|-------|-----|-----|-------|--------|-------|------------|
| $c$ | $0$ | $\frac{2}{3}$ | $1$ | $0$ | $0$ | $\sim 1.3$ |
| $b$ | $0$ | $-\frac{1}{3}$ | $0$ | $-1$ | $0$ | $\sim 4.2$ |
| $t$ | $0$ | $\frac{2}{3}$ | $0$ | $0$ | $1$ | $173$ |

and a baryon, composed of three quarks $(qqq)$, must be contained in

$$3 \otimes 3 \otimes 3 = 1 \oplus 8 \oplus 8 \oplus 10. \tag{1.3.2}$$

The quark content of the hadrons is indicated in figures 1.2–1.4. This simple model reproduces the representations seen prominently in experiments. It remains, of course, to understand why only these combinations of quarks and antiquarks are observed or to discover the circumstances under which more-complicated configurations (such as $q\bar{q}q\bar{q}$ or $qqqq\bar{q}$ or $6q$) might arise.

The existence of three flavors of quarks has thus been inferred from the quantum numbers of the light hadrons. Let us check the consistency of the properties attributed to the quarks.

If baryons, which are fermions, are to be made of three identical constituents, the constituents must themselves be fermions. The observed hadron spectrum corresponds to objects that can be formed as $(q\bar{q})$ or $(qqq)$ composites of spin-$\frac{1}{2}$ quarks. It is straightforward to work out the level structure in the meson sector:

$$(q\bar{q}) \to J^{PC} = \underbrace{0^{-+}, 1^{--}}_{L=0}; \underbrace{0^{++}, 1^{++}, 1^{+-}, 2^{++}}_{L=1}; \cdots \tag{1.3.3}$$

This conforms closely to the observed ordering of light-meson levels. Moving to the present, we now have identified three additional heavy quarks (*charm*, *bottom*, and *top*), with properties summarized in table 1.2. The expected order of levels is clearly shown in the spectrum of bound states of a charmed quark and antiquark depicted in figure 1.5 [7]. Combinations of spin, parity, and charge conjugation such as $J^{PC} = 0^{--}, 0^{+-}, 1^{-+}$, which cannot be formed from pairs of spin-$\frac{1}{2}$ quarks and antiquarks, have not been observed. The analysis of baryon multiplets is similar but

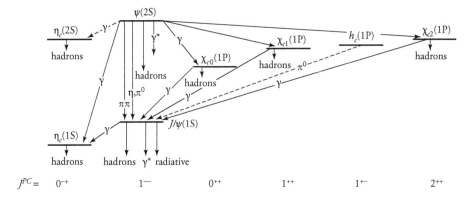

**Figure 1.5.** Spectrum of charm-anticharm bound states below the threshold for dissociation into charmed-particle pairs. (Adapted from Ref. [2].)

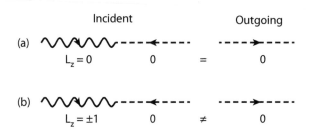

**Figure 1.6.** Photoabsorption by a scalar particle in the reference frame in which the three-momentum of the outgoing scalar is minus the incoming momentum. (a) Absorption of a longitudinal photon by a spinless particle is allowed by angular momentum conservation. (b) Absorption of a transverse photon is forbidden.

more tedious. Again, the observed spectrum is in agreement with the quark-model pattern.

In addition to this successful classification scheme, there are dynamical tests of the quark spin. Consider the cross sections for absorption of longitudinal or transverse virtual photons on point particles. In the Breit frame of the struck particle, illustrated in figure 1.6, it is easy to see that a spinless "quark" can absorb only a longitudinal (helicity = 0) photon, because angular momentum conservation forbids the absorption of a transverse (helicity = ±1) photon. Similarly (cf. problem 1.3), a spin-$\frac{1}{2}$ quark can absorb a transverse (helicity = ±1) photon but not a longitudinal photon. Within the parton model (cf. §7.4), deeply inelastic scattering of electrons from nucleon targets is analyzed as the scattering of electrons from noninteracting and structureless charged constituents. The relative size of the cross sections for absorption of longitudinal and transverse photons is a diagnostic for the spin of the charged constituents.

A related test is provided by the angular distribution of hadron jets in electron–positron annihilations, which is identical to the production angular distribution of muons in the reaction

$$e^+e^- \to \mu^+\mu^-. \tag{1.3.4}$$

That hadrons are emitted in well-collimated jets (cf. problem 1.5) supports the interpretation of particle production by means of the elementary process

$$e^+e^- \to q\bar{q}, \tag{1.3.5}$$

followed by the hadronization of the noninteracting quarks. The jet angular distribution then reflects the angular distribution of the spin-$\frac{1}{2}$ quarks.

Quarks have baryon number $\frac{1}{3}$; antiquarks have baryon number $-\frac{1}{3}$. This follows from the assertion that three quarks make up a baryon. The quarks also carry fractional electric charge [8]. The Gell-Mann–Nishijima formula [9] for displaced charge multiplets,

$$Q = I_3 + \tfrac{1}{2}Y = I_3 + \tfrac{1}{2}(B + S), \tag{1.3.6}$$

implies the charge assignments shown in table 1.1. The same assignments follow directly from examination of members of the baryon decimet:

$$\begin{aligned} \Delta^{++} &= uuu, \\ \Delta^+ &= uud, \\ \Delta^0 &= udd, \qquad \Omega^- = sss. \\ \Delta^- &= ddd, \end{aligned} \tag{1.3.7}$$

The characteristics of quarks that we have discussed until now—spin, flavor, baryon number, electric charge—are directly indicated by the experimental observations that originally motivated the quark model. Quarks have still another property, less obvious but of central importance for the strong interactions. This additional property is known as *color*. At first sight, the Pauli principle seems not to be respected by the wave function for the $\Delta^{++}$. This nucleon resonance is a ($uuu$) state with spin $= \frac{3}{2}$ and isospin $= \frac{3}{2}$, in which all the quark pairs are in relative $s$-waves. Thus, it is apparently a symmetric state of three identical fermions. Unless we are prepared to suspend the Pauli principle or to forgo the quark model, it is necessary [10] to invoke a new, hidden degree of freedom, which permits the $\Delta^{++}$ wave function to be antisymmetrized. In order that a ($uuu$) wave function can be antisymmetrized, each quark flavor must exist in no fewer than three distinguishable types, called colors. More than three colors would raise the unpleasant possibility of distinguishable (colored) species of protons, which is contrary to common experience. Thus motivated, the introduction of color may appear arbitrary and artificial—a "desperate remedy." However, a number of observables are sensitive to the number of distinct species of quarks, and subsequent measurements of these quantities have given strong support to the color hypothesis.

The inclusive cross section for electron–positron annihilation into hadrons is described, as earlier noted, by the elementary process $e^+e^- \to q\bar{q}$, where the quark and antiquark materialize with unit probability into the observed hadron jets. At a particular collision energy, the ratio

$$R \equiv \frac{\sigma(e^+ e^- \rightarrow \text{hadrons})}{\sigma(e^+ e^- \rightarrow \mu^+ \mu^-)} \qquad (1.3.8)$$

is then simply given as

$$R = \sum_{\substack{\text{quark} \\ \text{species}}} e_q^2. \qquad (1.3.9)$$

At barycentric energies between approximately 1.5 and 3.6 GeV, pairs of up, down, and strange quarks are kinematically accessible. In the absence of hadronic color, we would, therefore, expect a mean level

$$R_1 = e_u^2 + e_d^2 + e_s^2 = \tfrac{2}{3}, \qquad (1.3.10)$$

but if each quark flavor exists in three colors, we should have

$$R_3 = 3(e_u^2 + e_d^2 + e_s^2) = 2. \qquad (1.3.11)$$

Experiment decisively favors the color-triplet hypothesis, as shown in the top panel of figure 1.7. At still higher energies the heavier $c$- and $b$-quarks may be produced in the semifinal state. Between the $c\bar{c}$ and $b\bar{b}$ thresholds, the color-triplet model thus predicts

$$R = 3(e_u^2 + e_d^2 + e_s^2 + e_c^2) = \tfrac{10}{3}, \qquad (1.3.12)$$

to be compared with data in the middle panel of figure 1.7. Above the $b\bar{b}$ threshold, we expect

$$R = 3(e_u^2 + e_d^2 + e_s^2 + e_c^2 + e_b^2) = \tfrac{11}{3}, \qquad (1.3.13)$$

which agrees well with the data shown in the lower panel of figure 1.7. (cf. the extension to higher energies shown in figure 8.25).

A similar count of the number of distinguishable quarks of each flavor is provided by the decay branching ratios of the tau lepton. Within the quark model, decays may be described as shown in figure 1.8, namely, by the decay of $\tau$ into $\nu_\tau$ plus a virtual intermediate boson $W^-$. The intermediate boson may then disintegrate into all kinematically accessible fermion–antifermion pairs: $(e^- \bar{\nu}_e)$, $(\mu^- \bar{\nu}_\mu)$, $(u\bar{d})$. (See the following section and §7.1 for the refinement of Cabibbo mixing, which does not alter the logic of this argument.) The universality of the charged-current weak interactions implies equal rates for each of these decays. Therefore, in the absence of color, we expect

$$B_1 = \frac{\Gamma(\tau \rightarrow e^- \bar{\nu}_e \nu_\tau)}{\Gamma(\tau \rightarrow \text{all})} = \frac{1}{3}. \qquad (1.3.14)$$

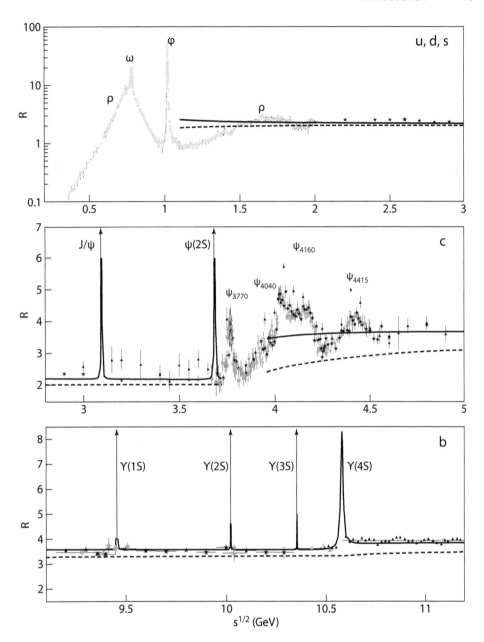

**Figure 1.7.** The ratio $R \equiv \sigma(e^+e^- \to \text{hadrons})/\sigma(e^+e^- \to \mu^+\mu^-)$ in the light-flavor, charm-threshold, and beauty-threshold regions. The dashed lines represent the parton-model expectations given by (1.3.11), (1.3.12), and (1.3.13). The solid curves show the influence of perturbative QCD corrections. (Adapted from Ref. [2].)

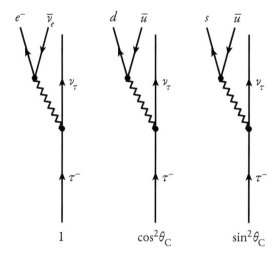

**Figure 1.8.** Leptonic decays (weight 1) and Cabibbo-favored (weight $\cos^2 \theta_C$) and Cabibbo-suppressed (weight $\sin^2 \theta_C$) semileptonic decays of the tau lepton.

If the quarks are color triplets, $(u\bar{d})$ is increased to $3(u\bar{d})$, and the leptonic branching ratio becomes [11]

$$B_3 = \tfrac{1}{5}. \tag{1.3.15}$$

The experimentally measured branching ratio,

$$B_{\text{exp}} = (17.44 \pm 0.85)\%, \tag{1.3.16}$$

is in accord with the color hypothesis.

The other important measure of the number of quark colors is the $\pi^0 \to \gamma\gamma$ decay rate [12]. A calculation for $\pi^0$-decay via a quark–antiquark loop is straightforward to carry out, although the required justification is somewhat subtle. The result is that

$$\Gamma(\pi^0 \to \gamma\gamma) = \left(\frac{\alpha}{2\pi}\right)^2 [N_c(e_u^2 - e_d^2)]^2 \frac{M_\pi^3}{16\pi f_\pi^2}, \tag{1.3.17}$$

where $N_c$ is the number of colors and $f_\pi \approx 92$ MeV is the pion decay constant, determined from the charged pion lifetime. The predicted rate is then

$$\Gamma(\pi^0 \to \gamma\gamma) = \begin{cases} 0.86 \text{ eV}, & N_c = 1, \\ 7.75 \text{ eV}, & N_c = 3, \end{cases} \tag{1.3.18}$$

to be compared with the measured rate of $(7.74 \pm 0.37)$ eV.

From all these experimental indications and from a further theoretical argument to be given in section 6.8, we may conclude that the hidden color degree of freedom is indeed present. It is then tempting to identify color as the property that distinguishes quarks from leptons and might play the role of a strong-interaction charge. This insight will eventually lead to the gauge theory of strong interactions, quantum chromodynamics.

The observation of hadrons with various internal quantum numbers, such as isospin, strangeness, and charm, has led to the application of flavor symmetries SU(2), SU(3), ..., to the strong interactions. These internal symmetry groups serve both for the classification of hadrons (whence the inspiration for the quark model) and for dynamical relations among strong-interaction amplitudes. Both isospin and SU(3) are excellent, but not exact, strong-interaction symmetries. Isospin invariance holds within a few percent, and flavor SU(3) is respected at the 10% to 20% level. All the same, the outcome of the preceding discussion has been to minimize the direct importance of flavors in the strong interactions and to emphasize the significance of color. Reasons for dismissing a strong-interaction theory based on flavor symmetry will be developed in chapter 4, but the problem of accounting for the existence and the breaking of the flavor symmetries will remain. According to an evolving view, the breaking of SU(N) flavor symmetry is a consequence of the quark mass differences

$$m_u < m_d < m_s < m_c < m_b < \cdots , \qquad (1.3.19)$$

which themselves follow, in a manner not fully understood, from the spontaneous symmetry breaking of the weak and electromagnetic interactions. The goodness of flavor SU(3) symmetry is then owed to the smallness and near degeneracy of the up, down, and strange quark masses.

This line of thought has not yet led to a complete understanding of the strong-interaction symmetries. It is ironic that isospin invariance, the oldest and most exact of strong-interaction symmetries, should seem merely coincidental. We do not know why the pattern of quark masses and mixing angles should be as it is or even why so many "fundamental" fermions should exist. Within the strong interactions, flavor appears not to have any essential dynamical role, its only role being to contribute a richness.

## 1.4 THE FUNDAMENTAL INTERACTIONS

The elementary interactions of the quarks and leptons can be understood as consequences of gauge symmetries. The notions of local gauge invariance and gauge theories and the construction and application of specific theories will be developed in logical sequence in succeeding chapters. As prologue, let us merely recall here some superficial aspects of the familiar gauge theories and their practical consequences.

The gauge theory known as quantum chromodynamics, in which colored quarks interact by means of massless colored gauge bosons named *gluons*, is now established as a comprehensive theory of the strong interactions among quarks. The three quark colors are regarded as the basis of the color-symmetry group SU(3)$_c$, distinct from the flavor SU(3) group that relates up, down, and strange quarks. An SU(3)$_c$ octet of vector gluons mediate the interactions among all colored objects, including the quarks and the gluons themselves.

It will be found appealing to argue that only color-singlet objects may exist in isolation. Color confinement, as it is called, would then imply quark confinement, which would, in turn, explain the fact that free quarks have never been observed: free quarks and gluons simply could not exist. We will show in chapter 8 that the coupling "constant" of the color interaction decreases at short distances and increases at long distances. This property of QCD, which is called asymptotic freedom, helps us understand why permanently confined quarks behave within hadrons as if they are free particles.

The experimental evidence for the existence of the color gauge bosons, the gluons, is multifaceted. We note two examples here, with more evidence to follow in chapter 8. First, energy-momentum sum rules in lepton–nucleon scattering indicate that the partons that interact electromagnetically or weakly, namely, the quarks, carry only about half the momentum of a nucleon. Something else, electrically neutral and inert with respect to the weak interactions, must carry the remainder. This is a role for which the gluons are ideally suited. Second, at barycentric energies exceeding about 17 GeV, a fraction of hadronic events produced in electron–positron annihilations display a three-jet structure instead of the familiar two-jet ($e^+e^- \rightarrow q\bar{q}$) structure [13]. We interpret the three-jet structure as evidence for the process

$$e^+e^- \rightarrow q\bar{q} + \text{gluon}, \qquad (1.4.1)$$

in which the gluon is radiated from the outgoing quark in a hadronic analogue of electromagnetic bremsstrahlung. This interpretation survives further scrutiny.

The electroweak theory is built upon the weak-isospin symmetry suggested by the $SU(2)_L$ doublets abstracted from the systematics of radioactive $\beta$-decay and other charged-current interactions, plus a $U(1)_Y$ phase symmetry associated with weak hypercharge. A key element is that the $SU(2)_L \otimes U(1)_Y$ gauge symmetry must be spontaneously broken to $U(1)_{EM}$, the phase symmetry that underlies quantum electrodynamics. We develop the theory and a plausible mechanism for spontaneous symmetry breaking in chapters 6 and 7. It will be useful here to anticipate some elementary features of the charged-current and neutral-current interactions [14]. We will see that, whereas flavor is incidental to the strong interactions, it has an intrinsic importance to the weak interactions.

Consider first a world populated by electron and muon doublets. The electromagnetic current is given by

$$J_\lambda^{(EM)} = -\bar{e}\gamma_\lambda e - \bar{\mu}\gamma_\lambda \mu, \qquad (1.4.2)$$

which evidently leaves all additive quantum numbers unchanged. The charged weak current is indicated by the weak-isospin doublets (1.2.1) as

$$J_\lambda^{(+)} = \bar{\nu}_e \gamma_\lambda (1 - \gamma_5) e + \bar{\nu}_\mu \gamma_\lambda (1 - \gamma_5) \mu. \qquad (1.4.3)$$

If the idea of weak-isospin symmetry is to be taken seriously, these long-studied leptonic charged currents must be supplemented by another current that completes the weak isovector:

$$J_\lambda^{(3)} = \tfrac{1}{2} \sum_i \bar{\psi}_i \tau_3 \gamma_\lambda (1 - \gamma_5) \psi_i$$

$$= \tfrac{1}{2} [\bar{\nu}_e \gamma_\lambda (1 - \gamma_5) \nu_e - \bar{e} \gamma_\lambda (1 - \gamma_5) e$$
$$+ \bar{\nu}_\mu \gamma_\lambda (1 - \gamma_5) \nu_\mu - \bar{\mu} \gamma_\lambda (1 - \gamma_5) \mu]. \qquad (1.4.4)$$

Like the electromagnetic current, the third component of weak isospin leaves additive quantum numbers unchanged. The weak neutral current that emerges after spontaneous symmetry breaking may be expressed as the linear combination

$$J_\lambda^{(0)} = J_\lambda^{(3)} - 2 \sin^2 \theta_W J_\lambda^{(EM)}, \qquad (1.4.5)$$

with the relative weights of the two terms governed by the weak mixing parameter $\sin^2 \theta_W$. The explicit form is

$$J_\lambda^{(0)} = \tfrac{1}{2} \bar{\nu}_e \gamma_\lambda (1 - \gamma_5) \nu_e + L_e \bar{e} \gamma_\lambda (1 - \gamma_5) e + R_e \bar{e} \gamma_\lambda (1 + \gamma_5) e + (e \to \mu), \qquad (1.4.6)$$

where the chiral coupling strengths $L_e$ and $R_e$ depend on $\sin^2 \theta_W$. Identical couplings occur in the muon sector because the lepton generations have the same $SU(2)_L \otimes U(1)_Y$ quantum numbers. Notice that the electron and muon sectors remain disconnected, as required by the separate conservation of electron number and muon number. The leptonic neutral current is said to be flavor conserving, or diagonal in flavors.

What of the hadronic sector? Consider the three quarks $u$, $d$, and $s$ that constitute the light hadrons. According to the Cabibbo hypothesis [15] of weak-interaction universality, the hadronic charged current is represented by the weak-isospin doublet

$$\begin{pmatrix} u \\ d_\theta \end{pmatrix}_L = \begin{pmatrix} u \\ d \cos \theta_C + s \sin \theta_C \end{pmatrix}_L. \qquad (1.4.7)$$

The charged current for quarks thus has the explicit form

$$J_\lambda^{(+)} = \bar{u} \gamma_\lambda (1 - \gamma_5) d \cdot \cos \theta_C + \bar{u} \gamma_\lambda (1 - \gamma_5) s \cdot \sin \theta_C. \qquad (1.4.8)$$

Even before entering the domain of the electroweak theory, we may ask why the hadron sector should have a superfluous quark, or, in other words, why the orthogonal combination of $s$- and $d$-quarks,

$$s_\theta = s \cos \theta_C - d \sin \theta_C, \qquad (1.4.9)$$

does not appear in the charged weak current. To pose the same question differently, why do quarks and leptons not enter more symmetrically?

These apparently idle questions become urgent in the framework of the electroweak, theory, where they also find their answers. In the three-quark theory, the weak neutral current would be

$$
\begin{aligned}
J_\lambda^{(0)} &= J_\lambda^{(3)} - 2\sin^2\theta_W J_\lambda^{(EM)} \\
&= \tfrac{1}{2}[\bar{u}\gamma_\lambda(1-\gamma_5)u - \bar{d}\gamma_\lambda(1-\gamma_5)d \cdot \cos^2\theta_C \\
&\quad -\bar{s}\gamma_\lambda(1-\gamma_5)s \cdot \sin^2\theta_C - \bar{s}\gamma_\lambda(1-\gamma_5)d \cdot \sin\theta_C \cos\theta_C \\
&\quad -\bar{d}\gamma_\lambda(1-\gamma_5)s \cdot \sin\theta_C \cos\theta_C] \\
&\quad -2\sin^2\theta_W(\tfrac{2}{3}\bar{u}\gamma_\lambda u - \tfrac{1}{3}\bar{d}\gamma_\lambda d - \tfrac{1}{3}\bar{s}\gamma_\lambda s).
\end{aligned}
\tag{1.4.10}
$$

Unlike the neutral leptonic current, this hadronic neutral current contains flavor-changing ($d \leftrightarrow s$) terms. This is experimentally unacceptable because of the small rates observed for the decays $K^+ \to \pi^+\nu\bar{\nu}$ and $K_L \to \mu^+\mu^-$. It was shown by Glashow, Iliopoulos, and Maiani [16] that the flavor-changing neutral currents could be eliminated and lepton–quark symmetry could be established by the addition of a second weak-isospin doublet

$$
\begin{pmatrix} c \\ s_\theta \end{pmatrix}_L
\tag{1.4.11}
$$

involving a then-hypothetical charmed quark [17]. The hadronic neutral current would then be

$$
\begin{aligned}
J_\lambda^{(0)} &= \tfrac{1}{2}[\bar{u}\gamma_\lambda(1-\gamma_5)u + \bar{c}\gamma_\lambda(1-\gamma_5)c - \bar{d}\gamma_\lambda(1-\gamma_5)d \\
&\quad -\bar{s}\gamma_\lambda(1-\gamma_5)s] - 2\sin^2\theta_W J_\lambda^{(EM)},
\end{aligned}
\tag{1.4.12}
$$

which is manifestly flavor diagonal. The discovery [18] of the family of charmonium bound states and the subsequent observation [19] of charmed particles that decay according to the $(c, s_\theta)_L$ pattern constituted a striking confirmation of the GIM hypothesis and an important validation of weak-electromagnetic unification. We shall have much more to say about the electroweak gauge bosons and the avatar of electroweak symmetry breaking.

This brief survey has presented some evidence that quarks and leptons may properly be regarded as elementary particles and has introduced some of the symmetries and relationships among them. In addition, we have recalled some of the basic ideas and properties of gauge theories and have indicated their connection with the current understanding of the fundamental interactions. With these concepts and aspirations as background, we now turn to the foundations that underlie gauge theories. The first of these is the notion of symmetry in Lagrangian field theory.

## PROBLEMS

1.1. Consider bound states composed of a $b$-quark and a $\bar{b}$-antiquark.
   (a) Show that a bound state with orbital angular momentum $L$ must have quantum numbers

$$
C = (-1)^{L+s}; \quad P = (-1)^{L+1},
$$

where $s$ is the spin of the composite system.

(b) Allowing for both orbital and radial excitations, construct a schematic mass spectrum of $(b\bar{b})$ bound states. Label each state with its quantum numbers $J^{PC}$.

1.2. Consider bound states composed of color-triplet scalars (denoted $\sigma$) and their antiparticles.

(a) Show that a $(\sigma\bar{\sigma})$ bound state with angular momentum $L$ (i.e., an orbital excitation) must have quantum numbers

$$C = (-1)^L; \; P = (-1)^L.$$

(b) Allowing for both orbital and radial excitation, construct a schematic mass spectrum of $(\sigma\bar{\sigma})$ bound states. Label each state with its quantum numbers $J^{PC}$. How does the spectrum differ from the quarkonium spectrum constructed in problem 1.1?

1.3. Analyze the absorption of a virtual photon by a spin-$\frac{1}{2}$ quark in the Breit frame (brick-wall frame) of the quark. Kinematics:

| Incident | | Outgoing |
|---|---|---|
| $\gamma^*$ | $q$ | |
| $(E; p_z) = (0; Q)$ | $\left(\frac{Q}{2}; -\frac{Q}{2}\right)$ | $\left(\frac{Q}{2}; \frac{Q}{2}\right)$ |

(a) Show that the squared matrix element for the absorption of a longitudinal photon vanishes.

(b) Compute the square of the matrix element for absorption of a photon with helicity $= +1$, that is, a transverse photon.

(c) How would your result for a longitudinal photon differ if the incident quark and photon were not precisely (anti)collinear?

1.4. Using the Feynman rules given in appendix B.5, compute the differential cross section $d\sigma/d\Omega$ and the total (integrated) cross section $\sigma \equiv \int d\Omega(d\sigma/d\Omega)$ for the reaction $e^+e^- \to \sigma^+\sigma^-$, where $\sigma^\pm$ is a charged scalar particle. Work in the center-of-momentum frame and in the high-energy limit (where all the masses may be neglected). Assume that the colliding beams are unpolarized.

1.5. (a) Again referring to Appendix B.5 for the Feynman rules, compute the differential cross section $d\sigma/d\Omega$ and the total (integrated) cross section $\sigma \equiv \int d\Omega(d\sigma/d\Omega)$ for the reaction $e^+e^- \to \mu^+\mu^-$. Work in the center-of-momentum frame and in the high-energy limit (where all the masses may be neglected). Assume that the colliding beams are unpolarized, and sum over the spins of the produced muons.

(b) Look up the original evidence for quark–antiquark jets in the inclusive reaction $e^+e^- \to$ hadrons [G. J. Hanson et al., *Phys. Rev. Lett.* **35**, 1609 (1975)]. Now recompute the differential cross section for the reaction $e^+e^- \to \mu^+\mu^-$, assuming the initial beams to be transversely polarized. [See also R. F. Schwitters et al., *Phys. Rev. Lett.* **35**, 1320 (1975).]

1.6. Define the requirements for an experiment to measure the gyromagnetic ratio of the tau lepton, taking into account the $\tau$ lifetime and the anticipated result

$g_\tau \approx 2$. For background, become acquainted with the methods used to measure the magnetic anomalies of the electron [A. Rich and J. Wesley, *Rev. Mod. Phys.* **44**, 250 (1972); R. S. Van Dyck, Jr., P. B. Schwinberg, and H. G. Dehmelt, *Phys. Rev. Lett.* **38**, 93 (1977); D. Hanneke, S. Fogwell, and G. Gabrielse, *Phys. Rev. Lett.* **100**, 120801 (2008)] and muon [F. Combley, F. J. M. Farley, and E. Picasso, *Phys. Rep.* **68**, 93 (1981); G. W. Bennett et al. [Muon $g - 2$ Collaboration], *Phys. Rev. D* **73**, 072003 (2006)], and the magnetic dipole moments of the nucleons [N. F. Ramsey, *Molecular Beams,* Oxford University Press, Oxford, 1956] and unstable hyperons [L. Schachinger et al., *Phys. Rev. Lett.* **41**, 1348 (1978); L. G. Pondrom, *Phys. Rept.* **122**, 57 (1985). A novel technique is described in D. Chen et al. [E761 Collaboration], *Phys. Rev. Lett.* **69**, 3286 (1992).] For the current state of the art, see J. Abdallah et al. [DELPHI Collaboration], *Eur. Phys. J.* **C35**, 159 (2004).

1.7. Assume that the charged weak current has the left-handed form of equation (1.2.2) and that the interaction Hamiltonian is of the "current–current" form, $\mathcal{H}_\mathrm{W} \sim J J^\dagger + J^\dagger J$.

(a) Enumerate the kinds of interactions (i.e., terms) in the Hamiltonian that may occur in a world composed of the electron and muon doublets.

(b) List the leptonic processes such as $\nu_\mu e \to \nu_\mu e$ that are consistent with the known selection rules but do not appear in the charged-current Hamiltonian.

1.8. Derive the connection between $|\Psi(0)|^2$ and the leptonic decay rate of a $(q\bar{q})$ vector meson. It is convenient to proceed by the following steps:

(a) Compute the spin-averaged cross section for the reaction $q\bar{q} \to e^+ e^-$. Show that it is

$$\sigma = \frac{\pi \alpha^2 e_q^2}{12 E^2} \cdot \frac{\beta_e}{\beta_q} (3 - \beta_e^2)(3 - \beta_q^2),$$

where $E$ is the c.m. energy of a quark and $\beta_i$ is the speed of particle $i$.

(b) The annihilation rate in a ${}^3S_1$ vector meson is the density $\times$ relative velocity $\times$ 4/3 (to undo the spin average) $\times$ the cross section, or

$$\Gamma = |\Psi(0)|^2 \times 2\beta_q \times \tfrac{4}{3} \times \sigma.$$

(c) How is the result modified if the vector-meson wave function is

$$|V^0\rangle = \sum_i c_i |q_i \bar{q}_i\rangle ?$$

(d) Now neglect the lepton mass and the quark binding energy and assume that the quarks move nonrelativistically. Show that

$$\Gamma(V^0 \to e^+ e^-) = \frac{16\pi\alpha^2}{3 M_V^2} |\Psi(0)|^2 \left(\sum_i c_i e_i\right)^2.$$

(e) How is this result modified if quarks come in $N_c$ colors and hadrons are color singlets? [This result is due to R. Van Royen and V. F. Weisskopf, *Nuovo Cim.* **50**, 617 (1967); *Nuovo Cim.* **51**, 583 (1967); and to H. Pietschmann and W. Thirring, *Phys. Lett.* **21**, 713 (1966).]

1.9. Outline a "three-neutrino experiment" to establish that a neutral, penetrating beam of $\nu_\tau$ materializes into $\tau$ upon interacting in matter. [For background, look at the first two-neutrino experiment, J. Danby et al., *Phys. Rev. Lett.* **9**, 36 (1962). See also the Nobel lectures of M. Schwartz, J. Steinberger, and L. M. Lederman, reprinted in *Rev. Mod. Phys.* **61**, 527, 533, 547 (1989).] What would provide a copious source of $\nu_\tau$? What energy would be advantageous for the detection of the produced $\tau$? What characteristics would be required of the detector? What are the important backgrounds, and how would you handle them? [K. Kodama et al. [DONuT Collaboration], *Phys. Rev.* **D78**, 052002 (2008)].

### ⅠⅠⅠⅠⅠⅠⅠⅠⅠⅠⅠ FOR FURTHER READING ⅠⅠⅠⅠⅠⅠⅠⅠⅠⅠⅠ

**Generalities.** Among the standard textbooks on particle physics, the following contain excellent summaries of experimental systematics:

    A. Bettini, *Introduction to Elementary Particle Physics*, Cambridge University Press, Cambridge, 2008;

    D. J. Griffiths, *Introduction to Elementary Particles*, 2nd ed., Wiley-VCH, Weinheim, 2008;

    A. Seiden, *Particle Physics: A Comprehensive Introduction*, Addison-Wesley, Reading, MA, 2004;

    R. N. Cahn and G. Goldhaber, *The Experimental Foundations of Particle Physics*, 2nd ed., Cambridge University Press, Cambridge, 2009.

These articles in the Spring 1997 issue of the SLAC *Beam Line* (http://j.mp/ACcaWv) celebrate the 100th anniversary of the discovery of the electron and the beginning of particle physics:

    A. Pais, "The Discovery of the Electron";

    S. Weinberg, "What Is an Elementary Particle?";

    C. Quigg, "Elementary Particles: Yesterday, Today, Tomorrow."

**SU(3) Flavor Symmetry.** The essentials are explained in many places, including

    P. Carruthers, *Introduction to Unitary Symmetry*, Interscience, New York, 1966,

    S. Coleman, "An introduction to unitary symmetry," in *Aspects of Symmetry: Selected Erice Lectures*, Cambridge University Press, Cambridge, 1988, Chapter 1,

    M. Gell-Mann and Y. Ne'eman, *The Eightfold Way*, Benjamin, New York, 1964,

    M. Gourdin, *Unitary Symmetries and Their Applications to High Energy Physics*, North-Holland, Amsterdam, 1967;

    D. B. Lichtenberg, *Unitary Symmetry and Elementary Particles*, Academic Press, New York, 1978;

    H. J. Lipkin, *Lie Groups for Pedestrians*, 2nd ed., North-Holland, Amsterdam, 1966.

**Quark Model.** The standard reference for early applications is the reprint volume by

    J.J.J. Kokkedee, *The Quark Model*, Benjamin, Reading, MA, 1969.

The next wave of developments is treated in detail in the book by
  F. E. Close, *Introduction to Quarks and Partons*, Academic Press, New York, 1979,

in review articles by
  O. W. Greenberg, *Ann. Rev. Nucl. Part. Sci.* **28**, 327 (1978),
  A. W. Hendry and D. B. Lichtenberg, *Rep. Prog. Phys.* **41**, 1707 (1978),
  H. J. Lipkin, *Phys. Rep.* **8C**, 173 (1973),
  J. L. Rosner, *Phys. Rep.* **11C**, 189 (1974),

and in summer school lectures by
  R. H. Dalitz, in *Fundamentals of Quark Models*, Scottish Universities Summer School in
    Physics, 1976, ed. I. M. Barbour and A. T. Davies, SUSSP, Edinburgh, 1977, p. 151,
  C. Quigg, in *Gauge Theories in High Energy Physics*, 1981 Les Houches Summer School,
    ed. M. K. Gaillard and R. Stora, North-Holland, Amsterdam, 1983,
    http://j.mp/ywNnnd,
  J. L. Rosner, in *Techniques and Concepts of High-Energy Physics*, St. Croix Advanced
    Study Institute, 1980, ed. T. Ferbel, Plenum, New York, 1981, p. 1.

In addition, condensed summaries and exhaustive lists of references are to be found in
  S. Gasiorowicz and J. L. Rosner, *Am. J. Phys.* **49**, 954 (1981),
  O. W. Greenberg, *Am. J. Phys.* **50**, 1074 (1982),
  J. L. Rosner, *Am. J. Phys.* **48**, 90 (1980),

**Quark abundance.** Searches for free quarks are reviewed by
  L. Lyons, *Phys. Rept.* **129**, 225 (1985),
  P. F. Smith, *Ann. Rev. Nucl. Part. Sci.* **39**, 73 (1989),
  M. L. Perl, E. R. Lee, and D. Loomba, *Ann. Rev. Nucl. Part. Sci.* **59**, 47 (2009).

**Parton model.** Full treatments appear in F. E. Close, *Quarks and Partons*, and in
  R. P. Feynman, *Photon-Hadron Interactions*, Benjamin, Reading, MA, 1972.

**Color.** A thorough historical review appears in
  O. W. Greenberg and C. A. Nelson, *Phys. Rep.* **32C**, 69 (1977).

**Gauge theories.** Many useful articles are cited in the resource letter by
  T. P. Cheng and L.-F. Li, *Am. J. Phys.* **56**, 586, 1048(E) (1988).

Past, present, and future of gauge theories are evoked in the 1979 Nobel lectures by
  S. L. Glashow, *Rev. Mod. Phys.* **52**, 539 (1980),
  A. Salam, *Rev. Mod. Phys.* **52**, 525 (1980),
  S. Weinberg, *Rev. Mod. Phys.* **52**, 515 (1980).

The following popularizations may also be read with profit at this point:
  C. Quigg, "Particles and the Standard Model," in *The New Physics: for the Twenty-first
    Century*, ed. G. Fraser, Cambridge University Press, Cambridge, 2006, chap. 4;
  M. Riordan, *The Hunting of the Quark*, Simon & Schuster, New York, 1987;
  F. E. Close, *The New Cosmic Onion: Quarks and the Nature of the Universe*, 2nd ed.,
    Taylor & Francis, London, 2006;
  F. Wilczek, *The Lightness of Being: Mass, Ether, and the Unification of Forces*, Basic
    Books, New York, 2008;
  F. Close, *The Infinity Puzzle*, Oxford University Press, Oxford, 2011;
  G. 't Hooft, *In Search of the Ultimate Building Blocks*, Cambridge University Press,
    Cambridge and New York, 1997;

Y. Nambu, *Quarks: Frontiers in Elementary Particle Physics,* World Scientific, Singapore, 1985;

A. Watson, *The Quantum Quark,* Cambridge University Press, Cambridge, 2004;

S. Weinberg,. "Unified Theories of Elementary Particle Interactions," *Sci. Am.* **231**, 50 (July 1974);

S. L. Glashow, "Quarks with Color and Flavor," *Sci. Am.* **233**, 38 (October 1975);

Y. Nambu, "The Confinement of Quarks," *Sci. Am.* **235**, 48 (November 1976);

G. 't Hooft, "Gauge Theories of the Forces between Elementary Particles," *Sci. Am.* **242**, 104 (June 1980);

H. Georgi, "A Unified Theory of Elementary Particles and Forces," *Sci. Am.* **244**, 40 (April 1981);

C. Rebbi, "The lattice theory of quark confinement," *Sci. Am.* **248**, 54 (February 1983);

C. Quigg, "Elementary Particles and Forces," *Sci. Am.* **252**, (4) 84 (April 1985);

D. H. Weingarten, "Quarks by computer," *Sci. Am.* **274**, 116 (February 1996);

C. Quigg, "The Coming Revolutions in Particle Physics," *Sci. Am.* **298**, (2) 46 (February 2008).

**Hidden symmetries.** Spontaneous symmetry breaking is common in physics, and parallels to condensed-matter physics are drawn in

Y. Nambu, *Rev. Mod. Phys.* **81**, 1015 (2009).

**Quantum Chromodynamics.** An extensive annotated bibliography appears in

A. S. Kronfeld and C. Quigg, *Am. J. Phys.* **78**, 1081 (2010).

## ⅢⅢⅢⅢⅢⅢⅢ REFERENCES ⅢⅢⅢⅢⅢⅢⅢⅢⅢ

1. H. Hertz, "Über die Beziehungen zwischen Licht und Elektrizität," in *Gesammelte Werke I* S. 339–354, http://j.mp/xdGK9t and http://j.mp/zobS0W.

2. K. Nakamura et al. [Particle Data Group], *J. Phys.* **G37**, 075021 (2010). This is the source for otherwise unattributed experimental results throughout this volume. Consult J. Beringer et al. [Particle Data Group], *Phys. Rev.* **D86**, 010001 (2012). for updates.

3. See K. Hagiwara, K. Hikasa, and M. Tanabashi "Searches for Quark and Lepton Compositeness" in Ref. [2] for a more detailed discussion.

4. C. S. Wu, E. Ambler, R. W. Hayward, D. D. Hoppes, and R. P. Hudson, *Phys. Rev.* **105**, 1413 (1957).

5. Perturbation theory conventions are given in appendix A. For a guided tour of simple Feynman-diagram calculations, see R. P. Feynman, *Quantum Electrodynamics,* and *The Theory of Fundamental Processes,* Westview Press, Boulder, CO, 1998. The prediagram approach is represented in W. Heitler, *The Quantum Theory of Radiation,* 3rd ed., Dover Publications, Mineola, NY, 2010.

6. M. Gell-Mann, *Phys. Lett.* **8**, 214 (1964); G. Zweig, CERN Report 8182/TH.401 (1964), http://j.mp/AFfsaG; CERN Report 8419/TH.412 (1964), http://j.mp/AvBGHM, reprinted in *Developments in the Quark Theory of Hadrons, Vol. I: 1964–1978,* ed. D. B. Lichtenberg and S. P. Rosen, Hadronic Press, Nonantum, MA, 1980, p. 22. Some historical perspective is provided by G. Zweig, in *Baryon 1980,* IV International Conference on Baryon Resonances, ed. N. Isgur, University of Toronto, Toronto, 1980, p. 439.

7. See the listing of $b\bar{b}$ mesons in Ref. [2], p. 1100, for the $b\bar{b}$ spectrum [http://j.mp/wGW7Kn]. The top-quark lifetime, $\tau_t \lesssim 10^{-24}$ s, is too short to allow the formation of top hadrons and, specifically, toponium states. See, for example, V. M. Abazov et al. [D0 Collaboration], *Phys. Rev. D* **85**, 091104 (2012).

8. Many predictions are in fact shared by a model of integrally charged colored quarks due originally to M. Y. Han and Y. Nambu, *Phys. Rev.* **139**, B1005 (1965), and it is notoriously difficult to draw distinctions. A straightforward experimental test will be developed in section 8.7 and problem 8.27. Properties of the $\eta(549)$ and $\eta'(958)$ strongly favor the fractional charge assignment, as explained in the articles cited in problem 8.27.

9. M. Gell-Mann, *Phys. Rev.* **92**, 833 (1953); T. Nakano and K. Nishijima, *Prog. Theor. Phys. (Kyoto)* **10**, 581 (1955).

10. O. W. Greenberg, *Phys. Rev. Lett.* **13**, 598 (1964).

11. Refined theoretical estimates for the leptonic branching ratio (17.75%) have been given by F. J. Gilman and D. H. Miller, *Phys. Rev. D* **17**, 1846 (1978) and by N. Kawamoto and A. I. Sanda, *Phys. Lett.* **76B**, 446 (1978).

12. A clear and thorough discussion of the calculation of the $\pi^0$ lifetime is given in J. F. Donoghue, E. Golowich, and B. R. Holstein, *Dynamics of the Standard Model*, Cambridge University Press, Cambridge, 1992.

13. The first thorough discussions of this phenomenon are to be found in the reports by H. Newman (Mark-J Collaboration), Ch. Berger (PLUTO Collaboration), G. Wolf (TASSO Collaboration), and S. Orito (JADE Collaboration), in *Proceedings of the 1979 International Symposium on Lepton and Photon Interactions at High Energies*, ed. T.B.W. Kirk and H.D.I. Abarbanel, Fermilab, Batavia, Illinois, 1979, pp. 3, 19, 34, 52, http://j.mp/AEG8vn.

14. Today's electroweak theory developed from a proposal by S. Weinberg, *Phys. Rev. Lett.* **19**, 1264 (1967) and by A. Salam, in *Elementary Particle Theory: Relativistic Groups and Analyticity* (8th Nobel Symposium), ed. N. Svartholm, Almqvist and Wiksell International, Stockholm, 1968, p. 367. The theory is built on the $SU(2)_L \otimes U(1)_Y$ gauge symmetry investigated by S. L. Glashow, *Nucl. Phys.* **22**, 579 (1961). For a meticulous intellectual history see M. Veltman, in *Proceedings of the 6th International Symposium on Electron and Photon Interactions at High Energies*, ed. H. Rollnik and W. Pfeil, North-Holland, Amsterdam, 1974, p. 429. A later historical perspective is that of S. Coleman, *Science* **206**, 1290 (1979).

15. N. Cabibbo, *Phys. Rev. Lett.* **10**, 531 (1963).

16. S. L. Glashow, J. Iliopoulos, and L. Maiani, *Phys. Rev. D* **2**, 1285 (1970).

17. B. J. Bjørken and S. L. Glashow, *Phys. Lett.* **11**, 255 (1964).

18. J. J. Aubert et al., *Phys. Rev. Lett.* **33**, 1404 (1974); J.-E. Augustin et al., *Phys. Rev. Lett.* **33**, 1406 (1974).

19. Definitive evidence for charmed mesons was presented by G. Goldhaber et al., *Phys. Rev. Lett.* **37**, 255 (1976), and by I. Peruzzi, et al., *Phys. Rev. Lett.* **37**, 569 (1976). The first example of a charmed baryon was reported by E. G. Cazzoli et al., *Phys. Rev. Lett.* **34**, 1125 (1975).

# Two

## Lagrangian Formalism and Conservation Laws

There are many ways to formulate the relativistic quantum field theory of interacting particles, each with its own set of advantages and shortcomings or inconveniences. The Lagrangian formalism has a number of attributes that make it particularly felicitous for our rather utilitarian purposes. Not to be neglected among its assets are that it is a familiar construct in classical mechanics and that many of its practical advantages can be understood already at the classical level. The Lagrangian approach is characterized by a simplicity in that field theory may be regarded as the limit of a system with $n$ degrees of freedom as $n$ tends toward infinity. Perhaps more to the point, it provides a formalism in which relativistic covariance is manifest, because the four coordinates of spacetime enter symmetrically. This is a decided, though by no means indispensable, advantage for the construction of a relativistic theory.

Lagrangian field theory is also particularly suited to the systematic discussion of invariance principles and the conservation laws to which they are related. In addition, a variational principle provides a direct link between the Lagrangian and the equations of motion. The foregoing properties are of particular value for the development of gauge theories, in which the interactions arise as consequences of local gauge symmetries. Finally, the path that leads from a Lagrangian to a quantum field theory by the method of canonical quantization is extremely well traveled. This makes it possible to conduct much of the discussion of the formulation of gauge theories at what is essentially the classical level and to make the leap to quantum field theory simply by using standard results, without repeating developments that are to be found in every field theory textbook. Thus equipped with the Feynman rules for a theory, we may proceed to calculate the consequences [1].

We do three things in this brief chapter. First, we summarize the basic elements of the Lagrangian formulation of classical mechanics and of field theory and the derivation of the equations of motion. Second, we evoke the intimate connection between continuous symmetries of the Lagrangian and constants of the motion, which is embodied in Noether's theorem. Third, in the course of these discussions, we take the opportunity to recall several elementary free-particle Lagrangians and the associated equations of motion. These will recur in more interesting physical contexts throughout this text.

## 2.1  HAMILTON'S PRINCIPLE

The Lagrangian function $L$ may be regarded as the fundamental object in classical mechanics. [2] From it may be constructed the classical action

$$S \equiv \int_{t_1}^{t_2} dt\, L(q, \dot{q}), \tag{2.1.1}$$

where $q(t)$ is the generalized coordinate and $\dot{q}(t)$, the generalized velocity. The equations of motion follow from Hamilton's principle of least action, according to which the variation

$$\delta S = \delta \int_{t_1}^{t_2} dt\, L(q, \dot{q}) = 0, \tag{2.1.2}$$

subject to the condition that variations of the generalized coordinates vanish at the endpoints $t_1$ and $t_2$. Thus the physical path is the particular trajectory joining $q_1 \equiv q_1(t)$ and $q_2 \equiv q_2(t)$, along which the action is stationary. An important generalization to quantum mechanics as a weighted sum over paths has been developed by Feynman [3].

In a large number of problems of physical interest, the Lagrangian depends only upon the generalized coordinates and their first derivatives. In this case, satisfaction of Hamilton's principle is guaranteed by the Euler–Lagrange equations in the form

$$\frac{\partial L}{\partial q} = \frac{d}{dt}\left(\frac{\partial L}{\partial \dot{q}}\right). \tag{2.1.3}$$

To show this, let us note that the variation in the action is given by

$$\delta S = \int_{t_1}^{t_2} dt\, \left[\frac{\partial L}{\partial q(t)}\delta q(t) + \frac{\partial L}{\partial \dot{q}(t)}\delta \dot{q}(t)\right]. \tag{2.1.4}$$

Because $\delta \dot{q}(t) = (d/dt)\delta q(t)$, the second term may be integrated by parts:

$$\int_{t_1}^{t_2} dt\, \frac{\partial L}{\partial \dot{q}} \cdot \frac{d}{dt}\delta q = \frac{\partial L}{\partial \dot{q}}\delta q\Big|_{t_1}^{t_2} - \int_{t_1}^{t_2} dt\, \left(\frac{d}{dt}\frac{\partial L}{\partial \dot{q}}\right)\delta q. \tag{2.1.5}$$

Because the endpoints are constrained,

$$\delta q(t_1) = 0 = \delta q(t_2), \tag{2.1.6}$$

the first term in (2.1.5) is identically zero, and the variation of the action becomes

$$\delta S = \int_{t_1}^{t_2} dt \left( \frac{\partial L}{\partial q} - \frac{d}{dt} \frac{\partial L}{\partial \dot{q}} \right) \delta q, \tag{2.1.7}$$

which vanishes provided that (2.1.3) holds. For the case of several (separately varied) generalized coordinates the same arithmetic leads to a set of Euler–Lagrange equations in the form (2.1.3), one for each coordinate.

The most familiar cases in classical mechanics are those for which the Lagrangian is simply the difference of kinetic and potential energies. Indeed, if we write in one dimension

$$L = \tfrac{1}{2}m\dot{x}^2 - V(x), \tag{2.1.8}$$

the Euler–Lagrange equation gives

$$\frac{-dV}{dx} = m\ddot{x}. \tag{2.1.9}$$

Upon identifying the applied force as

$$F \equiv \frac{-dV}{dx}, \tag{2.1.10}$$

we simply recover Newton's equation of motion (his second law).

The formalism can easily be generalized to include a Lagrangian that depends explicitly on the time $t$ or some other parameter. This is appropriate for an "open" system, which exchanges energy and momentum with its surroundings. A frequently encountered case is that of an external driving force $F(t)$, represented in the Lagrangian by a term $qF(t)$. The mathematical formalism can also be extended to treat Lagrangians that contain higher than first-order derivatives of the generalized coordinates. The ensuing Euler–Lagrange equations of motion are then higher than second order, but a "local" description in terms of local differential equations remains possible as long as the Lagrangian contains only derivatives of finite order. These generalizations will not be required for the study of gauge theories.

With the Lagrangian description of classical mechanics in hand, the transition to Lagrangian field theory, guided by the requirement of relativistic invariance, is straightforward. Construction of a field theory begins from the Lagrangian density $\mathcal{L}(\phi(x), \partial_\mu \phi(x))$, a functional of the field $\phi(x)$ and its four-gradient $\partial_\mu \phi(x) = \partial\phi(x)/\partial x^\mu$. The field may be regarded as a separate, generalized coordinate at each value of its argument, the spacetime coordinate $x$. Thus arises the description of field theory as a closed system with an infinite number of degrees of freedom.

The classical action is now defined as

$$S \equiv \int_{t_1}^{t_2} dt \int d^3\mathbf{x} \mathcal{L}(\phi(x), \partial_\mu \phi(x)), \tag{2.1.11}$$

where the spatial integral of the Lagrangian density takes the part of the Lagrangian in classical mechanics:

$$L \equiv \int d^3\mathbf{x} \mathcal{L}(\phi(x), \partial_\mu \phi(x)). \tag{2.1.12}$$

By the same reasoning followed in the classical problem, the action is required to be stationary,

$$\delta \int_{t_1}^{t_2} dt \int d^3\mathbf{x} \mathcal{L}(\phi(x), \partial_\mu \phi(x)) = 0, \tag{2.1.13}$$

subject as always to the constraint that the variations in the fields vanish at the endpoints characterized by $t_1$ and $t_2$. The requirement of least action is now ensured by the Euler–Lagrange equations for the Lagrangian density,

$$\frac{\partial \mathcal{L}}{\partial \phi(x)} = \partial_\mu \frac{\partial \mathcal{L}}{\partial(\partial_\mu \phi(x))}, \tag{2.1.14}$$

which in turn lead to explicit equations of motion for the fields. Evidently these equations will be covariant if the Lagrangian density itself transforms as a Lorentz scalar. Moreover, the equations of motion are unchanged if a total divergence is added to the Lagrangian density. For problems involving several fields $\phi_i$, the variational principle applied separately to each field leads to a set of equations (2.1.14), precisely as for mechanical problems with several generalized coordinates.

In what follows, it will frequently be convenient to refer in context to the Lagrangian density $\mathcal{L}$ by the shorter term Lagrangian.

## 2.2 FREE FIELD THEORY EXAMPLES

In many situations of physical interest, it is possible to establish, or to divine, the equations of motion for the fields directly. When this can be done, the Lagrangian is an afterthought that facilitates a systematic quantization or serves as a useful expedient for the investigation of invariance principles. For our purposes of constructing gauge theories from physical symmetries, it will often be the Lagrangian that comes first and the equations of motion that appear as consequences. Even when we proceed in this manner, however, the construction of the Lagrangian will be guided by the desire to reproduce certain characteristics of simpler equations of motion. It is, therefore, invaluable to be conversant with some of the most commonly occurring Lagrangians.

What is in many ways the simplest field theory is that of a field $\phi(x)$ that transforms as a Lorentz scalar or pseudoscalar. The Lagrangian

$$\mathcal{L} = \tfrac{1}{2}[(\partial^\mu \phi)(\partial_\mu \phi) - m^2 \phi^2] \tag{2.2.1}$$

leads, via the Euler–Lagrange equation (2.1.14), to the Klein–Gordon equation,

$$(\Box + m^2)\phi(x) = 0, \tag{2.2.2}$$

for a spinless free particle of mass $m$.

The complex scalar field

$$\phi(x) = \frac{a(x) + ib(x)}{\sqrt{2}}, \qquad (2.2.3)$$

which may be expressed in terms of the two independent real functions $a(x)$ and $b(x)$, is appropriate for the description of particles that carry an additive quantum number, such as electric charge. For most purposes it is convenient to write the Lagrangian in terms of $\phi(x)$ and the complex-conjugate field $\phi^*(x) = (a(x) - ib(x))/\sqrt{2}$ instead of $a(x)$ and $b(x)$. In special circumstances, still other parametrizations are more economical. The Lagrangian of the complex field is written by analogy with that of the real scalar field (2.2.1) as

$$
\begin{aligned}
\mathcal{L} &= \tfrac{1}{2}[(\partial^\mu a)(\partial_\mu a) - m^2 a^2] + \tfrac{1}{2}[(\partial^\mu b)(\partial_\mu b) - m^2 b^2] \\
&= (\partial^\mu \phi)^*(\partial_\mu \phi) - m^2 \phi^* \phi \\
&= |\partial^\mu \phi|^2 - m^2 |\phi|^2 .
\end{aligned}
\qquad (2.2.4)
$$

Independent variations of $\phi$ and $\phi^*$ (or, equivalently, of $a$ and $b$ then yield as equations of motion the two Klein–Gordon equations

$$(\square + m^2)\phi(x) = 0, \ (\square + m^2)\phi^*(x) = 0. \qquad (2.2.5)$$

We will see later that in the presence of appropriate interactions the fields $\phi$ and $\phi^*$ can be identified with particles of opposite charge.

The Dirac equation

$$(i\gamma^\mu \partial_\mu - m)\psi(x) = 0 \qquad (2.2.6)$$

for a free fermion follows from the Lagrangian

$$\mathcal{L} = \bar{\psi}(x)(i\gamma^\mu \partial_\mu - m)\psi(x), \qquad (2.2.7)$$

where the Dirac conjugate field is, as usual, defined as $\bar{\psi}(x) = \psi^\dagger(x)\gamma^0$. (Notations and conventions are reviewed in appendix A.)

Electromagnetic interactions will be discussed at some length in chapter 3, when the subject of gauge invariance is studied in detail. For the present, let us simply recall that a convenient formulation of the Maxwell theory may be obtained by expressing the electric and magnetic field strengths **E** and **B** in terms of a four-vector potential $A_\mu(x)$. In the absence of sources, an appropriate Lagrangian for the free electromagnetic field is

$$\mathcal{L} = -\tfrac{1}{4}(\partial_\nu A_\mu - \partial_\mu A_\nu)(\partial^\nu A^\mu - \partial^\mu A^\nu), \qquad (2.2.8)$$

from which Maxwell's equations may be readily verified. The factor of $\tfrac{1}{4}$ ensures the conventional normalization of the energy density.

Finally, for the case of a massive vector field with mass $M$, the Lagrangian is

$$\mathcal{L} = -\tfrac{1}{4}(\partial_\nu A_\mu - \partial_\mu A_\nu)(\partial^\nu A^\mu - \partial^\mu A^\nu) + \tfrac{1}{2}M^2 A^\mu A_\mu. \tag{2.2.9}$$

The difference in sign between the mass term for the vector field and that for the scalar field in (2.2.1) is simply a consequence of the fact that the physical degrees of freedom are represented by the spacelike components of $A_\mu$, which contribute with a minus sign to $A^\mu A_\mu$. The equation of motion, which follows from (2.2.9) upon regarding each component of $A_\mu$ as an independent field, is the Proca equation,

$$\partial_\nu(\partial^\nu A^\mu - \partial^\mu A^\nu) + M^2 A^\mu = 0. \tag{2.2.10}$$

Taking the divergence of this equation, we find that

$$M^2(\partial \cdot A) = 0, \tag{2.2.11}$$

which implies that the vector field is divergenceless,

$$\partial \cdot A = 0, \tag{2.2.12}$$

because $M^2 \neq 0$. The vanishing of the divergence $(\partial \cdot A)$ corresponds to the covariant elimination of one of the four apparent degrees of freedom of the vector field $A_\mu$. It also permits the Proca equation to be written in a more familiar form,

$$(\square + M^2)A^\mu = 0. \tag{2.2.13}$$

## 2.3 SYMMETRIES AND CONSERVATION LAWS

Much of the rest of this text will be devoted to recognizing symmetries in nature and deducing their implications for the fundamental particles and their interactions. Invariance principles and the associated symmetry transformations are of many types: continuous or discrete, geometrical or internal, and—as the following chapter will make apparent—global or local. They play many important roles in physics including that of restricting—and thus guiding—the formulation of theories, as we have already tacitly acknowledged in requiring the Lagrangian to be a Lorentz scalar.

When the equations of motion are derived from a variational principle, a general and systematic procedure is available for establishing conservation theorems and constants of the motion as a consequence of invariance properties. Thus, conservation laws and selection rules observed in nature may be imposed as symmetries of the Lagrangian, restricting or prescribing its form. The general framework for this program is provided by Noether's theorem [4], which correlates a conservation law with every continuous symmetry transformation under which the Lagrangian is invariant in form. Let us illustrate the utility of this theorem by considering two examples, one of a spacetime, or geometrical, invariance, and the other of an internal symmetry.

First consider, as an example of geometrical transformations of the spacetime variables, translations of the form

$$x_\mu \rightarrow x'_\mu = x_\mu + a_\mu, \tag{2.3.1}$$

where the infinitesimal displacement $a_\mu$ is independent of the coordinate $x_\mu$. A Lagrangian that is invariant in form under a transformation of this type will, therefore change by an amount

$$\delta\mathcal{L} = \mathcal{L}[x'] - \mathcal{L}[x] = a^\mu d\mathcal{L}/dx^\mu. \tag{2.3.2}$$

For a Lagrangian with no explicit dependence on the coordinates, we may equivalently compute the change as

$$\delta\mathcal{L} = \frac{\partial\mathcal{L}}{\partial\phi}\delta\phi + \frac{\partial\mathcal{L}}{\partial(\partial_\mu\phi)}\delta(\partial_\mu\phi), \tag{2.3.3}$$

where

$$\delta\phi = \phi(x') - \phi(x) = a^\mu\delta_\mu\phi(x) \tag{2.3.4}$$

and

$$\delta(\partial_\mu\phi) = \partial_\mu\phi(x') - \partial_\mu\phi(x) = a^\nu\partial_\nu\partial_\mu\phi(x). \tag{2.3.5}$$

Consequently, using the Euler–Lagrange equations to eliminate $\partial\mathcal{L}/\partial\phi$, we find

$$\delta\mathcal{L} = \left[\partial_\nu\frac{\partial\mathcal{L}}{\partial(\partial_\nu\phi)}\right]a^\mu\partial_\mu\phi(x) + \frac{\partial\mathcal{L}}{\partial(\partial_\nu\phi)}a^\mu\partial_\nu\partial_\mu\phi(x)$$
$$= \partial_\nu\frac{\partial\mathcal{L}}{\partial(\partial_\nu\phi)}a^\mu\partial_\mu\phi(x). \tag{2.3.6}$$

Equating the two expressions (2.3.2) and (2.3.6) for $\delta\mathcal{L}$ yields

$$a_\mu\partial_\nu\left[\frac{\partial\mathcal{L}}{\partial(\partial_\nu\phi)}\partial^\mu\phi - g^{\mu\nu}\mathcal{L}\right] = 0, \tag{2.3.7}$$

which is to be satisfied for arbitrary infinitesimal displacements $a_\mu$. We therefore conclude that the stress–energy–momentum flow characterized by the tensor

$$\Theta^{\mu\nu} \equiv \frac{\partial\mathcal{L}}{\partial(\partial_\nu\phi)}\partial^\mu\phi - g^{\mu\nu}\mathcal{L} \tag{2.3.8}$$

satisfies the local conservation law

$$\partial_\mu\Theta^{\mu\nu} = 0. \tag{2.3.9}$$

It is easily verified that $\Theta^{00}$ is the Hamiltonian density

$$\mathcal{H} = \frac{\partial \mathcal{L}}{\partial(\partial_0 \phi)} \partial^0 \phi - \mathcal{L}, \qquad (2.3.10)$$

so that the total energy

$$H = \int d^3 \mathbf{x} \Theta^{00} \qquad (2.3.11)$$

is a constant of the motion and the $\Theta^{0\nu}$ correspond to momentum densities. Thus translation invariance has led to four-momentum conservation.

It is also instructive—and more central to our later purposes—to consider by a simple example the consequences of an internal symmetry. If the neutron and proton are taken as elementary particles with a common mass $m$, the Lagrangian for free nucleons may be written in an obvious notation as

$$\mathcal{L} = \bar{p}(i\gamma^\mu \partial_\mu - m)p + \bar{n}(i\gamma^\mu \partial_\mu - m)n. \qquad (2.3.12)$$

In terms of the composite spinor

$$\psi \equiv \begin{pmatrix} p \\ n \end{pmatrix}, \qquad (2.3.13)$$

the Lagrangian may be rewritten more compactly as

$$\mathcal{L} = \bar{\psi}(i\gamma^\mu \partial_\mu - m)\psi. \qquad (2.3.14)$$

It is evidently invariant under global isospin rotations,

$$\psi \to \exp\left(\frac{i\boldsymbol{\alpha} \cdot \boldsymbol{\tau}}{2}\right)\psi, \qquad (2.3.15)$$

where $\boldsymbol{\tau} = (\tau_1, \tau_2, \tau_3)$ consists of the usual $2 \times 2$ Pauli isospin matrices, and $\boldsymbol{\alpha} = (\alpha_1, \alpha_2, \alpha_3)$ is an arbitrary constant (three-vector) parameter of the transformation. A global transformation is one that subjects the spinor $\psi$ to the same rotation everywhere in spacetime. Thus the parameter $\boldsymbol{\alpha}$ is independent of the coordinate $x$.

A continuous transformation such as an isospin rotation can be built up out of infinitesimal rotations. It is, therefore, interesting to consider the effect on the Lagrangian of an infinitesimal transformation

$$\psi(x) \to \psi(x) + \frac{i\boldsymbol{\alpha} \cdot \boldsymbol{\tau}}{2}\psi(x), \qquad (2.3.16)$$

which leads to the infinitesimal variations

$$\delta \psi = \frac{i \boldsymbol{\alpha} \cdot \boldsymbol{\tau}}{2} \psi,$$

$$\delta(\partial_\mu \psi) = \frac{i \boldsymbol{\alpha} \cdot \boldsymbol{\tau}}{2} (\partial_\mu \psi). \tag{2.3.17}$$

If the Lagrangian is to be invariant under such a transformation, we require that

$$\delta \mathcal{L} = 0. \tag{2.3.18}$$

Explicit computation yields

$$\begin{aligned}
\delta \mathcal{L} &= \frac{\partial \mathcal{L}}{\partial \psi} \delta \psi + \frac{\partial \mathcal{L}}{\partial(\partial_\mu \psi)} \delta(\partial_\mu \psi) + \frac{\partial \mathcal{L}}{\partial \bar{\psi}} \delta \bar{\psi} + \frac{\partial \mathcal{L}}{\partial(\partial_\mu \bar{\psi})} \delta(\partial_\mu \bar{\psi}) \\
&= \left[ \partial_\mu \frac{\partial \mathcal{L}}{\partial(\partial_\mu \psi)} \right] \frac{i \boldsymbol{\alpha} \cdot \boldsymbol{\tau}}{2} \psi + \frac{\partial \mathcal{L}}{\partial(\partial_\mu \psi)} \frac{i \boldsymbol{\alpha} \cdot \boldsymbol{\tau}}{2} (\partial_\mu \psi) \\
&= \partial_\mu \boldsymbol{\alpha} \cdot \left[ \frac{i}{2} \frac{\partial \mathcal{L}}{\partial(\partial_\mu \psi)} \boldsymbol{\tau} \psi \right],
\end{aligned} \tag{2.3.19}$$

where the second line follows from the equations of motion. The quantity in square brackets may be identified as a conserved current (density),

$$\mathbf{J}^\mu = \frac{i}{2} \frac{\partial \mathcal{L}}{\partial(\partial_\mu \psi)} \boldsymbol{\tau} \psi, \tag{2.3.20}$$

which satisfies the continuity equation

$$\partial_\mu \mathbf{J}^\mu = 0. \tag{2.3.21}$$

For the specific case at hand of the free-nucleon Lagrangian, the explicit form of the conserved current is

$$\mathbf{J}^\mu = \bar{\psi} \gamma^\mu \frac{\boldsymbol{\tau}}{2} \psi, \tag{2.3.22}$$

which is immediately recognizable as the isospin current, in analogy with the familiar electromagnetic current for Dirac particles. It is a standard exercise to show that these classical results carry over to quantum field theory.

In the following chapter we shall pursue further the consequences of internal symmetries and encounter the possibility of deriving interactions from symmetries. There we shall develop in detail the idea of local gauge invariance and study its implications for a theory of electrodynamics.

## PROBLEMS

2.1. Verify explicitly that changing the Lagrangian density by a total divergence leaves the Euler–Lagrange equations unchanged.

2.2. Using the freedom to add a total divergence to the Lagrangian, show that the Dirac Lagrangian for a free massless fermion may be written in the form

$$\mathcal{L}' = \tfrac{1}{2}\bar{\psi} i\gamma^\mu \partial_\mu \psi - \tfrac{1}{2}(\partial_\mu \bar{\psi}) i\gamma^\mu \psi.$$

2.3. Beginning from the Lagrangian for the free electromagnetic field (2.2.8), derive the equations of motion by independently varying each component of the electromagnetic field $A_\mu$.

2.4. Use the requirement that the Lagrangian be invariant under a continuous symmetry to deduce the conserved quantity that corresponds to a particular transformation. Show that invariance under (i) translations in space, (ii) translations in time, and (iii) spatial rotations implies conservation of (i) momentum, (ii) energy, and (iii) angular momentum.

2.5. Consider a gravitating system of $N$ particles in nonrelativistic motion, for which the Lagrangian is

$$L = T - V = \frac{1}{2}\sum_{i=1}^{N} m_i \dot{\mathbf{x}}_i^2 - G_N \sum_{i \neq j} \frac{m_i m_j}{(\mathbf{x}_i - \mathbf{x}_j)^2},$$

where $G_N$ is Newton's constant and $m_i$ and $\mathbf{x}_i$ represent, respectively, the masses and positions of the individual particles. Show that invariance of the Lagrangian under an infinitesimal boost implies that the center of mass $\mathbf{X} \equiv \sum_i m_i \mathbf{x}_i / \sum_i m_i$ of the $N$-particle system moves with constant velocity along the boost direction. [For a relativistic generalization, see L. D. Landau and E. M. Lifshitz, *The Classical Theory of Fields*, 4th ed., trans. M. Hamermesh, Butterworth–Heinemann, Oxford, 1980, §14.]

2.6. The Lagrangian for a free particle in one space dimension is $L = \tfrac{1}{2}m\dot{x}^2$. Show that the action $S$ corresponding to the classical motion of a free particle is

$$S = \frac{m(x_2 - x_1)^2}{2(t_2 - t_1)}.$$

2.7. For a harmonic oscillator with Lagrangian $L = \tfrac{1}{2}m(\dot{x}^2 - \omega^2 x^2)$, show that the classical action is

$$S = \frac{m\omega}{2\sin\omega T}\left[(x_1^2 + x_2^2)\cos\omega T - 2x_1 x_2\right],$$

where $T = (t_2 - t_1)$.

2.8. In the absence of the electroweak interactions, neither electric charge nor a small mass difference would distinguish neutron from proton. What observations might provide evidence for the existence of two nucleon species? For an isospin doublet of nucleons?

2.9. The near degeneracy of the neutron and proton masses, which differ by only 1.4 parts per mille [P. J. Mohr, B. N. Taylor, and D. B. Newell, *Rev. Mod. Phys.* **80**, 633 (2008)], is one hint that the neutron and proton may be taken as two states of a nucleon, with the *n*-*p* mass difference attributed to electroweak effects. A second element of the case for isospin invariance is the charge independence of nuclear forces. Compare the binding energies of the tritium ($^3$H) and $^3$He ground states [D. R. Tilley, H. R. Weller, and H. H. Hasan, *Nucl. Phys.* **A474**, 1 (1988)]. Use the $^3$He charge radius [D. C. Morton, Q. Wu, and G. W. F. Drake, *Phys. Rev.* **A73**, 034502 (2006)] to estimate the Coulomb repulsion in $^3$He, and so test the charge independence of forces in the three-nucleon system.

## ⅢⅢⅢⅢⅢⅢⅢ FOR FURTHER READING ⅢⅢⅢⅢⅢⅢⅢⅢⅢ

**Lagrangian mechanics and electrodynamics.** The Lagrangian description of classical mechanics is developed in many places, including

> H. Goldstein, C. P. Poole, and J. L. Sefko, *Classical Mechanics*, third edition, Addison-Wesley, Reading, MA, 2001,
>
> L. Landau and E. M. Lifshitz, *Mechanics,* 3rd ed., trans. by J. B. Sykes and J. S. Bell, Butterworth-Heinemann, Oxford, 1982,
>
> L. Landau and E. M. Lifshitz, *The Classical Theory of Fields,* 4th ed., trans. by M. Hamermesh, Butterworth-Heinemann, Oxford, 1980,
>
> D. E. Soper, *Classical Field Theory,* Dover, New York, 2008.

For treatments of the Lagrangian description of the classical electromagnetic field, see

> J. D. Jackson, *Classical Electrodynamics,* 3rd ed., Wiley, New York, 1998, chap. 12,
>
> F. Rohrlich, *Classical Charged Particles,* 3rd ed., World Scientific, Singapore, 2007.

Many topics of current interest in particle physics are approached from a (semi)classical perspective in

> V. A. Rubakov, *Classical Theory of Gauge Fields,* Princeton University Press, Princeton, 2002.

**Lagrangian formulation of quantum field theory.** The development of the vision that relativistic quantum field theory might serve as a description of all the fundamental interactions is surveyed in

> F. Wilczek, *Rev. Mod. Phys.* **71**, S85 (1999).

Elaboration of the material in this chapter may be found in many of the standard field theory textbooks, among which the following are especially accessible:

> I.J.R. Aitchison and A.J.G. Hey, *Gauge Theories in Particle Physics*, 4th ed., CRC Press, Boca Raton, 2012;
>
> L. Álvarez-Gaumé and M. Á. Vázquez-Mozo, *An Invitation to Quantum Field Theory,* Springer, Heidelberg, 2012;
>
> S. Coleman, *Notes from Sidney Coleman's Physics 253a,* http://j.mp/eCpRvP,
>
> M. E. Peskin and D. V. Schroeder, *An Introduction to Quantum Field Theory,* Westview Press, Boulder, CO, 1995;
>
> L. H. Ryder, *Quantum Field Theory,* 2nd ed., Cambridge University Press, Cambridge, 1996;
>
> S. Weinberg, *The Quantum Theory of Fields, Volume 1: Foundations,* Cambridge University Press, Cambridge, 2005.

Among classics that predate the primacy of gauge theories, it is worth consulting

J. D. Bjorken and S. D. Drell, *Relativistic Quantum Fields,* McGraw-Hill, New York, 1965,

N. N. Bogoliubov and D. V. Shirkov, *Introduction to the Theory of Quantized Fields,* trans. by G. M. Volkoff, Interscience, New York, 1959,

C. Itzykson and J.-B. Zuber, *Quantum Field Theory,* Dover, New York, 2006,

F. Mandl and G. Shaw, *Quantum Field Theory,* rev. ed., Wiley, New York 1993,

J. J. Sakurai, *Advanced Quantum Mechanics,* Addison-Wesley Reading, MA, 1967.

**Symmetry.** The physicist's notion of symmetry is exemplified by

H. Weyl, *Symmetry,* Princeton University Press, Princeton, NJ, 1983.

For an authoritative popular introduction, see

L. M. Lederman and C. T. Hill, *Symmetry and the Beautiful Universe,* Prometheus Books, Amherst NY, 2004.

A perceptive account of symmetries in field theory appears in

H. Georgi, *Weak Interactions and Modern Field Theory,* Dover Press, New York, 2009, chap. 1.

**Conservation Laws.** The classic reference on Noether's theorem and Hamilton's principle is

E. L. Hill, *Rev. Mod. Phys.* **23**, 253 (1953).

A thorough exegesis of Noether's work, accompanied by translations of key articles, is given in

Y. Kosmann-Schwarzbach, *The Noether Theorems: Invariance and Conservation Laws in the 20th Century,* Springer, New York, 2011.

For an informal history of Noether's discovery, see

N. Byers, "E. Noether's discovery of the deep connection between symmetries and conservation laws," in *The Heritage of Emmy Noether in Algebra, Geometry and Physics,* ed. M. Teicher, Bar-Ilan University, Ramat-Gan, Israel, 1999, p. 67, arXiv:physics/9807044.

The physical basis of numerous invariance principles is admirably explained in

J. J. Sakurai, *Invariance Principles and Elementary Particles,* Princeton University Press, Princeton, NJ, 1964.

**Path integrals.** The path–integral formulation of nonrelativistic quantum mechanics is developed thoroughly in

R. P. Feynman, A. R. Hibbs, and D. F. Styer, *Quantum Mechanics and Path Integrals,* emended ed., Dover, Mineola, NY, 2010.

The path–integral approach to quantum field theory is pursued in two recent books:

P. Ramond, *Field Theory: A Modern Primer,* 2nd ed., Westview Press, Boulder, CO, 1990;

A. Zee, *Quantum Field Theory in a Nutshell,* 2nd ed., Princeton University Press, Princeton, NJ, 2010.

See also

E. S. Abers and B. W. Lee, *Phys. Rep.* **9C**, 1 (1973).

**Variational principles.** The development of field theory from an action principle is expounded by

R. E. Peierls, *Proc. Roy. Soc. (London)* **A214**, 143 (1952),

J. Schwinger, *Phys. Rev.* **82**, 914 (1951); *Phys. Rev.* **91**, 91 (1953).

For a comprehensive tour, see

B. DeWitt, *The Global Approach to Quantum Field Theory,* 2nd ed., Oxford University Press, Oxford, 2003.

**Isospin invariance.** The historical path from Heisenberg to Wigner is chronicled, with illustrative examples, in

A. Bohr and B. R. Mottelson, *Nuclear Structure,* World Scientific, Singapore, 1998, §1.3.

It is instructive to look closely at the nuclear level schemes of the $A = 14$ nuclides, which may be taken as two nucleons outside a closed core, and at the $A = 7$ and $A = 11$ mirror nuclei, which can be seen as isospin doublets. Additional "isobaric analogue states" give evidence of larger multiplets. Consult

F. Ajzenberg-Selove, *Nucl. Phys.* **A490**, 1 (1988) for $A = 5$–10; *Nucl. Phys.* **A506**, 1 (1990) for $A = 11$–12; *Nucl. Phys.* **A523**, 1 (1991) for $A = 13$–15.

## ||||||||||||||| REFERENCES |||||||||||||||||||

1. To follow one of the true masters of the art, see M. Veltman, *Diagrammatica: The Path to Feynman Diagrams,* Cambridge University Press, Cambridge, 1994.
2. An instructive intellectual history of classical mechanics appears in R. B. Lindsay and H. Margenau, *Foundations of Physics,* Ox Bow Press, Woodbridge, CT., 1981.
3. The method of path integrals has its roots in the work of P. A. M. Dirac, *Phys. Zeits. Sowjetunion* **3**, 64 (1933) and was brought to its modern form by R. P. Feynman, *Rev. Mod. Phys.* **20**, 367 (1948); *Phys. Rev.* **80**, 440 (1950). These papers are reprinted in *Selcted Papers on Quantum Electrodynamics,* ed. by J. Schwinger, Dover, New York, 1958.
4. E. Noether, "Invariante Variationsprobleme," *Nachr. Ges. Wiss. Göttingen, Math-phys. Klasse* (1918), 235, http://j.mp/RZR9Iw. An English translation by M. A. Tavel, *Transport Theory and Statistical Mechanics* **1** (3), 183 (1971), is available as arXiv:physics/0503066 and http://j.mp/ROlPgs.

# Three

||||||||||||||||||||||||||||||||||||||||||||||||||||||||||||||||||||||||||||||||||||||||||||||||||||||||||||||||

## The Idea of Gauge Invariance

### 3.1 HISTORICAL PRELIMINARIES

We now turn to a discussion of the theory of electrodynamics, which is both the simplest gauge theory and the most familiar. The foundations for our present understanding of the subject were laid down by Maxwell in 1864 in his equations unifying the electric and magnetic interactions. The electromagnetic potential that we are led to introduce in order to generate fields that comply with Maxwell's equations by construction is not uniquely defined. The resulting freedom to choose many potentials that describe the same electromagnetic fields has come to be called *gauge invariance*. We shall see that the gauge invariance of electromagnetism can be phrased in terms of a continuous symmetry of the Lagrangian, which leads, through Noether's theorem, to the conservation of electric charge and to other important consequences.

Although it is clearly possible to regard gauge invariance as simply an outcome of Maxwell's unification, we may wonder whether a greater importance might not attach to the symmetry itself and thus be led to investigate the degree to which Maxwell's equations might be seen to follow from the symmetry. Indeed, we may trace the idea of symmetry as a dynamical principle to efforts by Hermann Weyl [1] to find a geometric basis for both gravitation and electromagnetism, a few years before the discovery of quantum mechanics. Weyl's attempt to unify the fundamental interactions of his day through the requirement of invariance under a spacetime-dependent change of scale were unsuccessful. His terminology, *Eichinvarianz* (*Eich* = gauge, or standard of calibration), has nevertheless survived, and his original program is worth recalling.

Consider the change in a function $f(x)$ between the point $x_\mu$ and the point $x_\mu + dx_\mu$. In a space with uniform scale, it is simply

$$f(x + dx) = f(x) + \partial^\mu f(x) dx_\mu. \tag{3.1.1}$$

But if, in addition, the scale, or unit of measure, for $f$ stretches by a factor $(1 + \mathcal{S}^\mu dx_\mu)$ in going from $x_\mu$ to $x_\mu + dx_\mu$, the value of the function becomes

$$f(x + dx) = (f(x) + \partial^\mu f(x)dx_\mu)(1 + \mathcal{S}^\nu dx_\nu)$$
$$= f(x) + (\partial^\mu f(x) + f(x)\mathcal{S}^\mu)dx_\mu + \mathrm{O}(dx)^2. \qquad (3.1.2)$$

To first order in the infinitesimal translation $dx_\mu$, the increment in the function $f$ is, therefore,

$$\Delta f = (\partial^\mu + \mathcal{S}^\mu)f dx_\mu. \qquad (3.1.3)$$

Weyl wished to base a theory upon the modified differential operator $(\partial^\mu + \mathcal{S}^\mu)$ and to identify the four-vector potential $A^\mu(x)$ of electromagnetism with a spacetime-dependent generator of scale changes, $\mathcal{S}^\mu$. Thus electromagnetism would find a basis in geometry. In an afterword to the first paper of Ref. [1], Einstein objected that if this were so, the rate at which a clock ticks—equivalently, the frequency at which an atom emits radiation—would be influenced not only by its current state, but also by its history. No such effect has been observed, then or now.

From a quantum-mechanical perspective, it is easy to see how the spirit of Weyl's program might be realized. Recall that the classical four-momentum

$$p^\mu = (E; p_x, p_y, p_z) \qquad (3.1.4)$$

goes over to the quantum-mechanical operator

$$\mathsf{p}^\mu = i\partial^\mu = (i\partial/\partial t; -i\nabla). \qquad (3.1.5)$$

For a particle with electric charge $q$, the canonical replacement is

$$(p^\mu - qA^\mu) \to i(\partial^\mu + iqA^\mu), \qquad (3.1.6)$$

in natural units with $\hbar = c = 1$. This suggests that Weyl's program could be implemented successfully if we identified

$$\mathcal{S}^\mu = iqA^\mu \qquad (3.1.7)$$

and required invariance of the laws of physics under a change of phase,

$$(1 + iqA^\mu dx_\mu) \simeq \exp(iqA^\mu dx_\mu), \qquad (3.1.8)$$

rather than under a change of scale. Following work by Vladimir Fock [2] and Fritz London [3], Weyl began in 1929 to study invariance under this phase rotation as a dynamical principle [4], but retained the terminology *gauge invariance* [5].

In the course of this chapter, we seek to develop an increasingly precise understanding of gauge invariance and its consequences. We begin by reviewing the manifestations of gauge invariance in the classical electrodynamics of Maxwell's equations. Next, we consider the implications of phase invariance in quantum mechanics and see for the first time how imposition of a local symmetry requires the existence of interactions. We then digress briefly on the importance of the potential and—above all—of the phase factor (3.1.8) in quantum theory. A systematic investigation of phase invariance in Lagrangian field theory follows. Finally, we close the chapter by deducing the Feynman rules for scalar and spinor electrodynamics from the Lagrangians for those theories.

## 3.2 GAUGE INVARIANCE IN CLASSICAL ELECTRODYNAMICS

We have already remarked on two of the motivations for examining gauge invariance in detail: the hope of finding an explanation, or at least a deeper understanding, of the conservation of electric charge and the desire (represented in the first instance by Weyl's attempts) to derive electrodynamics from some basic principle. As a necessary preliminary to these studies, let us review the manifestations of gauge arbitrariness in classical electrodynamics.

Maxwell's equation for magnetic charge,

$$\nabla \cdot \mathbf{B} = 0, \tag{3.2.1}$$

invites us to write the magnetic field as

$$\mathbf{B} = \nabla \times \mathbf{A}, \tag{3.2.2}$$

where $\mathbf{A}$ is called the vector potential. This identification ensures that $\mathbf{B}$ will be divergenceless by virtue of the identity

$$\nabla \cdot (\nabla \times \mathbf{A}) = 0. \tag{3.2.3}$$

If we add an arbitrary gradient to the vector potential

$$\mathbf{A} \to \mathbf{A} + \nabla \Lambda, \tag{3.2.4}$$

the magnetic field is unchanged because

$$\mathbf{B} = \nabla \times (\mathbf{A} + \nabla \Lambda) = \nabla \times \mathbf{A}. \tag{3.2.5}$$

In similar fashion, the curl equation (Faraday–Lenz) for the electric field,

$$\nabla \times \mathbf{E} = \frac{-\partial \mathbf{B}}{\partial t}, \tag{3.2.6}$$

which can be rewritten as

$$\nabla \times \left( \mathbf{E} + \frac{\partial \mathbf{A}}{\partial t} \right) = 0, \tag{3.2.7}$$

suggests the identification

$$\mathbf{E} + \frac{\partial \mathbf{A}}{\partial t} = -\nabla V, \tag{3.2.8}$$

where $V$ is known as the scalar potential. In order that the electric field remain invariant under the shift (3.2.4), we must also require

$$V \to V - \frac{\partial \Lambda}{\partial t}. \tag{3.2.9}$$

All this can be expressed compactly in covariant notation. The electromagnetic field-strength tensor

$$F^{\mu\nu} = -F^{\nu\mu} = \partial^\nu A^\mu - \partial^\mu A^\nu = \begin{pmatrix} 0 & E_1 & E_2 & E_3 \\ -E_1 & 0 & B_3 & -B_2 \\ -E_2 & -B_3 & 0 & B_1 \\ -E_3 & B_2 & -B_1 & 0 \end{pmatrix}, \tag{3.2.10}$$

built up from the four-vector potential

$$A^\mu = (V; \mathbf{A}), \tag{3.2.11}$$

is unchanged by the "gauge transformation"

$$A^\mu \to A^\mu - \partial^\mu \Lambda, \tag{3.2.12}$$

where $\Lambda(x)$ is an arbitrary function of the coordinate. The fact that many different four-vector potentials yield the same electromagnetic fields and thus describe the same physics is a manifestation of the gauge invariance of classical electrodynamics.

The Maxwell equations (3.2.1) and (3.2.6), which motivated the introduction of a potential, may be rewritten in covariant form as

$$\partial^\lambda F^{\mu\nu} + \partial^\mu F^{\nu\lambda} + \partial^\nu F^{\lambda\mu} = 0. \tag{3.2.13}$$

A more-compact expression follows upon introduction of the dual field-strength tensor:

$${}^*F^{\mu\nu} = -\tfrac{1}{2}\varepsilon^{\mu\nu\rho\sigma} F_{\rho\sigma} = \begin{pmatrix} 0 & B_1 & B_2 & B_3 \\ -B_1 & 0 & -E_3 & E_2 \\ -B_2 & E_3 & 0 & -E_1 \\ -B_3 & -E_2 & E_1 & 0 \end{pmatrix}, \tag{3.2.14}$$

which may be obtained formally from $F^{\mu\nu}$ by replacing $\mathbf{E} \to \mathbf{B}$ and $\mathbf{B} \to -\mathbf{E}$. Here we adopt the convention [cf. (A.2.19)] that Levi-Città's antisymmetric symbol $\varepsilon^{\mu\nu\rho\sigma}$ is equal to $\mp 1$ for even or odd permutations of (0123) and that $\varepsilon^{\mu\nu\rho\sigma} = -\varepsilon_{\mu\nu\rho\sigma}$. Evidently, the inverse relation is

$$F_{\mu\nu} = \tfrac{1}{2}\varepsilon_{\mu\nu\rho\sigma} {}^*F^{\rho\sigma}. \tag{3.2.15}$$

Taking the dual of (3.2.13), we find that the Maxwell equations (3.2.1) and (3.2.6) may be written as

$$\partial_\mu {}^*F^{\mu\nu} = 0. \tag{3.2.16}$$

The remaining Maxwell equations,

$$\nabla \cdot \mathbf{E} = \rho = -\nabla \cdot \dot{\mathbf{A}} - \nabla^2 V, \tag{3.2.17}$$

(where $\dot{\mathbf{A}} = \partial\mathbf{A}/\partial t$) and

$$\nabla \times \mathbf{B} = \mathbf{J} + \dot{\mathbf{E}} = \mathbf{J} - \ddot{\mathbf{A}} - \nabla \dot{V}$$
$$\parallel \tag{3.2.18}$$
$$\nabla \times (\nabla \times \mathbf{A}) = -\nabla^2\mathbf{A} + \nabla(\nabla \cdot \mathbf{A}),$$

[using (3.2.8) for $\mathbf{E}$ and (3.2.2) for $\mathbf{B}$] correspond, in covariant notation, to

$$\partial_\mu F^{\mu\nu} = -J^\nu, \tag{3.2.19}$$

with the electromagnetic current given by

$$J^\nu = (\rho; \mathbf{J}). \tag{3.2.20}$$

Two consequences are immediately apparent. First, the electromagnetic current is conserved:

$$\partial_\nu J^\nu = -\partial_\nu \partial_\mu F^{\mu\nu} = 0. \tag{3.2.21}$$

Second, the wave equation (3.2.19) may be expanded as

$$\Box A^\nu - \partial^\nu(\partial_\mu A^\mu) = J^\nu, \tag{3.2.22}$$

which becomes, in the absence of sources and in Lorenz [6] gauge ($\partial_\mu A^\mu = 0$),

$$\Box A^\nu = 0. \tag{3.2.23}$$

Each component of the vector potential, to be identified with the photon field, thus satisfies a Klein–Gordon equation for a massless particle.

In these familiar results we see a relationship between gauge invariance, current conservation, and massless vector fields. Let us now attempt to understand these connections more completely.

## 3.3 PHASE INVARIANCE IN QUANTUM MECHANICS

Suppose that we knew the Schrödinger equation but not the laws of electrodynamics. Would it be possible to derive—in other words, to guess—Maxwell's equations from a gauge principle? The answer is yes! It is worthwhile to trace the steps in the argument in detail.

A quantum-mechanical state is described by a complex Schrödinger wave function $\psi(x)$. Quantum-mechanical observables involve inner products of the form

$$\langle \mathcal{O} \rangle = \int d^n x \, \psi^* \mathcal{O} \psi, \tag{3.3.1}$$

where $\mathcal{O}$ is a Hermitian operator, which are unchanged under a global phase rotation:

$$\psi(x) \to e^{i\theta} \psi(x). \tag{3.3.2}$$

In other words, the absolute phase of the wave function cannot be measured and is a matter of convention. *Relative* phases between wave functions, as measured in interference experiments, are unaffected by such a global rotation.

This raises a question: Are we free to choose one phase convention in Paris and another in Batavia? Differently stated, can quantum mechanics be formulated to be invariant under local (position-dependent) phase rotations

$$\psi(x) \to \psi'(x) = e^{i\alpha(x)} \psi(x)? \tag{3.3.3}$$

We shall see that this can be accomplished, but at the price of introducing an interaction that will be constructed to be electromagnetism.

The quantum-mechanical equations of motion, such as the Schrödinger equation, always involve derivatives of the wave function $\psi$, as do many observables. Under local phase rotations, these transform as [7]

$$\partial_\mu \psi(x) \to \partial_\mu \psi' = e^{i\alpha(x)}[\partial_\mu \psi(x) + i(\partial_\mu \alpha(x))\psi(x)], \tag{3.3.4}$$

which involves more than a mere phase change. The additional gradient-of-phase term spoils local phase invariance. Local phase invariance may be achieved, however, if the equations of motion and the observables involving derivatives are modified by the introduction of the electromagnetic field $A_\mu(x)$. If the gradient $\partial_\mu$ is everywhere replaced by the *gauge-covariant derivative*

$$\mathcal{D}_\mu \equiv \partial_\mu + iq A_\mu, \tag{3.3.5}$$

where $q$ is the charge in natural units of the particle described by $\psi(x)$ and the field $A_\mu(x)$ transforms under phase rotations (3.3.3) as

$$A_\mu(x) \to A'_\mu(x) \equiv A_\mu(x) - \frac{1}{q} \partial_\mu \alpha(x), \tag{3.3.6}$$

it is easily verified that under local phase transformations

$$\mathcal{D}_\mu \psi(x) \to e^{i\alpha(x)} \mathcal{D}_\mu \psi(x). \tag{3.3.7}$$

Consequently, quantities such as $\psi^* \mathcal{D}_\mu \psi$ are invariant under local phase transformations. The required transformation law (3.3.6) for the four-vector potential $A_\mu$ is precisely the form (3.2.12) of a gauge transformation in electrodynamics. Moreover, the covariant derivative defined in (3.3.5) corresponds to the familiar replacement $p \to p - qA$ already noted in (3.1.6). Thus the form of the coupling $(\mathcal{D}_\mu \psi)$ between the electromagnetic field and matter is suggested, if not uniquely dictated, by local phase invariance.

This example has shown the possibility of using local gauge invariance as a dynamical principle, as promised in the discussion of Weyl's program. We should remark that the idea of modifying the equations of motion to accommodate an invariance principle is not without a successful precedent. It was precisely to accommodate local charge conservation that Maxwell modified Ampère's law by the addition of a "displacement current" $\dot{\mathbf{E}}$. Before proceeding to a more systematic application of the gauge principle, some commentary is in order on the import of the vector potential in quantum mechanics.

## 3.4 SIGNIFICANCE OF POTENTIALS IN QUANTUM THEORY

In classical electrodynamics, the field strengths $\mathbf{E}$ and $\mathbf{B}$ are regarded as the basic physical quantities and, as we saw in §3.2, the potential $A^\mu = (V; \mathbf{A})$ is introduced as a convenient calculational device. Because of the gauge ambiguity of electrodynamics, the potential corresponding to a given configuration of the fields is not uniquely defined. In other words, the potentials contain *too much* information. It may, therefore, appear that the potential is no more than an auxiliary mathematical quantity with no independent physical significance. Do the fields themselves then contain all the necessary physical information? This is decidedly not the case in quantum theory, as the simple but profound analysis of Aharonov and Bohm [8] made clear.

To see the effect of potentials in the absence of fields, consider a nonrelativistic charged particle moving through a static vector potential that corresponds to a vanishing magnetic field. If the wave function $\psi^0(\mathbf{x}, t)$ is a solution of the Schrödinger equation in the absence of a vector potential,

$$\frac{-\hbar^2}{2m} \nabla^2 \psi^0 = i\hbar \frac{\partial \psi^0}{\partial t}, \tag{3.4.1}$$

then the solution in the presence of the vector potential $\mathbf{A}$ will be

$$\psi(\mathbf{x}, t) = \psi^0(\mathbf{x}, t) e^{iS/\hbar}, \tag{3.4.2}$$

with

$$S = q \int d\mathbf{x} \cdot \mathbf{A} \tag{3.4.3}$$

the classical action and $q$ the particle's charge. The new solution follows from the Schrödinger equation

$$\frac{(-i\hbar\nabla - q\mathbf{A})^2}{2m} \psi = i\hbar \frac{\partial \psi}{\partial t} \tag{3.4.4}$$

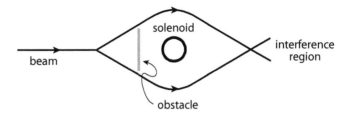

**Figure 3.1.** Schematic experiment to demonstrate interference influenced by a time-dependent vector potential.

and is a special case of the familiar fact that the phase shift experienced by a particle is given by the change in its classical action divided by $\hbar$. That the new solution differs from the old simply by a phase factor implies that there is no change in any physical result, as expected classically in the absence of electromagnetic fields.

Now suppose that a single coherent beam of charged particles is split into two parts, each passing on opposite sides of a solenoid but excluded from it, as shown in figure 3.1. After having passed the solenoid, the beams are recombined, and the resulting interference pattern is observed.

Now let

$$\psi^0(\mathbf{x}, t) = \psi_1^0(\mathbf{x}, t) + \psi_2^0(\mathbf{x}, t) \tag{3.4.5}$$

represent the wave function in the absence of a vector potential, where $\psi_1^0(\mathbf{x}, t)$ and $\psi_2^0(\mathbf{x}, t)$ denote the components of the beam that pass above and below the solenoid. When an electric current flows through the solenoid, it creates a magnetic field $\mathbf{B}$, which is essentially confined within the solenoid. In the experimental arrangement described, the beams pass only through field-free regions. The vector potential $\mathbf{A}$ cannot, however, be zero everywhere outside the solenoid, because the total flux through every loop containing the solenoid is a constant given by

$$\Phi = \int d\boldsymbol{\sigma} \cdot \mathbf{B} = \oint d\mathbf{x} \cdot \mathbf{A}. \tag{3.4.6}$$

By analogy with the single-beam configuration, the perturbed wave function is then

$$\psi = \psi_1^0 \, e^{iS_1/\hbar} + \psi_2^0 \, e^{iS_2/\hbar}, \tag{3.4.7}$$

where

$$S_i = \int_{\text{path } i} d\mathbf{x} \cdot \mathbf{A}. \tag{3.4.8}$$

Evidently the interference of the two components of the recombined beam will depend upon the phase difference $(S_1 - S_2)/\hbar$. Consequently, the vector potential does exert a physical effect, in spite of the fact that the beams have experienced no forces due to electromagnetic fields.

Some elementary remarks may help to reconcile this remarkable result with classical experience. First, because the effect of the potentials appears in an interference phenomenon, it is essentially quantum mechanical in nature. Second, the phase difference

$$\frac{S_1 - S_2}{\hbar} = \frac{q}{\hbar} \oint d\mathbf{x} \cdot \mathbf{A}(\mathbf{x}) = \frac{q\Phi}{\hbar} \qquad (3.4.9)$$

can be written in the form of an integral around a closed path. This form emphasizes that shifting the vector potential by a gradient, the familiar gauge freedom of classical electrodynamics, has no effect upon the result. Finally, we remark that the covariant generalization of (3.4.9) is simply

$$\frac{S_1 - S_2}{\hbar} = \frac{-q}{\hbar} \oint dx_\mu A^\mu. \qquad (3.4.10)$$

The experiment suggested by this analysis has been beautifully realized by Tonamura and collaborators using techniques of electron holography [9]. The observation that interference fringes shift as the enclosed flux is varied confirms that electromagnetic effects do occur in regions free of electric and magnetic fields. In its dependence on the enclosed flux, rather than forces encountered by the particles in transit, the Aharonov–Bohm effect is both nonlocal and topological. Thus, the field-strength tensor $F^{\mu\nu}$ is insufficient to determine all electromagnetic effects in quantum mechanics. A knowledge of the phase factor

$$\exp\left(\frac{-iq}{\hbar} \oint dx_\mu A^\mu\right), \qquad (3.4.11)$$

whose resemblance to (3.1.8) should not be overlooked, does make possible comprehensive predictions.

This brief digression has served several purposes: first, to call attention to the Aharonov–Bohm effect, which is interesting in its own right; second, to dispel the conviction, born of classical experience, that the electromagnetic vector potential is a mathematical artifice, without physical significance; and third, to introduce—albeit sketchily—the important role played in quantum theory by path-dependent phase factors.

## 3.5 PHASE INVARIANCE IN FIELD THEORY

In the preceding sections we have seen something of the connection between electromagnetic gauge invariance and the conservation of charge and have found that by generalizing the global phase invariance of quantum mechanics to a local phase symmetry, we may be led from a theory describing free particles to one in which the particles experience electromagnetic interactions. With this as background, it is now time to bring the logical structure into clear focus by making a more systematic investigation of phase invariance in the framework of Lagrangian field theory. To avoid becoming lost in formalism and because we shall take this opportunity to derive some specific results that will be of use later, we proceed by example.

Consider the Lagrangian for a free complex scalar field,

$$\mathcal{L} = |\partial^\mu \phi|^2 - m^2 |\phi|^2 , \qquad (3.5.1)$$

from which the Euler–Lagrange equations lead to the Klein–Gordon equations

$$(\Box + m^2)\phi(x) = 0, \; (\Box + m^2)\phi^*(x), \qquad (3.5.2)$$

as shown in section section 2.2. A global transformation on these fields,

$$\phi(x) \to e^{iq\alpha}\phi(x), \; \phi^*(x) \to e^{-iq\alpha}\phi^*(x), \qquad (3.5.3)$$

leads to the infinitesimal variations

$$\delta\phi = iq(\delta\alpha)\phi, \; \delta\phi^* = -iq(\delta\alpha)\phi^*; \qquad (3.5.4)$$

$$\delta(\partial_\mu \phi) = iq(\delta\alpha)\partial_\mu\phi, \; \delta(\partial_\mu\phi^*) = -iq(\delta\alpha)\partial_\mu\phi^*. \qquad (3.5.5)$$

The statement of global phase invariance is that such transformations leave the Lagrangian unchanged:

$$\delta\mathcal{L} = 0. \qquad (3.5.6)$$

Explicit computation yields

$$
\begin{aligned}
\delta\mathcal{L} &= \frac{\partial\mathcal{L}}{\partial\phi}\delta\phi + \frac{\partial\mathcal{L}}{\partial(\partial_\mu\phi)}\delta(\partial_\mu\phi) + \frac{\partial\mathcal{L}}{\partial\phi^*}\delta\phi^* + \frac{\partial\mathcal{L}}{\partial(\partial_\mu\phi^*)}\delta(\partial_\mu\phi^*) \\
&= \left[\partial_\mu\frac{\partial\mathcal{L}}{\partial(\partial_\mu\phi)}\right]iq(\delta\alpha)\phi + \frac{\partial\mathcal{L}}{\partial(\partial_\mu\phi)}iq(\delta\alpha)\partial_\mu\phi - (\phi \to \phi^*) \\
&= iq(\delta\alpha)\partial_\mu\left[\frac{\partial\mathcal{L}}{\partial(\partial_\mu\phi)}\phi - \frac{\partial\mathcal{L}}{\partial(\partial_\mu\phi^*)}\phi^*\right] \equiv 0, \qquad (3.5.7)
\end{aligned}
$$

where the equations of motion have been used in passing to the second line. Evidently we may identify a conserved Noether current,

$$
\begin{aligned}
j^\mu &= -iq\left[\frac{\partial\mathcal{L}}{\partial(\partial_\mu\phi)}\phi - \frac{\partial\mathcal{L}}{\partial(\partial_\mu\phi^*)}\phi^*\right] \\
&= iq[\phi^*\partial^\mu\phi - (\partial^\mu\phi)\phi^*] \equiv iq\phi^*\overleftrightarrow{\partial}^\mu\phi, \qquad (3.5.8)
\end{aligned}
$$

which satisfies

$$\partial_\mu j^\mu = 0. \qquad (3.5.9)$$

With the identification of $q$ as the electric charge, (3.5.8) is recognizable at once as the electromagnetic current of the charged scalar field. The connection between global phase invariance and current conservation is thus made explicit. Noether's theorem guarantees that this relation is in fact a general one.

What are the consequences of imposing invariance under *local* phase rotations, which transform the fields as

$$\phi(x) \rightarrow e^{iq\alpha(x)}\phi(x)? \tag{3.5.10}$$

Terms in the Lagrangian that depend only on the fields are left invariant, just as before, so there are no additional consequences beyond those of global gauge invariance. However, as we have seen in the quantum-mechanical discussion of section 3.3, gradient terms transform as

$$\partial_\mu\phi(x) \rightarrow e^{iq\alpha(x)}[\partial_\mu\phi(x) + iq(\partial_\mu\alpha(x))\phi(x)], \tag{3.5.11}$$

which necessitate the introduction of the gauge-covariant derivative

$$\mathcal{D}_\mu \equiv \partial_\mu + iq A_\mu(x). \tag{3.5.12}$$

Objects such as $\mathcal{D}_\mu\phi$ will then undergo the same phase rotation as the fields, namely,

$$\mathcal{D}_\mu\phi \rightarrow e^{iq\alpha(x)}\mathcal{D}_\mu\phi, \tag{3.5.13}$$

provided that the vector field $A_\mu$ transforms as

$$A_\mu(x) \rightarrow A_\mu(x) - \partial_\mu\alpha(x). \tag{3.5.14}$$

We remarked in section 3.3 that the replacement $\partial_\mu \rightarrow \mathcal{D}_\mu$ prescribes the form of the interaction between the gauge field $A_\mu$ and matter. To see how this comes about in Lagrangian field theory, let us look explicitly at the Dirac equation. The free-particle Lagrangian

$$\mathcal{L}_{\text{free}} = \bar{\psi}(i\gamma^\mu\partial_\mu - m)\psi \tag{3.5.15}$$

is replaced by the locally gauge-invariant expression

$$\begin{aligned} \mathcal{L} &= \bar{\psi}(i\gamma^\mu\mathcal{D}_\mu - m)\psi \\ &= \bar{\psi}(i\gamma^\mu\partial_\mu - m)\psi - q A_\mu\bar{\psi}\gamma^\mu\psi \\ &= \mathcal{L}_{\text{free}} - J^\mu A_\mu, \end{aligned} \tag{3.5.16}$$

where the (conserved) electromagnetic current has the familiar form

$$J^\mu = q\bar{\psi}\gamma^\mu\psi. \tag{3.5.17}$$

Precisely this form of the current follows from the requirement of global phase invariance. It is easy to verify that the Lagrangian (3.5.16) is indeed invariant under the combined transformations (3.5.10) and (3.5.14).

To arrive at the complete Lagrangian for quantum electrodynamics, it remains only to add a kinetic energy term for the vector field to describe the propagation of free photons. The Lagrangian (2.2.8) leads to Maxwell's equations and is manifestly invariant under local gauge transformations (3.5.14). Assembling all the pieces, we therefore have

$$\mathcal{L}_{\text{QED}} = \mathcal{L}_{\text{free}} - J^\mu A_\mu - \tfrac{1}{4} F_{\mu\nu} F^{\mu\nu}, \tag{3.5.18}$$

which is indeed the usual QED Lagrangian [10].

A photon mass term would have the form

$$\mathcal{L}_\gamma = \tfrac{1}{2} m^2 A^\mu A_\mu, \tag{3.5.19}$$

which obviously violates local gauge invariance because

$$A^\mu A_\mu \to (A^\mu - \partial^\mu \alpha)(A_\mu - \partial_\mu \alpha) \neq A^\mu A_\mu. \tag{3.5.20}$$

Thus, we find that local gauge invariance has led us to the existence of a massless photon. (Although this conclusion is not inescapable, it can be avoided only at a price, as will be explained in chapter 5.)

A restrictive direct limit on the photon mass has been deduced from a magnetohydrodynamic analysis of the solar wind data at Pluto's orbit collected by the Voyager 1 and 2 missions [11]. The upper limit is

$$M_\gamma < 1 \times 10^{-18} \text{ eV}, \tag{3.5.21}$$

corresponding to a modified Coulomb potential of the form

$$V \sim \frac{\exp(-r/r_0)}{r}, \tag{3.5.22}$$

with $r_0 > 2 \times 10^8$ km ($= 2 \times 10^{26}$ fm).

We have now seen how global phase invariance leads to the existence of a conserved charge. The more constraining requirement of local phase invariance necessitates the introduction of a massless gauge field and restricts the possible interactions of radiation with matter. For the Dirac equation, the interaction term was simply of the form $-J^\mu A_\mu$, with $J^\mu$ the conserved current of the free-fermion Lagrangian. In general, some care is required in making this identification because the structure of the current may be altered by the interactions. Such a modification occurs, for example, in the other case we have examined, that of the complex scalar field.

Replacing the gradient $\partial_\mu$ by the gauge-covariant derivative $\mathcal{D}_\mu$ in the Lagrangian (3.5.1) yields the locally gauge-invariant form

$$\begin{aligned} \mathcal{L} &= |\mathcal{D}^\mu \phi|^2 - m^2 |\phi|^2 \\ &= |\partial^\mu \phi|^2 - m^2 |\phi|^2 - iq A_\mu \phi^* \overset{\leftrightarrow}{\partial^\mu} \phi + q^2 A^2 \phi^* \phi \\ &= |\partial^\mu \phi|^2 - m^2 |\phi|^2 - j^\mu A_\mu + q^2 A^2 \phi^* \phi, \end{aligned} \tag{3.5.23}$$

where $j^\mu$ is the conserved current (3.5.8) for the free-particle Lagrangian. The final term in (3.5.23), which corresponds to a "contact interaction," can be identified as a $J \cdot A$ contribution when Noether's theorem is used to define a conserved current for the full Lagrangian, including interactions.

## 3.6 FEYNMAN RULES FOR ELECTROMAGNETISM

The passage from Lagrangian to Feynman rules is documented in many places. Here, we shall merely outline the final procedure without derivation, assuming the reader to be as familiar with the methods as he or she cares to be.

The recipe for extracting vertex factors from the interaction Lagrangian $\mathcal{L}_{\text{int}}$ is uncomplicated. In the quantity $i\mathcal{L}_{\text{int}}$, replace all the field operators by free-particle wave functions and evaluate the resulting expressions in momentum space. Omit all factors that correspond to the external lines. What remains is the vertex factor of the interaction.

Consider first the interaction in spinor electrodynamics. We have, from (3.5.16),

$$i\mathcal{L}_{\text{int}} = -iq A_\mu \bar{\psi} \gamma^\mu \psi, \tag{3.6.1}$$

into which we insert

$$\begin{aligned} \psi &= u_1 e^{-ip_1 \cdot x}, \\ \bar{\psi} &= \bar{u}_2 e^{ip_2 \cdot x}, \\ A_\mu &= \epsilon_\mu^*(k) e^{ik \cdot x}, \end{aligned} \tag{3.6.2}$$

so that

$$i\mathcal{L}_{\text{int}} = -iq \epsilon_\mu^*(k) \bar{u}_2 \gamma^\mu u_1 e^{i(p_2 + k - p_1)\cdot x}, \tag{3.6.3}$$

from which, after discarding the external factors, we retain

$$-iq\gamma^\mu. \tag{3.6.4}$$

For scalar electrodynamics, the interaction Lagrangian of (3.5.23) contains two terms. The trilinear term in $\mathcal{L}_{\text{int}}$ yields

$$i\mathcal{L}_{\text{int}}[3] = q A_\mu [\phi^* \partial^\mu \phi - (\partial^\mu \phi^*)\phi]. \tag{3.6.5}$$

Upon substitution of

$$\begin{aligned} \phi &= S_1 e^{-ip_1 \cdot x}, \\ \phi^* &= S_2 e^{ip_2 \cdot x} \end{aligned} \tag{3.6.6}$$

and the usual form for the photon, we find

$$i\mathcal{L}_{\text{int}}[3] \to \epsilon_\mu^*(k) S_1 S_2 e^{i(p_2 + k - p_1)\cdot x} q(-ip_1 - ip_2)^\mu, \tag{3.6.7}$$

which yields a vertex factor

$$-iq(p_1 + p_2)^\mu. \tag{3.6.8}$$

Finally, for the interaction term

$$i\mathcal{L}_{int}[4] = iq^2 A_\mu A_\nu g^{\mu\nu} \phi^* \phi, \tag{3.6.9}$$

we write

$$i\mathcal{L}_{int}[4] \to iq^2 g^{\mu\nu} (\epsilon_\mu^*(k_1)\epsilon_\nu^*(k_2) + \epsilon_\mu^*(k_2)\epsilon_\nu^*(k_1)) e^{i(p_2+k_1+k_2-p_1)\cdot x}, \tag{3.6.10}$$

from which the vertex factor is

$$2iq^2 g^{\mu\nu}. \tag{3.6.11}$$

The momentum-space propagators may be obtained with the aid of a similar mnemonic device. The general form of the wave functions is typified by that for the Klein–Gordon field,

$$(\Box + m^2)\phi(x) = J(x), \tag{3.6.12}$$

where the source term $J$ may itself depend on the fields. The propagator is simply related to the inverse of the momentum-space representation of the operator on the left-hand side. Thus for the Klein–Gordon field we have

$$G(p^2) = \frac{i}{p^2 - m^2 + i\varepsilon}, \tag{3.6.13}$$

whereas for the fermion field in the Dirac equation

$$(\not{p} - m)\psi(x) = J(x), \tag{3.6.14}$$

the propagator is

$$G(p) = \frac{i}{\not{p} - m + i\varepsilon} = i\frac{(\not{p} + m)}{p^2 - m^2 + i\varepsilon}. \tag{3.6.15}$$

To obtain the photon propagator, a choice of gauge is required. This is because the equation of motion,

$$\Box A^\nu - \partial^\nu(\partial_\mu A^\mu) = J^\nu, \tag{3.2.22}$$

does not uniquely determine the potential $A^\nu$ in terms of the conserved current $J^\nu$. (A massless photon has but two degrees of freedom.) To resolve the gauge ambiguity in the most pedestrian fashion we may choose the Lorenz gauge $\partial_\mu A^\mu = 0$, in which case (3.2.22) collapses to a set of massless Klein–Gordon equations, which lead to the propagator

$$G(k^2) = \frac{-ig^{\mu\nu}}{k^2 + i\varepsilon}. \tag{3.6.16}$$

Other gauge-fixing procedures are discussed briefly in section 6.3 and section 8.3. The gauge-independence of the theory guarantees that observable quantities will be independent of the choice of gauge. Graphical summaries of the Feynman rules for scalar and spinor electrodynamics are given in appendix B.5.

The analysis presented in this chapter has shown the theory of electromagnetism to be the gauge theory associated with the phase transformations that form the Abelian group U(1). We shall next investigate the generalization of these ideas to non-Abelian gauge theories, or Yang–Mills theories.

## PROBLEMS

3.1. Using the Feynman rules given in appendix B.5, compute the differential cross section $d\sigma/d\Omega$ and the total (integrated) cross section for the reaction $e^-\sigma^+ \rightarrow e^-\sigma^+$, where $\sigma^+$ is a charged scalar particle. Express the differential cross section in terms of (a) kinematic invariants, (b) c.m. variables, and (c) laboratory variables corresponding to the scalar target at rest. How would the results differ for $e^-\sigma^-$ scattering?

3.2. Making use of the Dirac equation, show that the most general parity-conserving form for the electromagnetic current of the proton is

$$J_\mu \sim \bar{u}(p')[\Gamma_1(q^2)\gamma_\mu + \Gamma_2(q^2)i\sigma_{\mu\nu}q^\nu + \Gamma_3(q^2)q_\mu]u(p),$$

where $\sigma_{\mu\nu} \equiv (i/2)[\gamma_\mu, \gamma_\nu]$ and $q \equiv p' - p$. What are the consequences of current conservation, $\partial^\mu J_\mu = 0$?

3.3. Calculate the differential cross section in the laboratory frame for elastic electron-proton scattering (a) for a structureless proton (i.e., a Dirac particle) and (b) for a real proton (using the results of problem 3.2). How do these results differ from the cross section for $e\sigma$ scattering?

3.4. Reformulate Maxwell's equations, taking into account the possibility that magnetic monopoles exist. Show that classical electrodynamics is invariant under the transformation

$$\mathbf{E} \rightarrow \mathbf{E}\cos\theta + \mathbf{B}\sin\theta,$$
$$\mathbf{B} \rightarrow -\mathbf{E}\sin\theta + \mathbf{B}\cos\theta,$$
$$q \rightarrow q\cos\theta + g\sin\theta,$$
$$g \rightarrow -q\sin\theta + g\cos\theta,$$

where $q$ and $g$ are the electric and magnetic charges, respectively. Show that if the ratio $g/q$ has the same value for all sources, the magnetic charge can be rotated away and the theory expressed in terms of electric charges only. What is the value of the new effective electric charge? Analyze the gauge invariance of the modified classical electrodynamics. [G. Wentzel, *Prog. Theoret. Phys. Suppl.* **37–38**, 163 (1966).]

3.5. Construct the O($e^4$) amplitude for the reaction $\gamma\gamma \rightarrow \gamma\gamma$ in QED, which represents the sum of six Feynman diagrams. Show that your final result is gauge invariant (in the sense that the amplitude vanishes upon replacement of $\epsilon_\mu(k)$ by $k_\mu$) and finite, whereas the contribution of each diagram separately is gauge dependent and divergent.

3.6. (a) Consider a nonrelativistic particle with charge $q$ moving along the axis of a cylindrical Faraday cage connected to an external generator, which causes the potential $V(t)$ on the cage to vary with time only when the particle is well within the cage. Show that, if the wave function $\psi^0(\mathbf{x}, t)$ is a solution of the Schrödinger equation for $V(t) \equiv 0$, the solution when the generator is operating will be $\psi(\mathbf{x}, t) = \psi^0(\mathbf{x}, t)e^{iS/\hbar}$, where $S = -\int^t dt'\, q\, V(t')$.

(b) Now suppose that a single coherent beam of charged particles is split into two parts, each of which is allowed to pass through its own long cylindrical cage of the kind just described. On emerging from the Faraday cages, the beams are recombined and the resulting interference pattern is observed. The beam is chopped into bunches that are long compared with the wavelength of an individual particle but short compared with the Faraday cages. The potentials on the two cages vary independently but are nonzero only when a bunch is well within the tubes. This ensures that the beam traverses a time-varying potential without experiencing electric or magnetic forces. Describe how the interference pattern depends upon the applied voltages [Aharonov and Bohm, Ref. [8]].

3.7. If baryon number is absolutely conserved, the conservation law may be a consequence of a global phase symmetry like that of electromagnetism, with the electric charge replaced by baryon number.
(a) How would Newton's law of gravitation be modified if the baryonic phase symmetry were a *local* gauge invariance?
(b) In view of the close equality of inertial and gravitational masses imposed by the Eöt-Wash experiment [S. Schlamminger et al., *Phys. Rev. Lett.* **100**, 041101 (2008); see earlier measurements by P. G. Roll, R. Krotkov, and R. H. Dicke, *Ann. Phys. (NY)* **26**, 442 (1967) and R. von Eötvös, D. Pekar, and E. Feteke, *Ann. Physik* **68**, 11 (1922)], what can be said about the strength of a hypothetical gauge interaction coupled to the baryon current? [T. D. Lee and C. N. Yang, *Phys. Rev.* **98**, 1501 (1955).]

3.8. How would a gauge boson $\gamma_B$ coupled to baryon number affect the properties of the $\Upsilon$ $(b\bar{b})$ resonances? Allow for the possibility that $\gamma_B$ has a nonzero mass. [C. D. Carone and H. Murayama, *Phys. Rev. Lett.* **74**, 3122 (1995).] On the possibility of a massless gauge boson coupled to lepton number, see L. B. Okun, *Yad. Fiz.* **10**, 358 (1969) [English translation: *Sov. J. Nucl. Phys.* **10**, 206 (1969)]. For an overview of the search for long-range forces, see A. D. Dolgov, *Phys. Rep.* **320**, 1 (1999).

## ⅢⅢⅢⅢⅢ FOR FURTHER READING ⅢⅢⅢⅢⅢ

**Classical electromagnetism.** The path to Maxwell's equations is described in the standard textbooks:

D. J. Griffiths, *Introduction to Electrodynamics,* 3rd ed., Addison-Wesley, Reading, MA, 1999.

J. D. Jackson, *Classical Electrodynamics*, 3rd ed., Wiley, New York, 1998.

and in the historical introduction to
M. Born and E. Wolf, *Principles of Optics*, 4th ed., Pergamon, New York, 1970.

The definitive account is that of
E. T. Whitaker, *A History of Theories of Æther and Electricity*, two volumes, Nelson, London, Vol. 1: *The Classical Theories*, 1910, revised and enlarged, 1951, Vol. 2: *The Modern Theories 1900–1926*, 1953; reprinted by Harper Torchbooks, New York, 1960.

**Gauge invariance.** The history of the concept of gauge invariance has been reviewed in meticulous detail by
J. D. Jackson and L. B. Okun, *Rev. Mod. Phys.* **73**, 663 (2001).

Weyl's contributions are highlighted in
N. Straumann, *Acta Phys. Polon. B* **37**, 575 (2006) [arXiv:hep-ph/0509116]
K. Chandrasekharan (ed.), *Hermann Weyl, 1885–1985: centenary lectures*, ETH Zürich and Springer-Verlag, Berlin, 1986.

A valuable collection of original papers, translated into English and with commentary, is
L. O'Raifeartaigh, *The Dawning of Gauge Theory*, Princeton University Press, Princeton, NJ, 1997.

For an inquiry into the historical origins and the path to recent developments, see
L. O'Raifeartaigh and N. Straumann, *Rev. Mod. Phys.* **72**, 1 (2000).

A very useful annotated bibliography is provided by
T. P. Cheng and L.-F. Li, "Resource Letter: GI-1 Gauge Invariance," *Am. J. Phys.* **56**, 586 (1988) [Erratum: ibid., p. 1048].

Perspectives of two of the founders of modern gauge theory may be found in
R. L. Mills, *Am. J. Phys.* **57**, 493 (1989).
C. N. Yang, "Magnetic monopoles, Fiber Bundles and Gauge Fields," in *Five Decades of Weak Interaction Theory*, ed. N.-P. Chang, *Ann. NY Acad. Sci.* **294**, 86 (1977).

For additional remarks on the connection between global gauge invariance and current conservation, see
E. P. Wigner, "Invariance in Physical Theory," in *Symmetries and Reflections*, Ox Bow Press, Woodbridge, Connecticut, 1979, p. 3.

**Aharonov–Bohm effect.** The nonclassical nature of the Aharonov–Bohm effect and the subtle elegance of the experiments that have been devised to observe it have spawned a vast literature, much of which can be sampled from
M. Peshkin and A. Tonomura, *The Aharonov–Bohm Effect*, Lecture Notes in Physics vol. 340, Springer-Verlag, New York, 1989,
*Quantum Phases: 50 Years of the Aharonov–Bohm effect and 25 Years of the Berry phase*, ed. M. Dennis, S. Popescu and L. Vaidman, *J. Phys. A: Math. Theor.* **43**, No. 35 (2010), See also the conference Web site, http://j.mp/gu4QuK,
H. Batelaan and A. Tonomura, "The Aharonov–Bohm Effects: Variations on a Subtle Theme," *Phys. Today* **62** (9), 38 (September 2009).

I would characterize the objections as homeopathic. For a dissident view, consider
T. H. Boyer, *Found. Phys.* **38**, 498 (2008).

The Aharonov–Bohm effect is an example of a "geometric phase," the result that the wave function of a quantum system may not return to its initial phase after its parameters cycle slowly around a circuit. In its modern form the geometric phase was discovered by

M. V. Berry, *Proc. Roy. Soc. London, Ser. A* **392**, 45 (1984),

M. V. Berry, "Anticipations of the Geometric Phase," *Phys. Today* **43**, 34 (December 1990).

For a collection of original papers and reprints, with commentary, consult

A. Shapere and F. Wilczek, *Geometrical Phases in Physics,* World Scientific, Singapore, 1989.

Manifestations of "Berry's phase" in molecular and condensed-matter physics are explored in

A. Böhm et al., *The geometric phase in quantum systems,* Springer, Berlin and New York, 2003,

D. J. Thouless, *Topological Quantum Numbers in Nonrelativistic Physics,* World Scientific, Singapore, 1998,

J. W. Zwanzinger, M. Koenig, and A. Pines, *Ann. Rev. Phys. Chem.* **41**, 601 (1990).

**Electromagnetism as a gauge theory.** Lucid introductions may be found in

E. S. Abers and B. W. Lee, *Phys. Rep.* **9C**, 1 (1973),

I.J.R. Aitchison and A.J.G. Hey, *Gauge Theories in Particle Physics,* Adam Hilger, Bristol, 1982.

**Photon mass.** The fascinating history of photon mass measurements is reviewed in many places, including

L. C. Tu, J. Luo and G. T. Gillies, *Rept. Prog. Phys.* **68**, 77 (2005),

J. C. Byrne, *Astrophys. Space Sci.* **46**, 115 (1977),

A. S. Goldhaber and M. M. Nieto, *Rev. Mod. Phys.* **43**, 277 (1971); *Sci. Am.* **234**, 84 (May 1976); *Rev. Mod. Phys.* **82**, 939 (2010),

I. Yu. Kobzarev and L. B. Okun, *Usp. Fiz. Nauk* **95**, 131 (1968) [English translation: *Sov. Phys.–Uspekhi* **11**, 338 (1968)],

J. D. Jackson, Classical Electrodynamics 3rd ed., Wiley, New York, 1988, section 1.2.

The upper bound on the mass of the photon can be improved by the analysis of galactic magnetic fields. The most restrictive limit, $M_\gamma < 3 \times 10^{-27}$ eV, has been inferred from the stability of the Magellanic clouds, as discussed by

G. V. Chibisov, *Usp. Fiz. Nauk* **119**, 551 (1976) [English translation: *Sov. Phys.–Uspekhi* **19**, 624 (1976)].

**Integral formulation of gauge theories.** The idea that electrodynamics can be derived from a path-dependent, or nonintegrable, phase, as evoked by the discussion of the Aharonov–Bohm effect, goes back to the celebrated "monopole paper"

P. A. M. Dirac, *Proc. Roy. Soc. (London)* **A133**, 60 (1931).

The idea has been implemented by

S. Mandelstam, *Ann. Phys. (NY)* **19**, 1 (1962),

C. N. Yang, *Phys. Rev. Lett.* **33**, 445 (1974),

T. T. Wu and C. N. Yang, *Phys. Rev. D* **12**, 3845 (1975).

The last work makes contact with the mathematical concepts of the theory of fiber bundles, which arose in the study of abstract problems in geometry. A brief nontechnical introduction to the mathematics has been given by

I. M. Singer, *Phys. Today* **35**, 41 (March, 1980).

An excellent resource for the links between gauge fields, gravitation, and differential geometry is

T. Eguchi, P. B. Gilkey, and A. J. Hanson, *Phys. Rep.* **66**, 213 (1980).

## ‖‖‖‖‖‖‖‖‖‖‖ REFERENCES ‖‖‖‖‖‖‖‖‖‖‖‖

1. H. Weyl, *Sitzber. Preuß. Akad. Wiss.* 465 (1918), in which the symmetry is called *Maßstab-Invarianz*. The terminology *Eichinvarianz* enters in *Ann. Phys. (Leipzig)* **59**, 101 (1919); and in *Space–Time–Matter*, trans. H. L. Brose, Dover, New York, 1952, Chapter IV, Section 35, p. 282. The German original dates from 1921.

2. V. Fock, *Z. Phys.* **39**, 226 (1927). For a modern English translation, see *Phys.-Usp.* **53**, 839 (2010). See also the commentary by L. B. Okun, *Phys.-Usp.* **53**, 835 (2010).

3. F. London, *Z. Phys.* **42**, 375 (1927).

4. H. Weyl, *Z. Phys.* **56**, 330 (1929).

5. For a summary of these early developments, see W. Pauli, *Handbuch der Physik*, edited by S. Flügge, Springer-Verlag, Bonn-Göttingen-Heidelberg, 1958, Vol. V/1, p. 1, which is largely identical with the 1933 edition.

6. J. D. Jackson, "Examples of the Zeroth Theorem of the History of Physics," *Am. J. Phys.* **76**, 704 (2008) [arXiv:0708.4249].

7. Covariant notation is adopted here in anticipation of the more general discussion to follow. Space and time components may be distinguished if an explicit analysis of the Schrödinger equation is desired.

8. Y. Aharonov and D. Bohm, *Phys. Rev.* **115**, 485 (1959). Similar observations were made earlier, in the context of electron microscopy, by W. Ehrenberg and R. E. Siday, *Proc. Roy. Soc. (London)* **62B**, 8 (1949).

9. A. Tonomura et al., *Phys. Rev. Lett.* **56**, 792 (1986); see also A. Tonomura, *PNAS* **102**, 14952-14959 (2005), published online with supporting movies as http://j.mp/ScFiud. The first experimental confirmation was given by R. G. Chambers, *Phys. Rev. Lett.* **5**, 3 (1960). A. Caprez, B. Barwick, and H. Batelaan, *Phys. Rev. Lett.* **99**, 210401 (2007), have demonstrated that an electron passing a solenoid experiences no force sufficient in magnitude to mimic the Aharonov–Bohm effect.

10. This is by no means the most general gauge-invariant Lagrangian that may be constructed. Although it does not arise from the minimal substitution $\partial_\mu \to \mathcal{D}_\mu$, a magnetic moment interaction involving the spin tensor $\sigma_{\mu\nu}$ is compatible with local gauge invariance, as problem 3.2 will show. This is but a single example. What does seem to distinguish (3.5.18), apart from its agreement with experiment, is that it defines a renormalizable theory.

11. D. D. Ryutov, *Plasma Phys. Control. Fusion* **49**, B429 (2007).

# Four

## Non-Abelian Gauge Theories

In this chapter we undertake the extension of our ideas about local gauge invariance to gauge groups that are more complicated than the group of phase rotations. We shall find that it is possible to enforce local gauge invariance by following essentially the same strategy that succeeded for electrodynamics. The principal difference, apart from algebraic complexity, will be the appearance of interactions among the gauge bosons as a consequence of the non-Abelian nature of the gauge symmetry. As before, we proceed by example, developing the SU(2)-isospin gauge theory put forward by Yang and Mills [1] and by Shaw [2]. The generalization to other gauge groups proceeds without complication.

### 4.1 MOTIVATION

The near degeneracy of the neutron and proton masses, the charge independence of nuclear forces, and many subsequent observations support the notion of isospin conservation in the strong interactions. What is meant by isospin conservation is that the laws of physics should be invariant under rotations in isospin space and that the proton and neutron should appear symmetrically in all equations. This means that if electromagnetism can be neglected, the isospin orientation is of no significance. The distinction between proton and neutron thus becomes entirely a matter of arbitrary convention. In such a world, the existence of two distinct kinds of nucleons could be inferred from the properties of the ground state of the $^4$He nucleus, in much the same manner as we have deduced from the spin-$\frac{3}{2}$ baryons $\Delta^{++}$ or $\Omega^-$ the need for three colors of quarks (cf. problems 4.6 and 4.7).

This sort of reasoning lay behind the introduction in section 2.3 of the free-nucleon Lagrangian

$$\mathcal{L}_0 = \bar{\psi}(i\gamma^\mu \partial_\mu - m)\psi, \tag{2.3.14}$$

written in terms of the composite fermion fields

$$\psi \equiv \begin{pmatrix} p \\ n \end{pmatrix}. \tag{2.3.13}$$

As the earlier discussion showed, the Lagrangian (2.3.14) has an invariance under global isospin rotations $\psi \to \exp(i\boldsymbol{\tau} \cdot \boldsymbol{\alpha}/2)\psi$, and the isospin current $\mathbf{J}^\mu = \bar{\psi}\gamma^\mu \frac{\boldsymbol{\tau}}{2}\psi$ is conserved. Thus, we have complete freedom in naming the proton and neutron (in the absence of electromagnetism); once freely chosen, the convention must be respected everywhere throughout spacetime.

This restriction may seem, as it did to Yang and Mills (Ref. [1]), at odds with the idea of local field theory. Furthermore, we have just seen in chapter 3 that electromagnetism possesses a *local* gauge invariance and that by imposing that local symmetry on a free-particle Lagrangian, it is possible to construct an interesting (and indeed correct) theory of electrodynamics. In analogy with electromagnetism we are led to ask whether we can require that the freedom to name the two states of the nucleon be available independently at every spacetime point. Can we, in other words, turn the global SU(2) invariance of the free field theory into a mathematically consistent local SU(2) invariance? If so, what are the physical consequences?

## 4.2 CONSTRUCTION

The formulation of the theory proceeds just as in the Abelian case. If, under a local gauge transformation, the field transforms as

$$\psi(x) \to \psi'(x) = G(x)\psi(x), \tag{4.2.1}$$

with

$$G(x) \equiv \exp\left(\frac{i\boldsymbol{\tau} \cdot \boldsymbol{\alpha}(x)}{2}\right), \tag{4.2.2}$$

then the gradient transforms as

$$\partial_\mu \psi \to G(\partial_\mu \psi) + (\partial_\mu G)\psi. \tag{4.2.3}$$

To ensure the local gauge invariance of the theory, we first introduce a gauge-covariant derivative

$$\mathcal{D}_\mu \equiv I\partial_\mu + ig B_\mu, \tag{4.2.4}$$

where

$$I = \begin{pmatrix} 1 & 0 \\ 0 & 1 \end{pmatrix} \tag{4.2.5}$$

serves as a reminder that the operators are $2 \times 2$ matrices in isospin space and $g$ will be seen to play the role of a strong-interaction coupling constant. The object $B_\mu$ is the $2 \times 2$ matrix defined by

$$B_\mu = \tfrac{1}{2}\boldsymbol{\tau} \cdot \mathbf{b}_\mu = \tfrac{1}{2}\tau^a b_\mu^a = \tfrac{1}{2}\begin{pmatrix} b_\mu^3 & b_\mu^1 - i b_\mu^2 \\ b_\mu^1 + i b_\mu^2 & -b_\mu^3 \end{pmatrix}, \tag{4.2.6}$$

where the three gauge fields are $\mathbf{b}_\mu = (b_\mu^1, b_\mu^2, b_\mu^3)$, boldface quantities denote isovectors, and the isospin index $a$ runs from 1 to 3.

The point of introducing the gauge fields and the gauge-covariant derivative is to obtain a generalization of the gradient that transforms as

$$\mathcal{D}_\mu \psi \to \mathcal{D}'_\mu \psi' = G(\mathcal{D}_\mu \psi). \tag{4.2.7}$$

Requiring this to be so will show us how $B_\mu$ must behave under gauge transformations. By explicit computation we have

$$\begin{aligned} \mathcal{D}'_\mu \psi' &= (\partial_\mu + ig B'_\mu)\psi' \\ &= G(\partial_\mu \psi) + (\partial_\mu G)\psi + ig B'_\mu (G\psi) \\ &\equiv G(\partial_\mu + ig B_\mu)\psi \\ &= G(\partial_\mu \psi) + ig G(B_\mu \psi), \end{aligned} \tag{4.2.8}$$

which may be solved to yield the condition

$$ig B'_\mu (G\psi) = ig G(B_\mu \psi) - (\partial_\mu G)\psi, \tag{4.2.9}$$

which must hold for arbitrary values of the nucleon field $\psi$. Inserting $I = G^{-1}G$ before each occurence of $\psi$ and regarding the transformation law as an operator equation, we obtain

$$B'_\mu = G B_\mu G^{-1} + \frac{i}{g}(\partial_\mu G)G^{-1} = G\left[B_\mu + \frac{i}{g}G^{-1}(\partial_\mu G)\right]G^{-1}. \tag{4.2.10}$$

Although this transformation law may appear formidable at first sight, it has a rather simple interpretation. Recall that in the case of electromagnetism the local gauge transformation was the phase rotation

$$G_{\text{EM}} = e^{iq\alpha(x)}, \tag{4.2.11}$$

where $\alpha(x)$ is an arbitrary function. A transcription of the general transformation law (4.2.10) is, therefore,

$$\begin{aligned} A'_\mu &= G_{\text{EM}} A_\mu G_{\text{EM}}^{-1} + \frac{i}{q}(\partial_\mu G_{\text{EM}})G_{\text{EM}}^{-1} \\ &= A_\mu - \partial_\mu \alpha, \end{aligned} \tag{4.2.12}$$

just as in (3.5.14). For the case of isospin gauge symmetry, the meaning of (4.2.10) is that $B_\mu$ is transformed by an isospin rotation plus a gradient term. To see this explicitly, consider an infinitesimal gauge transformation, represented by

$$G = 1 + \frac{i}{2}\boldsymbol{\tau} \cdot \boldsymbol{\alpha}, \quad |\alpha_i| \ll 1. \tag{4.2.13}$$

From the transformation law (4.2.10), we have

$$B'_\mu = B_\mu + \frac{i}{2}\boldsymbol{\tau} \cdot \boldsymbol{\alpha} B_\mu - \frac{i}{2} B_\mu \boldsymbol{\tau} \cdot \boldsymbol{\alpha} - \frac{1}{2g}\partial_\mu(\boldsymbol{\tau} \cdot \boldsymbol{\alpha}) + O(\alpha^2), \tag{4.2.14}$$

which, in view of the definition (4.2.6), is equivalent to

$$\boldsymbol{\tau} \cdot \mathbf{b}'_\mu = \boldsymbol{\tau} \cdot \mathbf{b}_\mu + \frac{i}{2}(\boldsymbol{\tau} \cdot \boldsymbol{\alpha}\,\boldsymbol{\tau} \cdot \mathbf{b}_\mu - \boldsymbol{\tau} \cdot \mathbf{b}_\mu\,\boldsymbol{\tau} \cdot \boldsymbol{\alpha}) - \frac{1}{g}\partial_\mu(\boldsymbol{\tau} \cdot \boldsymbol{\alpha}). \tag{4.2.15}$$

The middle term may be simplified at once with the aid of the familiar Pauli-matrix identity [*cf.* (A.2.13)]

$$\boldsymbol{\tau} \cdot \mathbf{a}\,\boldsymbol{\tau} \cdot \mathbf{b} = \mathbf{a} \cdot \mathbf{b} + i\boldsymbol{\tau} \cdot \mathbf{a} \times \mathbf{b}, \tag{4.2.16}$$

but some insight and the prospect of easy generalization are gained by proceeding more formally. Write the middle term in component form as

$$T_2 \equiv \frac{i}{2}\alpha^j b_\mu^k(\tau^j\tau^k - \tau^k\tau^j) = \frac{i}{2}\alpha^j b_\mu^k[\tau^j, \tau^k]. \tag{4.2.17}$$

For the isospin group SU(2), the commutator is given by

$$[\tau^j, \tau^k] = 2i\varepsilon_{jkl}\tau^l, \tag{4.2.18}$$

so that

$$T_2 = -\varepsilon_{jkl}\alpha^j b_\mu^k \tau^l = -\boldsymbol{\alpha} \times \mathbf{b} \cdot \boldsymbol{\tau} \tag{4.2.19}$$

and

$$\boldsymbol{\tau} \cdot \mathbf{b}'_\mu = \boldsymbol{\tau} \cdot \mathbf{b}_\mu - \boldsymbol{\alpha} \times \mathbf{b}_\mu \cdot \boldsymbol{\tau} - \frac{1}{g}\partial_\mu(\boldsymbol{\tau} \cdot \boldsymbol{\alpha}). \tag{4.2.20}$$

Because the three isospin components of the gauge field are linearly independent, we have as the transformation law for infinitesimal gauge transformations

$$b_\mu'^l = b_\mu^l - \varepsilon_{jkl}\alpha^j b^k - \frac{1}{g}\partial_\mu\alpha^l,$$

$$\mathbf{b}'_\mu = \mathbf{b}_\mu - \boldsymbol{\alpha} \times \mathbf{b}_\mu - \frac{1}{g}\partial_\mu\boldsymbol{\alpha}, \tag{4.2.21}$$

which has the claimed structure. The result in component form, by the way, shows that the transformation rule depends on the structure constants $\varepsilon_{jkl}$ and not on the representation of the isospin group. We learn from the intermediate steps (4.2.17) and (4.2.18) that the isospin rotation, which is the new feature compared with electromagnetism, arises from the noncommutativity of the gauge transformations or, in other words, from the non-Abelian nature of the symmetry group.

To this point in the construction of the isospin gauge theory of nucleons, we have a Lagrangian given by

$$
\begin{aligned}
\mathcal{L} &= \bar{\psi}(i\gamma^\mu \mathcal{D}_\mu - m)\psi \\
&= \mathcal{L}_0 - g\bar{\psi}\gamma^\mu B_\mu \psi \\
&= \mathcal{L}_0 - \frac{g}{2}\mathbf{b}_\mu \cdot \bar{\psi}\gamma^\mu \boldsymbol{\tau}\psi,
\end{aligned}
\tag{4.2.22}
$$

namely, a free Dirac Lagrangian plus an interaction term that couples the isovector gauge fields to the conserved isospin current of the nucleons. The structure of the interaction between the gauge fields and matter is precisely analogous to that found in the case of QED.

To proceed further, we must construct a field-strength tensor and, hence, a kinetic term for the gauge fields. Although elegant means are available for motivating the correct form, some understanding is to be gained from a pedestrian approach. In analogy with electromagnetism, we seek a field-strength tensor

$$
F_{\mu\nu} \equiv \tfrac{1}{2}\mathbf{F}_{\mu\nu} \cdot \boldsymbol{\tau} = \tfrac{1}{2}F_{\mu\nu}^a \tau^a
\tag{4.2.23}
$$

from which to construct a gauge-invariant kinetic term

$$
\mathcal{L}_{\text{gauge}} = -\tfrac{1}{4}\mathbf{F}_{\mu\nu} \cdot \mathbf{F}^{\mu\nu} = -\tfrac{1}{2}\text{tr}\left(F_{\mu\nu}F^{\mu\nu}\right),
\tag{4.2.24}
$$

where the last equality follows from the Pauli-matrix identity

$$
\text{tr}(\tau^a \tau^b) = 2\delta^{ab}.
\tag{4.2.25}
$$

Thus we wish to find a field-strength tensor that transforms under local gauge transformations $G$ as

$$
F'_{\mu\nu} = GF_{\mu\nu}G^{-1}.
\tag{4.2.26}
$$

Note first that a transcription of the QED form is not satisfactory:

$$
\begin{aligned}
\partial_\nu B'_\mu - \partial_\mu B'_\nu &= \partial_\nu\left[GB_\mu G^{-1} + \frac{i}{g}(\partial_\mu G)G^{-1}\right] - \partial_\mu\left[GB_\nu G^{-1} + \frac{i}{g}(\partial_\nu G)G^{-1}\right] \\
&= G(\partial_\nu B_\mu - \partial_\mu B_\nu)G^{-1} + [(\partial_\nu G)B_\mu - (\partial_\mu G)B_\nu]G^{-1} \\
&\quad + G[B_\mu(\partial_\nu G^{-1}) - B_\nu(\partial_\mu G^{-1})] \\
&\quad + \frac{i}{g}[(\partial_\mu G)(\partial_\nu G^{-1}) - (\partial_\nu G)(\partial_\mu G^{-1})] \\
&\neq G(\partial_\nu B_\mu - \partial_\mu B_\nu)G^{-1}.
\end{aligned}
\tag{4.2.27}
$$

This result may be cast in slightly more symmetric form by recalling that

$$GG^{-1} = G^{-1}G = I,$$  (4.2.28)

so that

$$\partial_\mu(GG^{-1}) = 0 = \partial_\mu(G^{-1}G),$$  (4.2.29)

whence

$$(\partial_\mu G)G^{-1} = -G(\partial_\mu G^{-1}) \text{ and } (\partial_\mu G^{-1})G = -G^{-1}(\partial_\mu G).$$  (4.2.30)

Judicious use of the identities (4.2.30) in (4.2.27) then yields

$$\begin{aligned}
\partial_\nu B'_\mu - \partial_\mu B'_\nu &= G(\partial_\nu B_\mu - \partial_\mu B_\nu)G^{-1} \\
&\quad + G\{[G^{-1}(\partial_\nu G), B_\mu] - [G^{-1}(\partial_\mu G), B_\nu]\}G^{-1} \\
&\quad + \frac{1}{ig}G\left[G^{-1}(\partial_\nu G), G^{-1}(\partial_\mu G)\right]G^{-1},
\end{aligned}$$  (4.2.31)

showing manifestly that the additional terms arise from the nonvanishing commutators owed to the non-Abelian group structure. It is natural to add a term to $\partial_\nu B_\mu - \partial_\mu B_\nu$, tuning the definition of $F_{\mu\nu}$ to recover the desired transformation properties.

For inspiration, we observe that the electromagnetic field-strength tensor (3.2.10) can also be written in the form

$$F_{\mu\nu} = \frac{1}{iq}\left[\mathcal{D}_\nu, \mathcal{D}_\mu\right],$$  (4.2.32)

with

$$\mathcal{D}_\mu = \partial_\mu + iqA_\mu,$$  (3.5.12)

because

$$\begin{aligned}
F_{\mu\nu} &= \frac{1}{iq}\left[(\partial_\nu + iqA_\nu), (\partial_\mu + iqA_\mu)\right] \\
&= \partial_\nu A_\mu - \partial_\mu A_\nu + iq\left[A_\nu, A_\mu\right],
\end{aligned}$$  (4.2.33)

where the commutator vanishes in an Abelian theory. This suggests that for the SU(2) gauge theory, a candidate field-strength tensor is the form

$$F_{\mu\nu} = \frac{1}{ig}\left[\mathcal{D}_\nu, \mathcal{D}_\mu\right] = \partial_\nu B_\mu - \partial_\mu B_\nu + ig\left[B_\nu, B_\mu\right].$$  (4.2.34)

QED:

 photon propagator

**Figure 4.1.** The photon propagator of quantum electrodynamics.

The commutator term transforms as

$$
\begin{aligned}
ig\left[B'_\nu, B'_\mu\right] &= ig\left[\left(GB_\nu G^{-1} + \frac{i}{g}(\partial_\nu G)G^{-1}\right), \left(GB_\mu G^{-1} + \frac{i}{g}(\partial_\mu G)G^{-1}\right)\right] \\
&= igG\left[B_\nu, B_\mu\right]G^{-1} \\
&\quad - G\left\{\left[G^{-1}(\partial_\nu G), B_\mu\right] - \left[G^{-1}(\partial_\mu G), B_\nu\right]\right\}G^{-1} \\
&\quad - \frac{1}{ig}G\left[G^{-1}(\partial_\nu G), G^{-1}(\partial_\mu G)\right]G^{-1},
\end{aligned}
\tag{4.2.35}
$$

where (4.2.30) has again been used to bring the result to a symmetric form. The terms after the first are precisely what is required to cancel the extra terms in (4.2.31). Thus, we find that the field-strength tensor given by (4.2.34) has the desired behavior (4.2.26) under local gauge transformations. The Yang–Mills Lagrangian,

$$
\mathcal{L}_{\mathrm{YM}} = \bar{\psi}(i\gamma^\mu \mathcal{D}_\mu - m)\psi - \tfrac{1}{2}\mathrm{tr}\left(F_{\mu\nu}F^{\mu\nu}\right),
\tag{4.2.36}
$$

is, therefore, invariant under local gauge transformations. Whereas a mass term $M^2 B_\mu B^\mu$ is incompatible with local gauge invariance, as in electromagnetism, a common nonzero mass for the nucleons is permitted, as the electron mass was.

It is of some interest to display the components of the field-strength tensor, because this will be of value for the generalization to other gauge groups. Using the definitions (4.2.6) and (4.2.23) and the commutation relations (4.2.18), it is straightforward to show that

$$
F^l_{\mu\nu} = \partial_\nu b^l_\mu - \partial_\mu b^l_\nu + g\varepsilon_{jkl}b^j_\mu b^k_\nu.
\tag{4.2.37}
$$

For a gauge group other than SU(2), the Levi-Cività symbol $\varepsilon_{jkl}$ will be replaced by the antisymmetric structure constants $f_{jkl}$.

## 4.3 SOME PHYSICAL CONSEQUENCES

It is now appropriate to consider briefly the differences between Yang–Mills theory and quantum electrodynamics and to investigate some of the experimental consequences of the Yang–Mills Lagrangian (4.2.36). First, let us note that sourceless QED is a free (or noninteracting) field theory. Photons, being electrically neutral, do not interact directly with other photons. Only bilinear combinations of the photon gauge field $A_\mu$ occur in the QED Lagrangian (3.5.18), and thus the only object for which Feynman rules are required is the photon propagator, as indicated in figure 4.1. The Yang–Mills theory has a richer structure, however. Even in the absence of fermion sources, there will be interactions as a consequence of the nonlinear term in $F_{\mu\nu}$. Trilinear and quadrilinear terms thus appear in $F_{\mu\nu}F^{\mu\nu}$,

SU(2):

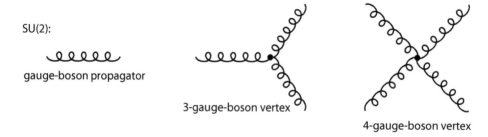

gauge-boson propagator

3-gauge-boson vertex

4-gauge-boson vertex

**Figure 4.2.** Gauge boson propagator and self-interactions in Yang–Mills theory.

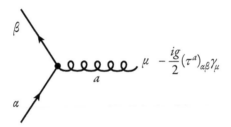

**Figure 4.3.** Feynman rule for the nucleon–gauge field interaction in Yang–Mills theory.

as well as the familiar bilinear term. In physical terms, the gauge bosons carry isospin and so interact among themselves. Thus, in addition to the gauge-field propagator, the theory contains the three- and four-gauge-boson vertices displayed in figure 4.2. These additional interactions have numerous physical consequences that will command our attention later on. For the moment, it suffices to remark that they exist and that they owe their existence to the non-Abelian structure of the gauge group.

Because the construction of Yang–Mills theory was motivated by the search for an isospin-conserving theory of nuclear interactions, let us analyze some consequences of the theory for the interactions among fermions. The interaction term

$$\mathcal{L}_{\text{int}} = -\frac{g}{2} b_\mu^a \bar{\psi} \gamma^\mu \tau^a \psi \tag{4.3.1}$$

leads by the usual procedure to the Feynman rule for the nucleon–nucleon–gauge-boson vertex. For the transition depicted in figure 4.3 of a nucleon with isospin label $\alpha (= p, n$ or $1, 2)$ into a nucleon with isospin label $\beta$ and a gauge boson with Lorentz index $\mu$ and isospin label $a$ ($= 1, 2, 3$), the vertex factor is simply

$$-\frac{ig}{2}(\tau^a)_{\alpha\beta} \gamma_\mu, \tag{4.3.2}$$

where $\alpha$ and $\beta$ label components of the $2 \times 2$ matrix $\tau^a$.

What does this say for the stability of two-nucleon systems? Suppose the two-body nuclear interaction were controlled by the exchange of a single gauge boson,

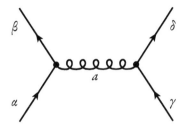

**Figure 4.4.** Model for the nucleon-nucleon interaction in SU(2) gauge theory.

as shown in figure 4.4. The interaction energy would then be characterized by

$$\mathcal{E} = \frac{g^2}{4} \sum_a \tau^a_{\alpha\beta} \tau^a_{\gamma\delta} \qquad (4.3.3)$$

times essentially kinematic factors. Instead of specifying the isospin labels $\alpha\beta\gamma\delta$ of the nucleons before and after interaction, it is more instructive to specify the two-nucleon state by its total isospin, 0 or 1, and to evaluate the average

$$\frac{1}{4}\langle \boldsymbol{\tau}^{(1)} \cdot \boldsymbol{\tau}^{(2)} \rangle = \langle \boldsymbol{I}^{(1)} \cdot \boldsymbol{I}^{(2)} \rangle = \frac{\langle I^2 \rangle - \langle I^{(1)\,2} \rangle - \langle I^{(2)\,2} \rangle}{2}, \qquad (4.3.4)$$

where the isospin operators $I^{(1,2)}$ refer to the individual nucleons with isospin $\frac{1}{2}$ and $I$ refers to the composite, with isospin 0 or 1. Using the familiar formula $\langle I^2 \rangle = I(I+1)$, we then find

$$\begin{aligned} \mathcal{E}(I=0) &= -\tfrac{3}{4}, \\ \mathcal{E}(I=1) &= +\tfrac{1}{4} \end{aligned} \qquad (4.3.5)$$

times common factors. The interaction is attractive in the isoscalar channel and repulsive in the isovector channel. If the Yang–Mills theory were an otherwise plausible theory of nuclear structure, this analysis would suggest an explanation for the fact that the only bound two-nucleon state is the (isoscalar) deuteron.

The applicability of lowest-order perturbation theory to a problem in strong-interaction dynamics is open to challenge because the coupling constant that would play the role of an expansion parameter is of order 1 at least; the nuclear forces are strong. It may be reasonable to hope that the systematics signaled by the Born term faithfully represent the pattern displayed by the full theory. At any rate, this hope is a recurrent one. The elementary calculation just concluded has been applied to other questions, such as the bound-state problem in quantum chromodynamics, under the name of a "maximally attractive channel analysis." It is generally true that the singlet channel emerges as maximally attractive.

Whether or not low-order perturbation theory can be taken as a reasonable guide is beside the point in the situation at hand, because there is a direct contradiction between the Yang–Mills theory and experimental nuclear physics. As a consequence of local gauge invariance, the Yang–Mills quanta $\mathbf{b}_\mu$ are massless vector bosons, which implies that the force mediated by them should be of infinite range, like the Coulomb interaction. On the contrary, a successful phenomenological description [3] of the nuclear force involves the exchange of massive particles: the light pseudoscalar pion and the heavier vector mesons $\rho^+$, $\rho^0$, $\rho^-$, and $\omega$. Thus, the theory cannot serve the purpose that inspired its conception.

## 4.4 ASSESSMENT

In this chapter and the preceding one, the examples of electromagnetism and the Yang–Mills theory have shown how gauge principles may be used to guide the construction of the theories. Global gauge invariance implies the existence of a conserved current, according to Noether's theorem. Local gauge invariance requires the introduction of massless vector gauge bosons, prescribes (or, more properly, restricts) the form of the interactions of gauge bosons with sources, and generates interactions among the gauge bosons if the symmetry is non-Abelian.

It is appealing to try to make use of the observed symmetries of nature as gauge symmetries, to exploit the potent idea that symmetries define interactions. This is, indeed, the course we shall follow in later applications. However, the example of the Yang–Mills theory shows that, whereas mathematical consistency is readily achieved, experimental success is not assured in advance for every would-be gauge group. Long-range nuclear forces mediated by massless vector quanta are not observed. Therefore, the Yang–Mills theory, based on the idea of isospin invariance, or flavor symmetry, as the strong-interaction gauge group is incorrect or at least incomplete.

If we are reluctant to abandon such an elegant concept, several courses are open: the choice of a new gauge group, the reinterpretation of the theory, and the evasion of massless gauge bosons. We shall make use of all three. By finding a different hadronic symmetry to which the mathematical structure of the Yang–Mills construction may be applied, we shall be led to quantum chromodynamics [4]. It is also natural to reinterpret the theory constructed in this chapter as a theory of weak interactions, based upon the "weak-isospin" symmetry apparent in nuclear $\beta$-decay, but this attempt would also founder on the prediction of massless gauge fields. Before we proceed to specific applications to the fundamental interactions, it will be important to understand how to circumvent the prediction of massless gauge bosons while preserving the local gauge invariance of the Lagrangian as a desirable restricting principle. This is the subject of the following chapter.

## PROBLEMS

4.1. Derive the Yang–Mills Lagrangian for a scalar field theory in which the three real scalar fields correspond to the triplet representation of SU(2). The free-particle Lagrangian is

$$\mathcal{L} = \tfrac{1}{2}[(\partial_\mu \phi)^2 - m^2 \phi^2],$$

with

$$\phi = \begin{pmatrix} \phi_1 \\ \phi_2 \\ \phi_3 \end{pmatrix}.$$

4.2. (a) By making the minimal substitution $\partial_\mu \to \mathcal{D}_\mu$ in the free-particle Lagrangian, construct a theory of the electrodynamics of a massive spin-1 boson $V^\pm$ and deduce the Feynman rules for the theory.

(b) Now, compute the differential cross section $d\sigma/d\Omega$ and the total cross section $\sigma \equiv \int d\Omega(d\sigma/d\Omega)$ for the reaction $e^+e^- \to V^+V^-$ for various helicities of the produced particles. Work in the c.m. frame and in the high-energy limit (where all the masses may be neglected). Assume the colliding beams are unpolarized. Compare the results with the cross sections for $e^+e^- \to \sigma^+\sigma^-$ and $e^+e^- \to \mu^+\mu^-$ derived in problems 1.4 and 1.5. [For Feynman rules, see J. D. Bjorken and S. D. Drell, *Relativistic Quantum Mechanics*, McGraw-Hill, New York, 1964, appendix B.]

4.3. Show that the transformations

$$B'_\mu = GB_\mu G^{-1} + \frac{i}{g}(\partial_\mu G)G^{-1} = G\left[B_\mu + \frac{i}{g}G^{-1}(\partial_\mu G)\right]G^{-1} \qquad (4.2.10)$$

of the gauge fields form a group—that is, under successive transformations on the matter field given by $\psi \to \psi' = G_1\psi \to \psi'' = G_2\psi'$, the transformation of the gauge field is characterized by $G = G_2G_1$.

4.4. Apply the analysis of nucleon–nucleon forces of section 4.3 to the nucleon–antinucleon case. Compare the Yang–Mills "predictions" with the spectrum of nonstrange mesons.

4.5. The flavor symmetry $SU(2)_{\text{isospin}}$ and the rotational symmetry $SU(2)_{\text{spin}}$ may be combined systematically in the group $SU(4)_{Is}$. In nuclear physics, this symmetry group provides the basis for classification into "Wigner supermultiplets" [E. P. Wigner, *Phys. Rev.* **51**, 106 (1937)]. For the quark model, the fundamental representation of $SU(4)_{Is}$ is

$$\mathbf{4} \equiv \begin{pmatrix} u_\uparrow \\ u_\downarrow \\ d_\uparrow \\ d_\downarrow \end{pmatrix}.$$

Using the notation $(\mathbf{2I+1}, \mathbf{2s+1})$ for the isospin $\times$ spin decomposition of $SU(4)_{Is}$ representations, we may write $\mathbf{4} = (\mathbf{2}, \mathbf{2})$, which recalls that the $\mathbf{4}$ transforms as a doublet under isospin rotations and under spin rotations.

(a) Work out the $SU(4)_{Is}$ content of the product $\mathbf{4} \otimes \mathbf{4^*}$. Characterize each representation by its Young tableau, symmetry properties, and dimension.

(b) Give the $(\mathbf{2I+1}, \mathbf{2s+1})$ content of each of the $SU(4)_{Is}$ representations in your decomposition of $\mathbf{4} \otimes \mathbf{4^*}$.

(c) Consider the implied $q\bar{q}$ bound states. Show that a bound state with orbital angular momentum $L$, isospin $I$, and total spin $S$ must have quantum numbers

$$C = (-1)^{L+S}; \quad P = (-1)^{L+1}; \quad G = (-1)^{L+S+I} \; .$$

Confirm that $\rho, \omega, \pi$ may be assigned to the $s$-wave **15**, whereas $\eta$ corresponds to **1**. (The physical $\eta$ contains an $s\bar{s}$ admixture.)

4.6. Now consider some implications of $SU(4)_{Is}$ for baryons.
(a) Work out the $SU(4)_{Is}$ content of the product $4 \otimes 4 \otimes 4$. Characterize each representation by its Young tableau, symmetry properties, and dimension.
(b) Give the $(2I + 1, 2s + 1)$ content of each of the $SU(4)_{Is}$ representations in your decomposition of $4 \otimes 4 \otimes 4$.
(c) Show that the nucleon and $\Delta(1232)$ resonances are plausibly members of the same flavor-spin supermultiplet, the **20**: ⊞ . Discuss the significance of the flavor × spin symmetry of the three-quark wave function.

4.7. To test the idea that the lowest-lying nucleon states make up a **20** of $SU(4)_{Is}$, let us examine in more detail the consequences for the proton and neutron.
(a) Construct *symmetric* flavor-spin wavefunctions for the nucleons. Note that there are two distinct paths to isodoublet three-quark states. The third quark (underlined) may be combined with an isoscalar pair to give

$$\varphi_A^p = |\tfrac{1}{2}, \tfrac{1}{2}\rangle_0 = (ud - du)\underline{u}/\sqrt{2} \; ,$$

which is antisymmetric under interchange of quarks 1 and 2; a similar situation holds for the neutron. Alternatively, the final quark may be added to an isovector pair, symmetric under interchange of the first two; show that this leads to

$$\varphi_S^p = |\tfrac{1}{2}, \tfrac{1}{2}\rangle_1 = -[(ud + du)\underline{u} - 2uu\underline{d}]/\sqrt{6} \; .$$

The corresponding spin parts of the wave functions $(\chi_{A,S}^{\uparrow})$ can be read from their flavor counterparts using the transcription

$$\text{flavor} \quad \left.\begin{matrix} u \\ d \end{matrix}\right\} \to \left\{\begin{matrix} \uparrow \\ \downarrow \end{matrix}\right. \quad \text{spin.}$$

Compute the fully symmetrized, normalized wave function for a proton with spin up, $|p \uparrow\rangle = (\varphi_A^p \chi_A^{\uparrow} + \varphi_S^p \chi_S^{\uparrow})/\sqrt{2}$, and show that

$$\begin{aligned}|p \uparrow\rangle = (&2u_\uparrow d_\downarrow u_\uparrow - u_\downarrow d_\uparrow u_\uparrow - u_\uparrow d_\uparrow u_\downarrow \\ &-d_\uparrow u_\downarrow u_\uparrow + 2d_\downarrow u_\uparrow u_\uparrow - d_\uparrow u_\uparrow u_\downarrow \\ &-u_\uparrow u_\downarrow d_\uparrow - u_\downarrow u_\uparrow d_\uparrow + 2u_\uparrow u_\uparrow d_\downarrow)/\sqrt{18} \; .\end{aligned}$$

The neutron counterpart, $|n \uparrow\rangle$, is obtained by interchanging $u \leftrightarrow d$.
(b) Take the quarks to be Dirac particles, with magnetic moments $\mu_q = e_q \hbar/2m_q c$, and assume that the constituent masses of up and down quarks are

equal: $m_u = m_d = m$, so that $\mu_u = \frac{2}{3}\mu$, $\mu_d = -\frac{1}{3}\mu$, with $\mu = e\hbar/2mc$. In terms of quark constituents, the magnetic moment of a nucleon $h = p, n$ is given by

$$\mu_h = \sum_i \langle h \uparrow |\mu_i \sigma_z^{(i)}|h \uparrow \rangle \ .$$

Show that $\mu_p = \mu$, and $\mu_n = -\frac{2}{3}\mu$, so that $\mu_n/\mu_p = -\frac{2}{3}$. Compare with data, and observe whether your prediction supports the $\mathbf{20}$ assignment, with symmetric flavor × spin wave functions.

## IIIIIIIIIIIIII FOR FURTHER READING IIIIIIIIIIIIIIII

**Non-Abelian gauge theories.** A general discussion, similar in tone to the one in this chapter, may be found in
  E. S. Abers and B. W. Lee, *Phys. Rep.* **9C**, 1 (1973).

A more abstract geometrical approach, which usefully complements this chapter, is followed by
  C. Itzykson and J.-B. Zuber, *Quantum Field Theory*, Dover Press, New York, 2006, Chap. 12.

A construction that does not make use of the Lagrangian formalism is given by
  I.J.R. Aitchison and A.J.G. Hey, *Gauge Theories in Particle Physics*, Adam Hilger, Bristol, 1982, §8.5.

For a wide-ranging collection, see
  G. 't Hooft, *50 Years of Yang–Mills Theories*, World Scientific, Singapore, 2005.

**General gauge groups.** The Yang–Mills strategy of constructing a theory from a local symmetry group was elaborated by
  R. Utiyama, *Phys. Rev.* **101**, 1597 (1956).

How to build a theory on a general gauge group is explained by
  S. L. Glashow and M. Gell-Mann, *Ann. Phys. (NY)* **15**, 437 (1961).

**Gauge theories of the strong interactions.** Following the work of Yang and Mills, there have been many attempts to construct phenomenologically acceptable strong-interaction theories based upon flavor symmetry groups. Among these, see
  Y. Fujii, *Prog. Theor. Phys. (Kyoto)* **21**, 232 (1959),
  J. J. Sakurai, *Ann. Phys. (NY)* **11**, 1 (1960),
  Y. Ne'eman, *Nucl. Phys.* **26**, 222 (1961).

The paper by Sakurai is quite remarkable, more for what is said than for what is done. The one by Ne'eman is better known for the proposal that the flavor symmetry be SU(3).

**Group theory.** A practical review of some essential manipulations appears in
  K. Nakamura et al. (Particle Data Group), *J. Phys.* G **37**, 075021 (2010), §38: "SU(*n*) Multiplets and Young Diagrams," by C. G. Wohl, available at http://pdg.lbl.gov.

Textbook treatments attuned to the needs of particle physicists include
  R. N. Cahn, *Semi-Simple Lie Algebras and Their Representations*, Dover Press, New York, 2006,

H. J. Lipkin, *Lie Groups for Pedestrians,* Dover Press, New York, 2002,
W.-K. Tung, *Group Theory in Physics,* World Scientific, Singapore, 1985.

**Flavor-spin SU(6).** The suggestion that Wigner's SU(4) could be extended to
$SU(3)_{flavor} \otimes SU(2)_{spin}$, with the baryon octet and decimet assigned to the symmetric **56**
representation, is due to
F. Gürsey and L. A. Radicati, *Phys. Rev. Lett.* **13**, 173 (1964).

Electromagnetic properties of the baryons, including the magnetic moments computed in
problem 4.7, are derived using tensor methods in
M. A. B. Bég, B. W. Lee, and A. Pais, *Phys. Rev. Lett.* **13**, 514 (1964),
B. Sakita, *Phys. Rev. Lett.* **13**, 643 (1964).

The path to the color degree of freedom is recounted in
O. W. Greenberg, "From Wigner's supermultiplet theory to quantum chromodynamics,"
*Acta Phys. Hung. A* **19**, 353 (2004), arXiv:hep-ph/0212174.

**Nonperturbative methods.** Quantum field theory defined on a spacetime lattice has
emerged as a powerful method for dealing with a variety of strong-coupling problems. For
elementary accounts of its application to the hadron spectrum, see
C. Rebbi, "The Lattice Theory of Quark Confinement," *Sci. Am.* **248**, 54 (February
1983),
D. Weingarten, "Quarks by Computer," *Sci. Am.* **274**, 116 (February 1996).

For a practical introduction to the notion of lattice QCD as a numerical path integral,
consult
G. P. Lepage, "Lattice QCD for Novices," in *Strong Interactions at Low and
Intermediate Energies,* ed. J. L. Goity, World Scientific, Singapore, 2000, pp. 49–90,
arXiv:hep-lat/0506036.

Additional readings on lattice QCD are suggested in chapter 8.

## |||||||||||||||| REFERENCES ||||||||||||||||

1. C. N. Yang and R. L. Mills, *Phys. Rev.* **95**, 631 (1954); *Phys. Rev.* **96**, 191 (1954).
2. R. Shaw, "The Problem of Particle Types and Other Contributions to the Theory
   of Elementary Particles," Cambridge University Thesis, 1955. The relevant chapter is
   reproduced in J. C. Taylor (ed.), *Gauge Theories in the Twentieth Century,* Imperial
   College Press, London, 2001, pp. 110–118, and is the subject of chapter 9 of
   L. O'Raifeartaigh, *The dawning of gauge theory,* Princeton University Press, Princeton,
   NJ, 1997.
3. General features of the nucleon–nucleon interaction are described in A. Bohr and
   B. R. Mottelson, *Nuclear Structure,* World Scientific, Singapore, 1998, Vol. 1, §2–5.
4. QCD predicts the properties of $\pi, \rho, \omega, \ldots$ as quark–antiquark-bound states, so it does
   account for the nuclear force particles. For progress toward deducing nuclear forces
   from QCD, see S. Aoki, "Lattice QCD and Nuclear Physics," arXiv:1008.4427, Lectures
   at the Summer School on *Modern perspectives in lattice QCD,* Les Houches, 2009;
   T. Inoue et al. [HAL QCD Collaboration], *Prog. Theor. Phys.* **124**, 591 (2010).

# Five

## Hidden Symmetries

$M$uch importance has been attached to symmetry principles in the preceding chapters. We have seen the connection between exact symmetries and conservation laws and have found that the requirement of local gauge invariance can serve as a dynamical principle to guide the construction of interacting field theories. Although a certain economy and mathematical elegance has thus been achieved, the results of the program to this point are unsatisfactory in several important respects. First, the gauge principle has led us to theories in which all the interactions are mediated by massless vector bosons, whereas only a single massless vector boson, the photon, is directly apparent in nature. Second, the algorithm for constructing gauge theories that we have developed applies only to exact symmetries, whereas nature exhibits numerous symmetries that are only approximate. Third, there are many situations in physics in which the exact symmetry of an interaction is concealed by circumstances. The canonical example, which will be elaborated shortly, is the Heisenberg ferromagnet, an infinite crystalline array of spin-$\frac{1}{2}$ magnetic dipoles. Below a critical temperature, the ground state is a completely ordered configuration in which all dipoles are ordered in some arbitrary direction, belying the rotation invariance of the underlying interaction. It is, thus, of interest to learn how to deal with symmetries that are not exact or not manifest in the hope of evading the conclusion that interactions must be mediated by massless gauge bosons.

In this chapter we shall, therefore, analyze the various types of symmetries more thoroughly than we have before. The distinction between internal symmetries and the geometrical symmetries that involve coordinate transformations (such as the Poincaré invariance of relativistic theories) has already been drawn in chapter 2. There, too, we observed that continuous symmetries imply local conservation laws through Noether's theorem, whereas discrete symmetries do not. A related difference between discrete and continuous symmetries will be uncovered in the discussion that follows. Most of our attention, however, will be concentrated on the distinction between exact and approximate symmetries. Among approximate symmetries, several different realizations are possible. The Lagrangian may display an imperfect or explicitly broken symmetry, or it may happen that the Lagrangian is symmetric but the physical vacuum does not respect the symmetry. In the latter case, the symmetry of the Lagrangian is said to be spontaneously broken. The various possibilities will be illustrated in what follows.

Our principal concerns in this chapter will be the conditions under which a symmetry is spontaneously broken and the consequences of spontaneous symmetry breakdown. We shall find that if the Lagrangian of a theory is invariant under an exact continuous symmetry that is not a symmetry of the physical vacuum, one or more massless spin-zero particles, known as Nambu–Goldstone bosons, must occur. From the point of view of unobserved massless particles, this would seem to double our trouble: gauge theories lead to unwanted massless vector bosons, and the spontaneous breakdown of a continuous symmetry implies the existence of unwanted spinless particles. However, if the spontaneously broken symmetry is a local gauge symmetry, a miraculous interplay between the would-be Nambu–Goldstone boson and the normally massless gauge bosons endows the gauge bosons with mass and removes the Nambu–Goldstone boson from the spectrum. The Higgs mechanism, by which this interplay occurs, is a central ingredient in our current understanding of the gauge bosons of the weak interactions.

The results of this chapter all can be codified and presented in a formal and even axiomatic manner. This was of some importance for the development of the subject because the precise statement of a theorem makes it possible to understand both its generality and its limitations. Our purposes will be better served, however, by considering a number of specific examples.

## 5.1   THE IDEA OF SPONTANEOUSLY BROKEN SYMMETRIES

The physical world manifests a number of apparently exact conservation laws, which we believe reflect the operation of exact symmetries of nature. These include the conservation of energy and momentum, of angular momentum, and of electric charge. In the language of Lagrangian field theory, exact symmetry is characterized by two conditions. The Lagrangian (density) is invariant under the symmetry in question,

$$\delta \mathcal{L} = 0, \tag{5.1.1}$$

and the unique physical vacuum is invariant under the symmetry transformations. From these requirements, a standard analysis demonstrates the mass degeneracy of particle multiplets.

Many of the useful internal symmetries, such as the flavor symmetries of isospin and SU(3) and the conservation of strangeness and charm, hold only approximately. It is usual to treat these approximate symmetries by writing the Lagrangian as

$$\mathcal{L} = \mathcal{L}_{\text{symmetric}} + \varepsilon \mathcal{L}_{\text{symmetry breaking}}. \tag{5.1.2}$$

This form is particularly useful if the symmetry-breaking term is small, in some sense, and can be treated as a perturbation upon the symmetric interaction.

The perturbation lifts the degeneracy of particle multiplets, with the resulting intermultiplet splitting, which is a function of the parameter $\varepsilon$, vanishing as $\varepsilon \to 0$. A familiar example (cf. problem 2.9 and the $A = 7, 11$, and 14 nuclear levels cited at the end of chapter 2) is

$$\mathcal{L} = \mathcal{L}_{\text{strong}} + \mathcal{L}_{\text{EM}}(\alpha), \tag{5.1.3}$$

in which the strong-interaction Lagrangian is isospin invariant and the responsibility for isospin violations resides in the electromagnetic term $\mathcal{L}_{\text{EM}}(\alpha)$. This characterization of explicit symmetry breaking lies behind the conventional view of a hierarchy of strong, electromagnetic, and weak interactions, which derives much of its utility from the fact that the dominant strong interaction respects the largest group of symmetries.

We saw by example in chapter 2 how continuous symmetries of the Lagrangian lead to exact conservation laws. Approximate conservation laws may arise if, as in the preceding paragraph, the Lagrangian is imperfectly symmetric. It may also happen that the Lagrangian $\mathcal{L}$ is exactly invariant under some symmetry, so that (5.1.1) holds, but that the dynamics determined by $\mathcal{L}$ imply a degenerate set of vacuum states that are not invariant under the symmetry. This leads to exact local conservation laws but conceals the symmetry of the theory—for example, by breaking the mass degeneracy of particle multiplets. Finally, we may imagine a hybrid situation in which an interaction that gives rise to a spontaneously broken symmetry is accompanied by an explicit symmetry-breaking interaction.

Each of these four situations can be illustrated by the infinite ferromagnet, to which we have already alluded. The nearest-neighbor interaction between spins, or magnetic dipole moments, is invariant under the group of spatial rotations SO(3). In the disordered, or paramagnetic, phase, which exists above the Curie temperature $T_C$, thermal fluctuations are more important than the dipole-dipole interactions, and the orientation of individual spins is *random*. The medium displays an unbroken SO(3) symmetry in the absence of an external field. The spontaneous magnetization of the system is zero, and there is no preferred direction in space, so the SO(3) invariance is manifest. A privileged direction may be selected by imposing an external magnetic field that biases the alignment of spins in the material. The SO(3) symmetry is thus broken down to an SO(2) symmetry of rotations about the external field direction. The full symmetry is restored when the external field is turned off.

For temperatures below $T_C$, when the system is in the ferromagnetic, or ordered, phase, the situation is rather different. In the absence of an impressed field, the configurations of lowest energy have a nonzero spontaneous magnetization $\langle \mathbf{M} \rangle \neq 0$, because the nearest-neighbor force favors the parallel alignment of spins. In these circumstances, the SO(3) symmetry is said to be spontaneously broken down to SO(2). The fact that the direction of the spontaneous magnetization is random can be demonstrated by repeating the experiment many times: averaged over many trials, $\langle\langle \mathbf{M} \rangle\rangle = 0$. The aleatory choice of direction and the fact that the measurable properties of the infinite ferromagnet do not depend upon its orientation are vestiges of the original SO(3) symmetry. The ground state is, thus, infinitely degenerate. A particular direction for the spontaneous magnetization may be chosen by imposing an external field that breaks the SO(3) symmetry explicitly. In contrast to the paramagnetic case, however, the spontaneous magnetization does not return to zero

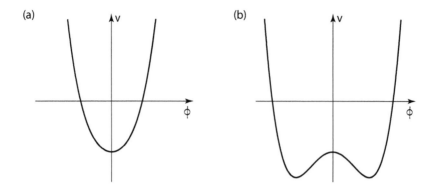

**Figure 5.1.** (a) Single-well potential with a unique minimum at $\phi = 0$. (b) Double-well potential with a degenerate vacuum, corresponding to a spontaneously broken symmetry.

when the applied field is turned off. For the rotational invariance to be broken spontaneously, it is crucial that the ferromagnet be infinite in extent, so that rotation from one degenerate ground state to another would require the impossible task of rotating an infinite number of elementary dipoles. This is the picturesque analogue of the statement in field theory that the degenerate vacua lie in distinct Hilbert spaces.

After this intuitive discussion, we are prepared to see how spontaneous symmetry breaking may arise in a simple mathematical model. Consider a Lagrangian for a self-interacting real scalar field $\phi$, which may be written in the form

$$\mathcal{L} = \tfrac{1}{2}(\partial_\mu \phi)(\partial^\mu \phi) - V(\phi). \tag{5.1.4}$$

How does the nature of the vacuum—and, thus, of the particle spectrum—depend upon the effective potential $V(\phi)$? If the potential is an even functional of the scalar field $\phi$,

$$V(\phi) = V(-\phi), \tag{5.1.5}$$

then the Lagrangian (5.1.4) is invariant under the parity transformation

$$\phi \to -\phi. \tag{5.1.6}$$

To explore the possibilities, it is convenient to consider an explicit potential

$$V(\phi) = \tfrac{1}{2}\mu^2 \phi^2 + \tfrac{1}{4}|\lambda|\,\phi^4. \tag{5.1.7}$$

The quadratic term has the familiar appearance of the mass term of scalar field theory. The positive coefficient of the quartic term is chosen to ensure stability against unbounded oscillations. Higher powers than the fourth are omitted in order that the theory be renormalizable.

Two cases, which correspond to manifest or spontaneously broken symmetry, can now be distinguished. If the parameter $\mu^2 > 0$, the potential (5.1.7) has a unique minimum at $\phi = 0$, as shown in figure 5.1(a), which corresponds to the vacuum state. This identification is perhaps most easily made in the Hamiltonian formalism. The Hamiltonian density is given by

$$\mathcal{H} = \pi\dot{\phi} - \mathcal{L}, \tag{5.1.8}$$

where

$$\dot{\phi} = \partial_0\phi \tag{5.1.9}$$

and the canonical momentum is

$$\pi \equiv \frac{\partial \mathcal{L}}{\partial \dot{\phi}}. \tag{5.1.10}$$

In the case at hand, we therefore have

$$\mathcal{H} = \tfrac{1}{2}[(\partial_0\phi)^2 + (\nabla\phi)^2] + V(\phi). \tag{5.1.11}$$

The state of lowest energy is thus seen to be one for which the value of the field $\phi$ is a constant, which we denote by $\langle\phi\rangle_0$. If the parameter $\mu^2$ is positive, the minimum of the potential (5.1.7) is at

$$\langle\phi\rangle_0 = 0. \tag{5.1.12}$$

The approximate form of the Lagrangian appropriate to the study of small oscillations about this minimum is then

$$\mathcal{L}_{so} = \tfrac{1}{2}[(\partial_\mu\phi)(\partial^\mu\phi) - \mu^2\phi^2], \tag{5.1.13}$$

which is that of a free particle with mass $= \mu$.

If $\mu^2 < 0$, the situation is that of a spontaneously broken symmetry. The potential

$$V(\phi) = -\tfrac{1}{2}|\mu|^2\phi^2 + \tfrac{1}{4}|\lambda|\phi^4, \tag{5.1.14}$$

shown in figure 5.1(b), has minima at

$$\langle\phi\rangle_0 = \pm\sqrt{-\frac{\mu^2}{|\lambda|}} \equiv \pm v, \tag{5.1.15}$$

which correspond to two degenerate lowest-energy states, either of which may be chosen to be the vacuum. Because of the parity invariance of the Lagrangian, the ensuing physical consequences must be independent of this choice. Whatever the choice, however, the symmetry of the theory is spontaneously broken; the parity transformation (5.1.6) is then an invariance of the Lagrangian but not of the vacuum state. Let us choose

$$\langle\phi\rangle_0 = +v \tag{5.1.16}$$

and define a shifted field

$$\phi' \equiv \phi - \langle\phi\rangle_0 = \phi - v, \tag{5.1.17}$$

so that the vacuum state corresponds to

$$\langle\phi'\rangle_0 = 0. \tag{5.1.18}$$

In terms of the shifted field, the Lagrangian is

$$\mathcal{L} = \tfrac{1}{2}(\partial_\mu\phi')(\partial^\mu\phi') - |\mu|^2 \left( \frac{\phi'^4}{4v^2} + \frac{\phi'^3}{v} + \phi'^2 - \frac{v^2}{4} \right), \tag{5.1.19}$$

which has no manifest symmetry properties with respect to the shifted field $\phi'$. For small oscillations about the vacuum, we have

$$\mathcal{L}_{\text{so}} = \tfrac{1}{2}[(\partial_\mu\phi')(\partial^\mu\phi') - 2\,|\mu|^2\,\phi'^2] \tag{5.1.20}$$

(plus an irrelevant constant), which describes the oscillation of a particle with $(\text{mass})^2 = 2\,|\mu|^2 = -2\mu^2 > 0$.

This simple example has illustrated how spontaneous symmetry breaking occurs when a symmetry of the Lagrangian is not respected by the vacuum state, defined as the state of lowest energy. The same methods of choosing a vacuum from among a degenerate set of vacua and discovering the particle spectra apply equally well to the more complicated and interesting physical situations, to which we now turn.

## 5.2 SPONTANEOUS BREAKING OF CONTINUOUS SYMMETRIES

The leap to the spontaneous breaking of continuous symmetries is easily made by considering a model first investigated by Goldstone [1], based on the Lagrangian for two scalar fields $\phi_1$ and $\phi_2$,

$$\mathcal{L} = \tfrac{1}{2}[(\partial_\mu\phi_1)(\partial^\mu\phi_1) + (\partial_\mu\phi_2)(\partial^\mu\phi_2)] - V(\phi_1^2 + \phi_2^2). \tag{5.2.1}$$

The Lagrangian is invariant under the group SO(2) of rotations in the plane

$$\boldsymbol{\phi} \equiv \begin{pmatrix} \phi_1 \\ \phi_2 \end{pmatrix} \rightarrow \begin{pmatrix} \cos\theta & \sin\theta \\ -\sin\theta & \cos\theta \end{pmatrix} \begin{pmatrix} \phi_1 \\ \phi_2 \end{pmatrix}. \tag{5.2.2}$$

As we have done before, we consider the effective potential

$$V(\boldsymbol{\phi}^2) = \tfrac{1}{2}\mu^2\boldsymbol{\phi}^2 + \tfrac{1}{4}\,|\lambda|\,(\boldsymbol{\phi}^2)^2, \tag{5.2.3}$$

where $\boldsymbol{\phi}^2 = \phi_1^2 + \phi_2^2$, and distinguish two cases.

(a)                                      (b)

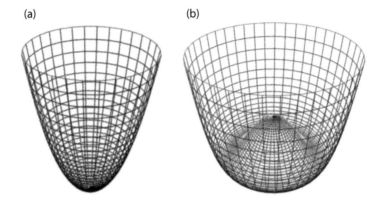

**Figure 5.2.** (a) Potential with a unique minimum at $\phi_1 = \phi_2 = 0$. (b) Potential with a degenerate vacuum at $\phi_1^2 + \phi_2^2 \neq 0$; selecting a vacuum spontaneously breaks the rotation symmetry.

A positive value of the parameter $\mu^2 > 0$ corresponds to the ordinary case of unbroken symmetry, illustrated by the potential in figure 5.2(a). The unique minimum, corresponding to the vacuum state, occurs at

$$\langle \boldsymbol{\phi} \rangle_0 = \begin{pmatrix} 0 \\ 0 \end{pmatrix}, \tag{5.2.4}$$

and so for small oscillations the Lagrangian takes the form

$$\mathcal{L}_{\text{so}} = \tfrac{1}{2}[(\partial_\mu \phi_1)(\partial^\mu \phi_1) - \mu^2 \phi_1^2] + \tfrac{1}{2}[(\partial_\mu \phi_2)(\partial^\mu \phi_2) - \mu^2 \phi_2^2]. \tag{5.2.5}$$

This is none other than the Lagrangian (2.2.4) for a pair of scalar particles with common mass $\mu$. Thus, the introduction of a symmetric interaction preserves the degenerate multiplet structure of the free theory, in accord with our simplest expectations.

The choice $\mu^2 < 0$ leads to a potential without a unique minimum, depicted in figure 5.2(b). The absolute minimum of the potential now occurs for

$$\langle \boldsymbol{\phi}^2 \rangle_0 = -\mu^2 / |\lambda| \equiv v^2, \tag{5.2.6}$$

which corresponds to a continuum of distinct vacuum states, degenerate in energy. The degeneracy is, of course, a consequence of the SO(2) symmetry of the potential (5.2.1). Designating one state as the vacuum selects a preferred direction in $(\phi_1, \phi_2)$ internal symmetry space and amounts to a spontaneous breakdown of the SO(2) symmetry. Let us select as the physical vacuum state the configuration

$$\langle \boldsymbol{\phi} \rangle_0 = \begin{pmatrix} v \\ 0 \end{pmatrix}, \tag{5.2.7}$$

as we may always do with a suitable definition of coordinates. Expanding about the vacuum configuration by defining

$$\boldsymbol{\phi}' \equiv \boldsymbol{\phi} - \langle \boldsymbol{\phi} \rangle_0 \equiv \begin{pmatrix} \eta \\ \zeta \end{pmatrix}, \tag{5.2.8}$$

we obtain the Lagrangian for small oscillations,

$$\mathcal{L}_{\text{so}} = \tfrac{1}{2}[(\partial_\mu \eta)(\partial^\mu \eta) + 2\mu^2 \eta^2] + \tfrac{1}{2}[(\partial_\mu \zeta)(\partial^\mu \zeta)], \tag{5.2.9}$$

plus an irrelevant constant. There are two particles in the spectrum. The $\eta$-particle, associated with radial oscillations, has $(\text{mass})^2 = -2\mu^2 > 0$, just as we found for the particle in the case of a spontaneously broken parity invariance. The $\zeta$-particle, however, is massless. The mass of the $\eta$-particle may be viewed as a consequence of the restoring force of the potential against radial oscillations. In contrast, the masslessness of the $\zeta$-particle is a consequence of the SO(2) invariance of the Lagrangian, which means that there is no restoring force against angular oscillations.

The splitting of the spectrum and the appearance of the massless particle are known as the Goldstone phenomenon. Such massless particles, which are referred to as Nambu–Goldstone bosons, are the zero-energy excitations that connect possible vacua. In the general case, one massless spin-zero particle will occur for each broken generator of the original symmetry group—that is, for each generator that connects distinct vacuum states. This is illustrated by further examples in problems 5.5 and 5.6.

The behavior exhibited by this example is a quite general consequence of the spontaneous breakdown of a continuous symmetry. This statement has attained the status of a theorem [2]. In any field theory that obeys the usual axioms, including locality, Lorentz invariance, and positive-definite norm on the Hilbert space, if an exact continuous symmetry of the Lagrangian is not a symmetry of the physical vacuum, then the theory must contain a massless spin-zero particle (or particles) whose quantum numbers are those of the broken group generator (or generators).

When coupled with the nonobservation of massless scalars or pseudoscalars (except perhaps for the pion), this strong statement would seem to preclude the use of spontaneous symmetry breaking in realistic theories of the fundamental interactions. However, gauge theories do not satisfy the assumptions on which the theorem is based, although they are respectable field theories. To quantize electrodynamics, for example, we must choose between the covariant Gupta–Bleuler formalism with its unphysical indefinite-metric states or quantization in a physical gauge wherein manifest covariance is lost. Without further ado, let us next investigate the spontaneous breakdown of a gauge symmetry.

## 5.3  SPONTANEOUS BREAKING OF A GAUGE SYMMETRY

We consider locally gauge-invariant Lagrangians that give rise to spontaneously broken symmetries. An unexpected cooperation will arise between the massless gauge fields (such as made the Yang–Mills theory an unacceptable description of the strong interactions of nucleons) and the massless Nambu–Goldstone bosons that have been seen to accompany spontaneous symmetry breaking. The simplest example of this interplay is provided by the Abelian Higgs model [3], which is nothing but the locally gauge-invariant extension of the Goldstone model discussed in the previous section: a U(1)-invariant theory that describes, in the absence of spontaneous symmetry breaking, the electrodynamics of charged scalars. The Lagrangian is simply

$$\mathcal{L} = |\mathcal{D}^{\mu}\phi|^2 - \mu^2\,|\phi|^2 - |\lambda|\,(\phi^*\phi)^2 - \tfrac{1}{4}F_{\mu\nu}F^{\mu\nu}, \tag{5.3.1}$$

where

$$\phi = \frac{\phi_1 + i\phi_2}{\sqrt{2}} \tag{5.3.2}$$

is a complex scalar field [4] and, as usual,

$$\mathcal{D}_{\mu} \equiv \partial_{\mu} + iq\,A_{\mu} \tag{5.3.3}$$

and

$$F_{\mu\nu} \equiv \partial_{\nu}A_{\mu} - \partial_{\mu}A_{\nu}. \tag{5.3.4}$$

The Lagrangian (5.3.1) is invariant under U(1) rotations

$$\phi \to \phi' = e^{i\theta}\phi \tag{5.3.5}$$

and under the local gauge transformations

$$\begin{aligned}
\phi(x) &\to \phi'(x) = e^{iq\alpha(x)}\phi(x),\\
A_{\mu}(x) &\to A'_{\mu}(x) = A_{\mu}(x) - \partial_{\mu}\alpha(x).
\end{aligned} \tag{5.3.6}$$

As usual, there are two cases, depending upon the parameters of the effective potential.

For $\mu^2 > 0$, the potential has a unique minimum at $\phi = 0$, and the exact symmetry of the Lagrangian is preserved. The spectrum is simply that of ordinary QED of charged scalars, with a single massless photon $A_{\mu}$ and two scalar particles, $\phi$ and $\phi^*$, with common mass $\mu$.

The situation when $\mu^2 = -\left|\mu^2\right| < 0$ is that of a spontaneously broken symmetry and requires a closer analysis. The potential has a continuum of absolute minima, corresponding to a continuum of degenerate vacua, at

$$\langle|\phi|^2\rangle_0 = \frac{-\mu^2}{2\,|\lambda|} \equiv \frac{v^2}{2}. \tag{5.3.7}$$

To explore the spectrum, we shift the fields in order to rewrite the Lagrangian (5.3.1) in terms of displacements from the physical vacuum. The latter may be chosen, without loss of generality, as

$$\langle\phi\rangle_0 = \frac{v}{\sqrt{2}}, \tag{5.3.8}$$

where $v > 0$ is a real number. We then define the shifted field

$$\phi' = \phi - \langle\phi\rangle_0, \tag{5.3.9}$$

which is conveniently parameterized in terms of

$$\phi = \frac{e^{i\zeta/v}(v + \eta)}{\sqrt{2}}$$
$$\approx \frac{(v + \eta + i\zeta)}{\sqrt{2}}. \tag{5.3.10}$$

Then the Lagrangian appropriate for the study of small oscillations is

$$\mathcal{L}_{so} = \tfrac{1}{2}[(\partial_\mu\eta)(\partial^\mu\eta) + 2\mu^2\eta^2] + \tfrac{1}{2}[(\partial_\mu\zeta)(\partial^\mu\zeta)]$$
$$- \tfrac{1}{4}F_{\mu\nu}F^{\mu\nu} + qvA_\mu(\partial^\mu\zeta) + \frac{q^2v^2}{2}A_\mu A^\mu + \cdots. \tag{5.3.11}$$

As we expect from out study of the Goldstone phenomenon, the $\eta$-field, which corresponds to radial oscillations, has a (mass)$^2 = -2\mu^2 > 0$. The gauge field $A_\mu$ appears to have acquired a mass, but it is mixed up in the penultimate term with the seemingly massless $\zeta$-field.

An astute choice of gauge will make it easier to sort out the spectrum of the spontaneously broken theory. To this end, it is convenient to rewrite the terms involving $A_\mu$ and $\zeta$ as

$$\frac{q^2v^2}{2}\left(A_\mu + \frac{1}{qv}\partial_\mu\zeta\right)\left(A^\mu + \frac{1}{qv}\partial^\mu\zeta\right), \tag{5.3.12}$$

a form that pleads for the gauge transformation

$$A_\mu \rightarrow A'_\mu = A_\mu + \frac{1}{qv}\partial^\mu\zeta, \tag{5.3.13}$$

which corresponds to the phase rotation on the scalar field

$$\phi \rightarrow \phi' = e^{-i\zeta(x)/v}\phi(x) = \frac{v + \eta}{\sqrt{2}}. \tag{5.3.14}$$

Knowing that the Lagrangian is locally gauge invariant, we may return to the original expression (5.3.1) to compute

$$\mathcal{L}_{so} = \tfrac{1}{2}[(\partial_\mu\eta)(\partial^\mu\eta) + 2\mu^2\eta^2] - \tfrac{1}{4}F_{\mu\nu}F^{\mu\nu} + \frac{q^2v^2}{2}A'_\mu A'^\mu, \tag{5.3.15}$$

plus an irrelevant constant. In this gauge the particle spectrum is manifest:

- an $\eta$-field, with $(\text{mass})^2 = -2\mu^2 > 0$ ;
- a massive vector field $A'_\mu$, with mass $= qv$ ;
- no $\zeta$-field.

By virtue of our choice of gauge, the $\zeta$-particle has disappeared from the Lagrangian entirely. Where has it gone? The gauge transformation (5.3.13) shows that what was formerly the $\zeta$-field has become the longitudinal component of the massive vector field $A'_\mu$. Before spontaneous symmetry breaking, the theory had four particle degrees of freedom: two scalars $\phi$ and $\phi^*$ plus two helicity states of the massless gauge field $A_\mu$. After spontaneous symmetry breaking, we are left with one scalar particle $\eta$ plus three helicity states of the massive gauge field $A'_\mu$, for a total of four particle degrees of freedom. In common parlance it is said that the massless photon "swallowed" the massless Nambu–Goldstone boson to become a massive vector boson. The remaining massive scalar ($\eta$) is known as the Higgs boson. The special gauge in which the particle spectrum became transparent is known as the unitary gauge, or U-gauge, because only physical states appear in the Lagrangian. This is not to say that the unitarity of the $S$-matrix, or the complete set of Feynman rules for the theory, is obvious in this gauge.

This is a truly remarkable result, suggesting as it does the possibility of constructing spontaneously broken gauge theories in which the interactions are mediated by massive vector bosons, rather than the phenomenologically unacceptable massless vector bosons of the unbroken theories. In a sense, each of the massless-particle diseases that we have encountered—gauge bosons and Nambu–Goldstone bosons—has provided the cure for the other [5]. Understanding the workings of spontaneous symmetry breaking in gauge theories has been the work of many hands, but the general strategy has come to be called the Higgs mechanism, in honor of one of the major contributors.

## 5.4 THE SIGMA MODEL

Before applying these lessons to non-Abelian gauge theories, it will be worthwhile to establish some insights that will serve us well when we examine the high-energy behavior of the electroweak theory and the low-energy behavior of quantum chromodynamics. Consider again the Dirac Lagrangian for free nucleons that we constructed in section 2.3. As before, we use the composite spinor

$$\psi \equiv \begin{pmatrix} p \\ n \end{pmatrix} \tag{2.3.13}$$

to write the free-nucleon Lagrangian compactly as

$$\mathcal{L} = \bar{\psi}(i\gamma^\mu \partial_\mu - m)\psi. \tag{2.3.14}$$

The Lagrangian is invariant under global isospin rotations that take $\psi \to \exp(i\boldsymbol{\alpha} \cdot \boldsymbol{\tau}/2)\psi$, where $\boldsymbol{\tau}$ consists of the usual $2 \times 2$ Pauli isospin matrices and $\boldsymbol{\alpha} = (\alpha_1, \alpha_2, \alpha_3)$ is an arbitrary, $x$-independent (three-vector) parameter of the transformation.

Now, following Gell-Mann and Lévy [6], let us take the nucleons to be massless. The Lagrangian simplifies to

$$\mathcal{L} = \bar{\psi}(i\gamma^\mu \partial_\mu)\psi. \tag{5.4.1}$$

This form remains invariant under global isospin rotations, but it is also invariant under the group of *chiral symmetries*, $SU(2)_L \otimes SU(2)_R$. To develop this statement, we designate the left- and right-handed components of the nucleon spinor as

$$\psi_L = \tfrac{1}{2}(1-\gamma_5)\psi, \qquad \psi_R = \tfrac{1}{2}(1+\gamma_5)\psi, \tag{5.4.2}$$

from which

$$\bar{\psi}_L = \tfrac{1}{2}\bar{\psi}(1+\gamma_5), \qquad \bar{\psi}_R = \tfrac{1}{2}\bar{\psi}(1-\gamma_5). \tag{5.4.3}$$

Now we may decompose the Lagrangian as

$$\mathcal{L} = i\bar{\psi}\partial\!\!\!/\psi = i\bar{\psi}[\tfrac{1}{2}(1-\gamma_5) + \tfrac{1}{2}(1+\gamma_5)]\partial\!\!\!/[\tfrac{1}{2}(1-\gamma_5) + \tfrac{1}{2}(1+\gamma_5)]\psi$$
$$= i\bar{\psi}_L\partial\!\!\!/\psi_L + i\bar{\psi}_R\partial\!\!\!/\psi_R, \tag{5.4.4}$$

in which left-right cross terms vanish because $\gamma_\mu\gamma_5 = -\gamma_5\gamma_\mu$, according to (A.2.21). The absence of terms coupling left and right means that we have the freedom to make separate isospin rotations on the left-handed and right-handed fields:

$$\psi_L \rightarrow \psi'_L = \exp\left(i\boldsymbol{\alpha}_L \cdot \frac{\boldsymbol{\tau}}{2}\right)\psi_L \equiv G_L\psi_L,$$
$$\psi_R \rightarrow \psi'_R = \exp\left(i\boldsymbol{\alpha}_R \cdot \frac{\boldsymbol{\tau}}{2}\right)\psi_R \equiv G_R\psi_R. \tag{5.4.5}$$

The nucleon mass term that we have eschewed is

$$m\bar{\psi}\psi = m\bar{\psi}[\tfrac{1}{2}(1-\gamma_5) + \tfrac{1}{2}(1+\gamma_5)]\psi = m\left(\bar{\psi}_L\psi_R + \bar{\psi}_R\psi_L\right), \tag{5.4.6}$$

which mixes the L and R fields. The answer falls out once we recall that $[\tfrac{1}{2}(1 \pm \gamma_5)]^2 = \tfrac{1}{2}(1 \pm \gamma_5)$. It is sometimes convenient to recast the transformations $G_L$ and $G_R$ in terms of an $SU(2)_V$ isospin rotation,

$$G_I = \exp\left(i\boldsymbol{\alpha}\cdot\frac{\boldsymbol{\tau}}{2}\right), \quad \boldsymbol{\alpha} = \boldsymbol{\alpha}_R + \boldsymbol{\alpha}_L, \tag{5.4.7a}$$

and an $SU(2)_A$ chiral rotation,

$$G_5 = \exp\left(i\gamma_5\boldsymbol{\alpha}_5 \cdot \frac{\boldsymbol{\tau}}{2}\right), \quad \boldsymbol{\alpha}_5 = \boldsymbol{\alpha}_R - \boldsymbol{\alpha}_L. \tag{5.4.7b}$$

The nucleon mass term explicitly breaks the chiral symmetry and leaves only normal isospin invariance intact.

Gell-Mann and Lévy found that they could build a Lagrangian with chiral symmetry and a nucleon mass if the chiral symmetry were spontaneously broken.

To see how this comes about, let us introduce $\boldsymbol{\Sigma}$, a $2 \times 2$ matrix of spinless meson fields that transforms under chiral isospin rotations as

$$\boldsymbol{\Sigma} \to \boldsymbol{\Sigma}' = G_L \boldsymbol{\Sigma} G_R^\dagger \qquad (5.4.8)$$

and add Yukawa couplings of the mesons and nucleons, so that

$$\mathcal{L} = \bar{\psi}(i\gamma^\mu \partial_\mu)\psi - g\bar{\psi}_L \boldsymbol{\Sigma} \psi_R - g\bar{\psi}_R \boldsymbol{\Sigma}^\dagger \psi_L + \mathcal{L}_\Sigma \,, \qquad (5.4.9)$$

where the final term describes meson self-interactions. If we expand $\boldsymbol{\Sigma}$ in terms of four real fields as

$$\boldsymbol{\Sigma} = \sigma + i\boldsymbol{\pi} \cdot \boldsymbol{\tau}, \qquad (5.4.10)$$

the Lagrangian becomes

$$\begin{aligned} \mathcal{L} &= \bar{\psi}(i\gamma^\mu \partial_\mu)\psi - g\sigma \left(\bar{\psi}_L \psi_R + \bar{\psi}_R \psi_L\right) - ig\boldsymbol{\pi} \cdot \left(\bar{\psi}_L \boldsymbol{\tau} \psi_R - \bar{\psi}_R \boldsymbol{\tau} \psi_L\right) + \mathcal{L}_\Sigma \\ &= \bar{\psi}(i\gamma^\mu \partial_\mu)\psi - g\sigma \bar{\psi}\psi + ig\boldsymbol{\pi} \cdot \bar{\psi}\boldsymbol{\tau}\gamma_5 \psi + \mathcal{L}_\Sigma. \end{aligned} \qquad (5.4.11)$$

The form is suggestive: the second term has the look of a potential nucleon mass term if $\langle\sigma\rangle_0$ were nonzero, and the third has the right form to describe the $\pi N\bar{N}$ coupling if we identify $g^2/4\pi = 13.7$ [7].

How can we build meson self-interactions that are invariant under $SU(2)_L \otimes SU(2)_R$? The most general invariant without derivatives will be a function of

$$\boldsymbol{\Sigma}^\dagger \boldsymbol{\Sigma} = I\sigma^2 + \boldsymbol{\pi} \cdot \boldsymbol{\tau} \, \boldsymbol{\pi} \cdot \boldsymbol{\tau} = I(\sigma^2 + \boldsymbol{\pi} \cdot \boldsymbol{\pi}), \qquad (5.4.12)$$

so we write

$$\mathcal{L}_\Sigma = \tfrac{1}{2}(\partial_\mu \sigma)(\partial^\mu \sigma) + \tfrac{1}{2}(\partial_\mu \boldsymbol{\pi}) \cdot (\partial^\mu \boldsymbol{\pi}) - V(\sigma^2 + \boldsymbol{\pi} \cdot \boldsymbol{\pi}). \qquad (5.4.13)$$

Now consider meson self-interactions dictated by a potential

$$V(\sigma^2 + \boldsymbol{\pi} \cdot \boldsymbol{\pi}) = \frac{\mu^2}{2}(\sigma^2 + \boldsymbol{\pi} \cdot \boldsymbol{\pi}) + \frac{|\lambda|}{4}(\sigma^2 + \boldsymbol{\pi} \cdot \boldsymbol{\pi})^2. \qquad (5.4.14)$$

If the parameter $\mu^2 > 0$, there is a unique minimum-energy state at $\sigma^2 + \boldsymbol{\pi} \cdot \boldsymbol{\pi} = 0$ for which the full symmetry of the theory is manifest: the left-handed and right-handed components of the proton and neutron remain massless, and the mesons $\sigma$ and $\boldsymbol{\pi}$ have a common mass $\mu$. The case $\mu^2 < 0$ is more interesting: the mimimum of the potential now occurs along a continuum defined by $\sigma^2 + \boldsymbol{\pi} \cdot \boldsymbol{\pi} = -\mu^2/|\lambda| \equiv v^2$. To analyze the consequences, it is convenient to rewrite the potential as

$$V(\sigma^2 + \boldsymbol{\pi} \cdot \boldsymbol{\pi}) = \tfrac{1}{4}|\lambda| \left(\sigma^2 + \boldsymbol{\pi} \cdot \boldsymbol{\pi} - v^2\right)^2, \qquad (5.4.15)$$

and to exploit the freedom to make $SU(2)_L \otimes SU(2)_R$ rotations to bring the vacuum expectation value to the $\sigma$-direction,

$$\langle \sigma \rangle_0 = v, \quad \langle \boldsymbol{\pi} \rangle_0 = \begin{pmatrix} 0 \\ 0 \\ 0 \end{pmatrix}. \tag{5.4.16}$$

Evidently the vacuum state is unchanged by isospin rotations but breaks $SU(2)_A$ symmetry (cf. problem 5.3).

To explore small perturbations about the vacuum state, we set $\sigma' = \sigma - v$, so that $\langle \sigma' \rangle_0 = 0$. Now the Lagrangian is

$$\mathcal{L} = \bar{\psi}(i\gamma^\mu \partial_\mu)\psi - gv\bar{\psi}\psi - g\sigma'\bar{\psi}\psi + ig\boldsymbol{\pi} \cdot \bar{\psi}\boldsymbol{\tau}\gamma_5\psi \tag{5.4.17}$$
$$+ \tfrac{1}{2}(\partial_\mu \sigma')(\partial^\mu \sigma') + \tfrac{1}{2}(\partial_\mu \boldsymbol{\pi}) \cdot (\partial^\mu \boldsymbol{\pi}) - \tfrac{1}{4}|\lambda| \left( \sigma'^2 + \boldsymbol{\pi} \cdot \boldsymbol{\pi} + 2\sigma'v \right)^2,$$

which describes nucleons with mass $M_N = gv$ coupled to a scalar field, $\sigma'$, with mass$^2$ $m_{\sigma'}^2 = 2|\lambda|v^2 = -2\mu^2$, and to *massless* pseudoscalar pions. The physical pions are far lighter than the other hadrons: $m_\pi^2/m_p^2 \approx 1/50$. Here they emerge, thanks to spontaneous breaking of the chiral symmetry $SU(2)_L \otimes SU(2)_R \to SU(2)_V$, as Nambu–Goldstone bosons, one for each broken generator of $SU(2)_A$.

What are the dynamical implications of the model? The conserved currents associated with the Lagrangian (5.4.11) are

$$V_\mu = \bar{\psi}\gamma_\mu \frac{\boldsymbol{\tau}}{2}\psi + \boldsymbol{\pi} \times (\partial_\mu \boldsymbol{\pi}),$$
$$A_\mu = \bar{\psi}\gamma_\mu \frac{\boldsymbol{\tau}}{2}\gamma_5\psi + \boldsymbol{\pi}(\partial_\mu \sigma) - \sigma(\partial_\mu \boldsymbol{\pi}). \tag{5.4.18}$$

In terms of the shifted field $\sigma' = \sigma - v$, the axial current has the form

$$A_\mu = -v(\partial_\mu \boldsymbol{\pi}) + \text{ bilinears in } \boldsymbol{\pi}, \sigma', \psi. \tag{5.4.19}$$

The first term is interesting, because it couples the pion to the vacuum through

$$\langle 0 | A_\mu^i | \pi^i(p) \rangle = ivp_\mu \delta^{ij} \equiv if_\pi p_\mu \delta^{ij}, \tag{5.4.20}$$

where the last form gives a conventional definition of the pseudoscalar decay constant $f_\pi \approx 92$ MeV [8]. Identifying $f_\pi = v$, we see that the spontaneous breakdown of chiral symmetry may be seen as inducing pion decay. Earlier we were led to identify $g$ as the pion–nucleon coupling. We may combine all our information and test for consistency:

$$939 \text{ MeV} = M_N = gv \stackrel{?}{=} g_{\pi NN} f_\pi = 1215 \text{ MeV}. \tag{5.4.21}$$

This is not a quantitative triumph, but the connection (5.4.21) is a simplified version of the celebrated Goldberger–Treiman relation [9],

$$1192 \text{ MeV} = g_A M_N = g_{\pi NN} f_\pi = 1215 \text{ MeV}, \tag{5.4.22}$$

where $g_A \approx 1.2695$ is the nucleon axial charge.

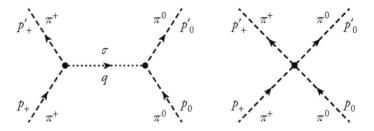

**Figure 5.3.** $\pi\pi$ scattering in the sigma model.

Finally, let us look more closely at the meson self-interactions, which appear in the final term of (5.4.17),

$$\mathcal{L}_{\text{int}} = -\tfrac{1}{4}|\lambda|\left(\sigma^2 + \boldsymbol{\pi}\cdot\boldsymbol{\pi} + 2\sigma v\right)^2$$
$$= -\tfrac{1}{4}|\lambda|\left[\sigma^4 + (\boldsymbol{\pi}\cdot\boldsymbol{\pi})^2 + 4\sigma^2 v^2 + 2\sigma^2\boldsymbol{\pi}\cdot\boldsymbol{\pi} + 4\sigma^3 v + 4\sigma v\boldsymbol{\pi}\cdot\boldsymbol{\pi}\right], \quad (5.4.23)$$

where we have dropped the primes to streamline notation. We have already considered the mass term, $4\sigma^2 v^2$. Putting it aside and expanding in components, the interactions are given by

$$\mathcal{L}'_{\text{int}} = -|\lambda|\, v\sigma\,(2\pi^+\pi^- + \pi^0\pi^0 + \sigma\sigma) - \tfrac{1}{4}|\lambda|\,(2\pi^+\pi^- + \pi^0\pi^0 + \sigma\sigma)^2. \quad (5.4.24)$$

The Feynman diagrams that contribute at tree level to $\pi^+\pi^0$ scattering are shown in figure 5.3. Being attentive to combinatorial factors, we can read off the Feynman rules for the $\pi\pi\sigma$ vertex $(-2i\,|\lambda|\,v)$ and for the four-pion contact term $(-2i\,|\lambda|)$. Accordingly, the scattering amplitude is given by

$$\mathcal{M}(\pi^+\pi^0 \to \pi^+\pi^0) = (-2i\,|\lambda|\,v)^2\,\frac{i}{q^2 - m_\sigma^2} - 2i\,|\lambda|$$
$$= -\frac{i m_\sigma^2}{v^2}\,\frac{q^2}{q^2 - m_\sigma^2}\,, \quad (5.4.25)$$

where we have used $m_\sigma^2 = 2\,|\lambda|\,v^2$ to simplify the result. Recalling our identification of the vacuum expectation value $v = f_\pi$, we can recast the scattering amplitude as

$$\mathcal{M} = \frac{i q^2}{f_\pi^2}\,\frac{1}{1 - q^2/m_\sigma^2} \approx \frac{i q^2}{f_\pi^2}\left(1 + \frac{q^2}{m_\sigma^2} + \cdots\right), \quad (5.4.26)$$

where the last expression holds for $q^2 \ll m_\sigma^2$. Two features are noteworthy: first, to leading order in $q^2$, the amplitude depends only on the symmetry-breaking parameter $f_\pi$, not on $m_\sigma$; second, the amplitude vanishes as $q^2 \to 0$. The vanishing of the amplitude at zero momentum is a universal feature of theories built on chiral symmetry.

To close this excursion, we remark that the interaction Lagrangian (5.4.24) is identical in form to the interaction Lagrangian for the Higgs sector of the electroweak theory, with the identification $(\pi^+, \pi^0, \pi^-) \to (w^+, z, w^-)$, the three massless scalars that become longitudinal components of $W^+$, $Z^0$, $W^-$, and $\sigma \to h$, the massive neutral scalar that becomes the Higgs boson [10].

## 5.5 SPONTANEOUS BREAKING OF A NON-ABELIAN SYMMETRY

To approach the additional complications that attend the spontaneous breakdown of a non-Abelian gauge symmetry, we choose as a useful prototype an SU(2) gauge theory and study scalar fields that make up the triplet (or isovector) representation

$$\boldsymbol{\phi} = \begin{pmatrix} \phi_1 \\ \phi_2 \\ \phi_3 \end{pmatrix}. \tag{5.5.1}$$

Construction of the unbroken gauge theory was the subject of problem 4.1. We require invariance under the gauge transformation

$$\boldsymbol{\phi} \to \boldsymbol{\phi}' = \exp(i\mathbf{T} \cdot \boldsymbol{\alpha})\boldsymbol{\phi}, \tag{5.5.2}$$

where the exponential factor is a $3 \times 3$ matrix. The operator $T_i$ generates isospin rotations about the $i$-axis and satisfies the usual SU(2) algebra

$$\left[T^j, T^k\right] = i\varepsilon_{jkl}T^l. \tag{5.5.3}$$

The explicit matrix representation is

$$(T^j)_{kl} = -i\varepsilon_{jkl}. \tag{5.5.4}$$

As usual, the covariant derivative takes the form

$$\mathcal{D}_\mu = I\partial_\mu + ig\mathbf{T} \cdot \mathbf{b}_\mu, \tag{5.5.5}$$

or, in isospin-component form,

$$(\mathcal{D}_\mu)_{kl} = \delta_{kl}\partial_\mu + g\varepsilon_{jkl}b^j_\mu. \tag{5.5.6}$$

In the presence of an effective potential, the interaction is

$$\mathcal{L} = \tfrac{1}{2}(\mathcal{D}_\mu\boldsymbol{\phi}) \cdot (\mathcal{D}^\mu\boldsymbol{\phi}) - V(\boldsymbol{\phi} \cdot \boldsymbol{\phi}) - \tfrac{1}{4}\mathbf{F}_{\mu\nu} \cdot \mathbf{F}^{\mu\nu}. \tag{5.5.7}$$

When $\boldsymbol{\phi} = 0$ is a unique minimum of the effective potential $V$, the spectrum is that of an ordinary, isospin-conserving Yang–Mills field theory: three massive scalar mesons, each with mass $\mu$, and three massless gauge bosons $\mathbf{b}_\mu$. The number of particle degrees of freedom is thus $3 \times 1 + 3 \times 2 = 9$.

Of more interest in the present context is the spontaneously broken case, in which the field configuration that minimizes $V(\boldsymbol{\phi}^2)$ may be chosen as

$$\langle \boldsymbol{\phi} \rangle_0 = \begin{pmatrix} 0 \\ 0 \\ v \end{pmatrix}. \tag{5.5.8}$$

Notice (cf. problem 5.5) that the vacuum state (5.5.8) is invariant under the action of $T_3$ (in the sense that $\exp(i\,T_3\alpha_3)\,\langle \boldsymbol{\phi} \rangle_0 = \langle \boldsymbol{\phi} \rangle_0$, which is true provided that $T_3\,\langle \boldsymbol{\phi} \rangle_0 = 0$) but not under $T_1$ and $T_2$.

We shift the scalar fields and expand about the minimum configuration using

$$\boldsymbol{\phi} = \exp\left[\frac{i}{v}(\zeta_1 T_1 + \zeta_2 T_2)\right] \begin{pmatrix} 0 \\ 0 \\ v + \eta \end{pmatrix}, \tag{5.5.9}$$

It is unnecessary to retrace step by step the arithmetic of the Abelian Higgs model of section 5.3. We may immediately exploit the gauge invariance of the theory by transforming to U-gauge by letting

$$\boldsymbol{\phi} \to \boldsymbol{\phi}' = \exp\left[\frac{-i}{v}(\zeta_1 T_1 + \zeta_2 T_2)\right] \boldsymbol{\phi} = \begin{pmatrix} 0 \\ 0 \\ v + \eta \end{pmatrix}, \tag{5.5.10}$$

with the implied transformation on the gauge fields. In the new gauge, the Lagrangian appropriate to the description of small oscillations about the minimum is

$$\mathcal{L}_{\text{so}} = \tfrac{1}{2}[(\partial_\mu \eta)(\partial^\mu \eta) + 2\mu^2\eta^2] - \tfrac{1}{4}\mathbf{F}_{\mu\nu} \cdot \mathbf{F}^{\mu\nu}$$
$$+ \frac{g^2 v^2}{2}(b_\mu^1 b^{1\mu} + b_\mu^2 b^{2\mu}) + \cdots. \tag{5.5.11}$$

In this form the Lagrangian reveals that

- $\eta$ has become a massive Higgs scalar, with (mass)$^2 = -2\mu^2 > 0$;
- the would-be Nambu–Goldstone bosons $\zeta_1$ and $\zeta_2$ have disappeared entirely, that is, they have been "gauged away";
- the vector bosons $b_\mu^1$ and $b_\mu^2$ corresponding to the (broken symmetry) generators $T_1$ and $T_2$ acquire a common mass $gv$;
- the gauge boson $b_\mu^3$ remains massless, reflecting the invariance of the vacuum under the generator $T_3$.

After spontaneous symmetry breaking, the number of particle degrees of freedom remains 9, now given by $1 \times 1 + 2 \times 3 + 1 \times 2$.

## 5.6 PROSPECTS

The analysis of this chapter has made available to us a large variety of field theories that exhibit the spontaneous breaking of internal symmetries. The breaking of a discrete symmetry poses no particular problems. If the broken symmetry is continuous, however, the spontaneous breakdown of the symmetry is accompanied by the appearance of massless spin-zero particles known as Nambu–Goldstone bosons. These appear as an impediment to the use of spontaneously broken symmetries to describe the physics of elementary particles, because—apart from the pions—no such massless scalars have yet been observed. If we go further and require the continuous symmetry to be a local gauge symmetry, the degrees of freedom associated with the Nambu–Goldstone bosons are absorbed into the gauge bosons that correspond to the broken generators of the gauge group. Thus, the disappearance of unwanted massless scalars and the acquisition of mass by the gauge bosons are coupled phenomena. The way is now open to the construction of theories in which interactions restricted by gauge principles are mediated by massive vector particles.

All the discussion of this chapter has been at the level of classical field theory. In particular situations, it will be important to verify that the elementary analysis of the minimum of the effective potential remains valid in the presence of quantum corrections. Generally speaking, this is the case unless the interactions represented by the effective potential are in some measure too strong. A brief discussion of one specific case will be presented in section 7.10, where we shall see nontrivial constraints on the Higgs-boson mass. It is also worthwhile to remark that the scalar fields that interact among themselves to effect the spontaneous symmetry breaking need not be elementary, but might arise as dynamical bound states of elementary fermions. This is the case, for example, in the BCS theory of superconductivity [11], where Cooper pairs of electrons are the objects of interest. This is a possibility to which we shall return.

We have developed the essential ideas of gauge theories and most of the tools needed to apply these ideas to the fundamental interactions. As a first application, we now turn to a unified description of the weak and electromagnetic interactions, the theory beyond QED that has been most extensively tested by experiment.

## PROBLEMS

5.1. The two-dimensional Ising model is an instructive caricature of a ferromagnet that displays a second-order phase transition—spontaneous magnetization—at a critical (inverse) temperature $\beta_c = 0.4407$. It is easily studied by Monte Carlo simulation. For this exercise, take advantage of the applets that are freely available on the Internet.

(a) Heat the system above the critical temperature and observe the disordered state.

(b) Cool the system just below the critical temperature to see the formation of domains.

(c) Cool the system further and watch the evolution of the spontaneous magnetization.

(d) Beginning from a zero-temperature (ordered) configuration, heat the system and observe the formation of small bubbles of flipped spins. When the temperature exceeds the critical temperature, the bubbles expand and merge, and up-down symmetry is restored. Repeat the experiments enough times to develop some intuition about the occurrence of configurations far from the average. [E. Ising, *Z. Phys.* **31**, 253 (1925); for an exhaustive treatment, see B. M. McCoy and T. T. Wu, *The Two-Dimensional Ising Model,* Harvard University Press, Cambridge, 1973.]

5.2. Using the transformation laws (5.4.6) and (5.4.8) for the behavior of the nucleon fields $\psi$ and meson fields $\Sigma$ under chiral isospin rotations, verify that the Yukawa terms in the $\sigma$-model Lagrangian (5.4.11) are left invariant by chiral isospin rotations.

5.3. Starting from the transformation law (5.4.8) for the composite meson field $\Sigma$, analyze the behavior of the components $\sigma$ and $\pi$ under chiral isospin rotations. Show in particular that $SU(2)_V$ isospin rotations leave $\sigma$ unchanged while rotating the components of $\pi$, whereas $SU(2)_A$ chiral rotations mix the $\sigma$ and $\pi$ fields.

5.4. In the four-fermion theory of the charged-current weak interactions, the invariant amplitude for the decay $\pi^+ \to \ell^+ \nu_\ell$ is given by

$$\mathcal{M} = G_F \cos\theta_C \, f_\pi \, p_\mu \bar{u}(\nu, q_\nu) \gamma^\mu (1 - \gamma_5) v(\ell, q_\ell),$$

where $\cos\theta_C \approx 0.9746$, the pion momentum is $p = q_\nu + q_\ell$, and the hadronic part of the transition matrix element is $\langle 0| A_\mu^i | \pi^j(p) \rangle \equiv i f_\pi p_\mu \delta^{ij}$.

(a) Evaluate the $\pi^+ \to \mu^+ \nu_\mu$ decay rate and use your result to determine the pion pseudoscalar decay constant, $f_\pi$.

(b) Predict the ratio $\Gamma(\pi^+ \to e^+ \nu_e)/\Gamma(\pi^+ \to \mu^+ \nu_\mu)$ and compare with observations. What accounts for the suppression of the $\pi^+ \to e^+ \nu_e$ decay rate? [E. D. Commins and P. H. Bucksbaum, *Weak Interactions of Leptons and Quarks,* Cambridge University Press, Cambridge, 1983, §4.4; J. F. Donoghue, E. Golowich, and B. R. Holstein, *Dynamics of the Standard Model,* Cambridge University Press, Cambridge, 1994, §6-1.]

5.5. Analyze the spontaneous breakdown of a global $SU(2)$ symmetry. Consider the case of three real scalar fields $\phi_1$, $\phi_2$, and $\phi_3$, which constitute an $SU(2)$ triplet, denoted

$$\phi = \begin{pmatrix} \phi_1 \\ \phi_2 \\ \phi_3 \end{pmatrix}.$$

The Lagrangian density is

$$\mathcal{L} = \tfrac{1}{2}(\partial_\mu \phi) \cdot (\partial^\mu \phi) - V(\phi \cdot \phi),$$

where, as usual,

$$V(\phi \cdot \phi) = \tfrac{1}{2}\mu^2 \phi \cdot \phi + \tfrac{1}{4}|\lambda| (\phi \cdot \phi)^2.$$

Assume that for $\mu^2 < 0$, the potential has a minimum at

$$\langle \boldsymbol{\phi} \rangle_0 = \begin{pmatrix} 0 \\ 0 \\ v \end{pmatrix}.$$

Then show that (a) the vacuum remains invariant under the action of the generator $T_3$ but not under $T_1$ or $T_2$; (b) the particles associated with $T_1$ and $T_2$ become massless (Nambu–Goldstone) particles; (c) the particle associated with $T_3$ acquires a mass $= \sqrt{-2\mu^2}$.

5.6. Generalize the preceding example to a Lagrangian that describes the interactions of $n$ scalar fields and is invariant under global transformations under the group $O(n)$. After spontaneous symmetry breaking and the choice of a vacuum state, show that the vacuum is invariant under the group $O(n-1)$. Verify that the number of Nambu–Goldstone bosons corresponds to the number of broken generators of the original symmetry group—that is, to the difference between the number of generators of $O(n)$ and $O(n-1)$.

5.7. The Ginzburg–Landau theory of superconductivity provides a phenomenological understanding of the Meissner effect: the observation that an external magnetic field does not penetrate the superconductor. Ginzburg and Landau introduce an "order parameter" $\psi$, such that $|\psi|^2$ is related to the density of superconducting charge carriers with effective charge $e^\star$ and effective mass $m^\star$. (The effective charge $e^\star$ turns out to be $-2e$ and the effective mass, twice the electron mass, because $|\psi|^2$ represents the density of Cooper pairs.) In the absence of an impressed magnetic field, expand the free energy of the superconductor as

$$G_{\text{super}}(0) = G_{\text{normal}}(0) + \alpha \, |\psi|^2 + \beta \, |\psi|^4,$$

where $\alpha$ and $\beta$ are phenomenological parameters.

(a) Minimize $G_{\text{super}}(0)$ with respect to the order parameter and discuss the circumstances under which spontaneous symmetry breaking occurs. Compute $\langle |\psi|^2 \rangle_0$, the value at which $G_{\text{super}}(0)$ is minimized.

(b) In the presence of an external magnetic field $\mathbf{B}$, a gauge-invariant expression for the free energy is

$$G_{\text{super}}(\mathbf{B}) = G_{\text{super}}(0) + \frac{\mathbf{B}^2}{2} + \frac{1}{2m^\star} \psi^* (-i\nabla - e^\star \mathbf{A})^2 \psi.$$

Derive the field equations that follow from minimizing $G_{\text{super}}(\mathbf{B})$ with respect to $\psi$ and $\mathbf{A}$. Show that in the slowly varying weak-field approximation (for which $\nabla \psi \approx 0$, $\psi \approx \langle \psi \rangle_0$), the photon acquires a mass $\lambda^{-1} = e^\star |\langle \psi \rangle_0| / \sqrt{m^\star c^2}$ within the superconductor. The parameter $\lambda$, the London penetration depth [F. London and H. London, *Proc. R. Soc. Lond.* A149, 71–88 (1935)], signals the exclusion of magnetic flux from the superconducting medium. [V. L. Ginzburg and L. D. Landau, *Zh. Eksp. Teor. Fiz.* 20, 1064 (1950); English translation: see *Men of Physics: Landau*, Vol. II, ed. D. ter Haar,

Pergamon, New York, 1965. For further information, see M. P. Marder, *Condensed Matter Physics,* Wiley-Interscience, New York, 2000, §27.2; and S. Weinberg, *The Quantum Theory of Fields,* Cambridge University Press, Cambridge, 1996, §21.6.]

5.8. Derive the equation of motion for the photon field $A_\nu$ in the Abelian Higgs model and show that it amounts to a relativistic generalization of the Ginzburg–Landau description of a superconductor.

## ⅢⅢⅢⅢⅢⅢ FOR FURTHER READING ⅢⅢⅢⅢⅢⅢⅢ

**Symmetry and the spectrum.** Implications of an exact quantum-mechanical symmetry are discussed in many places. Particularly clear presentations appear in

  K. Gottfried and T.-M. Yan, *Quantum Mechanics: Fundamentals*, second edition, Springer, Berlin, 2004, chap. VII,

and in The Source,

  E. P. Wigner, *Group Theory and Its Applications to the Quantum Mechanics of Atomic Spectra,* trans. J. J. Griffin, Academic Press, New York, 1959.

**Spontaneously broken gauge theories.** Fine introductory discussions at roughly the same technical level as this chapter are given by

  E. S. Abers and B. W. Lee, *Phys. Rep.* **9C**, 1 (1973),

  S. Coleman, "Secret Symmetry," in *Aspects of Symmetry: Selected Erice Lectures,* Cambridge University Press, Cambridge, 1988, chap. 5; *Phys. Rev. D* **15**, 2929 (1977) [Erratum: *Phys. Rev. D* **16**, 1248 (1977)],

  P. W. Higgs, "Spontaneous Symmetry Breaking," in *Phenomenology of Particles at High Energies,* 14th Scottish Universities Summer School in Physics, 1973, ed. R. L. Crawford and R. Jennings, Academic Press, New York, 1974, p. 529.

  C. Itzykson and J.-B. Zuber, *Quantum Field Theory,* McGraw-Hill, New York, 1980, chapters 11, 12,

  A. D. Linde, *Rep. Prog. Phys.* **42**, 389 (1979),

  J. C. Taylor, *Gauge Theories of the Weak Interactions,* Cambridge University Press, Cambridge, 1975, chap. 5, 6.

A number of specific examples are treated in

  H. Fritzsch and P. Minkowski, *Phys. Rep.* **73**, 67 (1981).

A comprehensive review at a slightly more formal level is presented in

  J. Bernstein, *Rev. Mod. Phys.* **46**, 7 (1974).

Still more formal is the article

  L. O'Raifeartaigh, *Rep. Prog. Phys.* **42**, 259 (1979).

Special attention to the interplay between spontaneous symmetry breaking and conservation laws is given in

  G. S. Guralnik, C. R. Hagen, and T.W.B. Kibble, in *Advances in Particle Physics,* ed. R. L. Cool and R. E. Marshak, Interscience, New York, 1968, vol. 2, p. 567.

A non-field-theoretic introduction, with reference to the physical analogy of the superconductor, is presented in

  I.J.R. Aitchison and A.J.G. Hey, *Gauge Theories in Particle Physics,* Adam Hilger, Bristol, 1982, chap. 9.

Among recollections of the early days, see

F. Englert, "Broken Symmetry and Yang–Mills Theory," in *50 Years of Yang–Mills Theories*, ed. G. 't Hooft, World Scientific, Singapore, 2005, p. 65, arXiv:hep-th/0406162.

G. S. Guralnik, "The History of the Guralnik, Hagen and Kibble Development of the Theory of Spontaneous Symmetry Breaking and Gauge Particles," *Int. J. Mod. Phys.* **24**, 2601 (2009), arXiv:0907.3466; "Gauge Invariance and the Goldstone Theorem," *Mod. Phys. Lett. A* **26**, 1381 (2011) [arXiv:1107.4592],

P. W. Higgs, "SBGT and all that," in *Discovery of Weak Neutral Currents: The Weak Interaction Before and After*, ed. A. K. Mann and D. B. Cline, *AIP Conf. Proc.* **300**, 159 (1994); "My Life as a Boson: The Story of 'The Higgs'," *Int. J. Mod. Phys. A* **17S1**, 86 (2002),

T.W.B. Kibble, "Englert-Brout-Higgs-Guralnik-Hagen-Kibble Mechanism (History)," *Scholarpedia* **4**(1), 8741 (2009); http://j.mp/yq8tqW.

**Spontaneous symmetry breaking in other physical contexts.** The phenomenon of spontaneously broken symmetries is a ubiquitous one, present in crystallization, the onset of turbulence, the Meissner effect, and many other circumstances. An interesting tour is given by

P. W. Anderson, in *Gauge Theories and Modern Field Theory*, ed. R. Arnowitt and P. Nath, MIT Press, Cambridge, MA, 1976, p. 311.

Goldstone and pseudo-Goldstone bosons in nuclear, particle, and condensed-matter physics are the subject of

C. P. Burgess, *Phys. Rept.* **330**, 193 (2000).

Nambu–Goldstone modes, spin waves, second sound, and their damping in magnets, superfluids, and liquid crystals are treated in

G. F. Mazenko, *Fluctuations, Order, and Defects*, Wiley, New York, 2003.

For more on the Heisenberg ferromagnet, see

D. C. Mattis, *The Theory of Magnetism Made Simple*, World Scientific, Singapore, 2006.

The relationship between the remarkable properties of superconductors and the broken electromagnetic gauge symmetry ($U(1)_{em} \to C_2$, phase rotations by $\pi$) in the superconducting ground state was illuminated by

Y. Nambu, Phys. Rev. **117**, 648 (1960),

which contains the suggestion that massless spin-zero particles must arise when any continuous symmetry is spontaneously broken.

Brief accounts of spontaneous symmetry breaking in superconductors appear in

E. A. Lynton, *Superconductivity*, Methuen, London, 1962, chap. 5,

J. R. Schrieffer, *Theory of Superconductivity*, Westview Press, Boulder, CO, 1999,

M. Tinkham, *Introduction to Superconductivity*, second edition, Dover Press, New York, 2004.

An analogy between ferromagnetism and the laser effect is described by

M. Sargent III, M. O. Scully, and W. E. Lamb, Jr., *Laser Physics*, Westview Press, Boulder, CO, 1978, §21-3.

Spontaneous symmetry breaking in the Bogoliubov superfluid is reviewed by Higgs (Scottish Universities Summer School lectures cited before) and in

D. Forster, *Hydrodynamic Fluctuations, Broken Symmetry, and Correlation Functions*, Westview Press, Boulder, CO, 1995.

Both classical and quantum systems are treated in
    F. Strocchi, *Symmetry Breaking,* Springer-Verlag, Berlin, 2008.

Evidence for a "Higgs boson" in superconductors is examined in
    C. M. Varma, *Journal of Low Temperature Physics* **126**, 901 (2002); P. B. Littlewood
        and C. M. Varma, *Phys. Rev. Lett.* **47**, 811 (1981),
    P. Higgs, "A Brief History of the Higgs Mechanism," http://j.mp/GNJA5Y,
    L. Dixon, "From Superconductors to Supercolliders," *SLAC Beam Line* **26**, 23 (1996)
        http://j.mp/GN4uAY.

See also the comments by P. Higgs (p. 509) and Y. Nambu (p. 514) in
    *The Rise of the Standard Model,* ed. L. Hoddeson, L. Brown, M. Riordan, and
        M. Dresden, Cambridge University Press, Cambridge, 1997.

**Spontaneously broken symmetries in particle physics.** Attempts to account for the
breaking of hadronic symmetries in terms of spontaneous symmetry breaking were initiated
by
    Y. Nambu, *Phys. Rev. Lett.* **4**, 380 (1960),
    Y. Nambu and G. Jona-Lasinio, *Phys. Rev.* **122**, 345 (1961); *Phys. Rev.* **124**, 246 (1961).

These papers are notable for combining tiny explicit violations of chiral symmetry with
spontaneous symmetry breaking. The light particles that arise when an approximate
symmetry is spontaneously broken are called pseudo-Nambu–Goldstone bosons.

**Gauge invariance and mass.** That dynamical effects could circumvent the prediction of
massless gauge bosons was first recognized by
    J. Schwinger, *Phys. Rev.* **124**, 397 (1962),

who presented a concrete example of a massive photon in the exactly solvable model of the
quantum electrodynamics of massless fermions in two dimensions, in
    J. Schwinger, *Phys. Rev.* **128**, 2425 (1962).

For the three-dimensional case, see
    J. F. Schonfeld, *Nucl. Phys.* **B185**, 157 (1981).

A review, emphasizing topological considerations, is given by
    R. Jackiw, "Gauge Invariance and Mass, III" in *Asymptotic Realms of Physics,* edited by
        A. Guth, K. Huang, and R. L. Jaffe, MIT Press, Cambridge, MA, 1983.

The link between gauge invariance and the Meissner effect is made in
    P. W. Anderson, *Phys. Rev.* **110**, 827 (1958).

For a well-chosen reprint collection, back to the prehistory of spontaneous symmetry
breaking, plus a perceptive introduction, see
    E. Farhi and R. Jackiw (eds.), *Dynamical Gauge Symmetry Breaking,* World Scientific,
        Singapore, 1982.

An explicit demonstration that spontaneous symmetry breaking in the form of nonvanishing
expectation values of a multiplet of scalar fields endows gauge bosons with mass was given
by
    F. Englert and R. Brout, *Phys. Rev. Lett.* **13**, 321 (1964).

The gauge-invariant extension of the Goldstone model was studied nearly simultaneously by
    P. W. Higgs, *Phys. Rev. Lett.* **13**, 508 (1964); *Phys. Lett.* **12**, 132 (1964).

Similar conclusions were reached in a different context at nearly the same moment by
G. S. Guralnik, C. R. Hagen, and T.W.B. Kibble, *Phys. Rev. Lett.* **13**, 585 (1964).

See also the work of
A. A. Migdal and A. M. Polyakov, *Sov. Phys. JETP* **24**, 91 (1967) [*Zh. Eksp. Teor. Fiz.* **51**, 135 (1966)]; included in the Farhi–Suskind reprint collection, p. 52.

The general group-theoretical analysis of this phenomenon is due to
T.W.B. Kibble, *Phys. Rev.* **155**, 1554 (1967).

For a valuable complement to the presentation in this chapter, see
J. Zinn-Justin, *Quantum Field Theory and Critical Phenomena,* 4th ed. Oxford University Press, Oxford, 2002, §18.10 and chap. 19.

For brief, but informative, intellectual histories, see
J. Bernstein, *Am. J. Phys.* **79**, 25-31 (2011),
J. Zinn-Justin, "From QED to the Higgs Mechanism: a Short Review," talk given at the conference on *Higgs Hunting,* Orsay, July 2010, available at http://j.mp/zzXWHw.

**The sigma model.** Excellent summaries from a modern perspective are in
J. F. Donoghue, E. Golowich, and B. R. Holstein, *Dynamics of the Standard Model,* Cambridge University Press, Cambridge, 1994, chap. 1,
H. Georgi, *Weak Interactions and Modern Field Theory,* Dover Press, New York, 2009, §2.6.

**Unitary gauge.** The demonstration that a U-gauge can be found for a general spontaneously broken gauge theory was given by
S. Weinberg, *Phys. Rev. D* **7**, 1068 (1973).

## ⅲⅲⅲⅲⅲⅲⅲⅲⅲ REFERENCES ⅲⅲⅲⅲⅲⅲⅲⅲⅲⅲ

1. J. Goldstone, *Nuovo Cim.* **19**, 154 (1961).
2. J. Goldstone, A. Salam, and S. Weinberg, *Phys. Rev.* **127**, 965 (1962); S. Bludman and A. Klein, *Phys. Rev.* **131**, 2364 (1963); W. Gilbert, *Phys. Rev. Lett.* **12**, 713 (1964); R. F. Streater, *Proc. Roy. Soc. (London)* **A287**, 510 (1965); D. Kastler, D. W. Robinson, and A. Swieca, *Commun. Math. Phys.* **2**, 108 (1966).
3. P. W. Higgs, *Phys. Rev.* **145**, 1156 (1966).
4. The analysis can be carried out equally well in terms of the two real fields $\phi_1$ and $\phi_2$ or in a two-dimensional vector notation with $\boldsymbol{\phi} = (\phi_1, \phi_2)^{\mathrm{T}}$. Pay attention to factors of 2.
5. That one zero-mass ill might cancel the other was suggested by analogy with the plasmon theory of the free-electron gas by P. W. Anderson, *Phys. Rev.* **130**, 439 (1963).
6. M. Gell-Mann and M. Lévy, *Nuovo. Cim.* **16**, 705 (1960).
7. D. V. Bugg, *Eur. Phys. J.* C **33**, 505 (2004).
8. Another convention in common usage defines the pseudoscalar decay constant to be $\sqrt{2}\times$ this value.
9. M. L. Goldberger and S. B. Treiman, *Phys. Rev.* **110**, 1178 (1958).
10. B. W. Lee, C. Quigg and H. B. Thacker, *Phys. Rev. D* **16**, 1519 (1977).
11. J. Bardeen, L. N. Cooper, and J. R. Schrieffer, *Phys. Rev.* **106**, 162 (1957).

# Six

# Electroweak Interactions of Leptons

In the preceding chapters we have developed a general strategy for the construction of exact or spontaneously broken gauge theories of the fundamental interactions. The point of departure for applications to particle physics is the recognition of a symmetry respected by the elementary fermions in the problem. What follows in the case of a spontaneously broken symmetry is an effort to conceal the exact symmetry in the same fashion that nature has chosen. There is much art in the selection of the gauge symmetry and of a pattern of symmetry breaking. Both experimental results and theoretical requirements offer constraints.

To illustrate the construction of a realistic gauge theory, we shall treat the model of leptons first given explicitly by Weinberg [1]. When completed by the addition of the quark sector, which is the subject of chapter 7, the $SU(2)_L \otimes U(1)_Y$ theory gives an excellent account of electroweak phenomena at the energies we have been able to explore. It has been validated in so many experimental tests that it deserves to be called—provisionally—a new law of nature. Experimental tools to search for, and characterize, the agent of electroweak symmetry breaking are now at hand.

Before constructing the theory, it is necessary to recount some basic elements of weak-interaction phenomenology that must be reproduced and to see the shortcomings of earlier descriptions. To this end, we shall summarize the effective Lagrangian approach originated by Fermi and the intermediate-boson extension of that picture. We then present the partially unified $SU(2)_L \otimes U(1)_Y$ gauge theory of weak and electromagnetic interactions and compare its predictions with the earlier descriptions. A central element of the $SU(2)_L \otimes U(1)_Y$ electroweak theory is the prediction of weak neutral-current phenomena. These will be discussed at some length.

We next consider the Higgs boson of the theory, the existence of which is a direct consequence of the mechanism of spontaneous symmetry breaking that we have developed. The properties of this scalar particle and prospects for its experimental detection are also discussed. We shall investigate the role of the Higgs boson in determining the high-energy behavior of the electroweak theory and the special status of the 1-TeV scale. A parenthetical section is devoted to neutrino mixing and neutrino mass.

Spontaneously broken gauge theories of the type we discuss here have the important asset of being renormalizable. Predictions are thus, in principle, calculable to all orders in perturbation theory. A proof of renormalizability is enormously complex and is beyond the scope of this book. We shall, however, look briefly at two aspects of the renormalization program: the definition of gauges better suited to higher-order calculations than the U-gauge, which was well adapted to the examination of particle spectra, and the issue of anomalies. We shall find that the standard model, when restricted to the lepton sector, behaves badly, but renormalizability can be achieved by introducing hadrons in a manner that will later be seen to have far-reaching implications. Throughout the chapter, many sample calculations are carried out in detail. An interim assessment concludes the chapter.

## 6.1 AN EFFECTIVE LAGRANGIAN FOR THE WEAK INTERACTIONS

Our contemporary view of the weak interactions is the result of a long evolution of the theoretical picture of radioactive $\beta$-decay introduced by Fermi [2] in 1933. The history of the subject up to the invention of spontaneously broken gauge theories and the low-energy phenomenology of nuclear $\beta$-decay, muon decay, and so on, are well documented in a number of excellent textbooks and monographs cited at the end of this chapter. We shall not reproduce this material but shall instead emphasize some aspects of the phenomenological theory that are not given prominence in the traditional textbook treatments. Our aims are to recall some calculational techniques, to point the way toward high-energy phenomena, and to expose the limitations of the historical approach.

We begin our exploration of weak-interaction phenomena with an investigation of neutrino–electron scattering. We shall briefly consider the implications of various possible spacetime forms for the charged-current interaction and then examine in greater detail the properties of the vector-minus-axial-vector structure favored by experiment.

The most general matrix element for the charged-current $\nu_e e$ interaction, free of derivative couplings, is

$$\mathcal{M} = \sum_i \mathcal{M}_i = \sum_i C_i \bar{\nu} \mathcal{O}_i e \bar{e} \mathcal{O}_i (1 - \gamma_5) \nu, \tag{6.1.1}$$

where the operators $\mathcal{O}_i$ are

$$\left. \begin{array}{lll} \mathcal{O}_S = 1 & \text{scalar} \\ \mathcal{O}_P = \gamma_5 & \text{pseudoscalar} \\ \mathcal{O}_T = \sigma_{\mu\nu} & \text{tensor} \\ \mathcal{O}_V = \gamma_\mu & \text{vector} \\ \mathcal{O}_A = \gamma_\mu \gamma_5 & \text{axial vector} \end{array} \right\}, \tag{6.1.2}$$

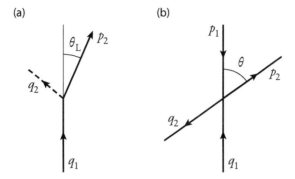

**Figure 6.1.** Kinematics of neutrino–electron scattering in the (a) laboratory and (b) center-of-momentum frame.

the summation runs over $i = $ S, P, T, V, A, and the explicit factor of $(1 - \gamma_5)$ appears in recognition of the fact that the neutrino sources of interest in nature yield left-handed neutrinos. For calculations, it will be convenient to employ the combinations $V \pm A$, which contribute incoherently to the absolute squares of matrix elements. Let us now compute the cross section that arises from each interaction in turn. Although much of this exercise is only of academic interest for the well-studied charged-current interactions, it will serve as an indication of what may be learned from scattering experiments. It will also be a useful reference when we turn to the study of neutral-current interactions or for applications to nonstandard neutrino interactions.

Denote the incoming and outgoing four-momenta as shown:

$$\left.\begin{array}{l} \nu_{\text{in}} : q_1 \\ \nu_{\text{out}} : q_2 \\ e_{\text{in}} : p_1 \\ e_{\text{out}} : p_2 \end{array}\right\}. \tag{6.1.3}$$

In the laboratory frame, depicted in figure 6.1(a), the four-momenta are explicitly given by

$$\left.\begin{array}{l} q_1^\mu = (E; 0, 0, E) \\ p_1^\mu = (m; 0, 0, 0) \\ p_2^\mu = (E'; P' \sin\theta_L, 0, P' \cos\theta_L) \\ q_2^\mu = (E + m - E'; -P' \sin\theta_L, 0, E - P' \cos\theta_L) \end{array}\right\}, \tag{6.1.4}$$

where the energy of the recoiling electron is

$$E' = \sqrt{P'^2 + m^2} \equiv yE, \tag{6.1.5}$$

with $m$ the electron mass. With these definitions, the invariants of interest may be expressed as

$$
\left.\begin{array}{l}
p_1 \cdot q_1 = p_2 \cdot q_2 = mE \\
p_1 \cdot q_2 = p_2 \cdot q_1 = m(E + m - E') \approx mE(1 - y) \\
p_1 \cdot p_2 = mE' = mEy \\
q_1 \cdot q_2 = m(E' - m) \approx mEy
\end{array}\right\}. \tag{6.1.6}
$$

In the c.m. frame, indicated in figure 6.1(b), the four-momenta are conveniently written in terms of

$$
p^* = \frac{s - m^2}{2\sqrt{s}} = \frac{mE}{\sqrt{s}} \approx \frac{\sqrt{s}}{2} \tag{6.1.7}
$$

and

$$
\omega^* = \sqrt{p^{*2} + m^2} = \frac{s + m^2}{2\sqrt{s}} \tag{6.1.8}
$$

as

$$
\left.\begin{array}{l}
q_1^\mu = (p^*; 0, 0, p^*) \\
p_1^\mu = (\omega^*; 0, 0, -p^*) \\
p_2^\mu = (\omega^*; p^* \sin\theta, 0, p^* \cos\theta) \\
q_2^\mu = (p^*; -p^* \sin\theta, 0, -p^* \cos\theta)
\end{array}\right\}. \tag{6.1.9}
$$

The invariants are then given by

$$
\left.\begin{array}{l}
p_1 \cdot q_1 = p_2 \cdot q_2 = p^*(p^* + \omega^*) \approx 2p^{*2} \\
p_1 \cdot q_2 = p_2 \cdot q_1 = p^*(\omega^* - p^* \cos\theta) \approx p^{*2}(1 - \cos\theta) \\
p_1 \cdot p_2 = \omega^{*2} + p^{*2} \cos\theta \approx p^{*2}(1 + \cos\theta) \\
q_1 \cdot q_2 = p^{*2}(1 + \cos\theta)
\end{array}\right\}. \tag{6.1.10}
$$

We now compute the differential and integral cross sections for the various interaction terms contained in (6.1.1).

1. Scalar interaction: $\nu_e e \to \nu_e e$. The matrix element is

$$
\mathcal{M} = \bar{u}(\nu, q_2) u(e, p_1) \bar{u}(e, p_2)(1 - \gamma_5) u(\nu, q_1). \tag{6.1.11}
$$

Squaring the matrix element and summing over the spins of the outgoing particles, we have

$$
\begin{aligned}
|\mathcal{M}|^2 &= \text{tr}[u(e, p_1)\bar{u}(e, p_1)u(\nu, q_2)\bar{u}(\nu, q_2)] \\
&\quad \times \text{tr}[(1 - \gamma_5)u(\nu, q_1)\bar{u}(\nu, q_1)(1 + \gamma_5)u(e, p_2)\bar{u}(e, p_2)] \\
&= \text{tr}[(m + p\!\!\!/_1)q\!\!\!/_2]\text{tr}[(1 - \gamma_5)q\!\!\!/_1(1 + \gamma_5)(m + p\!\!\!/_2)]. \tag{6.1.12}
\end{aligned}
$$

The first trace is simply

$$\text{tr}[(m + \not{p}_1)\not{q}_2] = 4p_1 \cdot q_2, \tag{6.1.13}$$

by virtue of (A.3.3) and (A.3.5b). The second is conveniently rearranged as

$$\text{tr}[(1 - \gamma_5)\not{q}_1(1 + \gamma_5)(m + \not{p}_2)] = 2\text{tr}[(1 - \gamma_5)\not{q}_1(m + \not{p}_2)]$$
$$= 8q_1 \cdot p_2, \tag{6.1.14}$$

where (A.3.3), (A.3.5b), (A.3.8), and (A.3.9) have been used. We therefore have $|\mathcal{M}|^2 = 32\,p_1 \cdot q_2\,p_2 \cdot q_1$, or, averaging over the spin states of the incident electron,

$$\overline{|\mathcal{M}|^2} = 16\,p_1 \cdot q_2\;p_2 \cdot q_1 = 16\,p^{*4}(1 - \cos\theta)^2. \tag{6.1.15}$$

The differential cross section in the c.m. frame is easily computed with the aid of (B.1.7) as

$$\frac{d\sigma}{d\Omega_{\text{c.m.}}} = \frac{\overline{|\mathcal{M}|^2}}{64\pi^2 s} = \frac{mE(1 - \cos\theta)^2}{32\pi^2}, \tag{6.1.16}$$

so that

$$\sigma = \int d\Omega_{\text{c.m.}} \left( \frac{d\sigma}{d\Omega_{\text{c.m.}}} \right) = \frac{mE}{16\pi} \int_{-1}^{1} dz(1 - z)^2 = \frac{mE}{6\pi}. \tag{6.1.17}$$

Comparing (6.1.6) and (6.1.10), we find that

$$(1 - \cos\theta) = 2(1 - y). \tag{6.1.18}$$

Consequently we may write the differential cross section in the useful form

$$\frac{d\sigma}{dy} = 4\pi \frac{d\sigma}{d\Omega_{\text{c.m.}}} = \frac{mE(1 - y)^2}{2\pi}, \tag{6.1.19}$$

which leads, of course, to the same integrated cross section,

$$\sigma = \int_0^1 dy \frac{d\sigma}{dy} = \frac{mE}{6\pi}. \tag{6.1.20}$$

2. Scalar interaction: $\bar{\nu}_e e \to \bar{\nu}_e e$. The matrix element is

$$\mathcal{M} = \bar{v}(\nu, q_1)u(e, p_1)\bar{u}(e, p_2)(1 - \gamma_5)v(\nu, q_2), \tag{6.1.21}$$

so that

$$
\begin{aligned}
|\mathcal{M}|^2 &= \mathrm{tr}[u(e, p_1)\bar{u}(e, p_1)v(v, q_1)\bar{v}(v, q_1)] \\
&\quad \times \mathrm{tr}[(1 - \gamma_5)v(v, q_2)\bar{v}(v, q_2)(1 + \gamma_5)u(e, p_2)\bar{u}(e, p_2)] \\
&= \mathrm{tr}[(m + \not{p}_1)\not{q}_1]\mathrm{tr}[(1 - \gamma_5)\not{q}_2(1 + \gamma_5)(m + \not{p}_2)],
\end{aligned} \tag{6.1.22}
$$

which may be obtained from the matrix element squared for $v_e e$ scattering by interchanging $q_1$ and $q_2$. As a result, we find

$$
\overline{|\mathcal{M}|^2} = 16 p_1 \cdot q_1 \, p_2 \cdot q_2, \tag{6.1.23}
$$

so that

$$
\frac{d\sigma}{dy} = \frac{mE}{2\pi}, \tag{6.1.24}
$$

and

$$
\sigma = \frac{mE}{2\pi}. \tag{6.1.25}
$$

It is straightforward to verify that the pseudoscalar interaction yields the same cross sections.

3. Tensor coupling: $v_e e \to v_e e$. The matrix element is

$$
\mathcal{M} = \bar{u}(v, q_2)\sigma_{\mu\nu}u(e, p_1)\bar{u}(e, p_2)\sigma^{\mu\nu}(1 - \gamma_5)u(v, q_1), \tag{6.1.26}
$$

so we are required to evaluate

$$
|\mathcal{M}|^2 = \mathrm{tr}[\sigma_{\mu\nu}(m + \not{p}_1)\sigma_{\kappa\lambda}\not{q}_2]\mathrm{tr}[\sigma^{\mu\nu}(1 - \gamma_5)\not{q}_1(1 + \gamma_5)\sigma^{\kappa\lambda}(m + \not{p}_2)]. \tag{6.1.27}
$$

To evaluate these traces, it is convenient to exploit the identity

$$
\sigma^{\mu\nu} = i(\gamma^\mu\gamma^\nu - g^{\mu\nu}) \tag{A.2.14}
$$

and to note that, by virtue of the anticommutation relations (A.2.1), a product of six $\gamma$-matrices may be simplified as

$$
\begin{aligned}
\mathrm{tr}(\not{a}\not{b}\not{c}\not{d}\not{e}\not{f}) &= a \cdot b \, \mathrm{tr}(\not{c}\not{d}\not{e}\not{f}) - a \cdot c \, \mathrm{tr}(\not{b}\not{d}\not{e}\not{f}) + a \cdot d \, \mathrm{tr}(\not{b}\not{c}\not{e}\not{f}) \\
&\quad - a \cdot e \, \mathrm{tr}(\not{b}\not{c}\not{d}\not{f}) + a \cdot f \, \mathrm{tr}(\not{b}\not{c}\not{d}\not{e}).
\end{aligned} \tag{6.1.28}
$$

A slightly tedious calculation then yields

$$
\begin{aligned}
|\mathcal{M}|^2 &= 256(2q_1 \cdot q_2 \, p_1 \cdot p_2 + 2q_1 \cdot p_1 \, q_2 \cdot p_2 - q_1 \cdot p_2 \, q_2 \cdot p_1) \\
&= 256(mE)^2(1 + y)^2,
\end{aligned} \tag{6.1.29}
$$

so that

$$\frac{d\sigma}{dy} = \frac{4mE}{\pi}(1+y)^2 \qquad (6.1.30)$$

and

$$\sigma = \frac{28mE}{3\pi}. \qquad (6.1.31)$$

4. Tensor coupling: $\bar{\nu}_e e \to \bar{\nu}_e e$. Again, the result may be obtained by interchanging $q_1$ and $q_2$. We have

$$\begin{aligned}
|\mathcal{M}|^2 &= 256(2q_1 \cdot q_2 \, p_1 \cdot p_2 + 2q_2 \cdot p_1 q_1 \cdot p_2 - q_2 \cdot p_2 q_1 \cdot p_1) \\
&= 256(mE)^2(1-2y)^2,
\end{aligned} \qquad (6.1.32)$$

from which

$$\frac{d\sigma}{dy} = \frac{4mE}{\pi}(1-2y)^2 \qquad (6.1.33)$$

and

$$\sigma = \frac{4mE}{3\pi}. \qquad (6.1.34)$$

5. S – T interference: $\nu_e e \to \nu_e e$. The matrix element is

$$\begin{aligned}
\mathcal{M} &= C_S \bar{u}(\nu, q_2)u(e, p_1)\bar{u}(e, p_2)(1-\gamma_5)u(\nu, q_1) \\
&\quad + C_T \bar{u}(\nu, q_2)\sigma_{\mu\nu}u(e, p_1)\bar{u}(e, p_2)\sigma^{\mu\nu}(1-\gamma_5)u(\nu, q_1),
\end{aligned} \qquad (6.1.35)$$

so the form of the interference terms is

$$\begin{aligned}
\mathcal{I} &= C_S C_T^* \mathrm{tr}[\bar{u}(\nu, q_2)u(e, p_1)\bar{u}(e, p_1)\sigma_{\mu\nu}u(\nu, q_2)] \\
&\quad \times \mathrm{tr}[\bar{u}(e, p_2)(1-\gamma_5)u(\nu, q_1)\bar{u}(\nu, q_1)(1+\gamma_5)\sigma^{\mu\nu}u(\nu, q_2)] \\
&\quad + C_S^* C_T \mathrm{tr}[\bar{u}(\nu, q_2)\sigma_{\mu\nu}u(e, p_1)\bar{u}(e, p_1)u(\nu, q_2)] \\
&\quad \times \mathrm{tr}[\bar{u}(e, p_2)\sigma^{\mu\nu}(1-\gamma_5)u(\nu, q_1)\bar{u}(\nu, q_1)(1+\gamma_5)u(e, p_2)] \\
&= C_S C_T^* \mathrm{tr}[(m+\not{p}_1)\sigma_{\mu\nu}\not{q}_2]\mathrm{tr}[(1-\gamma_5)\not{q}_1(1+\gamma_5)\sigma^{\mu\nu}(m+\not{p}_2)] \\
&\quad + C_S^* C_T \mathrm{tr}[\sigma_{\mu\nu}(m+\not{p}_1)\not{q}_2]\mathrm{tr}[\sigma^{\mu\nu}(1-\gamma_5)\not{q}_1(1+\gamma_5)(m+\not{p}_2)] \\
&= 2C_S C_T^* \mathrm{tr}[\sigma_{\mu\nu}\not{q}_2(m+\not{p}_1)]\mathrm{tr}[(1+\gamma_5)\sigma^{\mu\nu}(m+\not{p}_2)\not{q}_1] \\
&\quad + 2C_S^* C_T \mathrm{tr}[\sigma_{\mu\nu}(m+\not{p}_1)\not{q}_2]\mathrm{tr}[(1-\gamma_5)\sigma^{\mu\nu}\not{q}_1(m+\not{p}_2)].
\end{aligned} \qquad (6.1.36)$$

The commutation relations show the two products of traces to be equal. Again using the identity (A.2.14), we easily compute

$$\mathcal{I} = -128\mathrm{Re}(C_S^* C_T)(p_2 \cdot q_2 \, p_1 \cdot q_1 - q_1 \cdot q_2 \, p_1 \cdot p_2)$$
$$= -128\mathrm{Re}(C_S^* C_T)(mE)^2(1 - y^2), \tag{6.1.37}$$

so that

$$\frac{d\sigma_{\mathrm{ST}}}{dy} = -\frac{2mE}{\pi}(1 - y^2)\mathrm{Re}(C_S^* C_T), \tag{6.1.38}$$

and

$$\sigma_{\mathrm{ST}} = -\frac{4mE}{3\pi}\mathrm{Re}(C_S^* C_T). \tag{6.1.39}$$

The cross section corresponding to an interference term need not, of course, be positive definite. A parallel calculation shows that the same result is obtained for P – T interference.

6. S – T interference: $\bar{\nu}_e e \to \bar{\nu}_e e$. On interchanging $q_1 \leftrightarrow q_2$ in (6.1.37), we find

$$\mathcal{I} = -128 \, \mathrm{Re}(C_S^* C_T)(p_2 \cdot q_1 \, p_1 \cdot q_2 - q_1 \cdot q_2 \, p_1 \cdot p_2)$$
$$= -128 \, \mathrm{Re}(C_S^* C_T)(mE)^2(1 - 2y), \tag{6.1.40}$$

from which

$$\frac{d\sigma_{\mathrm{ST}}}{dy} = \frac{2mE}{\pi}(1 - 2y)\mathrm{Re}(C_S^* C_T) \tag{6.1.41}$$

and

$$\sigma_{\mathrm{ST}} = 0. \tag{6.1.42}$$

7. V – A interaction: $\nu_e e \to \nu_e e$. The matrix element is

$$\mathcal{M} = \bar{u}(\nu, q_2)\gamma_\mu(1 - \gamma_5)u(e, p_1)\bar{u}(e, p_2)\gamma^\mu(1 - \gamma_5)u(\nu, q_1). \tag{6.1.43}$$

Squaring the matrix element and summing over the spins of the outgoing particles, we obtain

$$|\mathcal{M}|^2 = \mathrm{tr}[\gamma_\mu(1 - \gamma_5)(m + \not{p}_1)(1 + \gamma_5)\gamma_\nu \not{q}_2]$$
$$\times \mathrm{tr}[\gamma^\mu(1 - \gamma_5)\not{q}_1(1 + \gamma_5)\gamma^\nu(m + \not{p}_2)]$$
$$\equiv A_{\mu\nu}B^{\mu\nu}. \tag{6.1.44}$$

The first factor is then

$$A_{\mu\nu} = 2\mathrm{tr}[(1 + \gamma_5)\gamma_\nu \not{q}_2 \gamma_\mu(m + \not{p}_1)]$$
$$= 8(q_{2\nu}p_{1\mu} - g_{\mu\nu}q_2 \cdot p_1 + q_{2\mu}p_{1\nu}) - 8i\varepsilon_{\mu\nu\rho\sigma}q_2^\rho p_1^\sigma, \tag{6.1.45}$$

and the second is

$$B^{\mu\nu} = 2\text{tr}[(1 - \gamma_5)\slashed{q}_1\gamma^\nu(m + \slashed{p}_2)\gamma^\mu]$$
$$= 8(q_1^\nu p_2^\mu - g^{\mu\nu}q_1 \cdot p_2 + q_1^\mu p_2^\nu) + 8i\varepsilon^{\mu\nu\kappa\lambda}q_{1\kappa}p_{2\lambda}. \qquad (6.1.46)$$

As a consequence,

$$|\mathcal{M}|^2 = A_{\mu\nu}B^{\mu\nu} = 128(q_1 \cdot q_2\, p_1 \cdot p_2 + q_1 \cdot p_1\, q_2 \cdot p_2)$$
$$- 64i\varepsilon_{\mu\nu\rho\sigma}q_2^\rho p_1^\sigma(q_1^\nu p_2^\mu + q_1^\mu p_2^\nu)$$
$$+ 64i\varepsilon^{\mu\nu\kappa\lambda}q_{1\kappa}p_{2\lambda}(q_{2\nu}p_{1\mu} + q_{2\mu}p_{1\nu})$$
$$+ 64\varepsilon_{\mu\nu\rho\sigma}\varepsilon^{\mu\nu\kappa\lambda}q_2^\rho p_1^\sigma q_{1\kappa}p_{2\lambda}. \qquad (6.1.47)$$

The second and third terms vanish because there are only three independent momenta in the problem. The last term may be evaluated with the aid of (A.3.14); it is simply

$$128(q_1 \cdot p_1\, q_2 \cdot p_2 - q_1 \cdot q_2\, p_1 \cdot p_2), \qquad (6.1.48)$$

so that

$$|\mathcal{M}|^2 = 256\, q_1 \cdot p_1\, q_2 \cdot p_2$$
$$= 256(mE)^2, \qquad (6.1.49)$$

whereupon

$$\frac{d\sigma}{dy} = \frac{4mE}{\pi} \qquad (6.1.50)$$

and

$$\sigma = \frac{4mE}{\pi}. \qquad (6.1.51)$$

8. V – A interaction: $\bar{\nu}_e e \rightarrow \bar{\nu}_e e$. The replacement $q_1 \leftrightarrow q_2$ yields

$$|\mathcal{M}|^2 = 256\, q_2 \cdot p_1\, q_1 \cdot p_2$$
$$= 256(mE)^2(1 - y)^2, \qquad (6.1.52)$$

from which

$$\frac{d\sigma}{dy} = \frac{4mE}{\pi}(1 - y)^2 \qquad (6.1.53)$$

and

$$\sigma = \frac{4mE}{3\pi}. \qquad (6.1.54)$$

9. V + A interaction: $\nu_e e \to \nu_e e$. We write the matrix element as

$$\mathcal{M} = \bar{u}(\nu, q_2)\gamma_\mu(1 + \gamma_5)u(e, p_1)\bar{u}(e, p_2)\gamma^\mu(1 - \gamma_5)u(\nu, q_1). \tag{6.1.55}$$

The calculation differs from the V − A calculation only by a change in the sign of the $\varepsilon_{\mu\nu\rho\sigma}$ term. Thus, we find that

$$\begin{aligned}|\mathcal{M}|^2 &= 256\, q_1 \cdot q_2\, p_1 \cdot p_2 \\ &= 256(mE)^2 y^2,\end{aligned} \tag{6.1.56}$$

which implies

$$\frac{d\sigma}{dy} = \frac{4mE}{\pi}y^2 \tag{6.1.57}$$

and

$$\sigma = \frac{4mE}{3\pi}. \tag{6.1.58}$$

10. V + A interaction: $\bar{\nu}_e e \to \bar{\nu}_e e$. The preceding results are unchanged by the interchange $q_1 \leftrightarrow q_2$, so we have at once

$$\frac{d\sigma}{dy} = \frac{4mE}{\pi}y^2 \tag{6.1.59}$$

and

$$\sigma = \frac{4mE}{3\pi}. \tag{6.1.60}$$

The foregoing calculations of

$$\frac{d\sigma}{dy} = \frac{d\sigma}{dy}(\text{STP}) + \frac{d\sigma}{dy}(\text{V} - \text{A}) + \frac{d\sigma}{dy}(\text{V} + \text{A}) \tag{6.1.61}$$

are summarized in table 6.1. We see that energy-loss distributions (or, equivalently, angular distributions) provide a means of probing the spacetime structure of the interaction. It is, however, easy to verify the following "confusion theorem" [3]: for the pair of spin-averaged $(\nu, \bar{\nu})$ cross sections, any combination of V − A and V + A interactions can be reproduced by a suitably chosen combination of S, T, and P interactions. Spin dependences are, in general, different for the two sets of interactions.

A number of experiments in the late 1950s established the structure of the weak charged current as V − A. Let us now concentrate more closely on this form, in the context of high-energy $\nu e$ scattering. It is conventional to write the effective

TABLE 6.1
Differential Cross Sections for Charged-Current Processes

| Coupling | $\dfrac{\pi}{mE}\dfrac{d\sigma}{dy}(\nu_e e \rightarrow \nu_e e)$ | $\dfrac{\pi}{mE}\dfrac{d\sigma}{dy}(\bar{\nu}_e e \rightarrow \bar{\nu}_e e)$ |
|---|---|---|
| $|C_S|^2 + |C_P|^2$ | $\frac{1}{2}(1-y)^2$ | $\frac{1}{2}$ |
| $|C_T|^2$ | $4(1+y)^2$ | $4(1-2y)^2$ |
| $\mathrm{Re}[(C_S^* + C_P^*)C_T]$ | $-2(1-y^2)$ | $-2(1-2y)$ |
| $|C_{V-A}|^2$ | $4$ | $4(1-y)^2$ |
| $|C_{V+A}|^2$ | $4y^2$ | $4y^2$ |

Lagrangian of the leptonic weak interaction as a product of charged currents,

$$\mathcal{L}_{\text{eff}} = -\frac{G_F}{\sqrt{2}}\, \bar{\nu}\gamma_\mu(1-\gamma_5)e\,\bar{e}\gamma^\mu(1-\gamma_5)\nu, \tag{6.1.62}$$

where the Fermi constant has been measured [4] to be

$$G_F = 1.166\,3787(6) \times 10^{-5} \text{ GeV}^{-2}, \tag{6.1.63}$$

so that

$$G_F^2 = 5.297 \times 10^{-38} \text{ cm}^2 \text{ GeV}^{-2}. \tag{6.1.64}$$

The cross sections for $\nu_e e$ and $\bar{\nu}_e e$ scattering are then given, in the V – A theory, by

$$\frac{d\sigma}{dy}(\nu_e e) = \frac{2G_F^2 mE}{\pi}, \tag{6.1.65}$$

$$\sigma(\nu_e e) = \frac{2G_F^2 mE}{\pi} \approx 1.72 \times 10^{-41} \text{ cm}^2 \left(\frac{E}{1 \text{ GeV}}\right), \tag{6.1.66}$$

$$\frac{d\sigma}{dy}(\bar{\nu}_e e) = \frac{2G_F^2 mE}{\pi}(1-y)^2, \tag{6.1.67}$$

$$\sigma(\bar{\nu}_e e) = \frac{2G_F^2 mE}{3\pi} \approx 0.574 \times 10^{-41} \text{ cm}^2 \left(\frac{E}{1 \text{ GeV}}\right). \tag{6.1.68}$$

The cross sections are small but within the reach of modern high-intensity neutrino beams.

It is interesting to analyze the reason for the factor-of-three difference between neutrino and antineutrino cross sections or, equivalently, the difference in angular distributions plotted in figure 6.2. It is simply this: the operators $(1 \pm \gamma_5)$ are spin-projection operators in the limit of vanishing fermion mass. Thus, for a V – A interaction, we may view all light fermions as left-handed (with helicity $-\frac{1}{2}$) and all antifermions as right-handed (with helicity $+\frac{1}{2}$). Thus, as shown in figure 6.3(a), the initial state is a state of $J_z = 0$ for $\nu_e e$ scattering and $J_z = 1$ for $\bar{\nu}_e e$ scattering. The same holds for forward scattering (which we have labeled by $\cos\theta = -1$, $y = 0$).

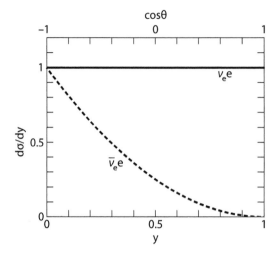

**Figure 6.2.** Angular and energy-loss distributions of $\nu_e e$ and $\bar{\nu}_e e$ scattering in the $V - A$ model.

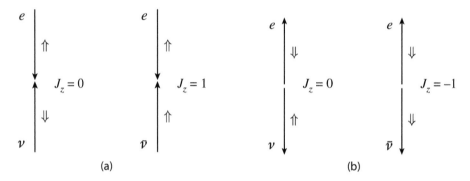

**Figure 6.3.** Spin configurations for $\nu e$ and $\bar{\nu} e$ scattering in the c.m. frame: (a) initial state; (b) final state for backward scattering.

For backward scattering (which we have described by $\cos\theta = +1$, $y = 1$), the situation is different, as shown in figure 6.3(b). The final state for $\nu_e e$ scattering is a state with total spin $J_z = 0$, as was the initial state, but the final state for $\bar{\nu}_e e$ scattering would correspond to $J_z = -1$, which would violate angular momentum conservation. Backward $\bar{\nu}_e e$ scattering is, therefore, forbidden, and the cross section must vanish at $y = 1$. This effect will occur many times in our study of the weak interactions.

One of the important observations of early experiments, which has been repeatedly confirmed and made more precise, is that the charged-current weak interactions are universal in strength: the same coupling constant applies to all leptonic and semileptonic interactions. As a consequence, the calculations we have just completed may be extended in straightforward fashion to the experimentally interesting case of inverse muon decay, the reaction

$$\nu_\mu(q_1)e(p_1) \rightarrow \mu(p_2)\nu_e(q_2). \tag{6.1.69}$$

Kinematic definitions in the laboratory frame are as in the case of $\nu_e e$ elastic scattering (6.1.4), except that the final muon energy is

$$E' = \sqrt{P'^2 + \mu^2}, \tag{6.1.70}$$

where $\mu$ is the muon mass. The invariants (6.1.6) become

$$\left.\begin{aligned}
q_1 \cdot p_1 &= me \\[1em]
q_2 \cdot p_2 &= mE - \frac{\mu^2 - m^2}{2} \\[1em]
p_1 \cdot p_2 &= mE \\[1em]
p_1 \cdot q_2 &= m(E + m - E') \\[1em]
q_1 \cdot q_2 &= mE' - \frac{\mu^2 + m^2}{2} \\[1em]
p_2 \cdot q_1 &= m(E - E') + \frac{\mu^2 + m^2}{2}
\end{aligned}\right\}. \tag{6.1.71}$$

The differential cross section is now

$$\frac{d\sigma}{d\Omega_{\text{c.m.}}} = \frac{\overline{|\mathcal{M}|^2}}{64\pi^2 s} \frac{p'_{\text{c.m.}}}{p_{\text{c.m.}}} = \frac{\overline{|\mathcal{M}|^2}}{64\pi^2 s}\left[1 - \frac{(\mu^2 - m^2)}{2mE}\right]. \tag{6.1.72}$$

Substitution of the matrix element [cf. (6.1.49)] yields

$$\frac{d\sigma}{d\Omega_{\text{c.m.}}} = \frac{G_F^2 mE}{2\pi^2}\left[1 - \frac{(\mu^2 - m^2)}{2mE}\right]^2 \tag{6.1.73}$$

and, therefore,

$$\sigma(\nu_\mu e \to \mu\nu_e) = \frac{2G_F^2 mE}{\pi}\left[1 - \frac{(\mu^2 - m^2)}{2mE}\right]^2 \tag{6.1.74}$$

$$= \sigma_{\text{V}-\text{A}}(\nu_e e \to \nu_e e)\left[1 - \frac{(\mu^2 - m^2)}{2mE}\right]^2. \tag{6.1.75}$$

The expected total cross section is plotted in figure 6.4. Because of its high threshold energy ($E \approx 10.9$ GeV), the reaction has been difficult to observe and has been studied only recently. From a signal of $15,758 \pm 324$ events, the CHARM-II collaboration [5] at CERN determined the asymptotic slope,

$$\sigma_\infty(\nu_\mu e \to \mu\nu_e)/E = (1.651 \pm 0.093) \times 10^{-41} \text{ cm}^2 \text{ GeV}^{-1}, \tag{6.1.76}$$

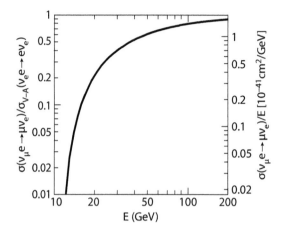

**Figure 6.4.** Cross section for inverse muon decay, $\nu_\mu e \to \mu \nu_e$, as a function of neutrino energy.

whereas from their sample of $1\,050 \pm 139$ events, the NuTeV collaboration [6] at Fermilab inferred

$$\sigma_\infty(\nu_\mu e \to \mu \nu_e)/E = (1.38 \pm 0.18) \times 10^{-41} \text{ cm}^2 \text{ GeV}^{-1}, \qquad (6.1.77)$$

both in agreement (within errors) with the predicted value [cf. (6.1.66)],

$$\sigma_\infty(\nu_\mu e \to \mu \nu_e)/E = 1.72 \times 10^{-41} \text{ cm}^2 \text{ GeV}^{-1}. \qquad (6.1.78)$$

Inverse muon decay is a particularly interesting process for experimental study because it is purely leptonic and, thus, free from some of the ambiguities that will present themselves in the case of semileptonic interactions. The small cross section was an early impediment to detailed study. However, the reaction is also well suited to exposing the limitations of the effective Lagrangian approach. The angular distribution is, according to (6.1.73), isotropic in the c.m. system, which is to say that the scattering is purely $s$-wave. Partial-wave unitarity constrains the modulus of an inelastic partial-wave amplitude to be

$$|\mathcal{M}_J| \le 1, \qquad (6.1.79)$$

where the partial-wave expansion for the scattering amplitude (here written for spinless external particles) is

$$f(\theta) = \left( 2 \frac{d\sigma}{d\Omega} \right)^{1/2} = \frac{1}{\sqrt{s}} \sum_{J=0}^{\infty} (2J+1) P_J(\cos\theta) \mathcal{M}_J, \qquad (6.1.80)$$

with $P_J(x)$ a Legendre polynomial. The explicit factor of two in (6.1.80) serves to undo the average over initial electron spins. The constraint (6.1.79) is equivalent to the familiar restriction

$$\sigma \leq \frac{\pi}{p_{\text{c.m.}}^2} \tag{6.1.81}$$

for inelastic $s$-wave scattering. For inverse muon decay, we easily compute

$$
\begin{aligned}
\mathcal{M}_0 &= \frac{G_F s}{\pi \sqrt{2}} \left[ 1 - \frac{(\mu^2 - m^2)}{s} \right] \\
&\xrightarrow[s \gg \mu^2]{} \frac{G_F s}{\pi \sqrt{2}} ,
\end{aligned}
\tag{6.1.82}
$$

which implies the unitarity constraint,

$$\frac{G_F s}{\pi \sqrt{2}} \leq 1 . \tag{6.1.83}$$

This means that the four-fermion theory can make sense only if [7]

$$s \leq \frac{\pi \sqrt{2}}{G_F}, \tag{6.1.84}$$

which is to say that

$$\sqrt{s} \leq 617 \text{ GeV} \tag{6.1.85}$$

or

$$p_{\text{c.m.}} \leq 309 \text{ GeV}. \tag{6.1.86}$$

Although the energy at which unitarity is violated by the point-coupling theory is elevated, it is not astronomical. We must, therefore, take seriously the objection that the effective point-coupling theory is unacceptable, as we have formulated it. Suppose now that (6.1.62) is taken as a true Lagrangian and not merely an effective one. Could the violation of unitarity that we have just exhibited be remedied in higher orders of perturbation theory? In second order we must compute the diagram shown in figure 6.5, for which the matrix element is readily seen to be

$$\mathcal{M} \propto \int \frac{d^4 k}{k^2}, \tag{6.1.87}$$

which is quadratically divergent. In fact, the divergence difficulty of the point-coupling theory grows more severe in each order of perturbation theory. In order to be saved—and this is evidently desirable because of its successes for low-energy phenomenology, the theory must be modified in a fundamental way. The simplest attempts to salvage the theory are the subject of the next section.

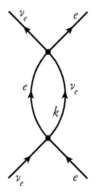

**Figure 6.5.** Second-order contribution to $\nu_e e$ elastic scattering in the four-fermion (point-coupling) theory.

## 6.2 INTERMEDIATE VECTOR BOSONS: A FIRST LOOK

Although Fermi's phenomenological interaction was inspired by the theory of electromagnetism, the analogy was not complete, and we may hope to obtain a more satisfactory theory by pushing the analogy further. An obvious device is to assume that the weak interaction, like quantum electrodynamics, is mediated by vector-boson exchange. The weak intermediate boson must have the following three properties:

1. It carries charge $\pm 1$, because the familiar manifestations of the weak interactions (such as $\beta$-decay) are charge-changing.
2. It must be rather massive to reproduce the short range of the weak force.
3. Its parity must be indefinite.

Furthermore, its couplings to the fermions are determined by the low-energy phenomenology. At each $\nu e W$ vertex will occur a factor

$$-i \left( \frac{G_F M_W^2}{\sqrt{2}} \right)^{1/2} \gamma_\mu (1 - \gamma_5), \tag{6.2.1}$$

where $M_W$ is the mass of the weak intermediate boson. The intermediate vector boson propagator has the standard form for a massive vector meson, derived in problem 4.2,

$$\frac{-i(g_{\mu\nu} - k_\mu k_\nu)}{k^2 - M_W^2}. \tag{6.2.2}$$

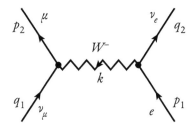

**Figure 6.6.** Lowest-order contribution to inverse muon decay, in the intermediate-boson model.

According to this picture, inverse muon decay proceeds by the $t$-channel exchange diagram shown in figure 6.6. The Feynman amplitude is

$$\mathcal{M} = \frac{i\,G_F M_W^2}{\sqrt{2}}\bar{u}(v, q_2)\gamma^\mu(1 - \gamma_5)u(e, p_1)\frac{g_{\mu\nu} - k_\mu k_\nu}{k^2 - M_W^2}$$

$$\times \bar{u}(e, p_2)\gamma^\nu(1 - \gamma_5)u(v, q_1), \tag{6.2.3}$$

with $k = p_2 - q_1 = p_1 - q_2$. By substituting $k_\mu = (p_1 - q_2)_\mu$ and $k_\nu = (p_2 - q_1)_\nu$ and using the Dirac equation, we find that the contribution of the $k_\mu k_\nu$ term in the propagator is of order $(m^2/M_W^2)$ and may be safely neglected. What remains is simply the amplitude we encountered in the point-coupling theory, times $M_W^2/(M_W^2 - k^2)$. The point-coupling result is recovered in the limit $M_W \to \infty$. Without further computation, we have

$$\frac{d\sigma}{d\Omega_{\text{c.m.}}}(\nu_\mu e \to \mu\nu_e) \approx \frac{G_F^2 mE\left[1 - (\mu^2 - m^2)/2mE\right]^2}{2\pi^2\left[1 + mE(1 - \cos\theta)/M_W^2\right]^2}, \tag{6.2.4}$$

where $m$ and $\mu$ have been neglected with respect to $M_W$ and $E$. The integrated cross section

$$\sigma(\nu_\mu e \to \mu\nu_e) \approx \frac{2G_F^2 mE}{\pi}\frac{\left[1 - (\mu^2 - m^2)/2mE\right]^2}{(1 + 2mE/M_W^2)} \tag{6.2.5}$$

reduces to the form (6.1.74) at modest energies but approaches a constant value at high energies:

$$\lim_{E\to\infty} \sigma(\nu_\mu e \to \mu\nu_e) = \frac{G_F^2 M_W^2}{\pi}. \tag{6.2.6}$$

This is a great improvement over the point-coupling theory, but a problem remains: the $s$-wave amplitude violates partial-wave unitarity. To see this, we write

$$
\begin{aligned}
\mathcal{M}_0 &= \frac{\sqrt{s}}{2} \int_{-1}^{1} d(\cos\theta) \, f(\theta) \\
&= \frac{G_F mE}{\pi\sqrt{2}} \int_{-1}^{1} dz \left[ 1 + \frac{mE}{M_W^2}(1-z) \right]^{-1} \\
&= \frac{G_F M_W^2}{\pi\sqrt{2}} \ln\left( 1 + \frac{2mE}{M_W^2} \right) \\
&= \frac{G_F M_W^2}{\pi\sqrt{2}} \ln\left( 1 + \frac{s}{M_W^2} \right).
\end{aligned}
\tag{6.2.7}
$$

In this instance the unitarity constraint (6.1.79) is respected, as long as

$$
s \leq M_W^2 \left[ \exp\left( \frac{\pi\sqrt{2}}{G_F M_W^2} \right) - 1 \right].
\tag{6.2.8}
$$

Notice that in the point-coupling limit $M_W \to \infty$ we recover the result (6.1.84). These violations of unitarity arise at incredibly high energies (for $M_W = 100$ GeV, for example, unitarity is respected up to beam energies of $3.5 \times 10^{23}$ GeV), and it might be hoped that the very mild violation of unitarity could be conquered in higher orders of perturbation theory. This is not the case. The $k_\mu k_\nu$ term in the $W$-boson propagator makes the theory nonrenormalizable by inducing new divergences in each order of perturbation theory. Equivalently, the introduction of the intermediate boson causes divergence problems in new physical processes, as we shall see presently, that make this ad hoc theory unacceptable. For the moment, however, let us continue to investigate the consequences of the $W$ in leptonic interactions.

In the crossed reaction,

$$
\bar{\nu}_e e \to \bar{\nu}_\mu \mu,
\tag{6.2.9}
$$

the intermediate boson appears as a direct-channel resonance, as shown in figure 6.7. The differential cross section is

$$
\frac{d\sigma}{d\Omega_{\text{c.m.}}}(\bar{\nu}_e e \to \bar{\nu}_\mu \mu) = \frac{G_F^2 mE}{8\pi^2} \frac{(1-\cos\theta)^2}{(1-2mE/M_W^2)^2} \left[ 1 - \frac{(\mu^2 - m^2)}{2mE} \right]^2,
\tag{6.2.10}
$$

which differs from the point-coupling result only by the factor $1/(1 - 2mE/M_W^2)^2$. The cross section becomes

$$
\sigma(\bar{\nu}_e e \to \bar{\nu}_\mu \mu) = \frac{2mE \, G_F^2 [1 - (\mu^2 - m^2)/2mE]^2}{3\pi (1 - 2mE/M_W^2)^2},
\tag{6.2.11}
$$

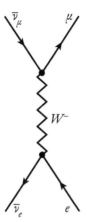

**Figure 6.7.** Lowest-order contribution to the reaction $\bar{\nu}_e e \to \bar{\nu}_\mu \mu$ in the intermediate-boson model.

which vanishes in the high-energy limit

$$\sigma(\bar{\nu}_e e \to \bar{\nu}_\mu \mu) \xrightarrow[E \to \infty]{} \frac{G_F^2 M_W^4}{6\pi m E}. \tag{6.2.12}$$

The apparent infinity in the cross section at $s = 2mE = M_W^2$ arises only because we have neglected the width of the unstable intermediate boson, an oversight we shall remedy at once. That aside, the unitarity problems for leptonic processes may be regarded as solved, at least in a practical sense, by the tree diagrams involving $W$ exchange.

Nothing that we have said fixes the mass of the intermediate boson, which remains a parameter. Once the mass is specified, however, the decay characteristics of the intermediate vector boson are essentially determined by the requirement that the low-energy phenomenology of the V – A theory be reproduced. In particular, we can predict definitely the rates for leptonic decays of the $W$.

In the rest frame of the $W^-$, the outgoing momenta for the decay

$$W^- \to e(p)\bar{\nu}_e(q) \tag{6.2.13}$$

are

$$\left. \begin{aligned} p &= \frac{M_W}{2}(1; \sin\theta, 0, \cos\theta) \\[2mm] q &= \frac{M_W}{2}(1; -\sin\theta, 0, -\cos\theta) \end{aligned} \right\}, \tag{6.2.14}$$

where the electron mass has been neglected. The matrix element for the decay is given by

$$\mathcal{M} = -i \left( \frac{G_F M_W^2}{\sqrt{2}} \right)^{1/2} \bar{u}(e, p) \gamma_\mu (1 - \gamma_5) v(\nu, q) \epsilon^\mu, \qquad (6.2.15)$$

where $\epsilon^\mu \equiv (0; \hat{\boldsymbol{\epsilon}})$ is the polarization vector of the decaying particle. Squaring the matrix element and summing over final-particle spins, we have, neglecting the electron mass,

$$
\begin{aligned}
|\mathcal{M}|^2 &= \frac{G_F M_W^2}{\sqrt{2}} \mathrm{tr}[\not{\epsilon}(1 - \gamma_5)\not{q}(1 + \gamma_5)\not{\epsilon}^* \not{p}] \\
&= \frac{G_F M_W^2}{\sqrt{2}} 2 \, \mathrm{tr}[(1 + \gamma_5)\not{\epsilon}\not{q}\not{\epsilon}^* \not{p}] \\
&= \frac{8 G_F M_W^2}{\sqrt{2}} (\epsilon \cdot q \, \epsilon^* \cdot p - \epsilon \cdot \epsilon^* \, p \cdot q + \epsilon \cdot p \, \epsilon^* \cdot q + i \varepsilon_{\mu\nu\rho\sigma} \epsilon^\mu q^\nu \epsilon^{*\rho} p^\sigma).
\end{aligned}
$$

$$(6.2.16)$$

The decay rate must be independent of the polarization of the decaying particle, so we choose first the simplest case of longitudinal polarization, helicity zero,

$$\epsilon^\mu = (0; 0, 0, 1) = \epsilon^{\mu*}, \qquad (6.2.17)$$

for which the $\varepsilon_{\mu\nu\rho\sigma}$ term vanishes. We then have, at once,

$$|\mathcal{M}|^2 = \frac{4 G_F M_W^4}{\sqrt{2}} \sin^2 \theta, \qquad (6.2.18)$$

so that, by virtue of (B.1.4), the differential decay rate is

$$\frac{d\Gamma}{d\Omega}(W_0^- \to e^- \bar{\nu}_e) = \frac{|\mathcal{M}|^2}{64 \pi^2 M_W} = \frac{G_F M_W^3}{16 \pi^2 \sqrt{2}} \sin^2 \theta. \qquad (6.2.19)$$

The leptonic decay rate is, therefore,

$$\Gamma(W^- \to e^- \bar{\nu}_e) = \frac{G_F M_W^3}{6 \pi \sqrt{2}} \approx 437 \text{ MeV} \left( \frac{M_W}{100 \text{ GeV}} \right)^3. \qquad (6.2.20)$$

The difficulty of the calculation is not greatly increased for other polarizations. For helicity $= +1$,

$$\epsilon^\mu = \frac{(0; -1, -i, 0)}{\sqrt{2}}, \qquad (6.2.21)$$

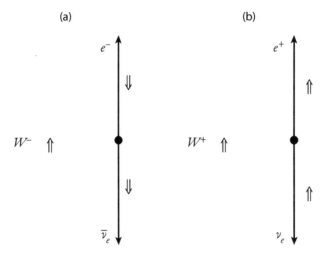

**Figure 6.8.** Spin configurations for the leptonic decay of intermediate bosons polarized along the positive $z$-axis: (a) $W^- \to e^- \bar{\nu}_e$; (b) $W^+ \to e^+ \nu_e$.

the square of the matrix element is

$$|\mathcal{M}|^2 = \frac{G_F M_W^4}{\sqrt{2}} \, 2 \left[ (1 + \cos^2 \theta) - 2 \cos \theta \right],\tag{6.2.22}$$

where the final term is the contribution from $\varepsilon_{\mu\nu\rho\sigma}$. Consequently, the differential decay rate is given by

$$\frac{d\Gamma}{d\Omega}(W_{+1}^- \to e^- \bar{\nu}_e) = \frac{G_F M_W^3}{32\pi^2 \sqrt{2}} (1 - \cos\theta)^2,\tag{6.2.23}$$

which corresponds, as it must, to the total rate given by (6.2.20). The result for helicity $= -1$ is simply obtained by replacing $\epsilon^\mu \leftrightarrow \epsilon^{\mu*}$ in the expression for $|\mathcal{M}|^2$. This yields

$$\frac{d\Gamma}{d\Omega}(W_{-1}^- \to e^- \bar{\nu}_e) = \frac{G_F M_W^3}{32\pi^2 \sqrt{2}} (1 + \cos\theta)^2.\tag{6.2.24}$$

The angular dependences merely reflect the helicity correlations of the V − A theory. This is illustrated in figure 6.8. Because the electron emitted the $W^-$ decay is left-handed, it is forbidden to travel parallel to the direction of $W^-$ polarization. For the decay $W^+ \to e^+ \nu_e$, the situation is just reversed: the emitted positron, being right-handed, tends to follow the direction of $W^+$ polarization. This effect is an example of C-violation in the weak interactions; the correlation played an important role in the discovery of intermediate bosons in high-energy proton–antiproton collisions, as we shall see in the discussion surrounding figure 7.18.

We have already mentioned that, whereas the introduction of the intermediate boson softens the divergence of the $s$-wave amplitude for the process $\nu_\mu e \to \mu \nu_e$, it

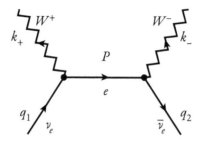

**Figure 6.9.** Lowest-order contribution to the reaction $\nu_e \bar{\nu}_e \to W^+ W^-$ in the intermediate boson model.

gives rise to new divergences in other processes. The most celebrated occurs in the reaction [8]

$$\nu_e(q_1)\bar{\nu}_e(q_2) \to W_0^+(k_+) + W_0^-(k_-), \qquad (6.2.25)$$

where the subscript 0 again denotes longitudinal polarization. According to the picture we have developed until now, we need only evaluate the electron-exchange diagram shown in figure 6.9. We define

$$P = q_1 - k_+ = k_- - q_2. \qquad (6.2.26)$$

The matrix element is

$$\mathcal{M} = -\frac{i\,G_F\,M_W^2}{\sqrt{2}}\,\bar{v}(v, q_2)\not{\epsilon}_-^*(1 - \gamma_5)\frac{(\not{P} + m)}{P^2 - m^2}\not{\epsilon}_+^*(1 - \gamma_5)u(v, q_1), \qquad (6.2.27)$$

and in the c.m. frame, the momentum four-vectors are

$$\left.\begin{aligned} q_1 &\equiv (Q; 0, 0, Q) \\ q_2 &= (Q; 0, 0, -Q) \\ k_+ &= (Q; K\sin\theta, 0, K\cos\theta) \\ k_- &= (Q; -K\sin\theta, 0, -K\cos\theta) \end{aligned}\right\}, \qquad (6.2.28)$$

with $K = \sqrt{Q^2 - M_W^2}$. If $\epsilon_\pm^\mu = (0; \hat{\epsilon}_\pm)$ represents the $W^\pm$ polarization in its rest frame, then after a boost to the c.m. frame, we have

$$\epsilon_\pm^\mu = \left[\frac{\mathbf{k}_\pm \cdot \hat{\boldsymbol{\epsilon}}_\pm}{M_W}; \hat{\boldsymbol{\epsilon}}_\pm + \frac{\mathbf{k}_\pm(\mathbf{k}_\pm \cdot \hat{\boldsymbol{\epsilon}}_\pm)}{M_W(Q + M_W)}\right]. \qquad (6.2.29)$$

For the case of longitudinal polarization, the c.m. polarization vector is, thus,

$$\epsilon_\pm^\mu = \left(\frac{K}{M_W}; \frac{Q\hat{\mathbf{k}}_\pm}{M_W}\right), \qquad (6.2.30)$$

which, in the limit of high energies $Q \approx K \gg M_W$, approaches the limiting form

$$\epsilon_{\pm}^{\mu} \xrightarrow[K \gg M_W]{} \frac{k_{\pm}}{M_W}. \tag{6.2.31}$$

The amplitude for the production of longitudinally polarized intermediate bosons at high energies thus becomes

$$\mathcal{M} = \frac{-i G_F}{\sqrt{2}} \bar{v}(\nu, q_2) \not{k}_{-}(1 - \gamma_5) \frac{(\not{P} + m)}{P^2 - m^2} \not{k}_{+}(1 - \gamma_5) u(\nu, q_1). \tag{6.2.32}$$

Using the Dirac equations

$$\left. \begin{array}{r} \not{q}_1 u(\nu, q_1) = 0 \\ \bar{v}(\nu, q_2) \not{q}_2 = 0 \end{array} \right\}, \tag{6.2.33}$$

we replace

$$\not{k}_{+} \rightarrow \not{k}_{+} - \not{q}_1 = -\not{P} \tag{6.2.34}$$

and

$$\not{k}_{-} \rightarrow \not{k}_{-} - \not{q}_2 = \not{P} \tag{6.2.35}$$

to obtain

$$\mathcal{M} = \frac{i G_F}{\sqrt{2}} \bar{v}(\nu, q_2) \not{P}(1 - \gamma_5) \frac{\not{P} \not{P}}{P^2}(1 - \gamma_5) u(\nu, q_1), \tag{6.2.36}$$

where we have everywhere discarded the electron mass $m$. Because $\not{P} \not{P} = P^2$, we now have

$$\mathcal{M} = i G_F \sqrt{2} \bar{v}(\nu, q_2) \not{P}(1 - \gamma_5) u(\nu, q_1). \tag{6.2.37}$$

To display the high-energy behavior, it is simplest to evaluate

$$\begin{aligned}
|\mathcal{M}|^2 &= 2 G_F^2 \, \mathrm{tr}[\not{P}(1 - \gamma_5) \not{q}_1 (1 + \gamma_5) \not{P} \not{q}_2] \\
&= 4 G_F^2 \, \mathrm{tr}[(1 - \gamma_5) \not{q}_1 \not{P} \not{q}_2 \not{P}] \\
&= 16 G_F^2 (2 q_1 \cdot P \, q_2 \cdot P - q_1 \cdot q_2 \, P^2) \\
&= 32 G_F^2 Q^2 K^2 \sin^2 \theta. \tag{6.2.38}
\end{aligned}$$

The angular dependence is again readily understood in terms of angular momentum conservation. At high energies, the cross section becomes

$$\sigma(\nu \bar{\nu} \rightarrow W_0^+ W_0^-) \xrightarrow[s \gg M_W^2]{} \frac{G_F^2 s}{3\pi}, \tag{6.2.39}$$

in gross violation of unitarity.

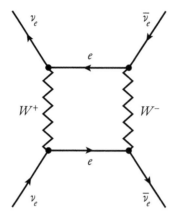

**Figure 6.10.** Second-order contribution to neutrino–antineutrino elastic scattering, in the intermediate boson model.

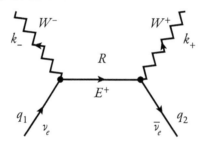

**Figure 6.11.** Paralepton-exchange contibution to the reaction $\nu\bar{\nu} \to W_0^+ W_0^-$.

Equivalently, we may remark that the second-order contribution to the neutrino–antineutrino scattering amplitude, given by the box diagram in figure 6.10, which is the square of the amplitude in figure 6.9, grows as $s^2$. Because the $k_\mu k_\nu$ piece of the $W$-boson propagator gives rise to more severe divergences in each order of perturbation theory, it seems imperative to find some means of eradicating the divergence order by order. We may accomplish this by inventing new physical particles, which will appear in new Feynman diagrams, in which the couplings are astutely chosen to cause cancellations of the offending divergences.

In the case of the reaction $\nu\bar{\nu} \to W^+W^-$, for which we have the diagram of figure 6.9, a possible solution [9] (though not the one chosen by nature) is to postulate a new *paralepton* $E^+$ with the same lepton number as $e^-$ and $\nu_e$, which gives rise to the diagram shown in figure 6.11. If the $\nu E W$ coupling is chosen to be $-i[G_E M_W^2/\sqrt{2}]^{\frac{1}{2}}\gamma_\mu(1 - \gamma_5)$, a V – A form with arbitrary strength, the amplitude corresponding to figure 6.11 is simply

$$\mathcal{M} = \frac{-i G_E M_W^2}{\sqrt{2}} \bar{v}(\nu, q_2) \not{\epsilon}_+^*(1 - \gamma_5) \frac{(\not{R} + m_E)}{R^2 - m_E^2} \not{\epsilon}_-^*(1 - \gamma_5) u(\nu, q_1), \qquad (6.2.40)$$

where

$$R \equiv q_1 - k_- = k_+ - q_2 = -P + q_1 - q_2 \qquad (6.2.41)$$

and $P$ was defined in (6.2.26). Neglecting fermion masses and taking both $W^+$ and $W^-$ to be longitudinally polarized, we have, using (6.2.31),

$$\mathcal{M} = \frac{-i\,G_E}{\sqrt{2}}\,\bar{v}(\nu, q_2)\not{k}_+(1 - \gamma_5)\frac{\not{R}\not{k}_-}{R^2}(1 - \gamma_5)u(\nu, q_1). \qquad (6.2.42)$$

Between spinors we may replace

$$\not{k}_- \rightarrow -\not{R} \qquad (6.2.43)$$

and

$$\not{k}_+ \rightarrow -\not{P}, \qquad (6.2.44)$$

so the amplitude becomes

$$\mathcal{M} = -i\,G_E\sqrt{2}\,\bar{v}(\nu, q_2)\not{P}(1 - \gamma_5)u(\nu, q_1), \qquad (6.2.45)$$

which is precisely the negative of the original amplitude (6.2.37), provided that the new coupling constant is chosen equal to the old,

$$G_E = G_F. \qquad (6.2.46)$$

This sort of order-by-order divergence cancellation is familiar in electrodynamics (see problems 3.5 and 6.10), where it is ensured by gauge invariance. Within the context of perturbation theory, such a cancellation is a prerequisite to renormalizability, because cancellations between different orders would be unstable against small changes in the coupling constant.

It is remarkable that a full program of divergence cancellation can be carried out systematically for the weak and electromagnetic interactions [10]. More remarkable still is that systematic pursuit of the program leads to theories with precisely the interaction structure of spontaneously broken gauge theories. This perhaps reflects the deep connection between gauge invariance, "safe" high-energy behavior, and renormalizability. It should not be left unsaid that for many physicists (the author among them), the intricate cancellations that we witness in the course of a gauge-theory calculation provide—together with the elegance of the gauge principle and the experimental support specific models have received, to be sure—an important reason for believing that gauge theories have something to do with reality. On that subjective note, let us now introduce the first elements of a gauge theory of the weak and electromagnetic interactions that has matured into a new law of nature.

## 6.3 THE STANDARD ELECTROWEAK THEORY OF LEPTONS

The $SU(2)_L \otimes U(1)_Y$ gauge theory that has emerged as "standard" is the result of a long development and many ingenious contributions. We first consider the electroweak theory in its purely leptonic form, which exhibits the motivation and the principal features.

We begin by designating the spectrum of fundamental fermions of the theory. It suffices for the moment to include only the electron and its neutrino, which form a left-handed "weak-isospin" doublet,

$$\mathsf{L}_e \equiv \begin{pmatrix} \nu \\ e \end{pmatrix}_L, \qquad (6.3.1)$$

where the left-handed states are

$$\begin{aligned} \nu_L &= \tfrac{1}{2}(1 - \gamma_5)\nu \\ e_L &= \tfrac{1}{2}(1 - \gamma_5)e. \end{aligned} \qquad (6.3.2)$$

The helicity of electron neutrinos emitted in charged-current interactions was determined to be negative in a classic experiment [11] that inferred the longitudinal polarization of the recoil nucleus in the reaction

$$\begin{aligned} e^- \, {}^{152}\mathrm{Eu}^m(J = 0) &\to {}^{152}\mathrm{Sm}^*(J = 1) \, \nu_e \\ &\quad\, \hookrightarrow \gamma \, {}^{152}\mathrm{Sm}. \end{aligned} \qquad (6.3.3)$$

A measurement of the photon polarization determines the neutrino helicity by angular momentum conservation. The electron neutrino is known to have a very small mass ($\lesssim 2$ eV), and all neutrino masses are too small to have been determined by kinematic constraints (Ref. [4]). It is convenient to idealize the neutrino as exactly massless,* in which case the right-handed state

$$\nu_R = \tfrac{1}{2}(1 + \gamma_5)\nu \qquad (6.3.4)$$

does not exist. Thus we designate only one right-handed lepton,

$$\mathsf{R}_e = e_R = \tfrac{1}{2}(1 + \gamma_5)e, \qquad (6.3.5)$$

which is a weak-isospin singlet. This completes a specification of the weak charged currents. To incorporate electromagnetism, we define a "weak hypercharge," Y. Requiring that the Gell-Mann–Nishijima relation for the electric charge,

$$Q = I_3 + \tfrac{1}{2}Y, \qquad (6.3.6)$$

---

* We will revisit the question of neutrino masses and mixings in section 6.7.

be satisfied leads to the assignments

$$Y_L = -1,$$
$$Y_R = -2. \tag{6.3.7}$$

By construction, the weak-isospin projection $I_3$ and the weak hypercharge $Y$ are commuting observables,

$$[I_3, Y] = 0. \tag{6.3.8}$$

We now take the (product) group of transformations generated by $I$ and $Y$ to be the gauge group $SU(2)_L \otimes U(1)_Y$ of the theory. To construct the theory, we introduce the gauge bosons

$$b_\mu^1, b_\mu^2, b_\mu^3 \qquad \text{for } SU(2)_L,$$
$$\mathcal{A}_\mu \qquad \text{for } U(1)_Y.$$

Evidently the Lagrangian for the theory may be written as

$$\mathcal{L} = \mathcal{L}_{\text{gauge}} + \mathcal{L}_{\text{leptons}} \tag{6.3.9}$$

where the kinetic term for the gauge fields is

$$\mathcal{L}_{\text{gauge}} = -\tfrac{1}{4} F_{\mu\nu}^l F^{l\mu\nu} - \tfrac{1}{4} f_{\mu\nu} f^{\mu\nu}, \tag{6.3.10}$$

and the field-strength tensors are

$$F_{\mu\nu}^l = \partial_\nu b_\mu^l - \partial_\mu b_\nu^l + g \varepsilon_{jkl} b_\mu^j b_\nu^k \tag{6.3.11}$$

for the $SU(2)_L$ gauge fields [cf. (4.2.37)] and

$$f_{\mu\nu} = \partial_\nu \mathcal{A}_\mu - \partial_\mu \mathcal{A}_\nu \tag{6.3.12}$$

for the $U(1)_Y$ gauge field [cf. (3.2.10)]. The matter term is

$$\mathcal{L}_{\text{leptons}} = \bar{R} i \gamma^\mu \left( \partial_\mu + \frac{ig'}{2} \mathcal{A}_\mu Y \right) R + \bar{L} i \gamma^\mu \left( \partial_\mu + \frac{ig'}{2} \mathcal{A}_\mu Y + \frac{ig}{2} \boldsymbol{\tau} \cdot \mathbf{b}_\mu \right) L. \tag{6.3.13}$$

The coupling of the weak-isospin group $SU(2)_L$ is called $g$, as in the Yang–Mills theory of chapter 4, and the coupling constant for the weak-hypercharge group $U(1)_Y$ is denoted as $g'/2$. the factor $\tfrac{1}{2}$ being chosen to simplify later expressions.

The theory of weak and electromagnetic interactions described by the Lagrangian (6.3.9) is not a satisfactory one for two immediately obvious reasons. It contains four massless gauge bosons $(b^1, b^2, b^3, \mathcal{A})$, whereas nature has but one, the photon. In addition, the expression (6.3.13) represents a massless electron; it lacks the $-m_e \bar{e}e$ term of the QED Lagrangian (2.2.7), and for good reason. A fermion mass term links left-handed and right-handed components:

$$\bar{e}e = \tfrac{1}{2}\bar{e}(1 - \gamma_5)e + \tfrac{1}{2}\bar{e}(1 + \gamma_5)e = \bar{e}_R e_L + \bar{e}_L e_R. \qquad (6.3.14)$$

The left-handed and right-handed components of the electron transform differently under $SU(2)_L$ and $U(1)_Y$, so an explicit fermion mass term would break the $SU(2)_L \otimes U(1)_Y$ gauge invariance of the theory: a mass term is forbidden. Our task is to modify the theory so that there will remain only a single conserved quantity (the electric charge) corresponding to one massless gauge boson (the photon), and the electron will acquire mass.

To accomplish these things, we introduce a complex doublet of scalar fields,

$$\phi \equiv \begin{pmatrix} \phi^+ \\ \phi^0 \end{pmatrix}, \qquad (6.3.15)$$

that transforms as an $SU(2)_L$ doublet and must, therefore, have weak hypercharge

$$Y_\phi = +1, \qquad (6.3.16)$$

by virtue of the Gell-Mann–Nishijima relation (6.3.6). We add to the Lagrangian a term

$$\mathcal{L}_{scalar} = (\mathcal{D}^\mu \phi)^\dagger (\mathcal{D}_\mu \phi) - V(\phi^\dagger \phi), \qquad (6.3.17)$$

where the gauge-covariant derivative is

$$\mathcal{D}_\mu = \partial_\mu + \frac{ig'}{2} A_\mu Y + \frac{ig}{2} \boldsymbol{\tau} \cdot \mathbf{b}_\mu \qquad (6.3.18)$$

and, as usual, the potential is

$$V(\phi^\dagger \phi) = \mu^2 (\phi^\dagger \phi) + |\lambda| (\phi^\dagger \phi)^2. \qquad (6.3.19)$$

We are also free to add an interaction term, which involves Yukawa couplings of the scalars to the fermions,

$$\mathcal{L}_{Yukawa} = -\zeta_e [\bar{R}(\phi^\dagger L) + (\bar{L}\phi)R], \qquad (6.3.20)$$

a Lorentz scalar symmetric under local $SU(2)_L \otimes U(1)_Y$ rotations.

Now let us imagine that $\mu^2 < 0$ and consider the consequences of spontaneous symmetry breaking. We choose as the vacuum expectation value of the scalar field

$$\langle \phi \rangle_0 = \begin{pmatrix} 0 \\ v/\sqrt{2} \end{pmatrix}, \qquad (6.3.21)$$

where $v = \sqrt{-\mu^2/|\lambda|}$, which breaks both $SU(2)_L$ and $U(1)_Y$ symmetries but preserves an invariance under the $U(1)_{EM}$ symmetry generated by the electric charge operator. Recall that a (would-be) Goldstone boson is associated with every generator of the gauge group that does not leave the vacuum invariant. The vacuum is left invariant by a generator $\mathcal{G}$ if

$$e^{i\alpha\mathcal{G}} \langle\phi\rangle_0 = \langle\phi\rangle_0 . \tag{6.3.22}$$

For an infinitesimal transformation, (6.3.22) becomes

$$(1 + i\alpha\mathcal{G}) \langle\phi\rangle_0 = \langle\phi\rangle_0 , \tag{6.3.23}$$

so that the condition for $\mathcal{G}$ to leave the vacuum invariant is simply

$$\mathcal{G} \langle\phi\rangle_0 = 0. \tag{6.3.24}$$

We easily compute that

$$\tau_1 \langle\phi\rangle_0 = \begin{pmatrix} 0 & 1 \\ 1 & 0 \end{pmatrix} \begin{pmatrix} 0 \\ v/\sqrt{2} \end{pmatrix} = \begin{pmatrix} v/\sqrt{2} \\ 0 \end{pmatrix} \neq 0, \quad \text{Broken!}$$

$$\tau_2 \langle\phi\rangle_0 = \begin{pmatrix} 0 & -i \\ i & 0 \end{pmatrix} \begin{pmatrix} 0 \\ v/\sqrt{2} \end{pmatrix} = \begin{pmatrix} -iv/\sqrt{2} \\ 0 \end{pmatrix} \neq 0, \quad \text{Broken!}$$

$$\tag{6.3.25}$$

$$\tau_3 \langle\phi\rangle_0 = \begin{pmatrix} 1 & 0 \\ 0 & -1 \end{pmatrix} \begin{pmatrix} 0 \\ v/\sqrt{2} \end{pmatrix} = \begin{pmatrix} 0 \\ -v/\sqrt{2} \end{pmatrix} \neq 0, \quad \text{Broken!}$$

$$Y \langle\phi\rangle_0 = Y_\phi \langle\phi\rangle_0 = +1 \langle\phi\rangle_0 = \begin{pmatrix} 0 \\ v/\sqrt{2} \end{pmatrix} \neq 0, \quad \text{Broken!}$$

However, if we examine the effect of the electric charge operator $Q$ on the (electrically neutral) vacuum state, we find that

$$Q\langle\phi\rangle_0 = \tfrac{1}{2}(\tau_3 + Y) \langle\phi\rangle_0 = \tfrac{1}{2} \begin{pmatrix} Y_\phi + 1 & 0 \\ 0 & Y_\phi - 1 \end{pmatrix} \langle\phi\rangle_0$$

$$= \begin{pmatrix} 1 & 0 \\ 0 & 0 \end{pmatrix} \begin{pmatrix} 0 \\ v/\sqrt{2} \end{pmatrix} = \begin{pmatrix} 0 \\ 0 \end{pmatrix}, \quad \text{Unbroken!} \tag{6.3.26}$$

This is promising! All the original four generators are broken, but the linear combination corresponding to electric charge is not. The photon will, therefore, remain massless, whereas three other gauge bosons will acquire mass.

We next expand the Lagrangian about the minimum of the Higgs potential $V$, by writing

$$\phi = \exp\left(\frac{i\boldsymbol{\zeta} \cdot \boldsymbol{\tau}}{2v}\right) \begin{pmatrix} 0 \\ (v+\eta)/\sqrt{2} \end{pmatrix} \tag{6.3.27}$$

and transforming at once to U-gauge:

$$\phi \to \phi' = \exp\left(-i\frac{\boldsymbol{\zeta} \cdot \boldsymbol{\tau}}{2v}\right) \phi = \begin{pmatrix} 0 \\ (v+\eta)/\sqrt{2} \end{pmatrix}, \tag{6.3.28}$$

$$\boldsymbol{\tau} \cdot \mathbf{b}_\mu \to \boldsymbol{\tau} \cdot \mathbf{b}'_\mu, \tag{6.3.29}$$

$$\mathcal{A}_\mu \to \mathcal{A}_\mu, \tag{6.3.30}$$

$$\mathsf{R} \to \mathsf{R}, \tag{6.3.31}$$

$$\mathsf{L} \to \mathsf{L}' = \exp\left(-i\frac{\boldsymbol{\zeta} \cdot \boldsymbol{\tau}}{2v}\right) \mathsf{L}. \tag{6.3.32}$$

Had we followed literally the procedure of section 5.5, we would have replaced the generator $\tau_3$ in the expansion (6.3.27) of $\phi$ about $\langle\phi\rangle_0$ by the combination $K = \frac{1}{2}(\tau_3 - Y)$ orthogonal to $Q$, which is, strictly speaking, the third broken generator. However, because $Q$ leaves the vacuum invariant and $\tau_3 = K + Q$, the effect is the same.

We may now reëxpress the Lagrangian in terms of the U-gauge fields (6.3.28)–(6.3.32), omitting primes to avoid notational clutter, and investigate the consequences of spontaneous symmetry breaking.

The scalar term in the Lagrangian now reads

$$\mathcal{L}_{\text{scalar}} = \frac{v^2}{8}\left[g^2 \left|b_\mu^1 - ib_\mu^2\right|^2 + (g'\mathcal{A}_\mu - gb_\mu^3)^2\right]$$
$$+ \frac{1}{2}\left[(\partial^\mu\eta)(\partial_\mu\eta) + 2\mu^2\eta^2\right] + \cdots \tag{6.3.33}$$

plus interaction terms. If we define the charged gauge fields

$$\mathsf{W}_\mu^\pm \equiv \frac{b_\mu^1 \mp ib_\mu^2}{\sqrt{2}}, \tag{6.3.34}$$

the term proportional to $g^2v^2$ is recognizable as a mass term for the charged vector bosons:

$$\frac{g^2v^2}{2}(\left|\mathsf{W}_\mu^+\right|^2 + \left|\mathsf{W}_\mu^-\right|^2), \tag{6.3.35}$$

corresponding to the charged intermediate boson masses

$$M_{\mathsf{W}^\pm} = \frac{gv}{2}. \tag{6.3.36}$$

Defining the orthogonal combinations

$$Z_\mu = \frac{-g' A_\mu + g b^3_\mu}{\sqrt{g^2 + g'^2}} \tag{6.3.37}$$

and

$$A_\mu = \frac{g A_\mu + g' b^3_\mu}{\sqrt{g^2 + g'^2}}, \tag{6.3.38}$$

we find that the neutral intermediate boson $Z^0$ has acquired a mass

$$M_{Z^0} = \sqrt{g^2 + g'^2} \, v/2 = M_W \sqrt{1 + g'^2/g^2}, \tag{6.3.39}$$

and that the field $A_\mu$ remains a massless gauge boson corresponding to the surviving $\exp[i\, Q\alpha(x)]$ symmetry.

Examining the second line of (6.3.33), we see at once [cf. (2.2.1)] that the $\eta$ field has acquired a (mass)$^2$,

$$M_H^2 = -2\mu^2 > 0; \tag{6.3.40}$$

it is the physical Higgs boson.

Finally, the Yukawa term in the Lagrangian has become

$$\mathcal{L}_{\text{Yukawa}} = -\zeta_e \frac{(v + \eta)}{\sqrt{2}} (\bar{e}_R e_L + \bar{e}_L e_R)$$

$$= -\frac{\zeta_e v}{\sqrt{2}} \bar{e}e - \frac{\zeta_e \eta}{\sqrt{2}} \bar{e}e, \tag{6.3.41}$$

so the electron has acquired a mass

$$m_e = \frac{\zeta_e v}{\sqrt{2}}. \tag{6.3.42}$$

As the originally massless nucleons in the $\sigma$ model (cf. §5.4) gained mass from their interactions with the massive scalar $\sigma$, the electron and other fermions acquire mass from their interactions with the $\eta$ field. We see that we have achieved, at least schematically, the desired particle content—plus a massive Higgs scalar we did not request.

Do the interactions correspond to those in nature? The interactions among the gauge bosons and leptons may be read from (6.3.13), $\mathcal{L}_{\text{leptons}}$. For the charged gauge bosons, we find

$$\mathcal{L}_{W:\ell} = -\frac{g}{\sqrt{2}} \left( \bar{\nu}_L \gamma^\mu e_L W^+_\mu + \bar{e}_L \gamma^\mu \nu_L W^-_\mu \right)$$

$$= -\frac{g}{2\sqrt{2}} \left[ \bar{\nu}\gamma^\mu (1 - \gamma_5) e \, W^+_\mu + \bar{e}\gamma^\mu (1 - \gamma_5)\nu \, W^-_\mu \right], \tag{6.3.43}$$

which reproduces the low-energy phenomenology of the ad hoc intermediate boson model of section 6.2, provided that we identify the coupling constants as

$$\frac{g^2}{8} = \frac{G_F M_W^2}{\sqrt{2}}. \tag{6.3.44}$$

With the aid of the expression (6.3.35) for the intermediate-boson mass, we find that the vacuum expectation value parameter $v$ is now determined as

$$v = (G_F \sqrt{2})^{-1/2} \approx 246 \text{ GeV}, \tag{6.3.45}$$

so that the vacuum expectation value of the scalar field is

$$\langle \phi^0 \rangle_0 = (G_F \sqrt{8})^{-1/2} \approx 174 \text{ GeV}. \tag{6.3.46}$$

Similarly, the neutral-gauge-boson couplings to leptons are given by

$$\begin{aligned}
\mathcal{L}_{0:\ell} = \; & \frac{gg'}{\sqrt{g^2 + g'^2}} \bar{e}\gamma^\mu e \, A_\mu \\
& - \frac{\sqrt{g^2 + g'^2}}{2} \bar{\nu}_L \gamma^\mu \nu_L \, Z_\mu \\
& + \frac{1}{\sqrt{g^2 + g'^2}} \left[ -g'^2 \bar{e}_R \gamma^\mu e_R + \frac{(g^2 - g'^2)}{2} \bar{e}_L \gamma^\mu e_L \right] Z_\mu.
\end{aligned} \tag{6.3.47}$$

Therefore, we may indeed identify $A_\mu$ as the photon, provided that we set

$$\frac{gg'}{\sqrt{g^2 + g'^2}} = e. \tag{6.3.48}$$

If, in analogy with the fine structure constant of electromagnetism, $\alpha \equiv e^2/4\pi\hbar c$, we define $\alpha_2 \equiv g^2/4\pi\hbar c$ and $\alpha_Y \equiv g'^2/4\pi\hbar c$, we may recast (6.3.48) as

$$1/\alpha = 1/\alpha_Y + 1/\alpha_2. \tag{6.3.49}$$

It is convenient to introduce a weak mixing angle $\theta_W$ to parametrize the mixing of the neutral gauge bosons. With the definition

$$g' \equiv g \tan \theta_W, \tag{6.3.50}$$

whence

$$\sqrt{g^2 + g'^2} = \frac{g}{\cos \theta_W}, \tag{6.3.51}$$

(6.3.37) and (6.3.38) may be rewritten as

$$Z_\mu = -\mathcal{A}_\mu \sin\theta_W + b_\mu^3 \cos\theta_W, \tag{6.3.52}$$

$$A_\mu = \mathcal{A}_\mu \cos\theta_W + b_\mu^3 \sin\theta_W, \tag{6.3.53}$$

which may be inverted to yield

$$\mathcal{A}_\mu = A_\mu \cos\theta_W - Z_\mu \sin\theta_W, \tag{6.3.54}$$

$$b_\mu^3 = A_\mu \sin\theta_W + Z_\mu \cos\theta_W. \tag{6.3.55}$$

In view of the identification (6.3.50), the coupling constants of the $SU(2)_L$ and $U(1)_Y$ gauge groups may be expressed as

$$g = \frac{e}{\sin\theta_W} \geq e, \tag{6.3.56}$$

$$g' = \frac{e}{\cos\theta_W} \geq e, \tag{6.3.57}$$

indicating that the previously disparate strengths of the weak and electromagnetic interactions are now related through a single parameter.

Taken together, the coupling constant identifications lead to

$$\begin{aligned}
M_W^2 &= \frac{g^2}{4G_F\sqrt{2}} = \frac{e^2}{4G_F\sqrt{2}\sin^2\theta_W} \\
&= \frac{\pi\alpha}{G_F\sqrt{2}\sin^2\theta_W} \\
&\approx \frac{(37.3 \text{ GeV})^2}{\sin^2\theta_W},
\end{aligned} \tag{6.3.58}$$

and

$$M_Z^2 = \frac{M_W^2}{\cos^2\theta_W} \geqslant M_W^2. \tag{6.3.59}$$

The feebleness of the weak interactions at low energies is thus laid to the large mass of the intermediate bosons and not to an intrinsically small coupling constant.

Note that the dimensionless coupling constant that endowed the electron with mass in (6.3.42) is both small,

$$\zeta_e = \frac{m_e\sqrt{2}}{v} = 2^{3/4} m_e G_F^{1/2} \approx 3 \times 10^{-6}, \tag{6.3.60}$$

and arbitrary.

It is convenient to rewrite the interaction Lagrangian (6.3.47) in terms of the weak mixing angle as

$$\mathcal{L}_{0:\ell} = e\,\bar{e}\gamma^{\mu}e\,A_{\mu} - \frac{g}{2\cos\theta_{W}}\,\bar{\nu}_{L}\gamma^{\mu}\nu_{L}\,Z_{\mu}$$

$$- \frac{g}{2\cos\theta_{W}}[2\sin^{2}\theta_{W}\bar{e}_{R}\gamma^{\mu}e_{R}\,Z_{\mu} + (2\sin^{2}\theta_{W} - 1)\bar{e}_{L}\gamma^{\mu}e_{L}\,Z_{\mu}]$$

$$= e\,\bar{e}\gamma^{\mu}e\,A_{\mu} - \frac{1}{\sqrt{2}}\left(\frac{G_{F}M_{Z}^{2}}{\sqrt{2}}\right)^{1/2}\bar{\nu}\gamma^{\mu}(1 - \gamma_{5})\nu\,Z_{\mu}$$

$$- \frac{1}{\sqrt{2}}\left(\frac{G_{F}M_{Z}^{2}}{\sqrt{2}}\right)^{1/2}[2\sin^{2}\theta_{W}\,\bar{e}\gamma^{\mu}(1 + \gamma_{5})e\,Z_{\mu}$$

$$+ (2\sin^{2}\theta_{W} - 1)\,\bar{e}\gamma^{\mu}(1 - \gamma_{5})e\,Z_{\mu}], \qquad (6.3.61)$$

from which Feynman rules for the elementary vertices may readily be deduced. These are shown in figure 6.12.

The low-energy phenomenology of the model has been constructed to reproduce that of the effective Lagrangian description, which has been the subject of sections 6.1 and 6.2. There are several novel elements as well: neutral weak-current phenomena (to be studied in section 6.4), some definite predictions for the properties of gauge bosons, and good high-energy behavior—indeed, renormalizability. Let us look briefly at the last two of these.

In the $SU(2)_{L} \otimes U(1)_{Y}$ model, the properties of the gauge bosons are correlated with those of the neutral-current interactions through the weak mixing angle $\theta_{W}$. The masses have already been expressed in terms of $\theta_{W}$ in (6.3.58) and (6.3.59). They are also plotted in the top pane of figure 6.13. The expression (6.2.20) for the partial decay rate of $W^{\pm}$ continues to hold because there has been no change in the charged-current phenomenology. By combining (6.3.58) and (6.2.20) we obtain the useful expression

$$\Gamma(W^{-} \to e^{-}\bar{\nu}_{e}) \approx \frac{23\text{ MeV}}{\sin^{3}\theta_{W}}. \qquad (6.3.62)$$

The leptonic decay rates of the neutral gauge boson may be computed almost by transcription. Evidently,

$$\left|\mathcal{M}(Z^{0} \to \nu\bar{\nu})\right|^{2} = \frac{1}{2} \cdot \frac{M_{Z}^{2}}{M_{W}^{2}}\left|\mathcal{M}(W^{-} \to e^{-}\bar{\nu})\right|^{2}, \qquad (6.3.63)$$

neglecting the electron mass, so that

$$\Gamma(Z^{0} \to \nu\bar{\nu}) = \frac{G_{F}M_{Z}^{3}}{12\pi\sqrt{2}} \approx \frac{11.4\text{ MeV}}{(\sin\theta_{W}\cos\theta_{W})^{3}}. \qquad (6.3.64)$$

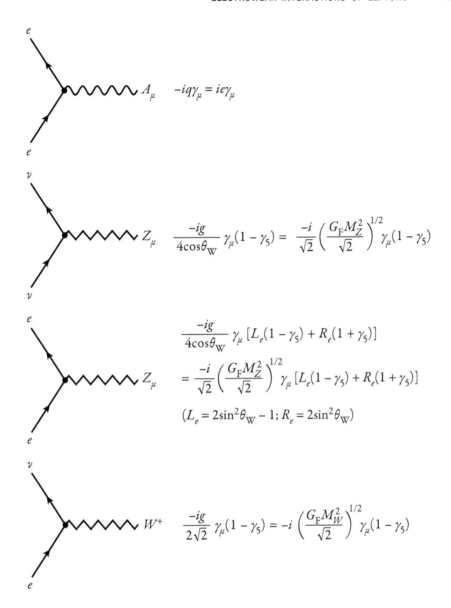

**Figure 6.12.** Feynman rules for lepton interactions in the $SU(2)_L \otimes U(1)_Y$ electroweak theory.

Similarly, we find for the decays into charged leptons

$$\Gamma(Z^0 \to e^+e^-) = \Gamma(Z^0 \to \nu\bar{\nu})[L_e^2 + R_e^2]$$
$$= \Gamma(Z^0 \to \nu\bar{\nu})[(2\sin^2\theta_W - 1)^2 + (2\sin^2\theta_W)^2], \quad (6.3.65)$$

where the two terms correspond to the contributions of left-handed and right-handed chiral couplings defined in figure 6.12. Our expectations for the partial decay rates are shown in the lower pane of figure 6.13.

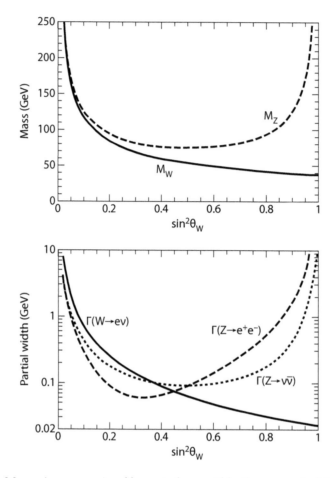

**Figure 6.13.** Masses (upper pane) and leptonic decay widths (lower pane) of the intermediate
bosons $W^\pm$ and $Z^0$ in the $SU(2)_L \otimes U(1)_Y$ electroweak theory, as functions of the
weak mixing parameter $\sin^2 \theta_W$.

We shall examine the predictions of the electroweak theory in some detail in
section 6.4 and chapter 7, but it is worth anticipating some of the consequences right
away. The observed mass of the $Z$-boson (Ref. [4]), $M_Z = (91.1876 \pm 0.0021)$ GeV,
and neutral current studies point to a value of the weak-mixing parameter $\sin^2 \theta_W \approx$
0.2. Using our lowest-order expression (6.3.59) for $M_Z$ to fix $\sin^2 \theta_W = 0.2122$, we
find (measured values in brackets)

$$M_W \approx 81 \text{ GeV} \quad [80.4 \text{ GeV}],$$
$$\Gamma(W \to e\nu) \approx 232 \text{ MeV} \quad [224 \text{ MeV}], \tag{6.3.66}$$

and

$$\Gamma(Z \to \nu\bar{\nu}) \approx 166 \text{ MeV} \quad [166 \text{ MeV}],$$
$$\Gamma(Z \to e^+e^-) \approx 85 \text{ MeV} \quad [84 \text{ MeV}], \tag{6.3.67}$$

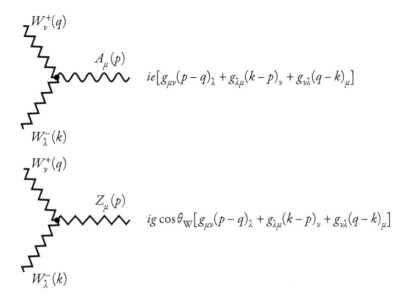

**Figure 6.14.** Trilinear interactions among electroweak gauge bosons. Recall that $g = e/\sin\theta_W$, $g' = e/\cos\theta_W = g\tan\theta_W$. Indicated charges and momenta flow toward the vertex.

which shows very promising agreement between theory and experiment. (Another choice [cf. (9.3.22)] would have been to fix the weak mixing parameter from the ratio $M_W^2/M_Z^2$, which yields $\sin^2\theta_W = 0.2226$.) We will complete the portrait of the electroweak gauge bosons in section 7.2, after incorporating quarks into the theory.

As a first look at the high-energy properties of the theory, let us reconsider the reaction

$$\nu_e\bar{\nu}_e \rightarrow W_0^+ W_0^-, \tag{6.2.25}$$

the comportment of which was symptomatic of the diseases of the ad hoc intermediate-boson theory. To do so, we must determine the Feynman rules for interactions among the gauge bosons. The trilinear and quadrilinear entries in the kinetic term (6.3.10) of the Lagrangian $\mathcal{L}_{\text{gauge}}$ give rise to the vertices shown in figures 6.14 and 6.15. The appearance of the $Z^0 W^+ W^-$ vertex, together with the $Z^0\nu\bar{\nu}$ vertex already discussed, signals that there will be a new contribution to the amplitude for the reaction $\nu_e\bar{\nu}_e \rightarrow W^+ W^-$: the $s$-channel $Z^0$-exchange graph shown in figure 6.16. Its contribution to the amplitude is

$$\mathcal{M}_Z = \frac{-ig^2}{4(S^2 - M_Z^2)}\bar{v}(v, q_2)\gamma_\mu(1 - \gamma_5)u(v, q_1)\left(g^{\mu\nu} - \frac{S^\mu S^\nu}{M_Z^2}\right)$$
$$\times\epsilon_+^{*\alpha}\epsilon_-^{*\beta}[g_{\alpha\beta}(k_- - k_+)_\nu + g_{\alpha\nu}[(k_+ + S)_\beta - g_{\beta\nu}(k_- + S)_\alpha], \tag{6.3.68}$$

where

$$S \equiv q_1 + q_2 = k_+ + k_- \tag{6.3.69}$$

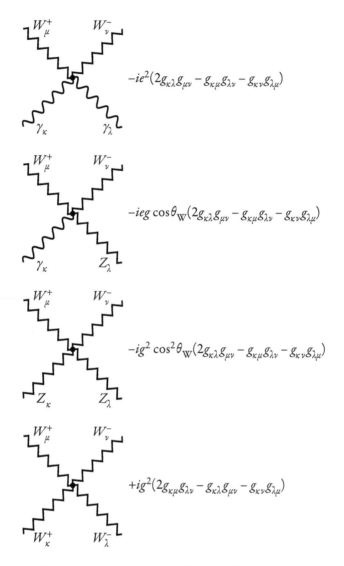

**Figure 6.15.** Quadrilinear interactions among electroweak gauge bosons. Recall that $g = e/\sin\theta_W$ and $g' = e/\cos\theta_W = g\tan\theta_W$. Indicated charges flow toward the vertex.

and the other kinematical quantities have been defined in (6.2.28)–(6.2.31). The $S^\mu S^\nu$ term is impotent between massless spinors. Thus we have only to evaluate

$$\mathcal{M}_Z = \frac{ig^2}{4(S^2 - M_Z^2)}\bar{v}(\nu, q_2)\gamma^\nu(1 - \gamma_5)u(\nu, q_1)[\epsilon_+^* \cdot \epsilon_-^* \ (k_- - k_+)_\nu$$

$$+ k_+ \cdot \epsilon_-^* \ \epsilon_{+\nu}^* - k_- \cdot \epsilon_+^* \ \epsilon_{-\nu}^* + \epsilon_-^* \cdot S \ \epsilon_{+\nu}^* - \epsilon_+^* \cdot S \ \epsilon_{-\nu}^*]. \qquad (6.3.70)$$

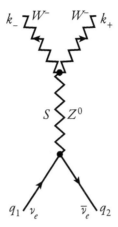

**Figure 6.16.** $Z^0$-pole contribution to the reaction $\nu_e\bar{\nu}_e \to W^+W^-$ in the $SU(2)_L \otimes U(1)_Y$ theory.

On substituting the asymptotic forms (6.2.31) of the polarization vectors for the longitudinally polarized intermediate bosons and using

$$k_+ \cdot S = k_- \cdot S = \frac{S^2}{2}, \qquad (6.3.71)$$

we find that

$$\mathcal{M}_Z = \frac{ig}{8\,M_W^2}\bar{v}(\nu,q_2)(\not{k}_+ - \not{k}_-)(1 - \gamma_5)u(\nu,q_1), \qquad (6.3.72)$$

for $S^2 \gg M_W^2$. Then, by virtue of (6.2.34) and (6.2.35) and the definition (6.2.26), we have

$$\begin{aligned}
\mathcal{M}_Z &= -\frac{ig^2}{4\,M_W^2}\bar{v}(\nu,q_2)\not{P}(1 - \gamma_5)u(\nu,q_1)\\
&= -i\,G_F\sqrt{2}\,\bar{v}(\nu,q_2)\not{P}(1 - \gamma_5)u(\nu,q_1),
\end{aligned} \qquad (6.3.73)$$

which is to be added to the contribution of the electron-exchange graph of figure 6.9,

$$\mathcal{M}_e = i\,G_F\sqrt{2}\,\bar{v}(\nu,q_2)\not{P}(1 - \gamma_5)u(\nu,q_1). \qquad (6.2.37)$$

The sum vanishes. Thus, the $p$-wave $s$-channel resonance, the $Z^0$, has canceled the divergence in the $p$-wave scattering amplitude, and the amplitude is "asymptotically safe."

Measurement of the cross section for the reaction

$$e^+e^- \to W^+W^-, \qquad (6.3.74)$$

(a)                                 (b)                             (c)

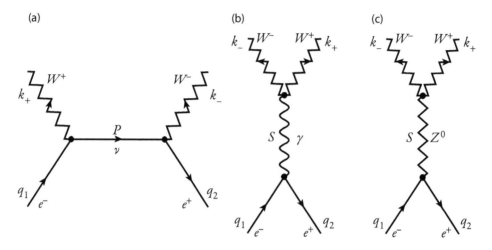

**Figure 6.17.** Diagrams that participate in the gauge cancellation in the reaction $e^+e^- \to W^+W^-$.

for which similar divergence cancellations occur, has been advocated [12] as a probe of the gauge structure of the electroweak theory. Neglecting the electron mass, this reaction is described by the three Feynman graphs in figure 6.17. The leading divergence in the neutrino-exchange diagram of figure 6.17(a) is canceled by the contributions of the direct-channel $\gamma$- and $Z^0$-exchange diagrams, figure 6.17(b) and (c), in complete analogy with the two graphs (figures 6.9 and 6.16) for $\nu\bar{\nu} \to W^+W^-$. We may verify the cancellation without carrying out the full calculation as follows. When fermion masses are neglected, the lepton-exchange diagrams of figures 6.9 and 6.17(a) represent equal amplitudes. In the very high energy approximation in which the $\gamma$ and $Z^0$ propagators are equal, it is easy to show that the sum of the contributions of diagram 6.17(b) and (c) is identical with the amplitude due to figure 6.16. The connection to the local gauge symmetry of the theory is easily traced.

A noteworthy achievement of the LEP experiments is the validation of the $SU(2)_L \otimes U(1)_Y$ symmetry for the interaction of gauge bosons with fermions and gauge bosons with gauge bosons in $e^+e^- \to W^+W^-$. This reaction is described by three Feynman diagrams that correspond to $s$-channel photon and $Z^0$ exchange and $t$-channel neutrino exchange. For the production of longitudinally polarized $W$-bosons, each diagram of figure 6.17 leads to a $J = 1$ partial-wave amplitude that grows as the square of the c.m. energy, but the gauge symmetry enforces a pattern of cooperation. The contributions of the direct-channel $\gamma$- and $Z^0$-exchange diagrams cancel the leading divergence in the $J = 1$ partial-wave amplitude of the neutrino-exchange diagram. The interplay is shown in figure 6.18. If the $Z$-exchange contribution is omitted (middle line) or if both the $\gamma$- and $Z$-exchange contributions are omitted (upper line), the calculated cross section grows unacceptably with energy. The measurements compiled by the LEP Electroweak Working Group [13] agree well with the benign high-energy behavior predicted by the full electroweak theory.

Hadron collider measurements do not directly determine the $W^+W^-$ invariant mass because of the missing energy carried by neutrinos, but they reach beyond

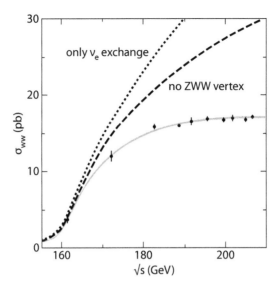

**Figure 6.18.** Cross section for the reaction $e^+e^- \to W^+W^-$ measured by the four LEP experiments, together with the full electroweak-theory simulation (solid line) and cross sections that would result from $\nu$-exchange alone (dots) and from $(\nu + \gamma)$ exchange (dashes). (From Ref. [13].)

the highest energy studied at LEP. The latest contributions, from the D0 [14] and CDF [15] collaborations, are in agreement with standard-model expectations [16] and tighten the bounds on anomalous couplings. The ATLAS and CMS experiments at the Large Hadron Collider will probe with increased sensitivity in $pp$ collisions at $\sqrt{s} = 7$ TeV and beyond [17].

This beautiful experimental validation of "gauge cancellation" is not the end of the story, however. As we shall see in section 6.6 and problem 6.12, an unacceptable growth with energy persists when a nonzero electron mass is restored. This is canceled not by the gauge symmetry, but by the contribution of the Higgs boson.

## 6.4 NEUTRAL-CURRENT INTERACTIONS AMONG LEPTONS

The prediction of neutral-current effects in gauge theories of the weak and electromagnetic interactions and the availability of high-energy neutrino beams spurred the search for experimental manifestations of the weak neutral current. In the summer of 1973, muonless events in deeply inelastic $\nu N$ and $\bar{\nu} N$ collisions were reported by the Gargamelle bubble chamber collaboration working at the CERN proton synchrotron [18]. Subsequently, these observations have been confirmed and extended [19], and the detailed study of neutral-current interactions has flowered so that an essentially complete determination of the neutral-current couplings has been carried out. We shall consider studies on the $Z^0$ pole and deeply inelastic scattering in the next chapter, after having incorporated hadrons into the electroweak theory and reviewed the essential features of the parton model. For the moment, let us investigate the implications of the $SU(2)_L \otimes U(1)_Y$ gauge theory for purely leptonic interactions, for which the comparison of theory and experiment does not require the intervention of parton-model assumptions. These reactions were historically important in establishing the neutral-current interaction and pinning down its characteristics.

As a prelude to our review of the experimental consequences of the leptonic neutral current derived in the preceding section, let us incorporate additional families of leptons into the model. In view of the universality of the electromagnetic and weak interactions, this is evidently to be done merely by cloning the existing fermion structure. We add further left-handed weak-isospin doublets

$$\mathsf{L}_\mu = \begin{pmatrix} \nu_\mu \\ \mu \end{pmatrix}_\mathrm{L}, \ \mathsf{L}_\tau = \begin{pmatrix} \nu_\tau \\ \tau \end{pmatrix}_\mathrm{L}, \ \dots \tag{6.4.1}$$

and right-handed singlets

$$\mathsf{R}_\mu = \mu_\mathrm{R}, \ \mathsf{R}_\tau = \tau_\mathrm{R}, \ \dots \tag{6.4.2}$$

with the same weak-hypercharge assignments (6.3.7) as their counterparts (6.3.1) and (6.3.5) in the electron family. By omitting right-handed neutrinos, we continue to idealize the neutrinos as massless. The Yukawa interaction term (6.3.20) in the electroweak Lagrangian is generalized to

$$\mathcal{L}_{\text{Yukawa}} = - \sum_{i=e, \mu\tau, \dots} \zeta_i [\bar{\mathsf{R}}_i(\phi^\dagger \mathsf{L}_i) + (\bar{\mathsf{L}}_i \phi)\mathsf{R}_i]. \tag{6.4.3}$$

This done, the Feynman rules for the interactions of the added leptons with gauge bosons are precisely those given for the electron family in figure 6.12, because universality is a direct consequence of the gauge symmetry. Mass terms and Higgs-boson interactions for the charged leptons are generated just as those given in (6.3.41) for the electron. We are now prepared to calculate the consequences of the theory.

Before we do so, it will be useful to repeat the exercise carried out for charged-current interactions with arbitrary spacetime structure in section 6.1. By now it is firmly established that the weak neutral current has the vector and axial vector structure of the electroweak theory. However, we may be presented with other novel interactions in the future, so it is important to see how we might begin to unravel their properties by considering the (now historical) neutral-current example. We consider the reactions

$$\begin{aligned} \nu_\mu e &\to \nu_\mu e, \\ \bar{\nu}_\mu e &\to \bar{\nu}_\mu e, \end{aligned} \tag{6.4.4}$$

which, though compatible with the general requirements of lepton-number conservation, are not mediated by charged currents. Upon writing down the matrix elements for these processes, we note immediately that the cross sections may be obtained from the earlier charged-current results by interchanging $p_2 \leftrightarrow q_2$ for $\nu_\mu e$ scattering or $p_1 \leftrightarrow q_2$ for $\bar{\nu}_\mu e$ scattering. Consequently there is no need to repeat the arithmetic in detail, and we may at once assemble the results in table 6.2. Recall that the energy-loss parameter is $y = E'_e/E_\nu$.

TABLE 6.2
Differential Cross Sections for $\nu_\mu e$ and $\bar{\nu}_\mu e$ Elastic Scattering

| Coupling | $\dfrac{\pi}{mE}\dfrac{d\sigma}{dy}(\nu_\mu e \to \nu_\mu e)$ | $\dfrac{\pi}{mE}\dfrac{d\sigma}{dy}(\bar{\nu}_\mu e \to \bar{\nu}_\mu e)$ |
|---|---|---|
| $\lvert C_S\rvert^2 + \lvert C_P\rvert^2$ | $\frac{1}{2}y^2$ | $\frac{1}{2}y^2$ |
| $\lvert C_T\rvert^2$ | $16(1 - y/2)^2$ | $16(1 - y/2)^2$ |
| $\mathrm{Re}[(C_S^* + C_P^*)C_T]$ | $-4y(1 - y/2)$ | $-4y(1 - y/2)$ |
| $\lvert C_{V-A}\rvert^2$ | $4$ | $4(1 - y)^2$ |
| $\lvert C_{V+A}\rvert^2$ | $4(1 - y)^2$ | $4$ |

TABLE 6.3
Mean Fractional Energy Loss for $\nu_\mu e$ and $\bar{\nu}_\mu e$ Elastic Scattering

| Coupling | $\langle y\rangle_{\nu_\mu e}$ | $\langle y\rangle_{\bar{\nu}_\mu e}$ |
|---|---|---|
| $\lvert C_S\rvert^2 + \lvert C_P\rvert^2$ | 0.75 | 0.75 |
| $\lvert C_T\rvert^2$ | 0.393 | 0.393 |
| $\mathrm{Re}[(C_S^* + C_P^*)C_T]$ | 0.625 | 0.625 |
| $\lvert C_{V-A}\rvert^2$ | 0.5 | 0.25 |
| $\lvert C_{V+A}\rvert^2$ | 0.25 | 0.5 |

As was the case for charge currents, the energy-loss distribution discriminates among the various forms for the interaction. Indeed, an interesting parameter of the neutral-current interaction is the mean value of the energy-loss parameter,

$$\langle y\rangle = \frac{\int_0^1 dy\, y\, (d\sigma/dy)}{\int_0^1 dy\, (d\sigma/dy)}, \tag{6.4.5}$$

the characteristic values of which are shown in table 6.3. Just as for the charged-current interactions, however, a "confusion theorem" holds: for the spin-averaged cross section, it is always possible to find a combination of S, P, and T couplings that reproduces the distribution due to an arbitrary superposition of $V \pm A$. The observation of $\gamma$-$Z^0$ interference effects in electron–positron annihilations and in electron–nucleon scattering provided early evidence for the $V \pm A$ interpretation, and this was generally assumed in the "model-independent" analyses of neutral-current interactions.

The leptonic neutral-current interactions for which experiments have been carried out include the neutrino reactions

$$\left.\begin{array}{l} \nu_\mu e \to \nu_\mu e \\ \bar{\nu}_\mu e \to \bar{\nu}_\mu e \\ \nu_e e \to \nu_e e \\ \bar{\nu}_e e \to \bar{\nu}_e e \end{array}\right\} \tag{6.4.6}$$

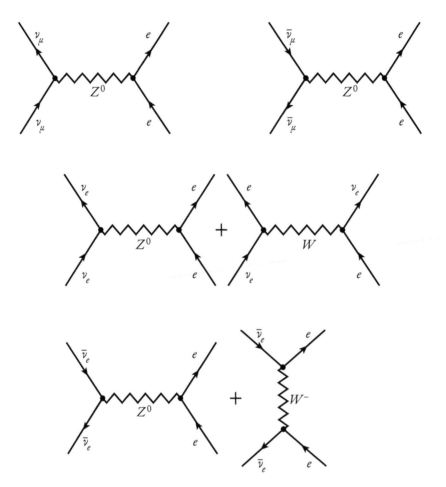

**Figure 6.19.** Lowest-order contributions to (anti)neutrino–electron elastic scattering in the electroweak theory.

and the electron–positron storage ring reactions

$$
\left.
\begin{aligned}
e^+ e^- &\rightarrow \mu^+ \mu^- \\
e^+ e^- &\rightarrow \tau^+ \tau^- \\
e^+ e^- &\rightarrow e^+ e^-
\end{aligned}
\right\} .
\tag{6.4.7}
$$

We deal with these two classes in turn.

Feynman diagrams for the neutrino–electron scattering reactions are shown in figure 6.19. As has already been remarked, the reactions initiated by muon neutrinos are pure neutral-current processes. At low energies (the condition $s \approx 2mE \ll M_Z^2$ is met for any conceivable fixed-target accelerator experiment), it is, therefore, possible to write down the cross sections at once by referring to the Feynman rules of figure 6.19 and the general results of table 6.2. They are

$$\frac{d\sigma}{dy}(\nu_\mu e \to \nu_\mu e) = \frac{4mE}{\pi} \frac{G_F^2}{8} \left[ L_e^2 + R_e^2(1-y)^2 \right]$$

$$= \frac{G_F^2 mE}{2\pi} \left[ (2x_W - 1)^2 + 4x_W^2(1-y)^2 \right], \qquad (6.4.8)$$

where we have introduced the useful notation

$$x_W \equiv \sin^2 \theta_W, \qquad (6.4.9)$$

and

$$\frac{d\sigma}{dy}(\bar{\nu}_\mu e \to \bar{\nu}_\mu e) = \frac{4mE}{\pi} \frac{G_F^2}{8} \left[ L_e^2(1-y)^2 + R_e^2 \right]$$

$$= \frac{G_F^2 mE}{2\pi} \left[ (2x_W - 1)^2(1-y)^2 + 4x_W^2 \right]. \qquad (6.4.10)$$

The total cross sections are, therefore,

$$\sigma(\nu_\mu e \to \nu_\mu e) = \frac{G_F^2 mE}{2\pi} \left( L_e^2 + \frac{R_e^2}{3} \right)$$

$$= \frac{G_F^2 mE}{2\pi} \left[ (2x_W - 1)^2 + \frac{4x_W^2}{3} \right] \qquad (6.4.11)$$

and

$$\sigma(\bar{\nu}_\mu e \to \bar{\nu}_\mu e) = \frac{G_F^2 mE}{2\pi} \left( \frac{L_e^2}{3} + R_e^2 \right)$$

$$= \frac{G_F^2 mE}{2\pi} \left[ \frac{(2x_W - 1)^2}{3} + 4x_W^2 \right]. \qquad (6.4.12)$$

These are plotted as functions of $x_W = \sin^2 \theta_W$ in figure 6.20.

We now investigate the changes to the charged-current processes $\nu_e e \to \nu_e e$ and $\bar{\nu}_e e \to \bar{\nu}_e e$ that are brought about by the introduction of neutral currents. The charged-current cross sections were given in (6.1.65)–(6.1.68). The full calculation is instructive and is left as an exercise (problem 6.6), but the steps are summarized here. At low energies corresponding to the point-coupling limit of $s \ll (M_W^2, M_Z^2)$ and, after a Fierz reordering of the $W$-boson exchange term, the Feynman amplitude for $\nu_e e$ scattering is seen to be identical in form to that for $\nu_\mu e$ scattering, with the replacement

$$L_e \to L_e + 2. \qquad (6.4.13)$$

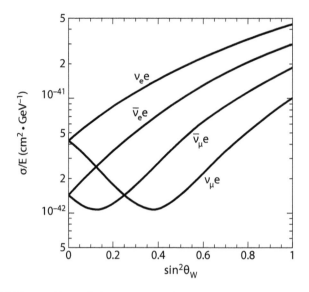

**Figure 6.20.** Cross sections for (anti)neutrino–electron scattering as functions of the weak mixing parameter.

Without further calculation, we may, therefore, write

$$\frac{d\sigma}{dy}(\nu_e e \to \nu_e e) = \frac{G_F^2 mE}{2\pi}\left[(L_e + 2)^2 + R_e^2(1-y)^2\right]$$

$$= \frac{G_F^2 mE}{2\pi}\left[(2x_W + 1)^2 + 4x_W^2(1-y)^2\right]. \qquad (6.4.14)$$

and

$$\sigma(\nu_e e \to \nu_e e) = \frac{G_F^2 mE}{2\pi}\left[(L_e + 2)^2 + \frac{R_e^2}{3}\right]$$

$$= \frac{G_F^2 mE}{2\pi}\left[(2x_W + 1)^2 + \frac{4x_W^2}{3}\right]. \qquad (6.4.15)$$

In precisely the same way, the cross section for $\bar{\nu}_e e$ scattering can be obtained from that for $\bar{\nu}_\mu e$ scattering, with the results

$$\frac{d\sigma}{dy}(\bar{\nu}_e e \to \bar{\nu}_e e) = \frac{G_F^2 mE}{2\pi}\left[(L_e + 2)^2(1-y)^2 + R_e^2\right]$$

$$= \frac{G_F^2 mE}{2\pi}\left[(2x_W + 1)^2(1-y)^2 + 4x_W^2\right] \qquad (6.4.16)$$

and

$$\sigma(\bar{\nu}_e e \to \bar{\nu}_e e) = \frac{G_F^2 mE}{2\pi} \left[ \frac{(L_e + 2)^2}{3} + R_e^2 \right]$$

$$= \frac{G_F^2 mE}{2\pi} \left[ \frac{(2x_W + 1)^2}{3} + 4x_W^2 \right]. \tag{6.4.17}$$

These cross sections are also plotted in figure 6.20.

Before proceeding to a discussion of the available data, it is profitable to exploit further the general manner in which we have arrived at the cross sections. It has become traditional to express the measured cross sections in terms of the neutral-current parameters

$$\left. \begin{array}{l} a_e = \frac{1}{2}(L_e - R_e) \\ v_e = \frac{1}{2}(L_e + R_e) \end{array} \right\}. \tag{6.4.18}$$

In the standard $SU(2)_L \otimes U(1)_Y$ theory, these parameters take on the values

$$\left. \begin{array}{l} a_e = -\frac{1}{2} \\ v_e = -\frac{1}{2} + 2x_W \end{array} \right\}, \tag{6.4.19}$$

whereas in the V–A picture of section 6.1 and 6.2, $a_e = v_e = 0$. In terms of the newly defined parameters, the neutrino–electron scattering cross sections are

$$\sigma(\nu_\mu e \to \nu_\mu e) = \frac{2G_F^2 mE}{\pi} \left( \frac{a^2 + av + v^2}{3} \right), \tag{6.4.20}$$

$$\sigma(\bar{\nu}_\mu e \to \bar{\nu}_\mu e) = \frac{2G_F^2 mE}{\pi} \left( \frac{a^2 - av + v^2}{3} \right), \tag{6.4.21}$$

$$\sigma(\nu_e e \to \nu_e e) = \frac{2G_F^2 mE}{\pi} \left( \frac{a^2 + av + v^2}{3} + a + v + 1 \right), \tag{6.4.22}$$

$$\sigma(\bar{\nu}_e e \to \bar{\nu}_e e) = \frac{2G_F^2 mE}{\pi} \left( \frac{a^2 - av + v^2 + a + v + 1}{3} \right), \tag{6.4.23}$$

where subscripts on $a$ and $v$ have been suppressed.

Experimental results may be represented as ellipses on the $a$-$v$ plane. The $\nu_\mu e$ and $\bar{\nu}_\mu e$ ellipses, which are centered at the origin, are perpendicular and intersect at four places. The four intersections reflect a $v \leftrightarrow a$ ambiguity and a sign ambiguity. The sign ambiguity can be resolved by considering the $\nu_e e$ or $\bar{\nu}_e e$ results as well, because expressions (6.4.22) and (6.4.21) contain terms linear in $(v + a)$. The vector–axial-vector (or $R_e \leftrightarrow -R_e$) ambiguity persists and cannot be resolved with neutrino cross sections alone. Experimental measurements of the $\bar{\nu}_e e$, $\nu_\mu e$, and $\bar{\nu}_\mu e$ cross sections from the early days of neutral-current studies are shown in figure 6.21. These results admit two solutions, corresponding approximately to $(a = 0, v = -\frac{1}{2})$ and to $(a = -\frac{1}{2}, v = 0)$. The latter is consistent with the expectations of the

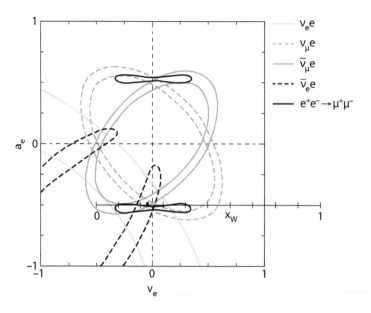

**Figure 6.21.** Constraints (ca. 1987) on the neutral-current parameters $a_e$ and $v_e$ from leptonic interactions (adapted from ref. [20]). The tiny dot near $x_W = 0.22$ shows the precision with which $e^+e^-$ experiments at the $Z^0$ pole determined the couplings.

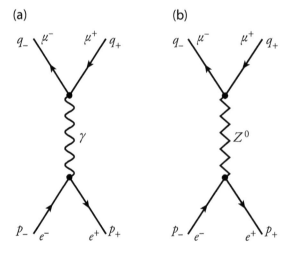

**Figure 6.22.** Lowest-order contributions to the reaction $e^+e^- \rightarrow \mu^+\mu^-$ in the electroweak theory.

electroweak theory, with $x_W \approx \frac{1}{4}$. Far more precise data are available now for all four $ve$ reactions [21].

The $e^+e^-$ storage-ring reactions (6.4.7) offer the possibility of resolving the $v \leftrightarrow a$ ambiguity. For electron-positron annihilations into $\mu$-pairs or $\tau$-pairs, the amplitude is given by the two Feynman graphs in figure 6.22, corresponding to

$s$-channel $\gamma$ and $Z^0$ poles. (The scattering amplitude for the Bhabha process $e^+e^- \rightarrow e^+e^-$ also receives contributions from $\gamma$ and $Z^0$ poles in the $t$-channel.) The matrix element for $e^+e^- \rightarrow \mu^+\mu^-$, which serves as a useful prototype for all the inelastic processes, is given by

$$\mathcal{M}(e^+e^- \rightarrow \mu^+\mu^-) = -ie^2\bar{u}(\mu, q_-)\gamma_\lambda Q_\mu v(\mu, q_+)\frac{g^{\lambda\nu}}{s}\bar{v}(e, p_+)\gamma_\nu u(e, p_-)$$

$$+ \frac{i}{2}\left(\frac{G_F M_Z^2}{\sqrt{2}}\right)\bar{u}(\mu, q_-)\gamma_\lambda[R_\mu(1 + \gamma_5) + L_\mu(1 - \gamma_5)]v(\mu, q_+)$$

$$\times \frac{g^{\lambda\nu}}{s - M_Z^2}\bar{v}(e, p_+)\gamma_\nu[R_e(1 + \gamma_5) + L_e(1 - \gamma_5)]u(e, p_-),$$

$$(6.4.24)$$

where (upon neglect of $m^2$ and $\mu^2$ compared to $s$) the kinematic invariants may be expressed in terms of c.m. variables as

$$\left.\begin{array}{l} p_+ \cdot p_- = q_+ \cdot q_- = \frac{1}{2}s \\[4pt] p_- \cdot q_- = p_+ \cdot q_+ = \frac{1}{4}s(1 - z) \\[4pt] p_+ \cdot q_- = p_- \cdot q_+ = \frac{1}{4}s(1 + z) \end{array}\right\}, \qquad (6.4.25)$$

with

$$z = \cos\theta_{\text{c.m.}}. \qquad (6.4.26)$$

The muon charge $Q_\mu = -1$ has been inserted explicitly to permit later generalizations. A straightforward calculation leads to the spin-averaged cross section,

$$\frac{d\sigma}{dz}(e^+e^- \rightarrow \mu^+\mu^-) = \frac{\pi\alpha^2 Q_\mu^2}{2s}(1 + z^2)$$

$$- \frac{\alpha Q_\mu G_F M_Z^2(s - M_Z^2)}{8\sqrt{2}[(s - M_Z^2)^2 + M_Z^2\Gamma^2]}$$

$$\times [(R_e + L_e)(R_\mu + L_\mu)(1 + z^2) + 2(R_e - L_e)(R_\mu - L_\mu)z]$$

$$+ \frac{G_F^2 M_Z^4 s}{64\pi[(s - M_Z^2)^2 + M_Z^2\Gamma^2]}$$

$$\times [(R_e^2 + L_e^2)(R_\mu^2 + L_\mu^2)(1 + z^2) + 2(R_e^2 - L_e^2)(R_\mu^2 - L_\mu^2)z],$$

$$(6.4.27)$$

where the $Z^0$ propagator has been replaced by the form appropriate for an unstable particle of total width $\Gamma$. Written in this form, the spin-averaged cross section can easily be transformed into the cross section for definite initial- or final-state helicities. The cross section for transversely polarized colliding beams is also readily obtained using density matrix techniques. The first term in (6.4.27) is simply the electromagnetic contribution, which was calculated in problem 1.5. The second is the weak-electromagnetic interference term, and the last represents the effect of the $Z^0$ diagram alone.

At energies far below the $Z^0$-boson mass, the final term is negligible. A quantity of particular experimental interest is the forward-backward asymmetry,

$$A \equiv \frac{\int_0^1 dz \, d\sigma/dz - \int_{-1}^0 dz \, d\sigma/dz}{\int_{-1}^1 dz \, d\sigma/dz}. \tag{6.4.28}$$

In the low-energy limit, the asymmetry is given approximately by

$$\lim_{s/M_Z^2 \to 0} A = \frac{3G_F s}{16\pi\alpha Q_\mu \sqrt{2}}(R_e - L_e)(R_\mu - L_\mu)$$

$$\approx -6.7 \times 10^{-5} \left(\frac{s}{1 \text{ GeV}^2}\right)(R_e - L_e)(R_\mu - L_\mu)$$

$$\approx -6.7 \times 10^{-5} \left(\frac{s}{1 \text{ GeV}^2}\right), \tag{6.4.29}$$

where the last line follows because in the $SU(2)_L \otimes U(1)_Y$ electroweak theory,

$$(R_e - L_e) = (R_\mu - L_\mu) = +1. \tag{6.4.30}$$

At an energy of $\sqrt{s} = 50$ GeV, we should expect an asymmetry $A \approx -17\%$ at this rough approximation.

Without specializing the result to the $SU(2)_L \otimes U(1)_Y$ theory, but assuming electron–muon universality, we may use the definition (6.4.18) to write

$$A(s \ll M_Z^2) = -\frac{3G_F s a^2}{4\pi\alpha\sqrt{2}}, \tag{6.4.31}$$

whereupon a measurement of the asymmetry may resolve the $v \leftrightarrow a$ ambiguity in the model-independent determination of neutral-current parameters from neutrino–electron scattering experiments. The fact that the axial coupling enters quadratically in (6.4.31) serves as a reminder, should one be needed, that the forward-backward asymmetry is not parity violating.

Extensive studies of the reactions $e^+e^- \to \mu^+\mu^-$ and $e^+e^- \to \tau^+\tau^-$ have been carried out in the continuum region well below the $Z^0$-pole [22]. These permit us to extract the quantities $a_e a_\mu$ and $a_e a_\tau$. According to the $SU(2)_L \otimes U(1)_Y$ electroweak theory [cf. (6.4.19)], we expect $a_e a_\mu = a_e a_\tau = 0.25$. From the data tabulated in Ref. [22], we find, for example at $\sqrt{s} = 29$ GeV,

$$a_e a_\mu = 0.23 \pm 0.03; \quad a_e a_\tau = 0.21 \pm 0.05, \tag{6.4.32}$$

in agreement with standard-model expectations. Assuming that the leptonic couplings of the $Z^0$ boson are universal, we conclude that $a_\ell \approx \pm 0.5$. When combined

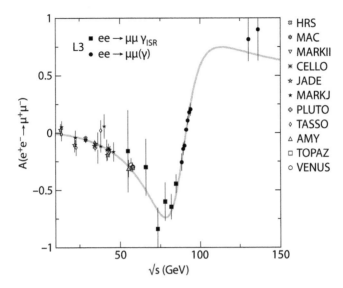

**Figure 6.23.** Forward-backward asymmetries of muon-pair production compared with the electroweak theory, from [23]. In addition to high-energy measurements by the L3 Collaboration at LEP, lower-energy measurements from experiments at the PEP, PETRA, and TRISTAN electron–positron storage rings are shown.

with the $\nu e$-scattering constraints displayed in figure 6.21, the asymmetry measurements rule decisively in favor of the $(a \approx -\frac{1}{2}, v \approx 0)$ solution for the $\nu e$ cross sections. Measurements of the forward-backward asymmetry in the reaction $e^+e^- \to \mu^+\mu^-$ are compared with the electroweak theory in figure 6.23.

It is amusing to note that the unique solution corresponds to the chiral couplings

$$L \approx -\tfrac{1}{2}, \quad R \approx \tfrac{1}{2}. \tag{6.4.33}$$

Now expand the electroweak gauge group to $SU(2)_L \otimes SU(2)_R \otimes U(1)_Y$, and interpret

$$L = 2x_W + \tau_3^{(L)}, \quad R = 2x_W + \tau_3^{(R)}. \tag{6.4.34}$$

Using the value $x_W \approx \frac{1}{4}$ (inferred from the value of $M_Z$, for example) we may measure the weak isospin of the charged leptons as

$$\tau_3^{(L)} = -1, \quad \tau_3^{(R)} = 0. \tag{6.4.35}$$

The $e^+e^-$ annihilation channel is an extremely rich area for the study of electroweak effects, and an enormous amount of physics is contained in the "master formula" (6.4.27). We shall return to a fuller discussion of its implications in section 7.3, following the incorporation of hadrons into the electroweak theory.

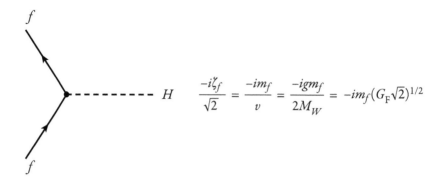

**Figure 6.24.** Feynman rule for Higgs-boson interaction with fermions.

## 6.5 THE HIGGS BOSON: A FIRST LOOK

The introduction of a complex doublet of self-interacting auxiliary scalar fields has given rise to spontaneous breaking of the local $SU(2)_L \otimes U(1)_Y$ gauge symmetry. This has led, by means of the Higgs mechanism, to several agreeable consequences: the gauge bosons associated with the weak interactions have acquired masses, as have the charged fermions [24], and a partial unification of the weak and electromagnetic interactions has been achieved. In place of three independent gauge coupling constants $e$, $g$, and $g'$ associated with the $U(1)_{EM}$, $SU(2)_L$, and $U(1)_Y$ gauge groups, there are now only two—the electric charge $e$ and the weak mixing parameter $\sin^2 \theta_W$. (Further reduction of the number of independent parameters will be discussed in chapter 9.) The resulting theory has an improved high-energy behavior, in that amplitudes corresponding to tree diagrams (Feynman diagrams without loops) are nondivergent, and indeed the theory can be made renormalizable (cf. §6.8). As we have begun to see, the low-energy phenomenology of the model is in excellent accord with experimental findings. An uninvited guest at these proceedings is the Higgs boson, the massive physical scalar particle that remains after spontaneous symmetry breaking. This section is devoted to a first look at the properties of this particle, which is an essential element of the standard electroweak theory. We shall discuss the interactions of the Higgs boson with fermions and gauge bosons, examine its role in divergence cancellation, consider prospects for detection of the Higgs boson, and seek constraints on the Higgs-boson mass. We defer a discussion of dynamical-symmetry-breaking alternatives to the standard Higgs scenario to section 8.9.

The interaction of the Higgs boson with fermions has already been given in the Yukawa Lagrangian (6.3.41), which leads to the Feynman rule shown in figure 6.24. We may at once calculate the rate for the decay

$$H \to f\bar{f}. \tag{6.5.1}$$

The invariant amplitude is given by

$$\mathcal{M} = -im_f (G_F \sqrt{2})^{1/2} \, \bar{u}(f, p_1) v(f, p_2), \tag{6.5.2}$$

where the c.m. momenta of the emitted fermions may be written as

$$p_1 = (\tfrac{1}{2} M_H; q \sin\theta, 0, q \cos\theta) \\ p_2 = (\tfrac{1}{2} M_H; -q \sin\theta, 0, -q \cos\theta) \Bigg\}, \tag{6.5.3}$$

where

$$q = \sqrt{\tfrac{1}{4} M_H^2 - m_f^2}. \tag{6.5.4}$$

Consequently, we compute

$$\begin{aligned} |\mathcal{M}|^2 &= G_F m_f^2 \sqrt{2} \, \mathrm{tr}[\not{p}_2 \not{p}_1 - m_f^2] \\ &= 4 G_F m_f^2 \sqrt{2} (p_1 \cdot p_2 - m_f^2) \\ &= 2 G_F M_H^2 m_f^2 \sqrt{2} (1 - 4 m_f^2 / M_H^2). \end{aligned} \tag{6.5.5}$$

The differential decay rate is, according to (B.1.4),

$$\frac{d\Gamma}{d\Omega} = \frac{|\mathcal{M}|^2}{64 \pi^2 M_H} = \frac{G_F m_f^2 M_H}{16 \pi^2 \sqrt{2}} \cdot N_c \cdot \left(1 - \frac{4 m_f^2}{M_H^2}\right)^{3/2}, \tag{6.5.6}$$

which is isotropic as it must be for the decay of a spinless particle. Here we have inserted the color factor $N_c = (1, 3)$ for (leptons, quarks), anticipating that the Higgs mechanism gives mass to quarks as well as leptons (cf. §7.1). The $f \bar{f}$ partial width is

$$\Gamma(H \to f \bar{f}) = \frac{G_F m_f^2 M_H}{4 \pi \sqrt{2}} \cdot N_c \cdot \left(1 - \frac{4 m_f^2}{M_H^2}\right)^{3/2}, \tag{6.5.7}$$

which is proportional to $N_c m_f^2 M_H$ as the Higgs-boson mass becomes large compared to the fermion mass. The dominant decay mode of a light ($M_H \lesssim 2 M_W$) Higgs boson is, therefore, into pairs of the most massive fermion that is kinematically accessible.

The Feynman rules for the interactions of Higgs bosons with gauge bosons and for Higgs-boson self-interactions may be obtained by expanding $\mathcal{L}_{\text{scalar}}$ (6.3.17) in terms of the U-gauge fields, as we did partially in (6.3.33). The Feynman rules are shown in figure 6.25. As a first application, the rates for decay of a heavy Higgs boson into a pair of intermediate vector bosons are given by (cf. problem 6.8)

$$\Gamma(H \to W^+ W^-) = \frac{G_F M_H^3}{32 \pi \sqrt{2}} \frac{(1-x)^{1/2}}{x} (3x^2 - 4x + 4), \tag{6.5.8}$$

$$\Gamma(H \to Z^0 Z^0) = \frac{G_F M_H^3}{64 \pi \sqrt{2}} \frac{(1-x')^{1/2}}{x'} (3x'^2 - 4x' + 4), \tag{6.5.9}$$

where $x = 4 M_W^2 / M_H^2$ and $x' = 4 M_Z^2 / M_H^2 = x / \cos^2 \theta_W$. The rates for decays into weak-boson pairs are asymptotically proportional to $M_H^3$ and $\tfrac{1}{2} M_H^3$, respectively.

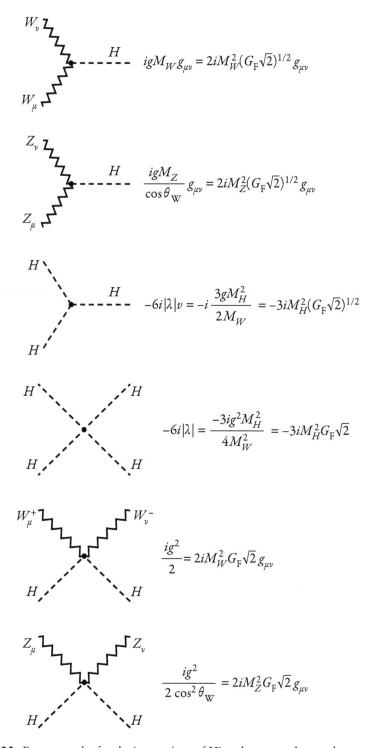

**Figure 6.25.** Feynman rules for the interactions of Higgs bosons and gauge bosons.

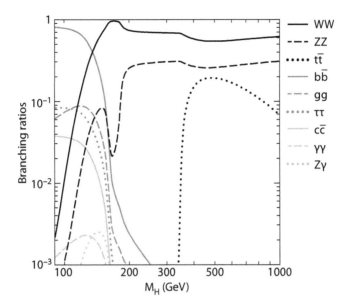

**Figure 6.26.** Branching fractions for prominent decay modes of the standard-model Higgs boson (from Ref. [25]).

In the final factors of (6.5.8) and (6.5.9), $2x^2$ and $2x'^2$, respectively, arise from decays into transversely polarized gauge bosons. The dominant decays for large $M_H$ are into pairs of longitudinally polarized weak bosons.

Branching fractions for decay modes that may hold promise for the detection of a Higgs boson are displayed in figure 6.26. In addition to the $f\bar{f}$ and $VV$ modes that arise at tree level, the plot includes the $\gamma\gamma$, $Z\gamma$, and two-gluon modes that proceed through loop diagrams. The rare $\gamma\gamma$ channel offers an important target for LHC experiments, if the Higgs boson is light, because the relatively benign backgrounds may be overcome by fine resolution.

Below the $W^+W^-$ threshold, the standard-model Higgs boson is rather narrow, with $\Gamma(H \to \text{all}) \lesssim 1$ GeV. Far above the threshold for decay into gauge-boson pairs, the total width is proportional to $M_H^3$. As its mass increases toward 1 TeV, the Higgs boson becomes highly unstable, with a perturbative width approaching its mass. It would, therefore, be observed as an enhanced rate rather than a distinct resonance. The Higgs-boson total width is plotted as a function of $M_H$ in figure 6.27.

Because a light Higgs boson would appear as a narrow resonance, it is worth asking whether it might be discovered or studied in detail at a "Higgs factory" in the formation reaction

$$e^+e^- \to H \to \text{all}, \tag{6.5.10}$$

shown in figure 6.28(a). The cross section is (cf. problem 6.7)

$$\sigma(e^+e^- \to H) = \frac{4\pi\Gamma(H \to e^+e^-)\Gamma(H \to \text{all})}{(s - M_H^2)^2 + M_H^2\Gamma(H \to \text{all})^2}, \tag{6.5.11}$$

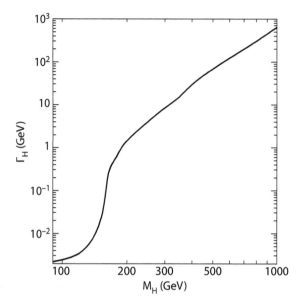

**Figure 6.27.** Total width of the standard-model Higgs boson vs. mass (from Ref. [25]).

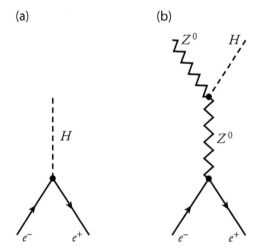

**Figure 6.28.** (a) Feynman diagram for direct-channel Higgs-boson formation in $e^+e^-$ annihilations. (b) "Higgs-strahlung" diagram for associated production of a Higgs boson and $Z^0$-boson in $e^+e^-$ annihilations.

where $\sqrt{s}$ is the $e^+e^-$ c.m. energy. The peak cross section (at $s = M_H^2$) is

$$\sigma_{\text{peak}}(e^+e^- \to H) = \frac{4\pi}{M_H^2} \cdot \frac{\Gamma(H \to e^+e^-)}{\Gamma(H \to \text{all})}$$

$$= 4.89 \times 10^{-31} \text{ cm}^2 \left[\frac{100 \text{ GeV}}{M_H}\right]^2 \cdot \frac{\Gamma(H \to e^+e^-)}{\Gamma(H \to \text{all})}.$$

$$(6.5.12)$$

The tiny branching fraction $\Gamma(H \to e^+e^-)/\Gamma(H \to \text{all}) \lesssim 10^{-8}$, which is owed to the small electron mass, makes this a dismal prospect, even for a relatively light Higgs boson. It is worth noting that cross sections for the reaction $\mu^+\mu^- \to H$ are larger by a factor $(m_\mu/m_e)^2 = 42{,}750$.

Because the Higgs-boson couplings with gauge bosons are not suppressed by small fermion masses, reactions involving gauge bosons provide favorable means for Higgs-boson production. The reaction

$$e^+e^- \to HZ^0, \qquad (6.5.13)$$

which proceeds via the $s$-channel formation of a virtual $Z^0$, as shown in figure 6.28(b), is a popular example. The cross section is

$$\sigma(e^+e^- \to HZ) = \frac{\pi\alpha^2}{24}\left(\frac{2K}{\sqrt{s}}\right)\frac{(K^2+3M_Z^2)}{(s-M_Z^2)^2+M_Z^2\Gamma_Z^2}\frac{(1-4x_W+8x_W^2)}{x_W^2(1-x_W)^2}, \qquad (6.5.14)$$

where, as usual, $x_W = \sin^2\theta_W$ and $K$ is the c.m. momentum of the emerging particles. At very high energies, for which $K \to \sqrt{s}/2$, the ratio

$$\frac{\sigma(e^+e^- \to HZ)}{\sigma(e^+e^- \to \mu^+\mu^-)} \to \frac{(1-4x_W+8x_W^2)}{128x_W^2(1-x_W)^2} \qquad (6.5.15)$$

approaches from below an asymptotic value of 0.142 for $x_W = 0.2122$. Other mechanisms for Higgs-boson production can be found in the resources cited at the end of this chapter, and we will revisit the search for the Higgs boson in section 7.9.

## 6.6 THE HIGGS BOSON, ASYMPTOTIC BEHAVIOR, AND THE 1-TeV SCALE

We saw at the end of section 6.3 that gauge symmetry enforces a cooperation among the $\nu$-. $\gamma$-, and $Z^0$-exchange contributions to the reaction $e^+e^- \to W_0^+ W_0^-$ that cancels the unacceptable ($\propto s$) high-energy behavior of the "normal-fermion-helicity" amplitudes that enter when the electron mass is neglected. This is not the whole story, however. Massive electrons may be found in the "wrong" helicity state, and the sum of the three amplitudes in figure 6.17 yields a softer divergence $\propto m_e\sqrt{s}$ for the production of longitudinal gauge bosons. The contribution to the $W^+W^-$ pair-production cross section is utterly negligible in the LEP energy range of the data shown in figure 6.18, but the unbounded growth with energy is a problem in principle. This residual divergence is precisely canceled by the Higgs-boson graph of figure 6.29, as problem 6.12 will show. If the Higgs boson did not exist, in a world with massive electrons, we should have to invent something very much like it. From the point of view of divergence cancellations in $S$-matrix theory, the $Hf\bar{f}$ coupling must be proportional to $m_f$, because "wrong-helicity" amplitudes are always proportional to $m_f$.

Let us note some of the interrelations that we have uncovered in our discussions of high-energy behavior. Without spontaneous symmetry breaking and the Higgs

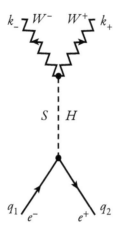

**Figure 6.29.** Higgs-boson-exchange contribution to the reaction $e^+e^- \to W^+W^-$.

boson, there would be no longitudinal gauge bosons and, thus, no extreme divergence difficulties. (Nor would there be a viable low-energy phenomenology of weak interactions.) The most severe divergences are eliminated by the gauge structure of the couplings among gauge bosons and leptons. A lesser, but still potentially fatal, divergence arises because the electron has acquired mass—thanks to the Higgs mechanism. Spontaneous symmetry breaking provides its own cure by supplying a Higgs boson to remove the last divergence. Here the common origin of the electron mass and the $He\bar{e}$ interaction in the Yukawa Lagrangian (6.3.41) is of crucial importance. We cannot help being impressed!

In spite of this tightly woven structure, some ambiguity persists. Nothing in the formulation of the $SU(2)_L \otimes U(1)_Y$ theory specifies the mass of the Higgs boson, and none of the applications to conventional processes that we have considered depends in any direct way upon the value of $M_H$. It may, therefore, appear that this is a completely free parameter of the theory, and this is indeed nearly the case. However, by imposing certain requirements of internal consistency upon the theory, we may narrow the range of possibilities somewhat. We shall identify an important general constraint at once and explore other restrictions in section 7.10.

It is relatively straightforward to obtain a sort of upper bound on $M_H$ by imposing the requirement that the theory make perturbative sense and that the amplitudes calculated in tree approximation satisfy partial-wave unitarity [26]. In gauge theories, the asymptotic growth of partial-wave amplitudes is regulated [27], and all amplitudes are *at worst* in logarithmic violation of unitarity in lowest order. When the calculable higher-order corrections are applied, these amplitudes become properly finite. Spontaneously broken theories present an exceptional case: interactions in which the Higgs boson plays an important role.

For example, the quartic Higgs-boson self-coupling is seen in figure 6.25 to be proportional to $G_F M_H^2$. Accordingly, it may happen that for large values of $M_H$ certain partial-wave amplitudes, although "asymptotically safe," may exceed the numerical bounds imposed by unitarity.

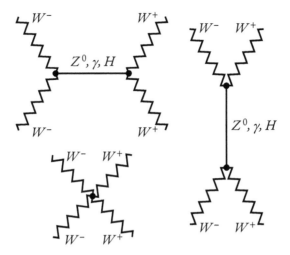

**Figure 6.30.** Feynman diagrams (in unitary gauge) for the reaction $W^+ W^- \to W^+ W^-$. A thick straight line represents the $Z^0$-boson, photon, or Higgs-boson propagator.

We can see how this comes about and also acquire a deeper understanding of how divergences are compensated in the electroweak theory, by considering the scattering of (longitudinal) gauge bosons and Higgs bosons. The most intricate case concerns the reaction $W_0^+ W_0^- \to W_0^+ W_0^-$, which proceeds at tree level by the seven Feynman diagrams sketched in figure 6.30. (The subscript 0 denotes the longitudinal polarization state.)

It is convenient to classify the cancellations among the tree graphs according to the power of $(q/M_W)$ that enters, where $q$ is the c.m. momentum of the gauge bosons. The high-energy behavior of the individual graphs in figure 6.30 is at worst $\sim (q/M_W)^4$. Accordingly, we may characterize the contribution of each graph to the $J$th partial wave in the form

$$a_J = A \left( \frac{q}{M_W} \right)^4 + B \left( \frac{q}{M_W} \right)^2 + C, \qquad (6.6.1)$$

where the partial-wave amplitude $a_J$ is defined through

$$\mathcal{M}(s, t) = 16\pi \sum_J (2J + 1) a_J(s) P_J(\cos\theta). \qquad (6.6.2)$$

We will refer to the coefficients in (6.6.1) as $A$-, $B$-, and $C$-forces.

The divergent behavior of the graphs in figure 6.30 is confined to the $J = 0$, 1, and 2 partial waves. In the $J = 2$ partial wave, an $A$-force from the contact term is canceled by an opposite contribution from the $t$-channel $(Z^0, \gamma)$ exchanges. In the $J = 1$ partial wave, an $A$-force from the contact graph is balanced by contributions from the $s$-channel and $t$-channel $(Z^0, \gamma)$ exchanges. In the $J = 0$ partial wave at order $(q/M_W)^4$, the contact-term contribution is canceled by an equal and opposite contribution from the $t$-channel $(Z^0, \gamma)$ exchanges. As we observed for the reaction $e^+e^- \to W_0^+ W_0^-$, gauge symmetry alone enforces the cancellation of the most-divergent pieces of the individual diagrams.

The cancellation of $B$-forces in the $J = 2$ partial wave again is due to a balance between the contact term and the $t$-channel $(Z^0, \gamma)$ exchanges. In the remaining partial waves, the Higgs boson plays an essential role. In the $J = 1$ partial wave, the $B$-forces from the contact term, the $t$-channel $(Z^0, \gamma)$ terms, and the $s$-channel $(Z^0, \gamma)$ exchanges leave a remainder that is erased by the contribution from the $t$-channel Higgs-boson exchange. In the $J = 0$ partial wave, $B$-forces from the contact term and $t$-channel $(Z^0, \gamma)$ exchange are compensated by the contributions from $s$-channel and $t$-channel Higgs-boson exchanges. Again in parallel to what we observed for the reaction $e^+e^- \to W_0^+ W_0^-$ at the beginning of this section, the elimination of subleading divergences involves the Higgs boson.

Because the longitudinal components of $W^\pm$ and $Z^0$ have their origin in the auxiliary scalar fields, it is informative to consider the problem of gauge-boson scattering at high energies as a phenomenon of the Higgs sector. Problem 6.13 provides a streamlined look at the interplay that leads to divergence cancellation.

In the limit of immediate interest, for $(s, M_H^2) \gg (M_W^2, M_Z^2)$, the full amplitude arises from the surviving $C$-force contributions of the $s$- and $t$-channel Higgs-boson exchange graphs,

$$\mathcal{M}(W_0^+ W_0^- \to W_0^+ W_0^-) = -\sqrt{2} G_F M_H^2 \left[ \frac{s}{s - M_H^2} + \frac{t}{t - M_H^2} \right]. \qquad (6.6.3)$$

An important clue resides in the factor $M_H^2$ on the right-hand side of (6.6.3), which suggests that the amplitude might become too large, for large values of the Higgs-boson mass. Let us project out the $J = 0$ partial wave, in the limit $(s, M_H^2) \gg (M_W^2, M_Z^2)$:

$$a_0(W_0^+ W_0^- \to W_0^+ W_0^-) = \frac{-G_F M_H^2}{8\pi\sqrt{2}} \left[ 2 + \frac{M_H^2}{s - M_H^2} - \frac{M_H^2}{s} \ln\left(1 + s/M_H^2\right) \right]. \qquad (6.6.4)$$

At energies far above the Higgs-boson pole, the partial-wave amplitude (6.6.4) approaches a constant value,

$$a_0(W_0^+ W_0^- \to W_0^+ W_0^-) \xrightarrow[s \gg M_H^2]{} \frac{-G_F M_H^2}{4\pi\sqrt{2}}. \qquad (6.6.5)$$

Now impose the partial-wave unitarity condition

$$\left| a_0(W_0^+ W_0^- \to W_0^+ W_0^-) \right| \le 1, \qquad (6.6.6)$$

which requires

$$\frac{G_F M_H^2}{4\pi\sqrt{2}} \le 1. \qquad (6.6.7)$$

The condition (6.6.7) is respected, provided that the Higgs-boson mass satisfies

$$M_H^2 \leq \frac{4\pi\sqrt{2}}{G_F} \lesssim 1.5 \text{ TeV}^2. \tag{6.6.8}$$

It is possible to refine the bound (6.6.8) somewhat by considering the consequences of partial-wave unitarity for the four-channel system of

$$W_0^+ W_0^-, \frac{Z_0^0 Z_0^0}{\sqrt{2}}, \frac{HH}{\sqrt{2}}, HZ_0^0, \tag{6.6.9}$$

where the factors of $\sqrt{2}$ account for identical-particle statistics. For these, the $s$-wave amplitudes are all asymptotically constant (i.e., well behaved) and proportional to $G_F M_H^2$. In the high-energy limit,

$$a_0 \xrightarrow[s \gg M_H^2]{} \frac{-G_F M_H^2}{4\pi\sqrt{2}} \cdot \begin{bmatrix} 1 & \frac{1}{\sqrt{8}} & \frac{1}{\sqrt{8}} & 0 \\ \frac{1}{\sqrt{8}} & \frac{3}{4} & \frac{1}{4} & 0 \\ \frac{1}{\sqrt{8}} & \frac{1}{4} & \frac{3}{4} & 0 \\ 0 & 0 & 0 & \frac{1}{2} \end{bmatrix}. \tag{6.6.10}$$

The matrix $a_0$ has eigenvalues $\frac{3}{2}, \frac{1}{2}, \frac{1}{2}, \frac{1}{2}$, in units of $-G_F M_H^2/4\pi\sqrt{2}$. Requiring that the largest eigenvalue respect the partial-wave unitarity condition $|a_0| \leq 1$ yields [28]

$$M_H \leq \left(\frac{8\pi\sqrt{2}}{3G_F}\right)^{1/2} \approx 1 \text{ TeV} \tag{6.6.11}$$

as a condition for perturbative unitarity.

If $M_H$ respects the bound, weak interactions remain weak at all energies, and perturbation theory is everywhere reliable. If the Higgs boson were heavier than 1 TeV, the weak interactions among $W^\pm$, $Z^0$, and $H$ would become strong on the 1-TeV scale, and perturbation theory would break down. At TeV energies, we might then observe multiple production of weak bosons, $W^+ W^-$ resonances, and other phenomena evocative of pion–pion scattering at GeV energies. One way or another, something new—a Higgs boson or strong scattering, if not some other new physics—is to be found in electroweak interactions at energies not much larger than 1 TeV. This conclusion provides a compelling motivation for experiments to be carried out at the Large Hadron Collider.

TABLE 6.4
Some Properties of the Leptons (Ref. [4])

| Lepton | Mass | Lifetime |
|--------|------|----------|
| $\nu_e$ | $< 2$ eV | |
| $e^-$ | $0.510\,998\,910(13)$ MeV | $> 4.6 \times 10^{26}$ y (90% CL) |
| $\nu_\mu$ | $< 0.19$ MeV (90% CL) | |
| $\mu^-$ | $105.658\,367(4)$ MeV | $2.197\,034(21) \times 10^{-6}$ s |
| $\nu_\tau$ | $< 18.2$ MeV (95% CL) | |
| $\tau^-$ | $1\,776.82 \pm 0.16$ MeV | $290.6 \pm 1.0 \times 10^{-15}$ s |

# 6.7 NEUTRINO MIXING AND NEUTRINO MASS

The electroweak theory's idealization that neutrinos are massless did not flow from any robust principle but was inferred from kinematical evidence against measurably large masses, summarized in table 6.4, where CL stands for confidence level. Because fermion mass normally requires linking left-handed and right-handed states, the presumed masslessness of the neutrinos could be captured by the omission of right-handed neutrinos from the theory, consistent with the evidence (Ref. [11]) that neutrinos produced in charged-current interactions are left-handed.

The preferred reaction for measuring the mass of the neutrino (mixture) associated with the electron is tritium $\beta$-decay,

$$^3\text{H} \rightarrow {}^3\text{He} \; e^- \; \bar{\nu}_e, \qquad (6.7.1)$$

for which the endpoint energy is $Q \approx 18.57$ keV. Sources of the spectral distortions that limited the sensitivity of early experiments are absent in modern experiments using free tritium. Nevertheless, detecting a small neutrino mass is enormously challenging: the fraction of counts in the beta spectrum for a massless neutrino that lie beyond the endpoint associated with a 1-eV neutrino is but $2 \times 10^{-13}$ of the total decay rate. The KATRIN experiment [29], which scales up the intensity of the tritium beta source as well as the size and precision of previous experiments by an order of magnitude, is designed to measure the mass of the electron neutrino directly with a sensitivity of 0.2 eV.

Cosmological arguments may also be used to constrain the sum of neutrino masses. If neutrinos are stable over the age of the universe, then according to the standard thermal history, neutrinos should be the most numerous particles in the universe, after the photons of the cosmic microwave background. At present, the number density of each (active) neutrino or antineutrino species would be [30]

$$n_{\nu_i 0} \approx 56 \text{ cm}^{-3}, \qquad (6.7.2)$$

up to a 1% correction from reheating effects [31]. Using the calculated number density (6.7.2), we can deduce the neutrino contribution to the mass density, expressed in units of the critical density

$$\rho_c \equiv \frac{3H_0^2}{8\pi G_N} = 1.05h^2 \times 10^4 \text{ eV cm}^{-3} \approx 5.78 \times 10^3 \text{ eV cm}^{-3}, \tag{6.7.3}$$

where $H_0$ is the Hubble constant, $G_N$ is Newton's constant, and the reduced Hubble constant is $h \equiv H_0/(100 \text{ km s}^{-1}/\text{Mpc}) = 0.742 \pm 0.036$ [32]. Combining (6.7.3) and (6.7.2), we deduce that the condition for neutrinos not to overclose the universe is [33]

$$\sum_i m_{v_i} \lesssim 50 \text{ eV}. \tag{6.7.4}$$

Neutrinos influence fluctuations in the cosmic microwave background, affect the development of density perturbations that set the pattern of large-scale structure, and modulate the baryon acoustic oscillations. Analyses of the interplay between neutrinos and astronomical observables provide bounds on the sum of neutrino masses, such as [34]

$$\sum_i m_{v_i} \lesssim 0.44 \text{ eV}. \tag{6.7.5}$$

Such constraints are less secure than the bound (6.7.4) derived from $\rho_v \lesssim \rho_c$, but they are highly suggestive.

If neutrinos are massless, we have the freedom to identify the mass eigenstates with flavor eigenstates, so the leptonic weak interactions are flavor preserving: $W^- \to \ell^- \bar{v}_\ell$ and $Z \to v_\ell \bar{v}_\ell$, where $\ell = e, \mu, \tau$. A neutrino that moves at the speed of light cannot change flavor between production and subsequent interaction, so massless neutrinos do not mix.

Time passes for massive neutrinos, which do not move at the speed of light. If neutrinos of definite flavor ($v_e, v_\mu, v_\tau$) are superpositions of different mass eigenstates ($v_1, v_2, v_3$), the mass eigenstates evolve in time with different frequencies, and so the superposition changes in time: a beam created as pure flavor $v_\alpha$ evolves into a flavor mixture. The essential phenomenological framework is well known [35]; we will review just enough to put the observations in context. For the initial discussion, it suffices to simplify to the case of two families.

Suppose that two flavor eigenstates $v_\alpha$ and $v_\beta$ are superpositions of the mass eigenstates $v_i$ and $v_j$, such that

$$v_\alpha = v_i \cos\theta + v_j \sin\theta; \quad v_\beta = -v_i \sin\theta + v_j \cos\theta . \tag{6.7.6}$$

An eventual theory of fermion masses should aspire to predict the mixing angle $\theta$; for now, it is to be determined experimentally. After propagating over a distance $L$, a beam created as $v_\alpha$ with energy $E \gg m_i$ has a probability to mutate into $v_\beta$ given by

$$P_{\alpha \to \beta} = \sin^2 2\theta \sin^2 \left( \frac{\Delta m^2 L}{4E} \right), \tag{6.7.7}$$

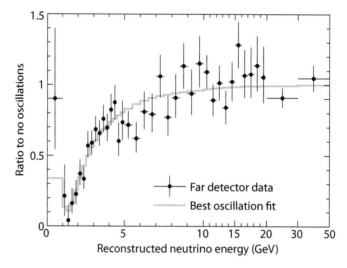

**Figure 6.31.** Ratio of the spectrum of neutrino-induced events in the MINOS far detector, with neutral-current events subtracted, to the null-oscillation prediction (points). The best-fit oscillation expectation is overlaid as the solid curve (from Ref. [39]).

whereas the (survival) probability to remain $\nu_\alpha$ is given by

$$P_{\alpha \to \alpha} = 1 - \sin^2 2\theta \sin^2 \left( \frac{\Delta m^2 L}{4E} \right), \tag{6.7.8}$$

where $\Delta m^2 = m_j^2 - m_i^2$. For practical applications to long-baseline experiments, it is convenient to recast the argument of the oscillatory factor as $1.27 \Delta m^2 L/E$, with $\Delta m^2$ measured in eV$^2$, $L$ in km, and $E$ in GeV.

Extending the observations of the KamiokaNDE experiment [36], Super-K has produced very compelling evidence [37] that $\nu_\mu$ produced in the atmosphere disappear (into other flavors, dominantly $\nu_\tau$) during propagation over long distances. Their evidence, in the form of a zenith-angle distribution and the $L/E$ plot, has been confirmed and refined by the long-baseline accelerator experiments K2K [38] and MINOS. The most comprehensive accelerator-based measurements come from MINOS, some 735 km distant from Fermilab at the Soudan mine in Minnesota. Figure 6.31 shows how their yield of $\nu_\mu$ events, compared to the no-oscillation expectation, varies with beam energy at fixed baseline—just as anticipated in the oscillation scenario. Current evidence is consistent with maximal mixing ($\theta \approx 45°$) and $|\Delta m^2| \approx 2.5 \times 10^{-3}$ eV$^2$ [39]. The OPERA experiment in the Gran Sasso Laboratory, which searches for examples of $\nu_\tau$ appearance in the $\nu_\mu$ beam generated 732 km away at CERN, has reported its first $\tau$ candidates [40].

No appreciable depletion of atmospheric $\nu_e$ has been observed, but flavor change has been observed in neutrinos created in the Sun. Analysis of the solar neutrino experiments is a bit involved, because neutrinos experience matter effects during their journey outward from the production region [41]. (Cf. problem 6.23 to see how resonant oscillation may occur.) The upshot is that the $E_\nu \approx 10$ MeV $^8$Be neutrinos emerge as a nearly pure $\nu_2$ mass eigenstate. Thus, these neutrinos do not oscillate during the passage from Sun to Earth; the flavor change has already happened within the Sun. The lower energy $pp$ ($E_\nu \approx 200$ keV) and $^7$Be ($E_\nu \approx 900$ keV) neutrinos are not strongly affected by matter in the Sun and do undergo oscillations on their Sun–Earth trajectory. The KamLAND reactor experiment [42] has seen a signal for vacuum neutrino oscillations of $\bar{\nu}_e$ that supports and refines the interpretation of the solar neutrino experiments, providing the tightest constraints on the "solar" mass-squared difference. The Borexino experiment [43] has detected the $^7$Be neutrinos, again supporting the oscillation interpretation and parameters.

A deficit of solar neutrinos observed as $\nu_e$, compared with the expectations of the standard solar model, had been a feature of data for some time [44]. Ruling in favor of the neutrino-flavor-change ("oscillation") hypothesis, as opposed to a defective solar model, required the combination of several measurements to demonstrate that the missing $\nu_e$ are present as $\nu_\mu$ and $\nu_\tau$ arriving from the Sun. The solar neutrinos are not energetic enough to initiate $\nu_\mu \to \mu$ or $\nu_\tau \to \tau$ transitions, but indirect means were provided by the Sudbury Neutrino Observatory (SNO), a heavy-water ($D_2O$) Cherenkov detector. The deuteron target makes it possible to distinguish three kinds of neutrino interactions sensitive to differently weighted mixtures of $\nu_e$, $\nu_\mu$. and $\nu_\tau$. The charged-current deuteron dissociation [CC] reaction,

$$\nu_e \mathrm{d} \to e^- \, p \, p \,, \tag{6.7.9}$$

proceeds by $W$-boson exchange and is sensitive only to the $\nu_e$ flux. Neutral-current dissociation [NC],

$$\nu_\ell \mathrm{d} \to \nu_\ell \, p \, n \,, \tag{6.7.10}$$

proceeds by $Z$ exchange and is sensitive to the total flux of active neutrino species $\nu_e + \nu_\mu + \nu_\tau$. Elastic scattering from electrons in the target [ES] is sensitive to a weighted average $\approx \nu_e + \frac{1}{7}(\nu_\mu + \nu_\tau)$ of the active-neutrino fluxes, as we can see by inspecting the cross sections

$$\sigma(\nu_{\mu,\tau} e \to \nu_{\mu,\tau} e) = \frac{G_F^2 m_e E_\nu}{2\pi} \left( \frac{L_e^2 + R_e^2}{3} \right) \quad (Z\text{-exchange only})$$

$$\tag{6.7.11}$$

$$\sigma(\nu_e e \to \nu_e e) = \frac{G_F^2 m_e E_\nu}{2\pi} \left[ (L_e + 2)^2 + \frac{R_e^2}{3} \right] \quad (W + Z\text{-exchange})$$

where the $Ze\bar{e}$ chiral couplings (cf. figure 6.12) are $L_e = 2\sin^2\theta_W - 1 \approx -\frac{1}{2}$ and $R_e = 2\sin^2\theta_W \approx \frac{1}{2}$. This reaction can be studied as well in ordinary water-Cherenkov detectors.

Super-K and SNO observations are summarized in figure 6.32. Taken together, they indicate that the *total* flux agrees with solar model, but only 30% of neutrinos arrive from the Sun as $\nu_e$. The nonzero value of $\phi_{\mu\tau}$ provides strong evidence that neutrinos created as $\nu_e$ are transformed into other active flavors. All the evidence to date is consistent with the conclusion that $\nu_e$ is dominantly a mixture of two mass eigenstates, designated $\nu_1$ and $\nu_2$, with a solar mass-squared difference $\Delta m_\odot^2 = m_2^2 - m_1^2 \approx 7.6 \times 10^{-5}$ eV$^2$ and mixing angle $\theta_\odot \approx 34°$.

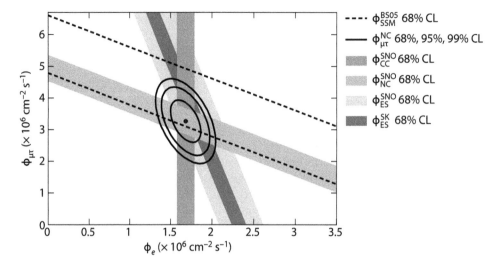

**Figure 6.32.** Flux of $\mu + \tau$ neutrinos versus flux of electron neutrinos determined in solar neutrino experiments. The filled bands indicate the CC, NC and ES flux measurements. The Standard Solar Model (Ref. [45]) predicts that the total $^8$Be solar neutrino flux lies between the dashed lines. The total flux measured with the NC channel is shown as the solid band parallel to the model prediction. The Super-Kamiokande ES result (Ref. [46]) is the dark narrow band within the SNO ES band. The intercepts of the experimental bands with the axes represent $\pm 1\sigma$ uncertainties. The point represents $\phi_e$ from the CC flux and $\phi_{\mu\tau}$ from the NC-CC difference; the contours represent 68%, 95%, and 99% CL allowed regions (from Ref. [47]).

In the presence of neutrino mixing, the charged-current interactions among the left-handed leptonic mass eigenstates $\nu = (\nu_1, \nu_2, \nu_3)$ and $\ell_L = (e_L, \mu_L, \tau_L)$ are specified by

$$\mathcal{L}_{\text{CC}}^{(q)} = -\frac{g}{\sqrt{2}} \, \bar{\nu} \, \gamma^\mu \mathcal{V}^\dagger \ell_L W_\mu^+ + \text{h.c.,} \tag{6.7.12}$$

where the neutrino mixing matrix [48], sometimes called the Pontecorvo–Maki–Nakagawa–Sakata (PMNS) matrix in tribute to neutrino-oscillation pioneers, is

$$\mathcal{V} = \begin{pmatrix} \mathcal{V}_{e1} & \mathcal{V}_{e2} & \mathcal{V}_{e3} \\ \mathcal{V}_{\mu 1} & \mathcal{V}_{\mu 2} & \mathcal{V}_{\mu 3} \\ \mathcal{V}_{\tau 1} & \mathcal{V}_{\tau 2} & \mathcal{V}_{\tau 3} \end{pmatrix} . \tag{6.7.13}$$

The three-flavor formalism for neutrino oscillations is developed in problem 6.22. It is conventional to factor the neutrino mixing matrix as

$$\mathcal{V} = \begin{bmatrix} 1 & 0 & 0 \\ 0 & c_{23} & s_{23} \\ 0 & -s_{23} & c_{23} \end{bmatrix} \begin{bmatrix} c_{13} & 0 & s_{13}e^{-i\delta} \\ 0 & 1 & 0 \\ -s_{13}e^{i\delta} & 0 & c_{13} \end{bmatrix} \begin{bmatrix} c_{12} & s_{12} & 0 \\ -s_{12} & c_{12} & 0 \\ 0 & 0 & 1 \end{bmatrix}, \tag{6.7.14}$$

where we abbreviate $s_{ij} = \sin\theta_{ij}$ and $c_{ij} = \cos\theta_{ij}$, and $\delta$ is a CP-violating phase.

**TABLE 6.5**
Neutrino Properties from a Global Fit to Oscillation Data (from Ref. [51])

| Parameter | Normal hierarchy | Inverted hierarchy |
|---|---|---|
| $\Delta m_{21}^2 \ [10^{-5} \ \mathrm{eV}^2]$ | $7.62 \pm 0.19$ | $7.62 \pm 0.19$ |
| $\Delta m_{31}^2 \ [10^{-3} \ \mathrm{eV}^2]$ | $2.53^{+0.08}_{-0.10}$ | $-2.40^{+0.10}_{-0.07}$ |
| $\sin^2 \theta_{12}$ | $0.320^{+0.015}_{-0.017}$ | $0.320^{+0.015}_{-0.017}$ |
| $\sin^2 \theta_{23}$ | $0.49^{+0.08}_{-0.05}$ | $0.53^{+0.05}_{-0.07}$ |
| $\sin^2 \theta_{13}$ | $0.026^{+0.003}_{-0.004}$ | $0.027^{+0.003}_{-0.004}$ |

True three-flavor mixing requires a finite value of $\theta_{13}$, which is also a prerequisite for CP violation in the neutrino sector. Adding to the established oscillation phenomena in the "atmospheric" and "solar" sectors, the Daya Bay and RENO experiments have observed the disappearance of reactor electron antineutrinos [49]. Establishing a finite value of $\theta_{13}$ qualifies as a milestone in neutrino physics [50]. Current evidence points to a value near $9°$.

The results of a recent global fit [51] are summarized in table 6.5. The atmospheric and solar neutrino experiments, with their reactor and accelerator complements, have partially characterized the neutrino spectrum in terms of a closely spaced solar pair $\nu_1$ and $\nu_2$ and a third neutrino, more widely separated in mass. We do not yet know whether $\nu_3$ lies above (*normal hierarchy*) or below (*inverted hierarchy*) the solar pair, and experiment has not yet set the absolute scale of neutrino masses. Indirect inferences for the CP phase are not yet conclusive. Figure 6.33 shows the normal and inverted spectra as functions of assumed values for the mass of the lightest neutrino.

With our (partial) knowledge of neutrino masses, we can estimate the contribution of neutrinos to the mass-energy density of the current universe. The left-hand scale of figure 6.34 shows the summed neutrino masses $m_1 + m_2 + m_3$ for the normal and inverted hierarchies, as functions of the lightest neutrino mass. The neutrino oscillation data imply that $\sum_i m_{\nu_i} \gtrsim 0.06$ eV in the case of the normal hierarchy and $\sum_i m_{\nu_i} \gtrsim 0.11$ eV in the case of the inverted hierarchy. If the neutrinos are stable on cosmological time scales, they contribute to the dark matter of the universe—most likely as minority, or even trace, components. Neutrinos are not, however, candidates for the cold dark matter (nonrelativistic at the epoch of structure formation) that is favored by scenarios for structure formation in the universe.

In contrast to massless neutrinos, which must be stable, massive neutrinos might decay. Over a distance $L$, decay would deplete the flux of extremely relativistic neutrinos of energy $E$, mass $m$, and lifetime $\tau$ by the factor $e^{-L/\gamma c\tau} = \exp[(-L/Ec)(m/\tau)]$, where $c$ is the speed of light and $\gamma$ is the Lorentz factor. A limit on depletion thus implies a bound on the reduced neutrino lifetime, $\tau/m$. The most stringent such bound, derived from solar $\gamma$- and x-ray fluxes, applies for radiative neutrino decay, $\tau/m > 7 \times 10^9$ s/eV [52]. The present bound on *nonradiative* decays, deduced from the survival of solar neutrinos, is far less constraining: $\tau/m \gtrsim 10^{-4}$ s/eV [53].

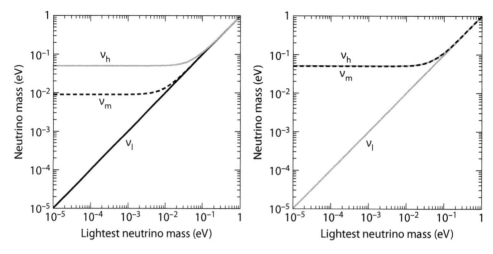

**Figure 6.33.** Favored values for the light, medium, and heavy neutrino masses $m_l$, $m_m$, $m_h$ as functions of the lightest neutrino mass in the three-neutrino oscillation scenario for the normal ($\nu_l = \nu_1$, $\nu_m = \nu_2$, $\nu_h = \nu_3$: left panel) and inverted hierarchy ($\nu_l = \nu_3$, $\nu_m = \nu_1$, $\nu_h = \nu_2$: right panel). The solar ($\Delta m_\odot^2 = m_2^2 - m_1^2$) and atmospheric ($\Delta m_{atm}^2 = m_3^2 - m_1^2$) mass-squared differences are taken from table 6.5. In the inverted hierarchy, $\nu_m$ and $\nu_h$ are nearly degenerate.

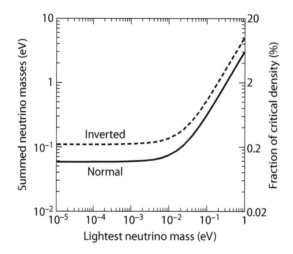

**Figure 6.34.** Summed masses (left-hand scale) of $\nu_1$, $\nu_2$, $\nu_3$, and contributions of stable relic neutrinos to the mass density of the universe (right-hand scale) vs. the lightest neutrino mass for the normal (solid line) and inverted (dashed line) mass hierarchies.

The mixing parameter ranges given in table 6.5 lead to the flavor content of the neutrino mass eigenstates depicted in figure 6.35. For both normal and inverted hierarchies, $\nu_3$ consists of nearly equal parts of $\nu_\mu$ and $\nu_\tau$, perhaps with a trace of $\nu_e$, whereas $\nu_2$ contains similar amounts of $\nu_e$, $\nu_\mu$, and $\nu_\tau$, and $\nu_1$ is rich in $\nu_e$,

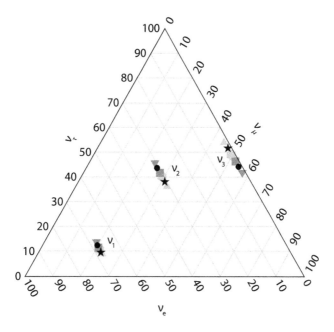

**Figure 6.35.** Current knowledge of the flavor ($\nu_e$, $\nu_\mu$, $\nu_\tau$) composition of the neutrino mass eigenstates $\nu_1$, $\nu_2$, $\nu_3$ for the normal mass hierarchy. The perpendicular distances from the axes to any interior point sum to 100%. The symbols range over $\pm 1\sigma$ variations in $\theta_{23}$, with $\theta_{12}$ and $\theta_{13}$ fixed at their central values, and the CP parameter $\delta$ set to 0.

with minority parts of $\nu_\mu$ and $\nu_\tau$. The observed structure of the neutrino-mixing matrix differs greatly from the pattern of the more familiar (Cabibbo–Kobayashi–Maskawa) quark-mixing matrix, which is displayed graphically in figure 7.1.

Might there be more than three neutrino species? Measurements of the invisible decay width of the $Z^0$ boson (also see the $Z^0$ line shape shown in figure 7.5) indicate that only three ($2.92 \pm 0.05$) light active neutrino flavors exist. Standard model fits to LEP data yield $2.984 \pm 0.008$ light active flavors (Ref. [4]). Accordingly, any additional light neutrinos would have to be "sterile," by which we mean inert with respect to the $SU(2)_L \otimes U(1)_Y$ interactions.

The experimental situation is unsettled. The Liquid Scintillator Neutrino Detector at Los Alamos reported evidence for $\bar\nu_\mu \to \bar\nu_e$ oscillations driven by a mass splitting $\Delta m^2_{LSND} \sim 1$ eV$^2$, which would necessitate a fourth neutrino species [54]. The MiniBooNE experiment at Fermilab tested the LSND claim by searching for $\nu_\mu \to \nu_e$ oscillations over a short baseline and found no corroborating evidence [55]. Subsequent running in a $\bar\nu_\mu$ beam yielded inconclusive evidence for an LSND effect [56]. A retrospective analysis of short-baseline reactor neutrino experiments hints at a deficit that might be explained by $\bar\nu_e \to \nu_{\text{sterile}}$ oscillations, also with $\Delta m^2 \sim 1$ eV$^2$ [57].

We probably do not know enough to specify a new ("$\nu$") standard model, but the inference from neutrino oscillations that neutrinos have mass makes it tempting to suppose, as a working hypothesis, that right-handed neutrinos do exist.

On current evidence, these right-handed neutrinos are sterile—inert with respect to the known electroweak interactions with the left-handed $W^{\pm}$, $Z^0$, and (of course) $\gamma$. Given the absence of detectable right-handed charged-current interactions, it is not surprising that what we surmise about the right-handed neutrinos is of little consequence for most studies of neutrino interactions.

We may seek to accommodate neutrino mass in the electroweak theory by adding to the spectrum a right-handed neutrino $N_R$ and constructing the (gauge-invariant) Dirac mass term

$$\mathcal{L}_D^{(\nu)} = -\zeta_\nu[(\bar{L}_\ell \tilde{\phi})N_R + \overline{N}_R(\tilde{\phi}^\dagger L_\ell)] \to -m_D\left[\bar{\nu}_L N_R + \overline{N}_R \nu_L\right], \tag{6.7.15}$$

where $\tilde{\phi} = i\sigma_2\phi^*$ is the complex conjugate of the Higgs doublet and $m_D = \zeta_\nu v/\sqrt{2}$. A Dirac mass term conserves the additive lepton number $L$ that takes on the value $+1$ for neutrinos and negatively charged leptons and $-1$ for antineutrinos and positively charged leptons. To account for the observed tininess of the neutrino masses, $m_\nu \lesssim 1$ eV, the Yukawa couplings must be extraordinarily small, $\zeta_\nu \lesssim 10^{-11}$. Whether they are qualitatively more puzzling than the factor of $3 \times 10^5$ that separates the electron and top-quark Yukawa couplings is for now a question for intuition.

Matters of taste aside, we have another reason to consider alternatives to the Dirac mass term: unlike all the other particles that enter the electroweak theory, the right-handed neutrinos are standard-model singlets. Alone among the standard-model fermions, they might be their own antiparticles, so-called Majorana fermions [58]. The charge conjugate of a right-handed field is left-handed, $\psi_L^c \equiv (\psi^c)_L = (\psi_R)^c$. Majorana mass terms connect the left-handed and right-handed components of conjugate fields,

$$-\mathcal{L}_A = A(\bar{\nu}_R^c \nu_L + \bar{\nu}_L \nu_R^c) = A\overline{\chi}\chi,$$
$$-\mathcal{L}_B = B(\overline{N}_L^c N_R + \overline{N}_R N_L^c) = B\bar{\omega}\omega. \tag{6.7.16}$$

Here the self-conjugate Majorana mass eigenstates are

$$\chi \equiv \nu_L + \nu_R^c = \chi^c,$$
$$\omega \equiv N_R + N_L^c = \omega^c. \tag{6.7.17}$$

A Majorana fermion cannot carry any additive quantum number. The mixing of particle and antiparticle fields means that the Majorana mass terms correspond to processes that violate lepton number by two units. Accordingly, the exchange of a Majorana neutrino can mediate neutrinoless double beta decay, $(Z, A) \to (Z+2, A) + e^- + e^-$. Detecting neutrinoless double beta decay would offer decisive evidence for the Majorana nature of the neutrino. If neutrinos are Majorana particles, the neutrino-mixing matrix acquires two additional phases,

$$\mathcal{V} \to \mathcal{V}\begin{bmatrix} e^{i\alpha_1/2} & 0 & 0 \\ 0 & e^{i\alpha_2/2} & 0 \\ 0 & 0 & 1 \end{bmatrix}. \tag{6.7.18}$$

The Majorana phases do not affect neutrino oscillations but are, in principle, accessible in neutrinoless double beta decay.

One possible source for a Majorana mass is an effective (i.e., nonrenormalizable) dimension-five operator of the form

$$\mathcal{L}_5^{(\nu)} = -\frac{\zeta_{\alpha\beta}}{M}(\mathsf{L}_\alpha\phi)^{\mathsf{T}}(\mathsf{L}_\beta\phi), \tag{6.7.19}$$

where $\mathsf{L}_\alpha$ [cf. (6.3.1)] is the left-handed weak-isospin doublet for flavor $\alpha$, $\phi$ is the Higgs doublet, $\zeta_{\alpha\beta}$ are dimensionless couplings, and $M$ specifies the cutoff scale. After electroweak symmetry breaking, characterized by

$$\langle\phi\rangle_0 = \begin{pmatrix} 0 \\ v/\sqrt{2} \end{pmatrix}, \tag{6.3.21}$$

the operator $\mathcal{L}_5^{(\nu)}$ generates Majorana neutrino masses

$$m_{\alpha\beta} = \frac{\zeta_{\alpha\beta}\, v^2}{2M}. \tag{6.7.20}$$

If we choose the "natural" values $\zeta_{\alpha\beta} = 1$ and set the cutoff scale at the Planck mass, $M = M_{\mathrm{Pl}} \equiv (\hbar c/G_{\mathrm{Newton}})^{1/2} = 1.22 \times 10^{19}$ GeV, then $m_{\alpha\beta} \sim 10^{-5}$ eV.

It is interesting to consider both Dirac and Majorana terms and specifically to examine the case in which Majorana masses corresponding to an active state $\chi$ and a sterile state $\omega$ arise from weak triplets and singlets, respectively, with masses $M_3$ and $M_1$. The neutrino mass matrix then has the form

$$(\bar{\nu}_{\mathrm{L}}\ \overline{N}_{\mathrm{L}}^c) \begin{pmatrix} M_3 & m_{\mathrm{D}} \\ m_{\mathrm{D}} & M_1 \end{pmatrix} \begin{pmatrix} \nu_{\mathrm{R}}^c \\ N_{\mathrm{R}} \end{pmatrix}. \tag{6.7.21}$$

In the highly popular seesaw limit [59], with $M_3 = 0$ and $m_{\mathrm{D}} \ll M_1$, diagonalizing the mass matrix (6.7.21) yields two Majorana neutrinos,

$$n_{1\mathrm{L}} \approx \nu_{\mathrm{L}} - \frac{m_{\mathrm{D}}}{M_1}N_{\mathrm{L}}^c, \qquad n_{2\mathrm{L}} \approx N_{\mathrm{L}}^c + \frac{m_{\mathrm{D}}}{M_1}\nu_{\mathrm{L}}, \tag{6.7.22}$$

with masses

$$m_1 \approx \frac{m_{\mathrm{D}}^2}{M_1} \ll m_{\mathrm{D}}, \qquad m_2 \approx M_1. \tag{6.7.23}$$

The seesaw produces one very heavy "neutrino" and one neutrino much lighter than a typical quark or charged lepton. Many alternative explanations of the small neutrino masses have been explored in the literature [60], including some in which collider experiments exploring the Fermi scale could reveal the origin of neutrino masses [61].

The discovery of neutrino mixing opened an exciting new area of study in particle physics. With the (nonzero!) value of $\theta_{13}$ tightly constrained, the sign of $\Delta m_{\mathrm{atm}}^2$, and the absolute scale of neutrino masses are near-term experimental targets.

A key question is whether the neutrino is its own antiparticle. The relatively large value of $\theta_{13}$ means that it may be possible to observe CP violation in the neutrino sector and test whether the neutrino-mixing matrix accounts for the observations. It is important to determine whether light sterile neutrinos exist and to learn whether $\nu_e$, $\nu_\mu$, or $\nu_\tau$ might experience any nonstandard (beyond $SU(2)_L \otimes U(1)_Y$) interactions. With respect to the electroweak theory and possible extensions, finding evidence for charged-lepton flavor violation and establishing the origin of neutrino mass would help set the direction for future research.

## 6.8 RENORMALIZABILITY OF THE THEORY

The quantization of non-Abelian gauge theories and the development of Feynman rules that are consistent when applied to diagrams containing loops is a highly nontrivial undertaking that we shall examine briefly in section 8.3. So, too, is the demonstration, first given by 't Hooft [62], that spontaneously broken gauge theories are renormalizable. A detailed discussion of quantization and renormalization lies outside the scope of this book, but two points have interesting repercussions and will, therefore, be developed briefly. These have to do with convenient choices of gauge and the role of anomalies.

To quantize a gauge theory and—particularly—to determine the vector-boson propagators requires a choice of gauge, as we have already remarked for the case of QED in section 3.6. The unitary gauge introduced for the first time in section 5.3 is extremely convenient for examining the particle content of the spontaneously broken gauge theory. However, the form of the massive vector-boson propagator that appears in this gauge,

$$\frac{-i(g_{\mu\nu} - q_\mu q_\nu/M^2)}{q^2 - M^2 + i\varepsilon}, \tag{6.8.1}$$

is ill suited to the calculation of higher-order diagrams in perturbation theory. Already at the level of tree diagrams for gauge-boson scattering, complicated cancellations among diagrams that are individually ill behaved are required for sensible results. The presence of the $q_\mu q_\nu/M^2$ term in the propagator means that finite amplitudes must result from the cancellation of contributions that are separately infinite, a procedure of some delicacy. Among 't Hooft's contributions was the insight that, because of the gauge invariance of the theory, we are empowered to choose one gauge to manifest the particle content and another to prove renormalizability, although each may be unsuited for the other's task.

The general method is nicely illustrated by the Abelian Higgs model of section 5.3. To the Lagrangian (5.3.11) of the Abelian Higgs model written in terms of shifted fields,

$$\mathcal{L}_{so} = \tfrac{1}{2}[(\partial_\mu \eta)(\partial^\mu \eta) + 2\mu^2 \eta^2] + \tfrac{1}{2}[(\partial_\mu \zeta)(\partial^\mu \zeta)]$$
$$- \tfrac{1}{4} F_{\mu\nu} F^{\mu\nu} + M A_\mu (\partial^\mu \zeta) + \tfrac{1}{2} M^2 A_\mu A^\mu + \cdots, \tag{6.8.2}$$

we add the gauge-fixing term

$$\mathcal{L}_{\text{gf}} = -\left(\frac{1}{2\xi}\right)(\partial^\mu A_\mu + \xi M \zeta)^2, \tag{6.8.3}$$

where the parameter $\xi$ determines the specific gauge. $\mathcal{L}_{\text{gf}}$ is so constructed as to cancel the unwanted mixing term $M A_\mu(\partial^\mu \zeta)$ in the original Lagrangian. In general, the unphysical $\zeta$ field is now present in the Feynman rules of the theory, with a propagator given by

$$\frac{i}{q^2 - \xi M^2}. \tag{6.8.4}$$

The vector boson propagator may also be found by the usual procedure. It is

$$\frac{-i[g_{\mu\nu} - (1 - \xi)q_\mu q_\nu/(q^2 - \xi M^2)]}{q^2 - M^2 + i\varepsilon}. \tag{6.8.5}$$

Because the $q_\mu q_\nu$ term appears in combination with $1/q^2$ for any finite value of $\xi$, there is no reason to expect nonrenormalizable behavior in loop integrals. The apparent poles in (6.8.4) and (6.8.5) at $q^2 = \xi M^2$ must cancel in any $S$-matrix element. Let us note that $\xi = 1$ corresponds to the Feynman gauge familiar in electrodynamics, whereas $\xi = 0$ yields the Landau (Lorenz) gauge, in which the numerator of (6.8.5) assumes the form

$$g_{\mu\nu} - \frac{q_\mu q_\nu}{q^2} \tag{6.8.6}$$

of a transverse projection operator. In the limit $\xi \to \infty$, the vector-boson propagator approaches the familiar unitary gauge form (6.8.1) and the unphysical scalar decouples as its mass tends to infinity.

The demonstrations of renormalizability make extensive use of the intricate cancellations implied by gauge invariance, such as those we have seen explicitly in operation at tree level. It is essential to the proofs that these cancellations, which are related to local current conservations and are codified for QED in the Ward–Takahashi identities, also take place in diagrams containing closed loops. However, situations are known in field theory in which a (classical) local conservation law derived from gauge invariance with the aid of Noether's theorem holds at tree level but is not respected by loop diagrams. Terms that violate the classical conservation laws are known as anomalies, and their properties have been studied extensively. The simplest example of a Feynman diagram leading to an anomaly is a fermion loop coupled to two vector currents and one axial current, as shown in figure 6.36. Because the weak interaction contains both vector and axial-vector currents, there is a danger that such diagrams may arise in the electroweak theory and destroy the renormalizability of the theory.

For a gauge theory, it is convenient to write the interaction Lagrangian in terms of left-handed and right-handed matter fields, as we have done in formulating the

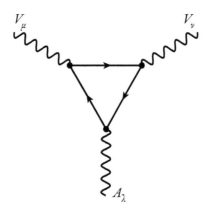

**Figure 6.36.** Triangle anomaly for the axial-vector current.

electroweak theory. The interaction terms may thus be written schematically as

$$\mathcal{L}_{\text{int}} = -g\, X_\mu^a (\bar{\mathsf{R}} \gamma^\mu T_+^a \mathsf{R} + \bar{\mathsf{L}} \gamma^\mu T_-^a \mathsf{L}), \tag{6.8.7}$$

where $T_\pm^a$ are generators of the appropriate representation of the gauge group and $X_\mu^a$ are the gauge fields. It may then be shown that the axial anomaly is proportional to

$$A^{abc} = A_+^{abc} - A_-^{abc}, \tag{6.8.8}$$

where the right-handed and left-handed contributions are given by

$$A_\pm^{abc} = \text{tr}\!\left(\{T_\pm^a,\, T_\pm^b\} T_\pm^c\right) \tag{6.8.9}$$

and do not depend on fermion masses. Evidently the theory will be anomaly free either if there is a cancellation among the left-handed and right-handed sectors,

$$A_+^{abc} = A_-^{abc}, \tag{6.8.10}$$

in which case the theory is called *vectorlike*, or if the two contributions vanish separately,

$$A_+^{abc} = 0 = A_-^{abc}. \tag{6.8.11}$$

For the electroweak theory, it is easy to verify that the only anomaly is proportional to

$$\text{tr}\!\left(\{\tau^a, \tau^b\} Y\right) = \text{tr}\!\left(\{\tau^a, \tau^b\}\right) \text{tr}(Y) \propto \sum_{\substack{\text{fermion}\\\text{doublets}}} Y. \tag{6.8.12}$$

By virtue of the Gell-Mann–Nishijima relation, we may, therefore, express the condition for the absence of anomalies in terms of the electric charge as

$$\Delta Q = Q_R - Q_L = \left( \sum_{\text{RH doublets}} Q - \sum_{\text{LH doublets}} Q \right) = 0. \tag{6.8.13}$$

In the model of leptons that we have considered until now, built upon left-handed doublets such as

$$L_e = \begin{pmatrix} \nu_e \\ e \end{pmatrix}_L, \tag{6.8.14}$$

we have

$$\Delta Q = -Q_L = 1 \tag{6.8.15}$$

for each lepton family.

To cancel the anomaly, we may consider two possibilities. One is to add doublets of right-handed fermions with appropriate charge assignments. This, of course, has the aesthetic appeal of restoring a measure of parity symmetry, but it lacks direct experimental support. The second is to add additional left-handed fermion doublets, with charges arranged to cancel the contribution of the leptons. This role falls naturally to the hadrons. Indeed, if we simply appended to the one-generation electroweak theory of leptons a left-handed doublet of nucleons,

$$L_N = \begin{pmatrix} p \\ n \end{pmatrix}_L, \tag{6.8.16}$$

the "left-handed charge" would become

$$Q_L = 0, \tag{6.8.17}$$

and the anomaly would be canceled. The nucleons are not elementary, however. One manifestation of the compositeness is that (even in the zero-momentum-transfer limit) the charged-current interaction is characterized by $\gamma^\mu(1 - g_A\gamma_5)$ with $g_A \approx 1.25 \neq 1$. We should not require any further encouragement to turn toward the quarks.

A single doublet of quarks,

$$L_q = \begin{pmatrix} u \\ d \end{pmatrix}_L, \tag{6.8.18}$$

contributes to the left-handed charge an amount

$$Q_L(q) = \tfrac{2}{3} - \tfrac{1}{3} = \tfrac{1}{3}, \tag{6.8.19}$$

which is insufficient to cancel the lepton-doublet contribution to the anomaly. But if for each lepton doublet we include three quark doublets,

$$
\begin{pmatrix} u_{\text{Red}} \\ d_{\text{Red}} \end{pmatrix}_{\text{L}}, \qquad \begin{pmatrix} u_{\text{Green}} \\ d_{\text{Green}} \end{pmatrix}_{\text{L}}, \qquad \begin{pmatrix} u_{\text{Blue}} \\ d_{\text{Blue}} \end{pmatrix}_{\text{L}}, \tag{6.8.20}
$$

corresponding to the three quark colors that had been suggested by numerous observations, the anomaly will be canceled [63]. The cancellation generalizes at once to many lepton (and, hence, quark) doublets.

The requirement that the theory of electroweak interactions be renormalizable thus contributes new support to the idea of quark color and makes the far-reaching suggestion that the spectra of leptons and quarks are related. We shall examine the consequences for the phenomenology of the weak interactions of quarks in the next chapter. Extended families of quarks and leptons will be central to a later attempt, in chapter 9, at a (grand) unification of the strong, weak, and electromagnetic interactions.

## 6.9 INTERIM ASSESSMENT

A spontaneously broken gauge theory based on $SU(2)_L \otimes U(1)_Y$ symmetry and leptons that occur in left-handed weak isospin doublets and right-handed singlets captures the essential features of low-energy charged-current weak interactions. In addition, it entails novel neutral-current weak interactions that (both at the first inspection we have made in this chapter and upon closer scrutiny) conform to what is observed. The force carriers $W^\pm$ of the charged current and $Z^0$ of the neutral current are predicted to be massive, as observed. A special feature of the Higgs mechanism that hides the electroweak symmetry is that it reveals a means by which fermions (the charged leptons, in this instance) can acquire mass—through their interactions with the Higgs boson. Should we discover a standard-model Higgs boson with the expected ($\propto m$) couplings to fermions, we will learn the origin of fermion masses. However, nothing in the electroweak theory empowers us to calculate the values of those masses; we merely attribute the values to the strength of Yukawa couplings of unknown origin. In that sense, *the electron mass represents physics beyond the standard model!* Neutrino mixing implies that neutrinos are massive and so requires some amendments to the electroweak theory—perhaps the incorporation of sterile, right-handed neutrinos or of Majorana neutrinos. The implications for charged-lepton flavor violation depend sensitively on the character of any new physics that might be responsible for neutrino mass. Renormalizability of the electroweak theory hangs on the absence of anomalies and so demands that each lepton family be accompanied by a quark family, a requirement that we shall implement in the following chapter. Once that is achieved, quantum corrections are calculable and the electroweak theory can be subjected to stringent tests as a quantum field theory.

# PROBLEMS

6.1. Compute the differential and total cross sections for the reaction $\nu_\mu e \to \nu_\mu e$, retaining the electron mass and the effect of the $Z^0$ propagator. In what kinematic regimes does the result differ noticeably from the point-coupling limit (6.4.8) and (6.4.11)? [Cf. equation (15) of R. Gandhi, C. Quigg, M. H. Reno, and I. Sarcevic, *Astropart. Phys.* **5**, 81 (1996), arXiv:hep-ph/9512364.]

6.2. Compute the cross section for the reaction $e^+ e^- \to \nu_\mu \bar{\nu}_\mu$ in the point-coupling limit, and compare it with that for $\nu_\mu e \to \nu_\mu e$. For the special case $\sin^2 \theta_W = \frac{1}{4}$, use angular momentum arguments to explain the ratio of the two cross sections.

6.3. Calculate the angular distribution $d\Gamma/d\Omega$ for the decay $Z^0 \to e^+ e^-$.

6.4. (a) Compute the decay rate for the disintegration of the tau lepton, $\tau^- \to e^- \bar{\nu}_e \nu_\tau$, neglecting the electron mass, and compare with the familiar rate for muon decay.
(b) How would the rate change if, by virtue of an unconventional lepton number assignment, $\nu_\tau \equiv \bar{\nu}_e$?

6.5. Derive the Fierz reordering theorem,

$$\bar{u}_3 \mathcal{O}_i u_2 \bar{u}_1 \mathcal{O}_i u_4 = \sum_{j=1}^{5} \lambda_{ij} \bar{u}_1 \mathcal{O}_j u_2 \bar{u}_3 \mathcal{O}_j u_4, \qquad (A.4.29)$$

where $\mathcal{O}_i = (1, \gamma_\mu, \sigma_{\mu\nu}, i\gamma_\mu\gamma_5, \gamma_5)$ and the $5 \times 5$ matrix $\lambda_{ij}$ is given by (A.4.30). [V. B. Berestetskii, E. M. Lifshitz, and L. P. Pitaevski, *Relativistic Quantum Theory*, part 1, trans. J. B. Sykes and J. S. Bell, Pergamon, Oxford, 1971, §22–28.]

6.6. Compute the differential and total cross sections for $\nu_e e$ and $\bar{\nu}_e e$ elastic scattering in the $SU(2)_L \otimes U(1)_Y$ electroweak theory. Work in the limit of large $M_W$ and $M_Z$, and neglect the electron mass with respect to large energies. The computation is done most gracefully by Fierz reordering one of the graphs, as indicated in section 6.4. [G. 't Hooft, *Phys. Lett.* **37B**, 195 (1971).]

6.7. Compute the cross section for the reaction $e^+ e^- \to H \to f\bar{f}$, where $f$ is a massive lepton, and thus show that the cross section for the reaction $e^+ e^- \to H \to$ all can be written in the form (6.5.11).

6.8. Show that for a heavy Higgs boson the rates for decay into a pair of intermediate vector bosons are given by (6.5.8) and (6.5.9). What fraction of the decays lead to longitudinally polarized gauge bosons? What accounts for the overall factor of two between the $WW$ and $ZZ$ rates?

6.9. Consider the decay of a standard-model Higgs boson into $W^+ W^-$ pairs near threshold.
(a) Show that the final states $W_{+1}^+ W_{+1}^-$, $W_L^+ W_L^-$, and $W_{-1}^+ W_{-1}^-$ are equally populated [cf. (6.5.8)], where the subscripts refer to the gauge-boson helicities.

(b) Now specialize to leptonic decays. Referring to the decay angular distributions (6.2.19), (6.2.23), and (6.2.24), what can you say about correlations between the outgoing charged leptons? [The generalization to below-threshold decays $H \to WW^*$ is given by V. D. Barger et al., *Phys. Rev.* D **49**, 79 (1994).]

6.10. Consider the process $e^+e^- \to \gamma\gamma$, which is described, in QED, by the two Feynman diagrams

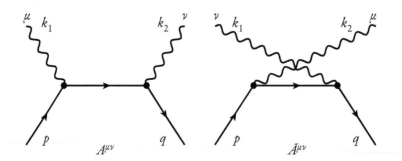

and for which the amplitude may be written $\mathcal{M} = \epsilon_{1\nu}^* \epsilon_{2\mu}^* (A^{\mu\nu} + \tilde{A}^{\mu\nu})$, where $\epsilon_i$ is the polarization vector of a photon and $A^{\mu\nu}(\tilde{A}^{\mu\nu})$ corresponds to the first (second) diagram.

(a) Calculate the tensors $A^{\mu\nu}$ and $\tilde{A}^{\mu\nu}$.

(b) Show that gauge invariance requires that

$$k_{1\nu}(A^{\mu\nu} + \tilde{A}^{\mu\nu}) = 0 = k_{2\mu}(A^{\mu\nu} + \tilde{A}^{\mu\nu}).$$

(c) Verify that these conditions are met, although the quantities $k_{1\nu}A^{\mu\nu}$, $k_{1\nu}\tilde{A}^{\mu\nu}$, $k_{2\mu}A^{\mu\nu}$, and $k_{2\mu}\tilde{A}^{\mu\nu}$ are all different from zero.

6.11. Evaluate the differential cross section $d\sigma/d\Omega_{\text{cm}}$ for the reaction $e^+e^- \to \gamma\gamma$ in the framework of quantum electrodynamics (cf. problem 6.10). Show that, neglecting the electron mass and away from the forward direction ($\sin\theta \neq 0$), the cross section is

$$\left. \frac{d\sigma_{\gamma\gamma}}{d\Omega_{\text{cm}}} \right|_{\text{QED}} = \frac{\alpha^2}{s} \left[ \frac{1 + \cos^2\theta}{\sin^2\theta} \right].$$

[For a comparison with experiment, see M. Derrick et al., *Phys. Rev.* D **34**, 3286 (1986).]

6.12. (a) Carry out the computation of the amplitudes for the reaction $e^+e^- \to W^+W^-$ described at the end of section 6.3, retaining the electron mass. Verify the role of the Higgs boson in the cancellation of divergences.

(b) Now suppose that there were no Higgs boson. Calculate the c.m. energy at which the residual ($\propto m_e E_{\text{c.m.}}$) behavior would violate $s$-wave unitarity.

(c) What is the corresponding critical energy for the top-quark-initiated reaction, $t\bar{t} \to W^+W^-$?

[T. Appelquist and M. S. Chanowitz, *Phys. Rev. Lett.* **59**, 2405 (1987) [Erratum: *Phys. Rev. Lett.* **60**, 1589 (1988)].]

6.13. Because the most serious high-energy divergences of a spontaneously broken gauge theory are associated with the longitudinal degrees of freedom of the gauge bosons, which arise from auxiliary scalars, it is instructive to study the Higgs sector in isolation. Consider, therefore, the Lagrangian for the Higgs sector of the Weinberg–Salam model before the gauge couplings are turned on,

$$\mathcal{L}_{\text{scalar}} = (\partial^\mu \phi)^\dagger (\partial_\mu \phi) - \mu^2 (\phi^\dagger \phi) - |\lambda| (\phi^\dagger \phi)^2.$$

(a) Choosing $\mu^2 < 0$, investigate the effect of spontaneous symmetry breaking. Show that the theory describes three massless scalars $(w^+, w^-, z^0)$ and one massive neutral scalar $(h)$, which interact according to

$$\mathcal{L}_{\text{int}} = -|\lambda| vh(2w^+w^- + z^2 + h^2) - (|\lambda|/4)(2w^+w^- + z^2 + h^2)^2,$$

where $v^2 = -\mu^2/|\lambda|$. In the language of the full electroweak theory, $1/v^2 = G_F\sqrt{2}$ and $\lambda = G_F M_H^2/\sqrt{2}$. Note the resemblance to the $\sigma$-model Lagrangian of (5.4.24). The correspondence is $\pi^\pm \leftrightarrow w^\pm$, $\pi^0 \leftrightarrow z^0$, $h \leftrightarrow \sigma$. In the $\sigma$ model, $v = f_\pi$ and $|\lambda| = m_\sigma^2/2 f_\pi^2$.

(b) Deduce the Feynman rules for interactions and compute the lowest-order (tree diagram) amplitude for the reaction $hz \to hz$.

(c) Compute the $s$-wave partial-wave amplitude in the high-energy limit and show that it respect partial-wave unitarity only if $M_H^2 < 8\pi\sqrt{2}/G_F$. [B. W. Lee, C. Quigg, and H. B. Thacker, *Phys. Rev. D* **16**, 1519 (1977).]

6.14. Compute the decay rate for the disintegration $Z^0 \to H\mu^+\mu^-$, which proceeds through the intermediate state $HZ^0_{\text{virtual}}$. [J. D. Bjorken, "Weak Interaction Theory and Neutral Currents," SLAC-PUB-1866, in *Weak Interactions at High Energies and the Production of New Particles*, 1976 SLAC Summer Institute, SLAC Report 198, p. 1. A factor of $\pi$ is missing from the denominator of equation (4.30).]

6.15. Glashow's $SU(2)_L \otimes U(1)_Y$ model of 1961 (see Ref. [1]) has the same structure as the Weinberg model of 1967 but makes less-definite predictions because the vector boson masses are not generated by a specific Higgs mechanism. To see how the choice of Higgs sector influences the predictions for vector meson masses, consider the following alternatives to the conventional weak-isospin doublet of complex scalars: (1) a weak isovector with $Y_\phi = 2$, which develops a vacuum expectation value $\langle\phi\rangle_0 = (0, 0, v/\sqrt{2})^T$; (2) a weak isovector with $Y_\phi = 0$, which develops a vacuum expectation value $\langle\phi\rangle_0 = (0, v/\sqrt{2}, 0)^T$; (3) a weak isotensor with $Y_\phi = 4$, which develops a vacuum expectation value $\langle\phi\rangle_0 = (0, 0, 0, 0, v/\sqrt{2})^T$. (The superscript T denotes transpose.) For each case, consider a spontaneously broken gauge theory, as constructed in section 6.3.

(a) Examine the symmetry properties of the vacuum state under the generators of $SU(2)_L \otimes U(1)_Y$ to determine which gauge bosons will acquire mass.

(b) Expand $\mathcal{L}_{\text{scalar}}$ following the procedure of (6.3.33)–(6.3.39), and determine for each the ratio $\rho \equiv M_W^2 / M_Z^2 \cos^2 \theta_W$.

(c) Show that the low-energy effective Lagrangian for the neutral-current interactions of leptons can be written in the current–current form,

$$\mathcal{L}_{\text{NC}} = -G_F \rho \sqrt{2} J_\lambda^0 J^{0\lambda},$$

where the neutral current is

$$J_\lambda^0 = \sum_i \bar{L}_i \gamma_\lambda \tau_3 L_i - 2 \sin^2 \theta_W J_\lambda^{(\text{EM})},$$

and the left-handed composite spinors have been defined in (6.3.1), (6.3.2), and (6.4.1). This is to be compared with the effective Lagrangian for charged-current interactions, which may be written as

$$\mathcal{L}_{\text{CC}} = -G_F \sqrt{2} J_{+\lambda} J_-^\lambda,$$

where

$$J_\pm^\lambda = \sum_i \bar{L}_i \gamma^\lambda \tau_\pm L_i.$$

6.16. Suppose that the $SU(2)_L \otimes U(1)_Y$ gauge symmetry of the Glashow model is broken by several sets of Higgs fields $\phi_i$, characterized by weak isospin $I_i$, weak hypercharge $Y_i$, and vacuum expectation value $\langle \phi_i \rangle_0$. Show that

$$\rho \equiv \frac{M_W^2}{M_Z^2 \cos^2 \theta_W} = \frac{\sum_i [I_i(I_i + 1) - \frac{1}{2} Y_i(\frac{1}{2} Y_i - 1)] \langle \phi_i \rangle_0^2}{\sum_i Y_i^2 \langle \phi_i \rangle_0^2}.$$

[D. A. Ross and M. J. G. Veltman, *Nucl. Phys.* **B95**, 135 (1975).]

6.17. At very low momentum transfers, as in atomic physics applications, the nucleon appears elementary. If the effective Lagrangian for nucleon $\beta$-decay can be written in the limit of zero momentum transfer as

$$\mathcal{L}_\beta = -\frac{G_F}{2\sqrt{2}} \bar{e} \gamma_\lambda (1 - \gamma_5) \nu \, \bar{p} \gamma^\lambda (1 - g_A \gamma_5) n,$$

where $g_A = 1.2695 \pm 0.0029$ is the axial charge of the nucleon (renormalized from unity by the strong interactions), show that the $eN$ neutral-current interactions may be represented by

$$\mathcal{L}_{ep} = \frac{G_F}{2\sqrt{2}} \bar{e} \gamma_\lambda (1 - 4x_W - \gamma_5) e \, \bar{p} \gamma^\lambda (1 - 4x_W - g_A \gamma_5) p$$

and

$$\mathcal{L}_{en} = -\frac{G_F}{2\sqrt{2}}\bar{e}\gamma_\lambda(1 - 4x_W - \gamma_5)e\bar{n}\gamma^\lambda(1 - g_A\gamma_5)n.$$

Now perform, for a nucleus regarded as a noninteracting collection of $Z$ protons and $N$ neutrons, the nonrelativistic reduction of the implied nuclear matrix elements. Show that, for a heavy nucleus, the dominant parity-violating contribution to the electron–nucleus amplitude will be of the form

$$\mathcal{M}_{p.v.} = -\frac{iG_F}{2\sqrt{2}}Q_W\bar{e}\rho_N(\mathbf{r})\gamma_5 e,$$

where $\rho_N$ is the nucleon density as a function of the electron coordinate $\mathbf{r}$, and the weak charge is

$$Q_W = Z(1 - 4x_W) - N.$$

6.18. The weak charge of $^{133}_{55}\text{Cs}$ has been measured through the 6S-7S transition probability to be

$$Q_W(\text{Cs}) = -73.16 \pm 0.29 \text{ (expt)} \pm 0.20 \text{ (theory)}$$

after reassessment of atomic structure effects. Using the first-order expression derived in problem 6.17, deduce the weak mixing parameter $x_W$ and its uncertainties. [S. C. Bennett and C. E. Wieman, *Phys. Rev. Lett.* **82**, 2484 (1999); S. G. Porsev, K. Beloy, and A. Derevianko, *Phys. Rev. Lett.* **102**, 181601 (2009).]

6.19. Neutrino observatories aim to detect ultrahigh-energy neutrinos from cosmic sources.
(a) At what neutrino energy $E_\nu$ would the $W^-$ be formed as a direct-channel resonance in $\bar{\nu}_e e$ scattering?
(b) Universality of the charged-current weak interactions implies that the decay rate of $W^\pm$ to a pair of light fermions in $N_c$ colors is $\Gamma(W \to f\bar{f}') = N_c G_F M_W^3 / 6\pi\sqrt{2}$ [cf. (6.2.20) and §7.2]. Neglecting the masses of all fermions but the top quark, estimate the total width of $W^\pm$ and compare with experiment.
(c) Compute the cross section for the reaction $\bar{\nu}_e e \to \bar{\nu}_\mu \mu$ and plot it for $10^{10} \text{ eV} \le E_\nu \le 10^{18} \text{ eV}$.
(d) What is the cross section at the resonance peak? [R. Gandhi, C. Quigg, M. H. Reno, and I. Sarcevic, *Astropart. Phys.* **5**, 81 (1996) [arXiv:hep-ph/9512364].]

6.20. Consider $\mu^- e^-$ scattering mediated by $t$-channel $(\gamma, Z^0)$ exchange.
(a) Calculate the differential cross section $d\sigma/dy$, where $y \equiv (E_i - E_f)/E_i$ is the fraction of the incoming muon's energy transferred to the electron target [cf. (7.4.94)].

(b) Show that the asymmetry that measures the difference between scattering of right-handed and left-handed muons on unpolarized electrons is

$$A_{\mu e} = \frac{\sigma_R - \sigma_L}{\sigma_R + \sigma_L} = \frac{G_F(4x_W - 1)ys}{2\pi\alpha\sqrt{2}[1 + (1-y)^2]},$$

where $s = 2m_e E_i$ is the square of the c.m. energy. [E. Derman and W. J. Marciano, *Ann. Phys.* **121**, 147 (1979).]

(c) Compute the asymmetry $A_{\bar{\mu}e}$ for $\mu^+ e^-$ scattering and explain its relation to the result for $A_{\mu e}$ derived in (b).

6.21. Show that if the couplings $Z\nu_\alpha\bar{\nu}_\beta$ of the neutral intermediate boson to neutrinos are flavor preserving, they also are diagonal when expressed in terms of mass eigenstates, $Z\nu_i\bar{\nu}_j$.

6.22. Beginning with the observation that neutrino flavor eigenstates $|\nu_\alpha\rangle$ and mass eigenstates $|\nu_i\rangle$ are related through $|\nu_\alpha\rangle = \sum_{i=1}^{3} \mathcal{V}_{\alpha i}^* |\nu_i\rangle$, show that the $\nu_\alpha \to \nu_\beta$ oscillation probability for ultrarelativistic neutrinos of energy $E$ over a baseline of length $L$ is given by

$$P(\nu_\alpha \to \nu_\beta) = \left| \sum_j \mathcal{V}_{\alpha j}^* \, \mathcal{V}_{\beta j} e^{-im_j^2 L/2E} \right|^2$$

$$= \delta_{\alpha\beta} - 4\sum_{i>j} \text{Re}(\mathcal{V}_{\alpha i}^* \mathcal{V}_{\alpha j} \mathcal{V}_{\beta i} \mathcal{V}_{\beta j}^*) \sin^2(\Delta m_{ij}^2 L/4E)$$

$$+ 2\sum_{i>j} \text{Im}(\mathcal{V}_{\alpha i}^* \mathcal{V}_{\alpha j} \mathcal{V}_{\beta i} \mathcal{V}_{\beta j}^*) \sin(\Delta m_{ij}^2 L/2E),$$

where $\Delta m_{ij}^2 = m_i^2 - m_j^2$ and the neutrino mixing matrix $\mathcal{V}$ is defined in (6.7.13). [For example, see C. Giunti and M. Laveder, "Neutrino mixing," arXiv:hep-ph/0310238.]

6.23. The propagation of $\nu_e$ and $\nu_\mu$ through matter is affected by interactions with the medium, taken here to be of uniform density. At low energies, inelastic scattering can be neglected, so only forward elastic scattering is relevant. An overall phase shift does not matter for flavor change, so only the difference $\Delta V = V_e - V_\mu$ between the potentials experienced by $\nu_e$ and $\nu_\mu$ is relevant.

(a) Demonstrate that in ordinary matter, $\Delta V$ arises from the charged-current contribution to $\nu_e e$ elastic scattering, so that $\Delta V = G_F \sqrt{2} n_e$, where $n_e$ is the number density of electrons in the medium.

(b) Show that the Hamiltonian governing the evolution of the $(\nu_e, \nu_\mu)$ system may be written as

$$\mathcal{H} = \begin{bmatrix} -\dfrac{\Delta m^2}{4E}\cos 2\theta + G_F\sqrt{2}n_e & \dfrac{\Delta m^2}{4E}\sin 2\theta \\[2ex] \dfrac{\Delta m^2}{4E}\sin 2\theta & \dfrac{\Delta m^2}{4E}\cos 2\theta \end{bmatrix},$$

where $\Delta m^2 \equiv m_2^2 - m_1^2$ and $\theta$ is the $(\nu_e, \nu_\mu)$ mixing angle in the vacuum.

(c) How does the Hamiltonian change in the case of antineutrino propagation?

(d) Compute the mixing parameter $\sin^2 2\theta_m$ in matter, and show that

$$\sin^2 2\theta_m = \frac{\sin^2 2\theta}{(\omega - \cos 2\theta)^2 + \sin^2 2\theta},$$

with $\omega = 2\sqrt{2}G_F n_e E / \Delta m^2$. Note that $\sin^2 2\theta_m$ is sensitive to the sign of $\Delta m^2$. (e) State the conditions for resonant oscillation. (Consult Refs. [35] and [41].)

## ⅢⅢⅢⅢⅢ FOR FURTHER READING ⅢⅢⅢⅢⅢ

**Weak interactions before gauge theories.** The universal $V - A$ interaction was proposed by

R. P. Feynman and M. Gell-Mann, *Phys. Rev.* **109**, 193 (1958),

S. S. Gershtein and Y. B. Zel'dovich, *Zh. Eksp. Teor. Fiz.* **29**, 698 (1955) [English translation: *Sov. Phys.–JETP* **2**, 576 (1956)],

E.C.G. Sudarshan and R. E. Marshak, *Phys. Rev.* **109**, 1860 (1958),

J. J. Sakurai, *Nuovo Cim.* **7**, 649 (1958).

Establishment of the $V - A$ interaction was a major achievement of the 1950s. Several of the key experiments are reported in

W.B. Hermannsfelt et al., *Phys. Rev. Lett.* **1**, 61 (1958),

T. Fazzini et al., *Phys. Rev. Lett.* **1**, 247 (1958),

G. Impeduglia et al., *Phys. Rev. Lett.* **1**, 249 (1958),

H. L. Anderson et al, *Phys. Rev. Lett.* **2**, 53 (1959),

M. Goldhaber, L. Grodzins, and A. W. Sunyar, *Phys. Rev.* **109**, 1015 (1958).

The history is reviewed in

P. K. Kabir (ed.), *The Development of Weak Interaction Theory*, Gordon and Breach, New York, 1963,

E. J. Konopinski, *The Theory of Beta Radioactivity*, Oxford University Press, London, 1966.

The status of searches for right-handed charged currents in muon decay is reported in

J. F. Bueno et al. [TWIST Collaboration], *Phys. Rev. D* **84**, 032005 (2011).

Tests of the electroweak theory in beta decay are reviewed in

N. Severijns, M. Beck, and O. Naviliat-Cuncic, *Rev. Mod. Phys.* **78**, 991 (2006).

Among introductions to the $V - A$ theory and its consequences, exemplary treatments appear in

E. D. Commins, *Weak Interactions*, McGraw-Hill, New York, 1973,

R. Marshak, Riazuddin, and C. P. Ryan, *Theory of Weak Interactions in Particle Physics*, Wiley-Interscience, New York, 1969,

L. B. Okun, *Weak Interactions of Elementary Particles*, Pergamon, Oxford, 1965.

Useful instruction in computational techniques is to be found in

R. P. Feynman, *The Theory of Fundamental Processes*, Westview Press, Boulder, CO, 1998,

M. Veltman, *Diagrammatica: The Path to Feynman Diagrams*, Cambridge University Press, Cambridge, 1994.

**Gauge theories of weak and electromagnetic interactions.** A lively account is given in
F. Close, *The Infinity Puzzle*, Oxford University Press, Oxford, 2011.

Informative intellectual histories have been presented by

M. Veltman, in *Proceedings of the 6th International Symposium on Electron and Photon Interactions at High Energies*, edited by H. Rollnik and W. Pfeil, North-Holland, Amsterdam, 1974, p. 429,
S. Coleman, *Science* **206**, 1290 (1979),
S. Weinberg, *Int. J. Mod. Phys. A* **23**, 1627 (2008).

Also of interest are the Nobel lectures of S. L. Glashow, A. Salam, and S. Weinberg cited in the bibliography to chapter 1, along with those of

G. 't Hooft, *Rev. Mod. Phys.* **72**, 333 (2000) [Erratum-ibid. **74**, 1343 (2003)],
M.J.G. Veltman, *Rev. Mod. Phys.* **72**, 341 (2000),

and the retrospective essays by

B. W. Lee, "Development of Unified Gauge Theories: Retrospect," in *Gauge Theories and Neutrino Physics*, ed. M. Jacob, North-Holland, Amsterdam, 1978, p. 147 [Fermilab-CONF-77-017/T] http://j.mp/xb5296,
S.A. Bludman, "The Role of Gauge Theory, Symmetry-Breaking, and Electroweak Unification in the Discovery of Weak Neutral Currents," *AIP Conf. Proc.* **300**, 164 (1994).

Among notable contributions in the development of the standard model, see

O. Klein, "On the Theory of Charged Fields," reprinted in L. O'Raifeartaigh, *The Dawning of Gauge Theory*, Princeton University Press, Princeton, 1997, p. 152, and in *Surveys in High Energy Physics* **5**, 269 (1986),
J. Schwinger, *Ann. Phys. (NY)* **2**, 407 (1957),
S. A. Bludman, *Nuovo Cim.* **9**, 443 (1958),
S. L. Glashow, *Nucl. Phys.* **22**, 579 (1961),
A. Salam and J. C. Ward, *Phys. Rev. Lett.* **13**, 168 (1964).

A number of these appear in the well-chosen reprint collections,

C. H. Lai (ed.), *Gauge Theories of the Weak and Electromagnetic Interactions*, World Scientific, Singapore, 1981,
C. H. Lai and R. N. Mohapatra (ed.), *Gauge Theories of the Fundamental Interactions*, World Scientific, Singapore, 1981.

In addition to the textbooks and monographs cited in earlier chapters, the articles by

G. Altarelli, *Nuovo Cim.* **123B**, 257 (2008),
S. Dawson, *Int. J. Mod. Phys. A* **21**, 1629 (2006),
P. Langacker, "Introduction to the Standard Model and Electroweak Physics," in *The Dawn of the LHC Era*, Proceedings of TASI 2008, ed. Tao Han, World Scientific, Singapore, 2010, p. 3 [arXiv:0901.0241],
C. Quigg, *Ann. Rev. Nucl. Part. Sci.* **59**, 505 (2009),

provide self-contained discussions of the electroweak model. Technical issues are superbly treated in

R. Balian and J. Zinn-Justin (ed.), *Methods in Field Theory*, 1975 Les Houches Lectures, North-Holland, Amsterdam, 1976; World Scientific, Singapore, 1981.

**$S$-matrix "derivation" of spontaneously broken gauge theories.** The program of divergence cancellation for the tree diagrams is thoroughly explained by

C. H. Llewellyn Smith, in *Phenomenology of Particles at High Energies,* 14th Scottish Universities Summer School in Physics, 1973, ed. R. L. Crawford and R. Jennings, Academic, New York and London, 1974, p. 459.

**Neutral currents.** A first-person account of the discovery of neutral currents is given by

D. Haidt, *Eur. Phys. J. C* **34**, 25 (2004).

For a historian's reconstruction of the search and discovery, see

P. Galison, *How Experiments End,* University of Chicago Press, Chicago, 1987, chap. 4; *Rev. Mod. Phys.* **55**, 477 (1983).

The outcome of (pre–$Z$-factory) "model-independent" determinations of the neutral current couplings is assessed in

P. Langacker, *Comments Nucl. Part. Phys.* **19**, 1 (1989).

A unit ratio between the neutral-current and charged-current couplings, expressed at tree level through $\rho \equiv M_W^2 / M_Z^2 \cos^2 \theta_W$, does not uniquely select Higgs doublets as the mechanism for electroweak symmetry breaking, but holds so long as a global "custodial" SU(2) symmetry is present. For discussions of custodial symmetry, see

M. Weinstein, *Phys. Rev. D* **8**, 2511 (1973),

P. Sikivie, L. Susskind, M. B. Voloshin, and V. I. Zakharov, *Nucl. Phys. B* **173**, 189 (1980).

**Neutral-current interactions at small momentum transfers.** For a prospectus of low-energy tests of the electroweak theory, see

J. Erler and M. J. Ramsey-Musolf, *Prog. Part. Nucl. Phys.* **54**, 351 (2005).

The search for new physics through measurements of the weak charge of the proton is treated in

J. Erler, A. Kurylov, and M. J. Ramsey-Musolf, *Phys. Rev. D* **68**, 016006 (2003).

Experiments to probe the neutral current in atomic physics are reviewed by

E. D. Commins and P. H. Bucksbaum, *Ann. Rev. Nucl. Part. Sci.* **30**, 1 (1980),

M.-A. Bouchiat and C. Bouchiat, *Rept. Prog. Phys.* **60**, 1351 (1997),

J. Guéna, M. Lintz, and M.-A. Bouchiat, *Mod. Phys. Lett. A* **20**, 375 (2005),

W. C. Haxton and C. E. Wieman, *Ann. Rev. Nucl. Part. Sci.* **51**, 261 (2001).

For a recent contribution, see

K. Tsigutkin et al., *Phys. Rev. Lett.* **103**, 071601 (2009). For context, see K. Jungmann, "Good Fortune from a Broken Mirror," *Physics* **2**, 68 (2009).

A new precision measurement of the parity-violating analyzing power in longitudinally polarized $ep$ scattering at very low $Q^2$ is described in

W. T. H. van Oers [Qweak Collaboration], *Nucl. Phys. A* **805**, 329 (2008).

For calculations, measurements, and commentary on studies of the weak neutral current in measurements of electron–electron scattering, see

A. Czarnecki and W. J. Marciano, *Phys. Rev. D* **53**, 1066 (1996),

P. L. Anthony et al. [SLAC E158 Collaboration], *Phys. Rev. Lett.* **95**, 081601 (2005),

A. Czarnecki and W. J. Marciano, *Nature* **435**, 437 (2005).

Prospects for high-precision studies of neutrino–electron scattering are presented in

W. J. Marciano and Z. Parsa, *J. Phys. G* **29**, 2629 (2003),

A. de Gouvêa and J. Jenkins, *Phys. Rev. D* **74**, 033004 (2006).

**The Higgs boson.** The expected properties of light Higgs bosons have been detailed in

J. F. Gunion, H. E. Haber, G. L. Kane, and S. Dawson, *The Higgs Hunter's Guide,* Westview Press, Boulder, CO, 2000,

A. Djouadi, *Phys. Rept.* **457**, 1 (2008).

The computer code HDECAY, by A. Djouadi, J. Kalinowski, M. C. Muehlleitner, and M. Spira, calculates the total decay widths and the branching ratios of the standard-model Higgs boson as well as those of the neutral, pseudoscalar, and charged Higgs bosons of the minimal supersymmetric extension of the standard model (MSSM). The program is available at http://j.mp/w5rxdJ. For documentation, see

A. Djouadi, J. Kalinowski, and M. Spira, *Comput. Phys. Commun.* **108**, 56 (1998), arXiv:hep-ph/9704448.

Extensive tables of partial widths for fermionic and bosonic decays of the Higgs boson are given in appendix B of

S. Dittmaier et al. [LHC Higgs Cross Section Working Group], "Handbook of LHC Higgs Cross Sections: 1. Inclusive Observables," arXiv:1101.0593.

A strategy for bounding the Higgs mass from above based on the calculability of radiative corrections, rather than the partial-wave unitarity methods described in the text, has been developed by

M. Veltman, *Acta Phys. Polon. B* **8**, 475 (1977); *Nucl. Phys.* **B123**, 89 (1977); *Phys. Lett.* **91B**, 95 (1980); and in *Quarks and Leptons,* Cargèse 1979, ed. M. Lévy et al., Plenum, New York, 1980, p. 1.

Related bounds on heavy-fermion masses are also implied by tree-unitarity considerations. For these, see

M. S. Chanowitz, M. A. Furman, and I. Hinchliffe, *Nucl. Phys.* **B153**, 402 (1979).

**Symmetries and gauge-boson interactions.** Low-energy theorems for the scattering of longitudinally polarized $W$ and $Z$ bosons that hold at a scale intermediate between $M_W$ and the scale of electroweak symmetry breaking are proved in

M. S. Chanowitz, M. Golden, and H. Georgi, *Phys. Rev. D* **36**, 1490 (1987).

These results correspond directly to the low-energy theorems for $\pi\pi$ scattering amplitudes exhibited in

S. Weinberg, *Phys. Rev. Lett.* **17**, 616 (1966).

**Strongly interacting gauge sector.** The phenomenology of strongly interacting electroweak gauge bosons is developed in

M. S. Chanowitz and M. K. Gaillard, *Nucl. Phys.* **B261**, 379 (1985),

G. F. Giudice et al., *JHEP* **0706**, 045 (2007).

**The idea of neutrinos.** For reflections on the early history of neutrino physics, see

B. Pontecorvo, "The Infancy and Youth of Neutrino Physics: Some Recollections," *J. Phys. Colloques* **43** C8-221-C8-236 (1982).

The evolution of the neutrino from conjecture to reality is chronicled in

A. Franklin, *Are There Really Neutrinos?,* Perseus, Cambridge, MA, 2004.

**Neutrino oscillations.** The possibility of neutrino oscillations was advanced by
> B. Pontecorvo, *Zh. Eksp. Teor. Fiz.* **34**, 247 (1958), [*Sov. Phys. JETP* **7**, 172 (1958)],
> Z. Maki, M. Nakagawa, and S. Sakata, *Prog. Theor. Phys.* **28**, 870 (1962).

The early history of neutrino oscillation studies is recounted in the Nobel lectures by
> R. Davis, *Rev. Mod. Phys.* **75**, 985 (2003),
> M. Koshiba, *Rev. Mod. Phys.* **75**, 1011 (2003).

An excellent short introduction, along with reprints of some classic papers, appears in
> R. Cahn and G. Goldhaber, *The Experimental Foundations of Particle Physics,* 2nd ed.,
> Cambridge University Press, Cambridge, 2009, chap. 16.

General references for the theory of neutrinos include
> R. N. Mohapatra and A. Y. Smirnov, *Ann. Rev. Nucl. Part. Sci.* **56**, 569 (2006),
> R. N. Mohapatra et al., *Rept. Prog. Phys.* **70**, 1757 (2007).

Explorations toward a picture of neutrino masses are reviewed by
> G. Altarelli and F. Feruglio, *Rev. Mod. Phys.* **82**, 2701 (2010).

Many derivations of the neutrino-oscillation formalism cut corners and so may seem paradoxical or internally inconsistent. Thoughtful treatments of the conceptual and practical issues are given in
> E. K. Akhmedov and A. Y. Smirnov, "Paradoxes of Neutrino Oscillations," *Phys. Atom. Nucl.* **72**, 1363 (2009); extended version at arXiv:0905.1903,
> E. K. Akhmedov and J. Kopp, "Neutrino Oscillations: Quantum Mechanics vs. Quantum Field Theory," *JHEP* **1004**, 008 (2010),
> E. K. Akhmedov and A. Y. Smirnov, "Neutrino Oscillations: Entanglement, Energy-Momentum Conservation and QFT," *Found. Phys.* **41**, 1279 (2011), arXiv:1008.2077.

In the quantum-mechanical picture, the neutrino is treated as a superposition of mass-eigenstate wave packets. Careful derivations appear in
> C. Giunti and C. W. Kim, *Phys. Rev. D* **58**, 017301 (1998);
> C. Giunti, *Found. Phys. Lett.* **17**, 103 (2004), hep-ph/0302026.

The quantum-field-theory approach treats the neutrino's interaction partners as wave packets and the neutrino as an internal line in a macroscopic Feynman diagram. For thorough presentations, see
> C. Giunti et al., *Phys. Rev. D* **48**, 4310 (1993),
> W. Grimus, S. Mohanty, and P. Stockinger, *Phys. Rev. D* **61**, 033001 (2000),
> A. G. Cohen, S. L. Glashow, and Z. Ligeti, *Phys. Lett. B* **678**, 191 (2009).

Neutrinos acquire effective masses from coherent scattering processes as they propagate through matter. The charged-current contribution to $\nu_e e$ elastic scattering differentiates the "index of refraction" of $\nu_e$ from the other neutrinos, as observed by
> L. Wolfenstein, *Phys. Rev. D* **17**, 2369 (1978), **20**, 2634 (1979),
> S. P. Mikheev and A. Y. Smirnov, *Sov. J. Nucl. Phys.* **42**, 913 (1985) [*Yad. Fiz.* **42**, 1441 (1985)]; *Nuovo Cim. C* **9**, 17 (1986).

The organic review
> A. Strumia and F. Vissani, "Neutrino Masses and Mixings and . . . ," http://j.mp/vx4S1t, see also arXiv:hep-ph/0606054

contains a wealth of experimental information and analysis.

**Sterile neutrinos.** For a brief review of additional neutrino flavors that do not experience interactions mediated by $SU(2)_L \otimes U(1)_Y$ gauge bosons, see

A. Kusenko, *New J. Phys.* **11**, 105007 (2009).

The role of sterile neutrinos in cosmology and astrophysics is explored in

A. Boyarsky, O. Ruchayskiy, and M. Shaposhnikov, *Ann. Rev. Nucl. Part. Sci.* **59**, 191 (2009).

A critical review of evidence for light sterile neutrinos is given by

J. Kopp, M. Maltoni, and T. Schwetz, *Phys. Rev. Lett.* **107**, 091801 (2011).

**Neutrinoless double beta decay.** Ordinary double-beta decay ($\beta\beta_{2\nu}$) was first discussed by

M. Goeppert-Mayer, *Phys. Rev.* **48**, 513 (1935).

The possibility of neutrinoless double-beta decay for "truly neutral" Majorana neutrinos was raised by

W. H. Furry, *Phys. Rev.* **56**, 1184 (1939).

Expectations for $SU(2)_L \otimes U(1)_Y$ (standard-model) contributions to $\beta\beta_{0\nu}$ decay are given in

J. Schechter and J.W.F. Valle, *Phys. Rev. D* **25**, 2951 (1982).

Evidence for $\beta\beta_{0\nu}$ in $^{76}$Ge is reported in

H. V. Klapdor-Kleingrothaus and I. V. Krivosheina, *Mod. Phys. Lett. A* **21**, 1547 (2006).

For recent improved limits in $^{136}$Xe, see

M. Auger et al. [EXO Collaboration], "Search for Neutrinoless Double-Beta Decay in $^{136}$Xe with EXO-200," arXiv:1205.5608,
A. Gando et al. [KamLAND-Zen Collaboration], *Phys. Rev. C* **85**, 045504 (2012).

A comprehensive introduction to $\beta\beta$ decay, with reprints of many important articles, is

H. V. Klapdor-Kleingrothaus, *Seventy Years of Double Beta Decay*, World Scientific, Singapore, 2010.

Recent reviews include

S. R. Elliott and P. Vogel, *Ann. Rev. Nucl. Part. Sci.* **52**, 115 (2002),
F. T. Avignone, S. R. Elliott, and J. Engel, *Rev. Mod. Phys.* **80**, 481 (2008),
W. Rodejohann, *Int. J. Mod. Phys. E* **20**, 1833 (2011),
J. J. Gomez-Cadenas et al., *Riv. Nuovo Cim.* **35**, 29 (2012).

**Majorana fermions.** For an informative general commentary, see

F. Wilczek, *Nature Physics* **5**, 614 (2009).

For a survey of the search for Majorana fermions in condensed-matter physics, see

T. L. Hughes, "Majorana Fermions Inch Closer to Reality," *Physics* **4**, 67 (2011).

Signatures of Majorana fermions in hybrid superconductor-semiconductor nanowire devices have been reported by

V. Mourik et al., *Science* **336**, 1003 (2012).

**Charged-lepton flavor violation.** For reviews of experimental searches for lepton flavor violation involving $e$, $\mu$, and $\tau$ leptons, and prospects in the context of theories beyond the standard model, see

W. J. Marciano, T. Mori, and J. M. Roney, *Ann. Rev. Nucl. Part. Sci.* **58**, 315 (2008),
A. de Gouvêa, *Nucl. Phys. Proc. Suppl.* **188**, 303 (2009),
Y. Okada, "Models of Lepton Flavor Violation," in *Lepton Dipole Moments,* ed. B. L. Roberts and W. J. Marciano, World Scientific, Singapore, 2009, p. 683,
Y. Kuno, "Search for the Charged Lepton-Flavor-Violating Transition Moments $\ell \rightarrow \ell'$," in *Lepton Dipole Moments*, p. 701.

Leptonic CP violation is reviewed in

G. C. Branco, R. González Felipe, and F. R. Joaquim, *Rev. Mod. Phys.* **84**, 515 (2012).

**Neutrinos in the universe.** The implications of neutrinos for cosmology are treated in

A. D. Dolgov, *Phys. Rept.* **370**, 333 (2002),
J. Lesgourgues and S. Pastor, *Phys. Rept.* **429**, 307 (2006).

**Renormalization of spontaneously broken gauge theories.** Among the classic papers on the subject are

G. 't Hooft, *Nucl. Phys.* **B33**, 173 (1971); **B35**, 167 (1971),
B. W. Lee and J. Zinn-Justin, *Phys. Rev. D* **5**, 3121, 3137, 3155 (1972); **7**, 1049 (1973),
G. 't Hooft and M. Veltman, *Nucl. Phys.* **B44**, 189 (1972); **B50**, 318 (1972),
D. A. Ross and J. C. Taylor, *Nucl. Phys.* **B51**, 125 (1973),
C. Becchi, A. Rouet, and R. Stora, *Ann. Phys. (NY)* **98**, 287 (1976).

Important for the renormalization program are the generalized Ward–Takahashi identities presented in

J. C. Taylor, *Nucl. Phys.* **B33**, 436 (1971),
A. A. Slavnov, *Teor. Mat. Fiz.* **10**, 153 (1972) [English translation: *Theor. Math. Phys.* **10**, 99 (1972)].

**Anomalies.** Surveys of the axial anomaly problem are given in

S. L. Adler, in Lectures on *Elementary Particles and Quantum Field Theory,* Brandeis Summer Institute 1970, ed. S. Deser, M. Grisaru, and H. Pendleton, MIT Press, Cambridge, MA, 1970, vol. 1, p. 1,
R. Jackiw, in *Lectures on Current Algebra and Its Applications,* by S. B. Treiman, R. Jackiw, and D. J. Gross, Princeton University Press, Princeton, NJ, 1972, p. 97. See also R. W. Jackiw, *Scholarpedia* **3** (10), 730 (2008) [http://j.mp/qsgylg]
J. F. Donoghue, E. Golowich, and B. R. Holstein, *Dynamics of the Standard Model,* Cambridge University Press, Cambridge, 1992, §III-3,
W. A. Bardeen, "Anomalies," in *Quantum Field Theory,* ed. P. Breitenlohner and D. Maison: *Lect. Notes Phys.* **558**, 3 (2000); *Prog. Theor. Phys. Suppl.* **167**, 44 (2007).

## REFERENCES

1. S. Weinberg, *Phys. Rev. Lett.* **19**, 1264 (1967). For a related development, see A. Salam, in *Elementary Particle Theory: Relativistic Groups and Analyticity* (Nobel Symposium No. 8), ed. N. Svartholm, Almqvist and Wiksell, Stockholm, 1968, p. 367, http://j.mp/r9dJOo. [See also A. Salam and J. C. Ward, *Phys. Lett.* **13**, 168 (1964).] The theory is built on the $SU(2)_L \otimes U(1)_Y$ gauge symmetry investigated by S. L. Glashow, *Nucl. Phys.* **22**, 579 (1961).

2. E. Fermi, *Ric. Sci.* **4**, 491 (1933); reprinted in *E. Fermi, Collected Papers,* edited by E. Segrè et al., University of Chicago Press, Chicago, 1962, vol. 1, p. 538; and *Z. Phys.* **88**, 161 (1934) [English translation: F. L. Wilson, *Am. J. Phys.* **36**, 1150 (1968)].

3. B. Kayser, G. T. Garvey, E. Fischbach, and S. P. Rosen, *Phys. Lett. B* **52**, 385 (1974).

4. K. Nakamura et al. [Particle Data Group], *J. Phys. G* **37**, 075021 (2010). Consult J. Beringer et al. [Particle Data Group], *Phys. Rev. D* **86**, 010001 (2012) for updates.

5. P. Vilain et al. [CHARM-II Collaboration], *Phys. Lett. B* **364**, 121 (1995).

6. J. A. Formaggio et al. [NuTeV Collaboration], *Phys. Rev. Lett.* **87**, 071803 (2001).

7. T. D. Lee and C. S. Wu, *Ann. Rev. Nucl. Sci.* **15**, 381 (1965).

8. M. Gell-Mann et al., *Phys. Rev.* **179**, 1518 (1969).

9. A model of this kind was proposed in a gauge-theory framework by H. Georgi and S. L. Glashow, *Phys. Rev. Lett.* **28**, 1494 (1972).

10. C. H. Llewellyn Smith, *Phys. Lett.* **46B**, 233 (1973); S. Joglekar, *Ann. Phys. (NY)* **83**, 427 (1974).

11. M. Goldhaber, L. Grodzins, and A. W. Sunyar, *Phys. Rev.* **109**, 1015 (1958). For a perceptive essay on neutrino helicity in light of neutrino mass, see A. S. Goldhaber and M. Goldhaber, *Phys. Today* **64** (5), 40 (May 2011). For determinations of the $\nu_\mu$ helicity, see M. Bardon, P. Franzini, and J. Lee, *Phys. Rev. Lett.* **7**, 23 (1961); A. Possoz et al., *Phys. Lett.* **B70**, 265 (1977); L. Ph. Roesch et al., *Am. J. Phys.* **50**, 931 (1982). The $\nu_\tau$ helicity was measured by K. Abe et al. [SLD Collaboration], *Phys. Rev. Lett.* **78**, 4691 (1997); A. Heister et al. [ALEPH Collaboration], *Eur. Phys. J. C* **22**, 217 (2001).

12. O. P. Sushkov, V. V. Flambaum, and I. B. Khriplovich, *Yad. Fiz.* **20**, 1016 (1974) [English translation: *Sov. J. Nucl. Phys.* **20**, 537 (1975)]; W. Alles, C. Boyer, and A.J. Buras, *Nucl. Phys.* **B119**, 125 (1977); K. Gaemers and G. Gounaris, *Z. Phys. C* **1**, 259 (1979).

13. LEP Electroweak Working Group, http://lepewwg.web.cern.ch/LEPEWWG/.

14. V. M. Abazov et al. [D0 Collaboration], *Phys. Rev. Lett.* **103**, 191801 (2009).

15. T. Aaltonen et al. [CDF Collaboration], *Phys. Rev. Lett.* **104**, 201801 (2010).

16. J. M. Campbell and R. K. Ellis, *Phys. Rev. D* **60**, 113006 (1999); S. Frixione and B. R. Webber, *JHEP* **0206**, 029 (2002). For theoretical expectations in the LHC energy regime, see for example J. M. Campbell, R. K. Ellis, and C. Williams, *JHEP* **1107**, 018 (2011).

17. Some early results include G. Aad et al. [ATLAS Collaboration], *Phys. Lett. B* **709**, 341 (2012); *Phys. Rev. Lett.* **108**, 041804 (2012). CMS Collaboration, "Measurement of the WW, WZ and ZZ cross sections at CMS," CMS Physics Analysis Summary EWK-11-010, http://j.mp/xBdSUL.

18. F. J. Hasert et al., *Phys. Lett.* **46B**, 121, 138 (1973); *Nucl. Phys.* **B73**, 1 (1974). See also the evidence from Fermilab Experiment E-1A, A. Benvenuti et al., *Phys. Rev. Lett.* **32**, 800 (1974).

19. B. Aubert et al., *Phys. Rev. Lett.* **34**, 1454, 1457 (1974); S. J. Barish et al., *Phys. Rev. Lett.* **33**, 468 (1974); B. C. Barish et al., *Phys. Rev. Lett.* **34**, 538 (1975).

20. The ALEPH, DELPHI, L3, OPAL, SLD Collaborations, the LEP Electroweak Working Group, the SLD Electroweak and Heavy Flavour Groups, *Phys. Rept.* **427**, 257 (2006), Figure 1.15.

21. $\sigma(\bar{\nu}_e e)$: M. Deniz et al. [TEXONO Collaboration], *Phys. Rev. D* **81**, 072001 (2010); $\sigma(\nu_e e)$: R. C. Allen et al., *Phys. Rev. D* **47**, 11 (1993); $\sigma(\nu_\mu e)$ and $\sigma(\bar{\nu}_\mu e)$: P. Vilain et al. [CHARM-II Collaboration], *Phys. Lett. B* **281**, 159 (1992).

22. J. Mnich, *Phys. Rept.* **271**, 181 (1996) contains very useful compilations of data on forward-backward asymmetries in $e^+e^- \to \mu^+\mu^-$ (table 20) and $e^+e^- \to \tau^+\tau^-$ (table 21).

23. M. Acciarri et al. [L3 Collaboration], *Phys. Lett. B* **374**, 331 (1996).

24. The masslessness of the neutrinos is a consequence of the absence of right-handed neutrinos in the original enumeration of fermions. We discuss neutrino mass briefly in §6.7.

25. S. Dittmaier et al. [LHC Higgs Cross Section Working Group], "Handbook of LHC Higgs Cross Sections: 1. Inclusive Observables," arXiv:1101.0593.

26. Logarithmic violations of unitarity that occur at exponentially high energies $\sim M_W \exp^{1/\alpha}$ need not concern us here.

27. J. M. Cornwall, D. N. Levin, and G. Tiktopoulos, *Phys. Rev. Lett.* **30**, 1268 (1973); *Phys. Rev. D* **10**, 1145 (1974).

28. B. W. Lee, C. Quigg, and H. B. Thacker, *Phys. Rev. D* **16**, 1519 (1977). See also D. A. Dicus and V. S. Mathur, *Phys. Rev. D* **7**, 3111 (1973).

29. KArlsruhe TRItium Neutrino experiment, http://www-ik.fzk.de/tritium/.

30. G. Steigman, *Ann. Rev. Nucl. Part. Sci.* **29**, 313 (1979) is a convenient primer.

31. R. E. Lopez et al., *Phys. Rev. Lett.* **82**, 3952 (1999).

32. A. G. Riess et al., *Astrophys. J.* **699**, 539 (2009).

33. This line of argument may be traced to S. S. Gershtein and Y. B. Zel'dovich, *JETP Lett.* **4**, 120 (1966); R. Cowsik and J. McClelland, *Phys. Rev. Lett.* **29**, 669 (1972), *Astrophys. J.* **180**, 7 (1973); A. S. Szalay and G. Marx, *Astron. Astrophys.* **49**, 437 (1976).

34. S. Hannestad et al., *JCAP* **1008**, 001 (2010).

35. A convenient reference is K. Nakamura and S. T. Petcov, "Neutrino Mass, Mixing, and Oscillations," in Ref. [4], p. 164.

36. K. S. Hirata et al. [Kamiokande-II Collaboration], *Phys. Lett. B* **280**, 146 (1992).

37. Y. Ashie et al. [Super-Kamiokande Collaboration], *Phys. Rev. D* **71**, 112005 (2005).

38. M. H. Ahn et al. [K2K Collaboration], *Phys. Rev. D* **74**, 072003 (2006).

39. P. Adamson et al. [MINOS Collaboration], *Phys. Rev. Lett.* **106**, 181801 (2011). Consistent results for $\bar{\nu}_\mu$ oscillations are reported in *Phys. Rev. Lett.* **108**, 191801 (2012).

40. N. Agafonova et al. [OPERA Collaboration], *Phys. Lett. B* **691**, 138 (2010). See also M. Nakamura, "Results from OPERA" presented at Neutrino 2012, http://j.mp/NNXGGS.

41. A. Y. Smirnov, "The MSW effect and solar neutrinos," X International Workshop on Neutrino Telescopes, Venice (2003), arXiv:hep-ph/0305106.

42. S. Abe et al. [KamLAND Collaboration], *Phys. Rev. Lett.* **100**, 221803 (2008).

43. C. Arpesella et al. [Borexino Collaboration], *Phys. Rev. Lett.* **101**, 091302 (2008).

44. R. Davis, Jr., D. S. Harmer, and K. C. Hoffman, *Phys. Rev. Lett.* **20**, 1205 (1968); B. T. Cleveland et al., *Astrophys. J.* **496**, 505 (1998); J. N. Abdurashitov et al. [SAGE Collaboration], *Phys. Rev. C* **80**, 015807 (2009). M. Cribier et al. [GALLEX Collaboration], *Nucl. Phys. Proc. Suppl.* **70**, 284 (1999).

45. J. N. Bahcall, A. M. Serenelli, and S. Basu, *Astrophys. J.* **621**, L85 (2005).

46. S. Fukuda et al. [Super-Kamiokande Collaboration], *Phys. Lett. B* **539**, 179 (2002).

47. B. Aharmim et al. [SNO Collaboration], *Phys. Rev. C* **72**, 055502 (2005); see also "Combined Analysis of all Three Phases of Solar Neutrino Data from the Sudbury Neutrino Observatory," arXiv:1109.0763.

48. B. W. Lee and R. E. Shrock, *Phys. Rev. D* **16**, 1444 (1977).

49. F. P. An et al. [Daya Bay Collaboration], *Phys. Rev. Lett.* **108**, 171803 (2012), arXiv:1210.6327; J. K. Ahn et al. [RENO Collaboration], *Phys. Rev. Lett.* **108**, 191802 (2012). For earlier evidence for $\theta_{13} \neq 0$ from $\nu_\mu \to \nu_e$ appearance over long baselines in beams produced at accelerators, see K. Abe et al. [T2K Collaboration], *Phys. Rev. Lett.* **107**, 041801 (2011); P. Adamson et al. [MINOS Collaboration], *Phys. Rev. Lett.* **107**, 181802 (2011). See also Y. Abe et al. [Double-CHOOZ Collaboration], *Phys. Rev. Lett.* **108**, 131801 (2012); A. Gando et al. [KamLAND Collaboration], *Phys. Rev. D* **83**, 052002 (2011).

50. H. Nunokawa, S. J. Parke, and J. W. F. Valle, *Prog. Part. Nucl. Phys.* **60**, 338 (2008).

51. D. V. Forero, M. Tortola, and J.W.F. Valle, "Global Status of Neutrino Oscillation Parameters after Recent Reactor Measurements," arXiv:1205.4018.

52. G. G. Raffelt, *Stars as Laboratories for Fundamental Physics*, University of Chicago Press, Chicago, 1996; A. Mirizzi, D. Montanino, and P. D. Serpico, *Phys. Rev. D* **76**, 053007 (2007).

53. J. F. Beacom and N. F. Bell, *Phys. Rev. D* **65**, 113009 (2002).

54. A. Aguilar et al. [LSND Collaboration], *Phys. Rev. D* **64**, 112007 (2001).

55. A. A. Aguilar-Arevalo et al. [MiniBooNE Collaboration], *Phys. Rev. Lett.* **98**, 231801 (2007).

56. A. A. Aguilar-Arevalo et al. [MiniBooNE Collaboration], *Phys. Rev. Lett.* **105**, 181801 (2010).

57. G. Mention et al., *Phys. Rev. D* **83**, 073006 (2011).

58. E. Majorana, *Nuovo Cim.* **14**, 171 (1937) [new English translation by L. Maiani, *Soryushiron Kenkyu* **63**, 149 (1981), http://j.mp/xTLCDU, and (with summary added) in *Ettore Majorana: Scientific Papers*, ed. G. F. Bassani et al., Società Italiana di Fisica, Bologna & Springer, Berlin Heidelberg New York, 2006, p. 218, http://j.mp/y9lnve].

59. P. Minkowski, *Phys. Lett. B* **67**, 421 (1977); T. Yanagida, in *Proceedings of the Workshop on the Baryon Number of the Universe and Unified Theories*, ed. O. Sawada and A. Sugamoto (KEK, Tsukuba, 1979), pp. 95Ð98, [reprinted in *SEESAW 25: Proceedings of the International Conference on the Seesaw Mechanism*, ed. J. Orloff, S. Lavignac, and M. Cribier, World Scientific, Singapore, 2005, p. 261]; M. Gell-Mann, P. Ramond, and R. Slansky, in *Supergravity*, ed. P. van Nieuwenhuizen and D. Z. Freedman, North-Holland, Amsterdam, 1979, p. 95; R. N. Mohapatra and G. Senjanovic, *Phys. Rev. Lett.* **44**, 912 (1980); J. Schechter and J.W.F. Valle, *Phys. Rev. D* **22**, 2227 (1980).

60. A. Y. Smirnov, *SEESAW 25: Proceedings of the International Conference on the Seesaw Mechanism*, ed. J. Orloff, S. Lavignac, and M. Cribier, World Scientific, Singapore, 2005, p. 221, arXiv:hep-ph/0411194.

61. M.-C. Chen, A. de Gouvêa, and B. A. Dobrescu, *Phys. Rev. D* **75**, 055009 (2007).

62. G 't Hooft, *Nucl. Phys.* **B33**, 173 (1971); **B35**, 167 (1971).

63. C. Bouchiat, J. Iliopoulos, and P. Meyer. *Phys. Lett.* **38B**, 519 (1972). A thorough analysis of the anomaly problem in the Weinberg–Salam model was made by D. J. Gross and R. Jackiw, *Phys. Rev. D* **6**, 477 (1972). General conditions for anomaly cancellation were investigated by H. Georgi and S. L. Glashow, *Phys. Rev. D* **6**, 429 (1972).

# Seven ‖‖‖‖‖‖‖‖‖‖‖‖‖‖‖‖‖‖‖‖‖‖‖‖‖‖‖‖‖‖‖‖‖‖‖‖‖‖‖‖‖‖‖‖‖‖‖‖‖‖‖‖‖‖‖‖‖‖‖‖‖‖‖‖

## Electroweak Interactions of Quarks

In this chapter, we extend the electroweak theory to the hadronic sector. This will be accomplished through the medium of the quark model, as is natural in view of the experimental suggestions that quarks and leptons are comparably elementary. Because of the similarity to leptons, construction of a theory at the quark model level is relatively straightforward, and yet nontrivial. We remarked in section 6.8 that an anomaly-free and, hence, renormalizable theory could be formulated if each weak-isospin doublet of leptons were accompanied by a color triplet of weak-isospin doublets of quarks. Now we shall encounter the necessity of enlarging the quark spectrum beyond $u$, $d$, and $s$ to eliminate flavor-changing neutral currents through the Glashow–Iliopoulos–Maiani mechanism. We will generalize the GIM construction to theories with many quark and lepton doublets.

The difficulty in describing completely the weak interactions of quarks is that the quarks are bound within hadrons. We lack a comprehensive theory of hadron structure and cannot yet give a full account of the influence of the strong interactions upon weak transition amplitudes [1]. For semileptonic processes, the parton model has been found to give a reliable approximate description. Its development here serves two purposes: the first, to provide a tool for analyzing hadronic weak interactions and testing weak-interaction models; and the second, to recognize the challenge to understanding that the success of the parton model poses for a theory of the strong interactions. The role of the parton model in the development of quantum chromodynamics will be taken up further in chapter 8.

In the parton-model approximation, we shall complete our description of the properties of the electroweak gauge bosons and then turn to the study of several high-energy processes. Electron–positron annihilation is treated at some length, both for the $Z^0$ formation reaction and for the study of $\gamma$-$Z^0$ interference already developed in section 6.4. This discussion prepares the way for a discussion of the tests that have validated the electroweak theory. We then develop both the general kinematics and the parton-model description of deeply inelastic lepton–nucleon scattering. This leads to an analysis of weak neutral-current effects in neutrino–nucleon and electron–nucleon scattering and to a brief discussion of electroweak studies carried out at the high-energy electron–proton collider HERA. Next, we discuss the production of electroweak gauge bosons in hadron–hadron collisions, where $W^{\pm}$ and $Z^0$ were first observed as real particles. The quantum corrections that enabled precision tests of the electroweak theory are treated by example. We then turn to the search for the Higgs boson, hitherto the missing agent of electroweak symmetry breaking in the standard model. Outside a restricted range of Higgs-boson masses, the electroweak theory cannot be complete up to the Planck scale at which gravity is expected to intervene. Two other challenges to the completeness of the electroweak theory—the problem of maintaining widely separated scales and the vacuum energy problem—point to new physics beyond the standard model. An assessment concludes the chapter.

## 7.1 THE STANDARD ELECTROWEAK THEORY: PRELIMINARIES

We shall now arrive at the minimal internally consistent theory of weak and electromagnetic interactions. Consider first the direct generalization of the theory of electrons developed in chapter 6. We noted in section 6.8 that an $SU(2)_L \otimes U(1)_Y$ theory based upon the electron doublet

$$L_e \equiv \begin{pmatrix} \nu \\ e \end{pmatrix}_L \tag{6.3.1}$$

and quark doublets

$$\begin{pmatrix} u_{\text{Red}} \\ d_{\text{Red}} \end{pmatrix}_L, \qquad \begin{pmatrix} u_{\text{Green}} \\ d_{\text{Green}} \end{pmatrix}_L, \qquad \begin{pmatrix} u_{\text{Blue}} \\ d_{\text{Blue}} \end{pmatrix}_L \tag{6.8.20}$$

would be free of anomalies and, therefore, have the benign ultraviolet behavior required for renormalizability. For many purposes in weak-interaction theory, color merely serves as a device for counting multiplets. Whenever it is unnecessary to retain the color indices, we shall suppress them with the understanding that the quarks are color triplets.

The hadronic sector of the theory is thus built upon a single (color triplet of) left-handed weak-isospin doublet(s),

$$L_q = \begin{pmatrix} u \\ d \end{pmatrix}_L, \tag{7.1.1}$$

with weak hypercharge

$$Y(q_L) = \tfrac{1}{3} \tag{7.1.2}$$

and two (color-triplet) right-handed weak-isospin singlets,

$$\left. \begin{aligned} R_u &= u_R = \tfrac{1}{2}(1 + \gamma_5)u \\ R_d &= d_R = \tfrac{1}{2}(1 + \gamma_5)d \end{aligned} \right\} , \tag{7.1.3}$$

with weak hypercharge

$$Y(u_R) = \tfrac{4}{3} \left.\vphantom{\begin{matrix}1\\1\end{matrix}}\right\}$$
$$Y(d_R) = -\tfrac{2}{3} \quad. \tag{7.1.4}$$

The complex doublet of Higgs bosons is, as usual,

$$\phi \equiv \begin{pmatrix} \phi^+ \\ \phi^0 \end{pmatrix} \tag{6.3.15}$$

with

$$Y_\phi = +1, \tag{6.3.16}$$

and its complex conjugate is the $SU(2)_L$ doublet

$$\bar{\phi} = i\sigma_2 \phi^* = \begin{pmatrix} \bar{\phi}^0 \\ -\phi^- \end{pmatrix} \tag{7.1.5}$$

with

$$Y_{\bar{\phi}} = -1. \tag{7.1.6}$$

The only change in form from the theory of leptons is in the Yukawa interaction term,

$$\mathcal{L}_{\text{Yukawa}} = -\zeta_u[(\bar{L}_q \bar{\phi})u_R + \bar{u}_R(\bar{\phi}^\dagger L_q)]$$
$$- \zeta_d[(\bar{L}_q \phi)d_R + \bar{d}_R(\phi^\dagger L_q)], \tag{7.1.7}$$

which will generate, upon spontaneous symmetry breaking, masses for both up and down quarks.

We need not repeat all the arithmetic of section 6.3 to complete the construction of the theory but may simply refer to our earlier results for the consequences. Evidently the charged-current interaction is, by construction,

$$\mathcal{L}_{W-q} = -\frac{g}{2\sqrt{2}} \left[ \bar{u}\gamma^\mu(1 - \gamma_5)d\,W_\mu^+ + \bar{d}\gamma^\mu(1 - \gamma_5)u\,W_\mu^- \right], \tag{7.1.8}$$

and the weak neutral-current interaction is of the now familiar form [cf. (6.3.61)]

$$\mathcal{L}_{Z-q} = -\frac{g}{4\cos\theta_W} \left\{ \bar{u}\gamma^\mu[(1 - \gamma_5)\tau_3 - 4x_W Q]u\,Z_\mu \right.$$
$$\left. + \bar{d}\gamma^\mu[(1 - \gamma_5)\tau_3 - 4x_W Q]d\,Z_\mu \right\}, \tag{7.1.9}$$

where, as usual, $x_W = \sin^2\theta_W$. The gauge-boson and Higgs-boson sectors are unchanged from the theory of leptons, and the Yukawa couplings $\zeta_u$ and $\zeta_d$ may be chosen to reproduce the quark masses.

The theory of the weak and electromagnetic interactions of $\nu_e$, $e$, $u$, and $d$ that we have described is formally neat and internally self-consistent. However, the low-energy phenomenology is incorrect in a small but significant respect: the hadronic charged current is described not by the quark doublet (7.1.1), but rather by the quark-model transcription of the Cabibbo current [2], namely,

$$\mathsf{L}_q = \begin{pmatrix} u \\ d_\theta \end{pmatrix}_\mathrm{L}, \tag{7.1.10}$$

with

$$d_\theta = d \cos \theta_\mathrm{C} + s \sin \theta_\mathrm{C}, \tag{7.1.11}$$

where $\theta_\mathrm{C}$ is the Cabibbo angle, given by [3]

$$\cos \theta_\mathrm{C} = 0.974\,25 \pm 0.000\,22. \tag{7.1.12}$$

The implied $SU(2)_\mathrm{L} \otimes U(1)_\mathrm{Y}$ model is defective in several ways. It seems peculiar that the combination of charge $-\frac{1}{3}$ quarks orthogonal to (7.1.11),

$$s_\theta = s \cos \theta_\mathrm{C} - d \sin \theta_\mathrm{C}, \tag{7.1.13}$$

is apparently superfluous. More urgently, the weak neutral-current interaction now contains terms proportional to

$$\sin \theta_\mathrm{C} \cos \theta_\mathrm{C} \left[ \bar{d} \gamma^\mu (1 - \gamma_5) s + \bar{s} \gamma^\mu (1 - \gamma_5) d \right] Z_\mu, \tag{7.1.14}$$

which correspond to flavor-changing (specifically, strangeness-changing) neutral currents. This is phenomenologically unacceptable because stringent experimental limits have been placed on the rates of decays mediated by strangeness-changing neutral currents. For example, the decay

$$K^+ \to \pi^+ \nu \bar{\nu}, \tag{7.1.15}$$

which may be interpreted in terms of the elementary transition

$$\bar{s} \to \bar{d} \nu \bar{\nu}, \tag{7.1.16}$$

is observed only at the highly suppressed rate of [4]

$$\frac{K^+ \to \pi^+ \nu \bar{\nu}}{K^+ \to \mathrm{all}} = (1.73^{+1.15}_{-1.05}) \times 10^{-10}. \tag{7.1.17}$$

Similarly, the observed rate [3],

$$\frac{\Gamma(K_\mathrm{L} \to \mu^+ \mu^-)}{\Gamma(K_\mathrm{L} \to \mathrm{all})} = (6.84 \pm 0.11) \times 10^{-9} \tag{7.1.18}$$

can be understood in terms of QED and the known $K_\mathrm{L} \to \gamma\gamma$ transition rate and leaves little room for an elementary $\bar{s}d \to \mu^+\mu^-$ transition. A similar conclusion may be drawn from the smallness of observables linked to $|\Delta S| = 2$ transition amplitudes, such as the mass difference (Ref. [3])

$$M(K_L) - M(K_S) = (0.5292 \pm 0.0009) \times 10^{10} \hbar \, s^{-1}$$
$$\approx 3.5 \times 10^{-6} \text{ eV}$$
$$\approx 7 \times 10^{-15} M(K^0). \tag{7.1.19}$$

Thus, in the incipient theory before us and, more generally, in models that entail neutral-current interactions proportional to the third component of weak isospin, it becomes important to guard against the appearance of strangeness-changing neutral currents.

An elegant solution to the problem of flavor-changing neutral currents was devised by Glashow, Iliopoulos, and Maiani [5]. The key observation is that by introducing a new "charmed" quark $c$ as the weak-isospin partner of $s_\theta$, we cancel the offending terms (7.1.14) in the neutral-current interaction with quarks. To follow the historical development, we consider an $SU(2)_L \otimes U(1)_Y$, in which the fundamental fermions are the leptons

$$\mathsf{L}_e = \begin{pmatrix} \nu_e \\ e \end{pmatrix}_L, \qquad \mathsf{L}_\mu = \begin{pmatrix} \nu_\mu \\ \mu \end{pmatrix}_L; \qquad \mathsf{R}_e = e_R, \; \mathsf{R}_\mu = \mu_R, \tag{7.1.20}$$

and the quarks

$$\mathsf{L}_u = \begin{pmatrix} u \\ d_\theta \end{pmatrix}_L, \qquad \mathsf{L}_c = \begin{pmatrix} c \\ s_\theta \end{pmatrix}_L; \; \mathsf{R}_u = u_R, \; \mathsf{R}_d = d_R, \; \mathsf{R}_c = c_R, \; \mathsf{R}_s = s_R. \tag{7.1.21}$$

The usual construction then yields an anomaly-free theory with flavor-preserving neutral currents, Cabibbo universality for the charged currents, and an agreeable lepton-hadron symmetry. The success of the remarkable inference that a not very massive charmed quark must exist is well known and thoroughly documented elsewhere [6].

The flavor-conserving property of the neutral-current interactions is easily generalized to the case of many quark generations. Suppose that there are $n$ left-handed doublets,

$$\begin{pmatrix} u \\ d' \end{pmatrix}_L, \quad \begin{pmatrix} c \\ s' \end{pmatrix}_L, \quad \begin{pmatrix} t \\ b' \end{pmatrix}_L, \cdots \tag{7.1.22}$$

where the primes denote generalized Cabibbo mixing among the quarks of charge $-\frac{1}{3}$. We write all the quarks in terms of a composite $2n$-component spinor

$$\Psi = \begin{pmatrix} u \\ c \\ t \\ \vdots \\ \hline d \\ s \\ b \\ \vdots \end{pmatrix} \tag{7.1.23}$$

and express the structure of the charged current as

$$J_+^\lambda = \frac{1}{\sqrt{2}}\bar{\Psi}\gamma^\lambda(1-\gamma_5)\mathcal{O}\Psi, \tag{7.1.24}$$

where the $2n \times 2n$ matrix $\mathcal{O}$ is of the form

$$\mathcal{O} = \begin{pmatrix} 0 & V \\ 0 & 0 \end{pmatrix} \tag{7.1.25}$$

and $V$ is the unitary $n \times n$ matrix that describes quark mixing. In the four-quark theory, the quark-mixing matrix assumes the familiar form

$$V = \begin{pmatrix} \cos\theta_C & \sin\theta_C \\ -\sin\theta_C & \cos\theta_C \end{pmatrix}. \tag{7.1.26}$$

In the six-quark theory, $V$ is the (Cabibbo–Kobayashi–Maskawa [7]) matrix, namely,

$$\begin{pmatrix} d' \\ s' \\ b' \end{pmatrix} = \begin{pmatrix} V_{ud} & V_{us} & V_{ub} \\ V_{cd} & V_{cs} & V_{cb} \\ V_{td} & V_{ts} & V_{tb} \end{pmatrix} \begin{pmatrix} d \\ s \\ b \end{pmatrix} \equiv V \begin{pmatrix} d \\ s \\ b \end{pmatrix}, \tag{7.1.27}$$

The magnitudes of the quark-mixing matrix elements, determined from a global fit to charged-current observables, are approximately (Ref. [3])

$$|V| = \begin{bmatrix} 0.9743 & 0.2253 & 0.0035 \\ 0.2252 & 0.9734 & 0.0410 \\ 0.0086 & 0.0403 & 0.9992 \end{bmatrix}. \tag{7.1.28}$$

The family relationships captured in the (CKM) quark-mixing matrix are displayed in the ternary plot in figure 7.1. The coordinates are given by the squares of the CKM mixing matrix elements in each row of (7.1.28). The $u$ quark couples mostly to $d$, $c$ couples mostly to $s$, and $t$ couples almost exclusively to $b$. The pattern of quark mixing, with its definite family structures, is very different from the pattern of neutrino mixing shown in figure 6.35.

In the general $n$-doublet case, the weak-isospin contribution to the neutral current is

$$J_3^\lambda = \tfrac{1}{2}\bar{\Psi}\gamma^\lambda(1-\gamma_5)[\mathcal{O}, \mathcal{O}^\dagger]\Psi; \tag{7.1.29}$$

but since

$$[\mathcal{O}, \mathcal{O}^\dagger] = \begin{pmatrix} I & 0 \\ 0 & -I \end{pmatrix}, \tag{7.1.30}$$

the neutral current will be flavor diagonal with a weak-isospin contribution simply proportional to the value of $\tau_3$, as expected:

$$J_3^\lambda = \tfrac{1}{2}\bar{\Psi}\gamma^\lambda(1-\gamma_5)\tau_3\Psi. \tag{7.1.31}$$

The resulting Feynman rules for gauge-boson–quark interactions are shown in figure 7.2.

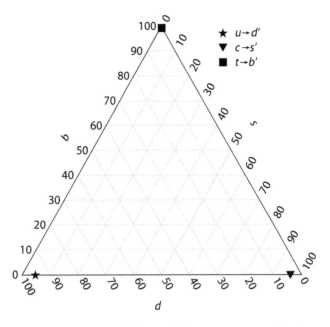

**Figure 7.1.** $d$, $s$, and $b$ composition of the quark flavor eigenstates $d'$ ($\bigstar$), $s'$ ($\blacktriangledown$), $b'$ ($\blacksquare$). The perpendicular distances from the axes to any interior point sum to 100%.

Let us take up once more the question of fermion masses. As was the case for the leptons, the generation of fermion masses by the Higgs mechanism is both possible (which is a virtue) and entirely ad hoc (which is not). Quark mixing, which is such an important aspect of charged-current phenomenology, also arises from the Yukawa interaction

$$\mathcal{L}_{\text{Yukawa}} = -\zeta_{ij}^{u}(\bar{\mathsf{L}}_{q_i}\tilde{\phi})u_{j\mathrm{R}} - \zeta_{ij}^{d}(\bar{\mathsf{L}}_{q_i}\phi)d_{j\mathrm{R}} + \text{h.c.}, \tag{7.1.32}$$

where $\zeta_{ij}^{f}$ are $3 \times 3$ (or, in general, $n \times n$) complex matrices, $i$, $j$ are generation indices, $f$ is a flavor index denoting uplike or downlike quarks, $\mathsf{L}_{q_i}$ are quark doublets in the weak-eigenstate basis [cf. (7.1.21)], and $\phi$ is the Higgs field. When $\phi$ acquires its vacuum expectation value

$$\langle\phi\rangle_0 = \begin{pmatrix} 0 \\ v/\sqrt{2} \end{pmatrix}, \tag{6.3.21}$$

the Yukawa interaction (7.1.32) generates mass terms for the quarks. Problem 7.1 is a concrete illustration in the relatively tractable two-generation case. To characterize the mass eigenstates, we diagonalize the Yukawa matrices by unitary transformations so that the diagonal mass matrices are

$$\mathcal{M}^{f} = V_{\mathrm{L}}^{f}\zeta^{f}V_{\mathrm{R}}^{f\dagger}, \tag{7.1.33}$$

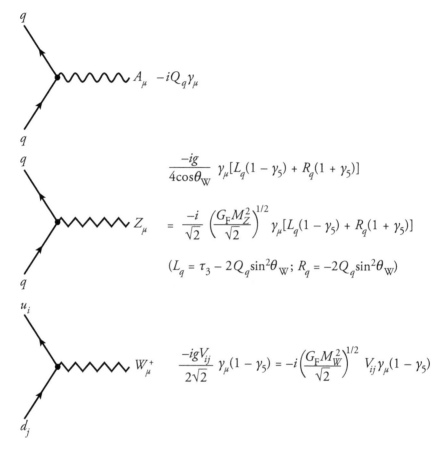

**Figure 7.2.** Feynman rules for interactions of quarks and gauge bosons in the $SU(2)_L \otimes U(1)_Y$ electroweak theory.

with $f = u, d$. Accordingly, the charged-current interaction couplings to the mass eigenstates $u_{iL}$ and $d_{jL}$ are given by the quark-mixing matrix

$$V = V_L^u V_L^{d\dagger}; \qquad (7.1.34)$$

the explicit form for three generations was given in (7.1.27).

As problem 7.2 reveals, the $3 \times 3$ quark-mixing matrix is specified in terms of three real mixing angles and one complex phase. The phase generates CP-violating effects in phenomena involving hadrons (Ref. [7]). Exhaustive experimental tests have shown that the phase in the quark-mixing matrix is the dominant source (indeed, the only established source) of CP-violating effects in processes studied in the laboratory [8]. Problems 7.3–7.5 illuminate further aspects of the CKM matrix.

## 7.2 ELECTROWEAK GAUGE BOSONS

Let us begin to explore the consequences of the electroweak theory built on three generations of quarks and leptons. The results that we derived in chapter 6 for the masses of the gauge bosons,

$$M_W^2 = \frac{\pi \alpha}{G_F \sqrt{2} \sin^2 \theta_W} \tag{6.3.58}$$

and

$$M_Z^2 = \frac{M_W^2}{\cos^2 \theta_W}, \tag{6.3.59}$$

are unaffected, to leading order, by the introduction of quarks into the theory.* To compute the nonleptonic decay rates of the intermediate bosons, we appeal to the parton model in analogy with the successful description of the reaction

$$e^+ e^- \to \text{hadrons} \tag{7.2.1}$$

in terms of the elementary process

$$e^+ e^- \to \gamma_{\text{virtual}} \to q\bar{q}, \tag{7.2.2}$$

in which the quark and antiquark in the semifinal state are assumed to materialize with probability one into the observed jets of hadrons. Hadronic decays are expected to occur as back-to-back jets in the $W$-boson rest frame. The normalized angular distribution of the product quark, the direction of which defines a jet axis, is the same as that of the electron described in (6.2.19), (6.2.23), and (6.2.24).

The partial widths for nonleptonic $W^\pm$ and $Z^0$ decays may then be related to those for the leptonic decays. For the charged intermediate boson, we have

$$\left. \begin{aligned} \Gamma(W^+ \to u\bar{d}) &= 3\,|V_{ud}|^2\,\Gamma(W^+ \to \ell^+ \nu_\ell) \\ \Gamma(W^+ \to u\bar{s}) &= 3\,|V_{us}|^2\,\Gamma(W^+ \to \ell^+ \nu_\ell) \\ \Gamma(W^+ \to c\bar{d}) &= 3\,|V_{cd}|^2\,\Gamma(W^+ \to \ell^+ \nu_\ell) \\ \Gamma(W^+ \to c\bar{s}) &= 3\,|V_{cs}|^2\,\Gamma(W^+ \to \ell^+ \nu_\ell) \end{aligned} \right\}, \tag{7.2.3}$$

where the factor of 3 accounts for quark colors. As in the discussion of leptonic decays, we have idealized the decay products as massless, an approximation that may readily be undone. As before [cf. (6.2.20)], the leptonic decay width is given by

$$\Gamma(W^+ \to \ell^+ \nu_\ell) = \frac{G_F M_W^3}{6\pi \sqrt{2}} \approx 227 \text{ MeV}, \tag{7.2.4}$$

where the numerical value is here derived from the observed mass (Ref. [3]),

$$M_W = (80.399 \pm 0.023) \text{ GeV}. \tag{7.2.5}$$

The experimental value (Ref. [3]) of the leptonic width is $225 \pm 11$ MeV, in excellent agreement with the estimate (7.2.4).

If $D_q$ is the number of color-triplet $SU(2)_L$ doublets into which the intermediate boson can decay, the rate for the inclusive decay of $W^+ \to$ hadrons is given by

$$\Gamma(W^+ \to \text{hadrons}) = 3 D_q \Gamma(W^+ \to \ell^+ \nu_\ell). \tag{7.2.6}$$

---

* In section 7.7, we consider radiative corrections to the gauge-boson masses that arise from quark loops.

Similarly, the total width is given by

$$\Gamma(W^+ \to \text{all}) = (D_\ell + 3D_q)\Gamma(W^+ \to \ell^+ \nu_\ell), \tag{7.2.7}$$

where $D_\ell$ is the number of kinematically accessible lepton doublets. Consequently, for the observed values of $D_\ell = 3$ and $D_q = 2$, we anticipate the branching fractions for each lepton flavor and for nonleptonic decays as

$$\frac{\Gamma(W^+ \to \mu^+ \nu_\mu)}{\Gamma(W^+ \to \text{all})} = \frac{1}{9},$$
$$\frac{\Gamma(W^+ \to \text{hadrons})}{\Gamma(W^+ \to \text{all})} = \frac{2}{3}, \tag{7.2.8}$$

to be compared with the experimental values of $(10.80 \pm 0.09)\%$ and $(67.60 \pm 0.27)\%$, respectively (Ref. [3]). These results imply a total width

$$\Gamma(W^+ \to \text{all}) = 2.05 \text{ GeV}, \tag{7.2.9}$$

in excellent agreement with the measured value (Ref. [3]), $2.085 \pm 0.042$ GeV.

Apart from the prediction (6.3.58) for the $W$-boson mass in terms of the weak mixing parameter $\sin^2 \theta_W$, none of the results for decays of the charged intermediate boson depends sensitively upon the specifics of the $SU(2)_L \otimes U(1)_Y$ theory of the weak and electromagnetic interactions. Rather, they all follow from the intermediate boson picture developed in section 6.2, augmented by the idea of quark-lepton universality and by the parton model for nonleptonic decays. It is worthwhile to remark that the latter has been tested for virtual $W^\pm$ decays by the successful estimate of the branching ratios of the $\tau$-lepton.

The properties of the neutral electroweak gauge boson, in contrast, are evidently quite specific to the $SU(2)_L \otimes U(1)_Y$ theory, although no less easy to compute. Comparing the Feynman rules for $Z^0 \to \nu\bar{\nu}$ and $Z^0 \to e^+ e^-$ given in figure 6.12 with those given for $Z^0 \to q\bar{q}$ given in figure 7.2, we have, at once,

$$\Gamma(Z^0 \to q\bar{q}) = 3(L_q^2 + R_q^2)\Gamma(Z^0 \to \nu\bar{\nu}) \tag{7.2.10}$$

for each quark flavor. Summing the kinematically allowed decays of $Z^0$ into $u\bar{u}$, $d\bar{d}$, $s\bar{s}$, $c\bar{c}$, and $b\bar{b}$, we find

$$\frac{\Gamma(Z^0 \to \text{hadrons})}{\Gamma(Z^0 \to \nu\bar{\nu})} \approx 3\left[2 \cdot \Gamma(Z^0 \to u\bar{u}) + 3 \cdot \Gamma(Z^0 \to d\bar{d})\right]$$
$$= 3\left[2\left(1 - \frac{8x_W}{3} + \frac{32x_W^2}{9}\right) + 3\left(1 - \frac{4x_W}{3} + \frac{8x_W^2}{9}\right)\right]$$
$$= 3\left(5 - \frac{28x_W}{3} + \frac{88x_W^2}{9}\right), \tag{7.2.11}$$

whereas [cf. (6.3.65)]

$$\Gamma(Z^0 \to e^+ e^-) = (1 - 4x_W + 8x_W^2)\Gamma(Z^0 \to \nu\bar{\nu}). \tag{7.2.12}$$

The scale of the partial widths is set by the neutrinic decay rate,

$$\Gamma(Z^0 \to \nu\bar{\nu}) = \frac{G_F M_Z^3}{12\pi\sqrt{2}} \approx 166 \text{ MeV}, \tag{6.3.64}$$

which we evaluated using the observed mass (Ref. [3]), $M_Z = 91.1876 \pm 0.0021$ GeV.

We may compute the total width of the neutral gauge boson as

$$\Gamma(Z^0 \to \text{all}) \approx 3\Gamma(Z^0 \to \nu\bar{\nu}) + 3\Gamma(Z^0 \to e^+e^-) + \Gamma(Z^0 \to \text{hadrons})$$

$$= \left(21 - 40x_W + \frac{160x_W^2}{3}\right)\Gamma(Z^0 \to \nu\bar{\nu}). \tag{7.2.13}$$

Now, using the value of the weak mixing parameter inferred from $M_Z$ in section 6.3, $\sin^2\theta_W = 0.2122$, we predict at lowest order

$$\Gamma(Z^0 \to \text{all}) = 2.474 \text{ GeV}, \tag{7.2.14}$$

to be compared with the experimental value, $(2.4952 \pm 0.0023)$ GeV;

$$\Gamma(Z^0 \to \text{hadrons}) = 1.722 \text{ GeV}, \tag{7.2.15}$$

to be compared with $(1.7444 \pm 0.0020)$ GeV, and

$$\Gamma(Z^0 \to e^+e^-) = 84.8 \text{ MeV}, \tag{7.2.16}$$

to be compared with $(83.984 \pm 0.086)$ MeV. It is also useful to observe the expected and observed branching fractions

$$\frac{\Gamma(Z^0 \to \ell^+\ell^-)}{\Gamma(Z^0 \to \text{all})} = 3.43\% \ [(3.3658 \pm 0.0023)\%],$$

$$\frac{\Gamma(Z^0 \to \text{hadrons})}{\Gamma(Z^0 \to \text{all})} = 69.59\% \ [(69.91 \pm 0.06)\%], \tag{7.2.17}$$

$$\frac{3\Gamma(Z^0 \to \nu\bar{\nu})}{\Gamma(Z^0 \to \text{all})} = 20.1\% \ [(20.00 \pm 0.06)\%].$$

The $SU(2)_L \otimes U(1)_Y$ electroweak theory successfully anticipates the existence and gross properties of the gauge bosons $W^\pm$ and $Z^0$. Indeed, the close agreement between lowest-order predictions and experiment is most remarkable. Although attention to quantum corrections is required to extract the most information from experiment and to test the theory with high precision, it is significant that lowest-order calculations suffice for a rather comprehensive appreciation of electroweak phenomena. We turn next to a number of more-detailed applications of the electroweak theory.

# 7.3 ELECTRON–POSITRON ANNIHILATIONS

Because of the simplicity and definiteness of the parton model in this situation, we begin our discussion of electroweak processes involving hadrons by considering electron–positron interactions. The contribution of direct-channel $\gamma$- and $Z^0$-exchange diagrams (figure 6.22) to the reaction

$$e^+e^- \to \mu^+\mu^- \tag{7.3.1}$$

has already been presented in (6.4.27). The differential cross section for the parton-model reaction

$$e^+e^- \to q\bar{q} \to \text{hadrons} \tag{7.3.2}$$

differs only by an overall factor of $N_c = 3$ for quark color and in the replacement of the electroweak couplings of the muon by those of the quark in question. It is thus

$$
\begin{aligned}
\frac{d\sigma}{dz}(e^+e^- \to q\bar{q}) = N_c \Bigg\{ & \frac{\pi\alpha^2 Q_q^2}{2s}(1+z^2) \\
& - \frac{\alpha Q_q G_F M_Z^2(s - M_Z^2)}{8\sqrt{2}[(s - M_Z^2)^2 + M_Z^2\Gamma^2]} \\
& \times [(R_e + L_e)(R_q + L_q)(1 + z^2) + 2(R_e - L_e)(R_q - L_q)z] \\
& + \frac{G_F^2 M_Z^4 s}{64\pi[(s - M_Z^2)^2 + M_Z^2\Gamma^2]} \\
& \times [(R_e^2 + L_e^2)(R_q^2 + L_q^2)(1 + z^2) + 2(R_e^2 - L_e^2)(R_q^2 - L_q^2)z] \Bigg\},
\end{aligned}
\tag{7.3.3}
$$

where $z = \cos\theta_{\text{c.m.}}$ measures the angle between incoming and outgoing fermions. We will be concerned mostly with the consequences that arise in the standard $\text{SU}(2)_\text{L} \otimes \text{U}(1)_\text{Y}$ theory, but it is worth emphasizing that the forms (6.4.27) and (7.3.3) are valid for any theory containing a single neutral weak boson with vector and axial-vector couplings. We may test the standard model by reverting to the notation of the model-independent analysis introduced in section 6.4.

The cross section expected for the production of various species in the standard model with $x_\text{W} = \sin^2\theta_\text{W} = 0.2122$ is plotted in figure 7.3 in terms of the quantity

$$R \equiv \frac{\sigma(e^+e^- \to f\bar{f})}{\sigma_{\text{QED}}(e^+e^- \to \mu^+\mu^-)} \tag{7.3.4}$$

The $Z^0$ has been assumed to decay with negligible kinematic suppression into three families of leptons and into $u$, $d$, $s$, $c$, and $b$ quark–antiquark pairs, and no radiative corrections have been made to the cross section. The $Z^0$ peak is quite unmistakable.

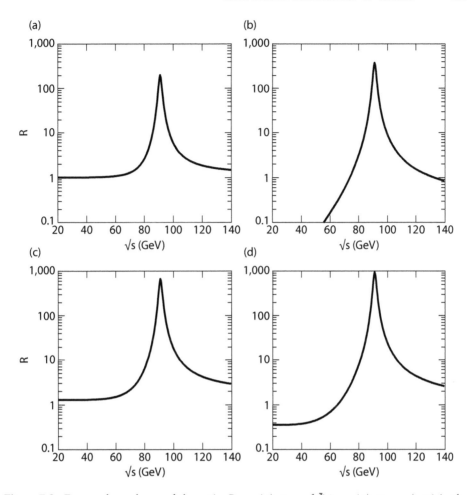

**Figure 7.3.** Energy dependence of the ratio $R \equiv \sigma(e^+e^- \rightarrow f\bar{f})/\sigma_{\mathrm{QED}}(e^+e^- \rightarrow \mu^+\mu^-)$ in the $\mathrm{SU(2)_L \otimes U(1)_Y}$ electroweak theory with $x_W = 0.2122$ and three generations of quarks and leptons: (a) $\ell^+\ell^-$ final state; (b) $\nu\bar{\nu}$ final state; (c) $u\bar{u}$ (charge $=\frac{2}{3}$ quarks) final state; $d\bar{d}$ (charge $= -\frac{1}{3}$ quarks) final state.

At the peak, the cross section for each fermion species is

$$\sigma_{\mathrm{peak}}(e^+e^- \rightarrow f\bar{f}) = \frac{G_F^2 M_Z^4 (L_e^2 + R_e^2)(L_f^2 + R_f^2)N_c}{24\pi\Gamma_Z^2}$$

$$= \frac{G_F^2 M_Z^4 (v_e^2 + a_e^2)(v_f^2 + a_f^2)N_c}{6\pi\Gamma_Z^2}, \qquad (7.3.5)$$

where $N_c$ is the number of colors of the produced fermion $f$ and, in the second line, we employ the notation

$$\left.\begin{aligned} a_i &= \tfrac{1}{2}(L_i - R_i) \\ v_i &= \tfrac{1}{2}(L_i + R_i) \end{aligned}\right\} \qquad (7.3.6)$$

used for model-independent analyses.

TABLE 7.1
Cross Sections for Electron–Positron Annihilation at the $Z^0$ Peak.

| Channel | $\sigma_{\text{peak}}$ (nb) | $R_{\text{peak}}$ |
|---------|------|------|
| $\nu\bar{\nu}$ | 4.06 | 389 |
| $\mu^+\mu^-$ | 2.08 | 199 |
| $u\bar{u}$ | 7.24 | 693 |
| $d\bar{d}$ | 9.22 | 883 |
| hadrons | 42.1 | 4034 |

Thus, we may write the ratio (7.3.4) at the neutral gauge-boson peak as

$$R_{\text{peak}}(e^+e^- \to f\bar{f}) = \frac{G_F^2 M_Z^6 (L_e^2 + R_e^2)(L_f^2 + R_f^2) N_c}{32\pi^2 \alpha^2 \Gamma_Z^2}. \tag{7.3.7}$$

In the standard electroweak theory, with $\Gamma_Z$ given by (7.2.13), we therefore have

$$R_{\text{peak}}^{(\nu\bar{\nu})} = \frac{9(1 - 4x_W + 8x_W^2)}{\alpha^2(21 - 40x_W + 160x_W^2/3)^2} \tag{7.3.8}$$

and

$$R_{\text{peak}}^{(\mu^+\mu^-)} = (1 - 4x_W + 8x_W^2) R_{\text{peak}}^{(\nu\bar{\nu})}, \tag{7.3.9}$$

$$R_{\text{peak}}^{(u\bar{u})} = 3\left(1 - \frac{8x_W}{3} + \frac{32x_W^2}{3}\right) R_{\text{peak}}^{(\nu\bar{\nu})}, \tag{7.3.10}$$

$$R_{\text{peak}}^{(d\bar{d})} = 3\left(1 - \frac{4x_W}{3} + \frac{8x_W^2}{9}\right) R_{\text{peak}}^{(\nu\bar{\nu})}. \tag{7.3.11}$$

The resulting peak cross sections and ratios are given in table 7.1 for $x_W = 0.2122$ and five quark flavors (neglecting masses). The visible cross section is quite enormous with respect to the QED reference cross section of 10.45 pb:

$$R_{\text{peak}}(e^+e^- \to \text{charged leptons or hadrons}) \approx 4630. \tag{7.3.12}$$

Although this is reduced somewhat in practice by radiative corrections [9] and uncertainty in beam energy, the copious yield contributed to the immense productivity of the "$Z^0$ factories," LEP (at CERN) and the Stanford Linear Collider. The measured cross sections for the reaction $e^+e^- \to$ hadrons over a wide range of energies are compiled in figure 7.4. Should one or more $Z'$ gauge bosons be found, they would provide inviting targets for a future lepton collider (cf. problem 7.11).

The peak cross section's sensitivity to the total width of the $Z^0$ boson makes possible a determination of the number of universally coupled light neutrino species that complements the invisible-width approach of problem 7.6. Figure 7.5 compares expectations and observations for the $Z^0$ line shape; the data decisively select three light neutrino species over two or four.

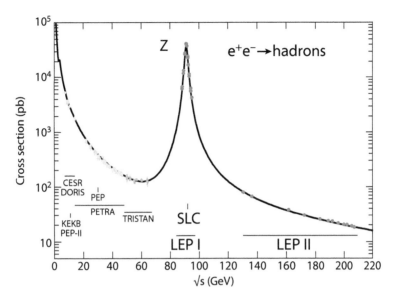

**Figure 7.4.** The hadronic cross section in electron–positron annihilations as measured at a number of colliders (Ref. [10]). The curve is the standard-model prediction that corresponds, at leading order, to (7.3.3). The cross sections have been corrected for the effects of photon radiation.

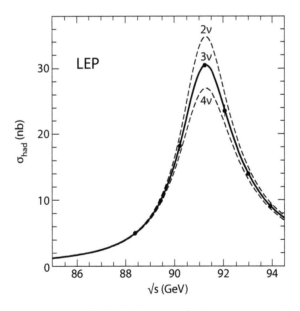

**Figure 7.5.** LEP measurements of the hadron production cross section around the $Z^0$ resonance (Ref. [10]). The points are averages of measurements by the ALEPH, DELPHI, L3, and OPAL experiments, with uncertainties magnified by a factor of ten. The curves represent the line-shape expectations for two, three, and four neutrino species with standard-model couplings and negligible masses.

The most stringent constraint on the number of light neutrino species, $N_\nu = 2.9840 \pm 0.0082$, comes from standard-model fits to the universe of LEP and SLC data [10]. It leaves little room for decays of $Z$ into exotic weakly interacting particles. The conclusion that only three active light species exist does not rule out a fourth generation of quarks and leptons, provided that the neutral leptons are heavy enough that their contributions to the invisible width would be negligible—if not zero! A fourth generation is constrained, but not excluded, by what we know of charged-current and neutral-current interactions [11].

A fruitful observable is the forward-backward asymmetry

$$A \equiv \frac{\int_0^1 dz \, d\sigma/dz - \int_{-1}^0 dz \, d\sigma/dz}{\int_{-1}^1 dz \, d\sigma/dz}, \tag{6.4.28}$$

already discussed for the charged leptons in section 6.4. In the low-energy approximation, the asymmetry is approximately given by

$$A(q\bar{q}) = \frac{3G_F s}{16\pi\alpha Q_q \sqrt{2}}(R_e - L_e)(R_q - L_q). \tag{7.3.13}$$

In the standard model, $(R_i - L_i) = \tau_3^{(i)}$, so that

$$R_e - L_e = R_d - L_d = -(R_u - L_u) = 1. \tag{7.3.14}$$

Therefore, comparing with (6.4.29), we find

$$\left. \begin{aligned} A(u\bar{u}) &= \tfrac{3}{2} A(\mu^+\mu^-) \\ A(d\bar{d}) &= 3 A(\mu^+\mu^-) \end{aligned} \right\}, \tag{7.3.15}$$

where $u$ and $d$ are to be understood as generic symbols for the quarks of charge $+\tfrac{2}{3}$ and $-\tfrac{1}{3}$, respectively.

Thus, in experiments at $\sqrt{s} = 40$ GeV, the leading-order expectation is

$$\left. \begin{aligned} A(u\bar{u}) &\approx -16\% \\ A(d\bar{d}) &\approx -32\% \end{aligned} \right\}, \tag{7.3.16}$$

in the standard model. Measurements of these asymmetries—for example by tagging the decays of short-lived particles containing $c$ or $b$ quarks—can enable independent determinations of the neutral-current couplings of the heavy quarks. In the notation (7.3.6) of vector and axial couplings, the forward-backward asymmetry at low energies may be written as

$$A(q\bar{q}) = \frac{3G_F s a_e a_q}{4\pi\alpha Q_q \sqrt{2}}. \tag{7.3.17}$$

Problem 7.12 illuminates the role of $A(b\bar{b})$ in determining the weak isospin of the $b$ quark.

Although the low-energy forward-backward asymmetry can be a sensitive probe of neutral-current couplings, it is not directly sensitive to the *position* of the

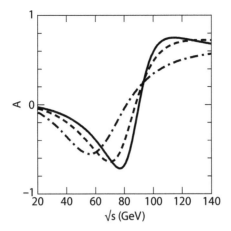

**Figure 7.6.** Energy dependence of the forward–backward asymmetry (6.4.28) in electron–positron annihilations in the $SU(2)_L \otimes U(1)_Y$ electroweak theory with $x_W = 0.2122$. The solid curve refers to the $\mu^+\mu^-$ final state, the dashed curve, to $u\bar{u}$, and the dot-dashed curve, to $d\bar{d}$. The $Z^0$ has been assumed to decay with negligible kinematic suppression into three families of leptons and into $u, d, s, c$, and $b$ quark–antiquark pairs, and no radiative corrections are included.

$Z^0$-boson pole, as the absence of $M_Z$ from (6.4.29), (6.4.31), (7.3.13), and (7.3.17) makes plain. The evolution of the asymmetry with increasing energy is, however, influenced by the $Z^0$ propagator, because the linear rise with $s$ of the magnitude of the asymmetry is damped. The behavior anticipated in the standard model is illustrated in figure 7.6.

On the $Z^0$ peak, the asymmetry may be expressed as

$$A_{\text{peak}}^{(f\bar{f})} = \frac{3(L_e^2 - R_e^2)(L_f^2 - R_f^2)}{4(L_e^2 + R_e^2)(L_f^2 + R_f^2)}$$

$$= \frac{3a_e v_e a_f v_f}{(v_e^2 + a_e^2)(v_f^2 + a_f^2)}. \tag{7.3.18}$$

In the standard model, we compute

$$A_{\text{peak}}^{(\mu^+\mu^-)} = \frac{3}{4}\left(\frac{1 - 4x_W}{1 - 4x_W + 8x_W^2}\right)^2, \tag{7.3.19a}$$

$$A_{\text{peak}}^{(u\bar{u})} = \frac{3}{4}\left(\frac{1 - 4x_W}{1 - 4x_W + 8x_W^2}\right)\left(\frac{1 - 8x_W/3}{1 - 8x_W/3 + 32x_W^2/9}\right), \tag{7.3.19b}$$

$$A_{\text{peak}}^{(d\bar{d})} = \frac{3}{4}\left(\frac{1 - 4x_W}{1 - 4x_W + 8x_W^2}\right)\left(\frac{1 - 4x_W/3}{1 - 4x_W/3 + 8x_W^2/9}\right). \tag{7.3.19c}$$

The $Z^0$-pole asymmetries vanish for $x_W = 0.25$, where the vector coupling of the electron to the weak neutral current is zero, and so are highly sensitive to the value

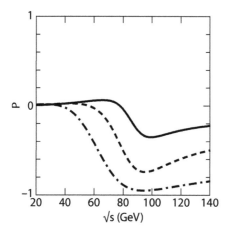

**Figure 7.7.** Energy dependence of the net polarization of charged fermions produced in electron–positron annihilations in the $SU(2)_L \otimes U(1)_Y$ electroweak theory with $x_W = 0.2122$. The solid curve refers to the $\mu^+\mu^-$ final state, the dashed curve, to $u\bar{u}$, and the dot-dashed curve, to $d\bar{d}$. The $Z^0$ has been assumed to decay with negligible kinematic suppression into three families of leptons and into $u, d, s, c$, and $b$ quark–antiquark pairs, and no radiative corrections are included.

of the weak mixing parameter. For our example value, $x_W = 0.2122$, we would anticipate peak asymmetries (0.07, 0.16, 0.21) for lepton pairs, uplike quarks, and downlike quarks, respectively. These are only in rough agreement with the measurements at LEP and SLC (Ref. [10]),

$$A_{\text{peak}}^{(\ell^+\ell^-)} = 0.0171 \pm 0.0010, \tag{7.3.20a}$$

$$A_{\text{peak}}^{(c\bar{c})} = 0.0707 \pm 0.0035, \tag{7.3.20b}$$

$$A_{\text{peak}}^{(b\bar{b})} = 0.0992 \pm 0.0016; \tag{7.3.20c}$$

but, as problem 7.9 will show, only a modest adjustment in the weak-mixing parameter, to $x_W \approx 0.232$, is required to reproduce the measurements. The full electroweak theory, including quantum corrections, is in impressive agreement with the entire suite of precision measurements. We introduce the loop corrections in section 7.7.

Parity violation in the neutral-current interaction is manifested as a net polarization of the produced fermions,

$$P = \frac{\sigma_R - \sigma_L}{\sigma}. \tag{7.3.21}$$

In the standard model with $x_W = 0.2122$, the polarization of all species is expected to be small at low energies, becoming appreciable in the neighborhood of the $Z^0$ resonance. This is shown in figure 7.7. On the $Z^0$ peak, the net polarization (integrated over $z = \cos\theta$) is given by

$$P_{\text{peak}}^{(f\bar{f})} = \frac{R_f^2 - L_f^2}{R_f^2 + L_f^2}$$

$$= \frac{-2a_f v_f}{v_f^2 + a_f^2}. \tag{7.3.22}$$

This means that the electron parameters could, in principle, be isolated in the ratio

$$\frac{A_{\text{peak}}^{(f\bar{f})}}{P_{\text{peak}}^{(f\bar{f})}} = \frac{-3a_e v_e}{2(v_e^2 + a_e^2)}, \tag{7.3.23}$$

without intervening universality assumptions. The $Z^0$-pole polarizations predicted at leading order are significant for our example value of $x_{\text{W}} = 0.2122$:

$$P_{\text{peak}}^{(\mu^+\mu^-)} = \frac{4x_{\text{W}} - 1}{1 - 4x_{\text{W}} + 8x_{\text{W}}^2} \rightarrow -0.30, \tag{7.3.24a}$$

$$P_{\text{peak}}^{(u\bar{u})} = \frac{8x_{\text{W}}/3 - 1}{1 - 8x_{\text{W}}/3 + 32x_{\text{W}}^2/9} \rightarrow -0.73, \tag{7.3.24b}$$

$$P_{\text{peak}}^{(d\bar{d})} = \frac{4x_{\text{W}}/3 - 1}{1 - 4x_{\text{W}}/3 + 8x_{\text{W}}^2/9} \rightarrow -0.95. \tag{7.3.24c}$$

The negative sign of the polarization follows from our defining the polarization of a right-handed fermion as positive in a world in which left-handed couplings dominate. The polarization of the $\tau$-lepton is accessible to experiment through the kinematic distributions of the decay products and the V − A form of the charged-current interaction that mediates the decays. Problem 7.10 explores the information contained in the angular distribution of $\tau$ polarization on the $Z^0$ pole, an observable studied at LEP. The availability of longitudinally polarized electron or positron beams greatly expands the range of possible measurements, as we may learn from the basic formulas (6.4.27) and (7.3.3).

The four LEP experiments (ALEPH [12], DELPHI [13], L3 [14], and OPAL [15]) together recorded 17 million examples of $Z^0$ decay. Using polarized beams, the SLD [16] experiment at SLAC studied 600 thousand $Z^0$. Supplemented by inputs from neutrino scattering, hadron collider experiments, and atomic physics experiments, these experiments determined standard-model parameters, including chiral couplings of the $Z^0$-boson, with very remarkable precision [10]. Operating above the $W^+W^-$ pair threshold, the LEP experiments also made precise measurements of the $W$-boson mass [17] ($M_W = 80.376 \pm 0.033$ GeV) and three-gauge-boson couplings [18].

## 7.4 DEEPLY INELASTIC LEPTON–HADRON SCATTERING

When hadron structure is examined on a very short time or distance scale, as is possible in large-momentum-transfer collisions, hadrons are found to behave as if they were composed of essentially noninteracting pointlike structures. The pointlike constituents of hadrons have been named *partons* by Feynman. The basic idea of the parton model is analogous to the well-traveled notion that for many purposes a nucleus may be regarded as a collection of structureless and noninteracting nucleons—but with a critical difference. Nucleons are rather easily liberated from nuclei, but the division of a hadron into its constituent partons has never been observed. The parton model has thus had a paradoxical aspect, which may be summarized in the question, How could partons be quasifree within hadrons if they interact so strongly that they cannot be separated? The question became more urgent with each success of the model. Only with the development of the theory of strong interactions known as quantum chromodynamics has a resolution emerged, as we shall see in chapter 8. Much progress has been made on the general problem of hadron structure, but we cannot yet claim a complete solution.

Our principal concern in this section is not with hadron structure, but with the interactions of weak and electromagnetic currents with the fundamental constituents. Thus, we shall develop and apply the parton model in an intuitive fashion that is close to the original spirit in which the model was formulated. Loftier formulations may be found among the suggested readings at the end of the chapter. The heuristic approach has the advantage of leading immediately to a transparent physical picture of hard collisions that is readily generalized to incorporate QCD refinements. The development of this section will have three primary goals: to formulate in a general language the kinematics and observables of inclusive lepton–hadron scattering, to specialize to the parton model and verify its approximate validity, and to investigate what can be learned about the electroweak interactions in deeply inelastic lepton–hadron scattering.

We wish to consider inclusive reactions of the generic form

$$\ell + N \rightarrow \ell' + \text{anything}, \tag{7.4.1}$$

for which the kinematic notation is indicated in figure 7.8. From the four-momenta designated there we may form the useful invariants

$$s = (\ell + P)^2, \tag{7.4.2}$$

$$Q^2 \equiv -q^2 = -(\ell - \ell')^2, \tag{7.4.3}$$

$$\nu = \frac{q \cdot P}{M}, \tag{7.4.4}$$

where $M$ is the target mass and

$$W^2 = 2M\nu + M^2 - Q^2 \tag{7.4.5}$$

is the square of the invariant mass of the produced hadronic system, "anything." In the laboratory frame, in which (neglecting the lepton mass)

$$\left.\begin{aligned}
\ell^\mu &= (E; 0, 0, E) \\
\ell'^\mu &= (E'; E' \sin\theta, 0, E' \cos\theta) \\
P^\mu &= (M; 0, 0, 0)
\end{aligned}\right\}, \tag{7.4.6}$$

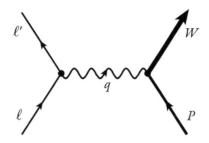

**Figure 7.8.** Kinematics of deeply inelastic scattering.

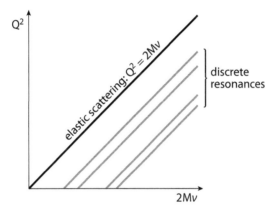

**Figure 7.9.** The $Q^2$-$2M\nu$ plane in deeply inelastic scattering.

we may write

$$Q^2 = 2EE'(1 - \cos\theta)$$
$$= 4EE'\sin^2\frac{\theta}{2} \tag{7.4.7}$$

and recognize as an energy-loss variable

$$\nu = E - E'. \tag{7.4.8}$$

From the connection (7.4.5) we see that the line

$$Q^2 = 2M\nu \tag{7.4.9}$$

in the $Q^2$-$2M\nu$ plane corresponds to elastic scattering, for which

$$W^2 = M^2. \tag{7.4.10}$$

Similarly, the excitation of a system with mass $M^*$, such as a discrete resonance, occurs along a line

$$Q^2 = 2M\nu + M^2 - M^{*2}. \tag{7.4.11}$$

These kinematic regimes are indicated in figure 7.9.

As a first specific example, consider the case of inelastic electron–proton scattering with an unpolarized beam and target. The square of the invariant amplitude, averaged over initial spins and summed over final spins, may be written as

$$\overline{|\mathcal{M}|^2} = \frac{e^4}{(q^2)^2} L^{\mu\nu} W_{\mu\nu}, \tag{7.4.12}$$

where the anticipated factors for the spin average have been absorbed into the definitions of the tensors $L^{\mu\nu}$ and $W_{\mu\nu}$, which describe the structures of the leptonic and hadronic vertices, respectively. The lepton tensor is prescribed by QED as

$$\begin{aligned} L^{\mu\nu} &= \tfrac{1}{2}\mathrm{tr}[\bar{u}(\ell')\gamma^\mu u(\ell)\bar{u}(\ell)\gamma^\nu u(\ell')] \\ &= \tfrac{1}{2}\mathrm{tr}[\gamma^\mu(m+\ell)\gamma^\nu(m+\ell')] \\ &= 2\left[\ell^\mu\ell'^\nu + \ell^\nu\ell'^\mu - g^{\mu\nu}(\ell\cdot\ell' - m^2)\right]. \end{aligned} \tag{7.4.13}$$

In the absence of any detailed information about the hadronic vertex, we may write the general form

$$\begin{aligned} W_{\mu\nu} &= V_1 g_{\mu\nu} + V_2 P_\mu P_\nu + V_3(P_\mu q_\nu + P_\nu q_\mu) \\ &\quad + V_4(P_\mu q_\nu - P_\nu q_\mu) + V_5 q_\mu q_\nu + V_6 \varepsilon_{\mu\nu\alpha\beta} P^\alpha q^\beta. \end{aligned} \tag{7.4.14}$$

Because $L^{\mu\nu}$ is symmetric under interchange of the indices $\mu$ and $\nu$, the antisymmetric terms proportional to $V_4$ and $V_6$ cannot contribute to the cross section. Current conservation requires that

$$q^\mu W_{\mu\nu} = 0 = q^\nu W_{\mu\nu}. \tag{7.4.15}$$

This implies that

$$V_1 q_\nu + V_2(q\cdot P)P_\nu + V_3(q\cdot P\, q_\nu + q^2 P_\nu) + V_5 q^2 q_\nu = 0, \tag{7.4.16}$$

from which the coefficients of $P_\nu$ and $q_\nu$ must vanish separately, so that

$$\left.\begin{aligned} V_1 + (q\cdot P)V_3 + q^2 V_5 &= 0 \\ (q\cdot P)V_2 + q^2 V_3 &= 0 \end{aligned}\right\}. \tag{7.4.17}$$

Eliminating

$$V_3 = -\frac{(q\cdot P)V_2}{q^2} \tag{7.4.18}$$

and

$$V_5 = \frac{(q\cdot P)^2 V_2}{q^4} - \frac{V_1}{q^2}, \tag{7.4.19}$$

we find that

$$W_{\mu\nu} = V_1\left(g_{\mu\nu} - \frac{q_\mu q_\nu}{q^2}\right) + V_2\left[P_\mu - \frac{(q\cdot P)q_\mu}{q^2}\right]\left[P_\nu - \frac{(q\cdot P)q_\nu}{q^2}\right], \tag{7.4.20}$$

which is conventionally written as

$$W_{\mu\nu} = -W_1\left(g_{\mu\nu} - \frac{q_\mu q_\nu}{q^2}\right) + \frac{W_2}{M^2}\left[P_\mu - \frac{(q \cdot P)q_\mu}{q^2}\right]\left[P_\nu - \frac{(q \cdot P)q_\nu}{q^2}\right]. \quad (7.4.21)$$

The objects $W_{1,2}$ are known as the structure functions for (unpolarized) inelastic electron–proton scattering. All that may be learned about hadron structure from such collisions is contained in these two structure functions, which depend upon $P$ and $q$ or, more precisely, upon the invariants $Q^2$ and $\nu$.

To make contact with observables, we neglect lepton masses, so that

$$\ell \cdot \ell' = q \cdot \ell' = -q \cdot \ell = \frac{Q^2}{2}, \quad (7.4.22)$$

and form

$$L^{\mu\nu} W_{\mu\nu} = 2 W_1(Q^2, \nu) Q^2 + W_2(Q^2, \nu)\left[\frac{4(\ell \cdot P)(\ell' \cdot P)}{M^2} - Q^2\right]. \quad (7.4.23)$$

In the laboratory frame, we may express

$$\begin{aligned} \ell \cdot P &= ME, \\ \ell' \cdot P &= ME', \end{aligned} \quad (7.4.24)$$

to obtain

$$L^{\mu\nu} W_{\mu\nu} = 4EE'\left[2 W_1(Q^2, \nu) \sin^2\left(\frac{\theta}{2}\right) + W_2(Q^2, \nu) \cos^2\left(\frac{\theta}{2}\right)\right]. \quad (7.4.25)$$

The differential cross section in the laboratory frame is given by

$$\frac{d^2\sigma}{dE' d\Omega'} = \frac{1}{16\pi^2}\frac{E'}{E}\overline{|\mathcal{M}|^2} = \frac{(4\pi\alpha)^2}{16\pi^2 Q^4}\frac{E'}{E} L^{\mu\nu} W_{\mu\nu} \quad (7.4.26)$$

$$= \frac{4\alpha^2 E'^2}{Q^4}\left[2 W_1(Q^2, \nu) \sin^2\left(\frac{\theta}{2}\right) + W_2(Q^2, \nu) \cos^2\left(\frac{\theta}{2}\right)\right].$$

It is often more convenient to express the differential cross section with respect to the invariants $\nu$ and $Q^2$ as

$$\frac{d^2\sigma}{dQ^2 d\nu} = \frac{\pi}{EE'}\frac{d^2\sigma}{dE' d\Omega'} = \frac{4\pi\alpha^2}{Q^4}\frac{E'}{E}\left[2 W_1(Q^2, \nu) \sin^2\left(\frac{\theta}{2}\right) + W_2(Q^2, \nu) \cos^2\left(\frac{\theta}{2}\right)\right]. \quad (7.4.27)$$

It is interesting to compare the cross section (7.4.26) with the more familiar results for elastic scattering. The Rosenbluth formula for elastic electron–proton scattering, which was derived in problem 3.3, may be written in the form

$$
\frac{d^2\sigma}{dE'd\Omega'} = \frac{4\alpha^2 E'^2}{Q^4} \left\{ 2Q^2 \left[ \frac{\Gamma_1(Q^2)}{2M} + \Gamma_2(Q^2) \right]^2 \sin^2\left(\frac{\theta}{2}\right) \right.
$$

$$
\left. + \left[ \Gamma_1^2(Q^2) + Q^2\Gamma_2^2(q^2) \right] \cos^2\left(\frac{\theta}{2}\right) \right\} \delta\left(\nu - \frac{Q^2}{2M}\right), \quad (7.4.28)
$$

which has precisely the same structure. For a structureless Dirac proton, also considered in problem 3.3, the form factors are simply

$$
\left.\begin{array}{l}
\Gamma_1(Q^2) = 1 \\
\Gamma_2(Q^2) = 0
\end{array}\right\}, \quad (7.4.29)
$$

so we may identify the structure functions for scattering from a pointlike spin-$\frac{1}{2}$ particle as

$$
\left.\begin{array}{l}
W_1^{pt}(Q^2, \nu) = (Q^2/4M^2)\delta(\nu - Q^2/2M) \\
W_2^{pt}(Q^2, \nu) = \delta(\nu - Q^2/2M)
\end{array}\right\}. \quad (7.4.30)
$$

The dimensionless combinations

$$
\left.\begin{array}{l}
2MW_1^{pt}(Q^2, \nu) = (Q^2/2M\nu)\delta(Q^2/2M\nu - 1) \\
\nu W_2^{pt}(Q^2, \nu) = \delta(Q^2/2M\nu - 1)
\end{array}\right\} \quad (7.4.31)
$$

are seen to depend upon the kinematic invariants only through the dimensionless ratio

$$
x = \frac{Q^2}{2M\nu}. \quad (7.4.32)
$$

Similarly, by examining the form of the differential cross section for electron–elementary-scalar scattering derived in problem 3.1 or by considering directly the structure of the photon–scalar vertex, we see at once that

$$
\left.\begin{array}{l}
W_2^{(0)}(Q^2, \nu) = \delta(\nu - Q^2/2M) \\
W_1^{(0)}(Q^2, \nu) = 0
\end{array}\right\}. \quad (7.4.33)
$$

To conclude these essentially kinematic developments, let us relate the structure functions to the cross sections for the absorption of virtual photons. Writing the flux of virtual photons as $\Phi$, we may define the cross section for absorption of photons of helicity $\lambda$ as

$$
\sigma_\lambda = \frac{4\pi^2\alpha}{\Phi} \epsilon_\lambda^{\mu*} W_{\mu\nu} \epsilon_\lambda^\nu, \quad (7.4.34)
$$

following the conventions appropriate to real photons. For photons of helicity $\pm 1$, for which

$$\epsilon^\mu_{\pm 1} = (0; -1, \mp i, 0)/\sqrt{2}, \tag{7.4.35}$$

we compute

$$\sigma_T \equiv \frac{1}{2}(\sigma_{+1} + \sigma_{-1}) = \frac{4\pi^2\alpha}{\Phi} W_1, \tag{7.4.36}$$

whereas for longitudinally polarized (scalar) photons, with

$$\epsilon^\mu_S = (\sqrt{Q^2 + \nu^2}; 0, 0, \nu)/\sqrt{Q^2}, \tag{7.4.37}$$

the cross section is

$$\sigma_S = \frac{4\pi^2\alpha}{\Phi} \left[ -W_1 + \left(1 + \frac{\nu^2}{Q^2}\right) W_2 \right]. \tag{7.4.38}$$

The requirement that $\sigma_T$ and $\sigma_S$ be nonnegative leads to the restrictions

$$\left.\begin{array}{c} W_1 \geq 0 \\[2mm] W_2 \left(1 + \dfrac{\nu^2}{Q^2}\right) \geq W_1 \end{array}\right\}. \tag{7.4.39}$$

Finally, we note that the ratio

$$\frac{\sigma_S}{\sigma_T} = \frac{W_2(1 + \nu^2/Q^2)}{W_1} - 1 \tag{7.4.40}$$

is sensitive to the spin of the target. For a structureless spin-$\frac{1}{2}$ target, with structure functions given by (7.4.30), the ratio vanishes as $\nu \to \infty$:

$$\frac{\sigma_S}{\sigma_T} = \frac{2M}{\nu} \xrightarrow[\nu \to \infty]{} 0, \quad \text{spin-}\tfrac{1}{2}. \tag{7.4.41}$$

In contrast, for a spinless target with structure functions given by (7.4.33), the transverse cross section vanishes, so that

$$\frac{\sigma_S}{\sigma_T} \to \infty, \quad \text{spin-zero}. \tag{7.4.42}$$

This completes our discussion of the kinematics of electron–proton scattering. We now turn to the parton model itself.

The transition to the parton model is most gracefully made in the infinite-momentum frame, in which the longitudinal momentum of the target (proton, for example) is extremely large. The target is regarded as a collection of $N$ free partons, each carrying a fraction $x_i$ $(i = 1, 2, \ldots, N)$ of the longitudinal momentum of the target, as shown in figure 7.10. Assuming the mass of a parton to be insignificant both before and after a collision and that the transverse momentum of an incident

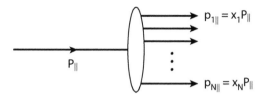

**Figure 7.10.** Parton-model depiction of the proton in the infinite-momentum frame.

parton is negligible, we may write the four-momentum of an individual parton as

$$p_i^\mu = x_i P^\mu. \tag{7.4.43}$$

Then if interactions among the partons can be neglected, so that the individual current–parton interactions may be treated incoherently, we may write the contribution to $W_2$ due to scattering from a single parton of charge $e_i$ as [19]

$$
\begin{aligned}
W_2^{(i)}(Q^2, \nu; x_i) &= x_i e_i^2\, \delta\left(\frac{q \cdot p_i}{M} - \frac{Q^2}{2M}\right) \\
&= x_i e_i^2\, \delta\left(\frac{x_i q \cdot P}{M} - \frac{Q^2}{2M}\right) \\
&= e_i^2\, \delta\left(\nu - \frac{Q^2}{2Mx_i}\right),
\end{aligned}
\tag{7.4.44}
$$

which reproduces, as it must, the Rutherford cross section

$$\frac{d\sigma}{dQ^2} = \frac{4\pi\alpha^2 e_i^2}{Q^4} \tag{7.4.45}$$

at high energies.

The incoherence assumption, or impulse approximation, means that the structure function for electron–proton scattering is simply the sum over the contributions of individual partons:

$$W_2(Q^2, \nu) = \sum_i \int_0^1 dx_i\, f_i(x_i)\, W_2^{(i)}(Q^2, \nu; x_i), \tag{7.4.46}$$

where $f_i(x_i)$ gives the probability of finding the $i$th parton with momentum fraction $x_i$. The integration over $dx_i$ is readily carried out using the rule

$$\int dx\, \delta(h(x)) = \frac{1}{\partial h/\partial x}\bigg|_{h(x)=0}. \tag{7.4.47}$$

We find at once that

$$W_2(Q^2, \nu) = \sum_i \frac{e_i^2 f_i(x)x}{\nu}, \tag{7.4.48}$$

where the scaling variable $x = Q^2/2M\nu$ has already been defined in (7.4.32). Consequently, the dimensionless quantity

$$\nu W_2(Q^2, \nu) = \sum_i e_i^2 f_i(x)x, \qquad (7.4.49)$$

obtained as an incoherent sum over the contributions of individual partons, is identified as a function of a single dimensionless variable $x$. It is convenient to define the scaling form as the combination

$$F_2(x) \equiv \nu W_2(Q^2, \nu). \qquad (7.4.50)$$

Knowing the relation between $W_1$ and $W_2$ for structureless particles of spin-zero or spin-$\frac{1}{2}$, we may construct from this result the general form of the cross section. It is again convenient to define a dimensionless form based on $W_1$, which will evidently depend only upon the scaling variable $x$. It is conventionally written as

$$F_1(x) \equiv MW_1(Q^2, \nu), \qquad (7.4.51)$$

which is given by

$$F_1(x) = \begin{cases} 0, & \text{spin-zero partons,} \\ \left(\dfrac{1}{2x}\right) F_2(x), & \text{spin-}\frac{1}{2} \text{ partons.} \end{cases} \qquad (7.4.52)$$

That $\nu W_2(Q^2, \nu)$ should become independent of $Q^2$ for fixed values of $x$ as $\nu, Q^2 \to \infty$ was anticipated by Bjorken [20] and demonstrated in classic experiments by the SLAC–MIT Collaboration [21]. This behavior, which we have seen to be quite natural within the parton model, is in sharp contrast to the comportment of elastic scattering or resonance excitation, both characterized by rapidly falling form factors. Experiments have also shown the ratio $\sigma_S/\sigma_T$ to be small [22], suggesting that the charged partons carry spin-$\frac{1}{2}$. The fact that deeply inelastic electron scattering is readily interpreted as scattering from structureless, spin-$\frac{1}{2}$ constituents suggests that the charged partons be identified as quarks. What are the implications of this identification?

To fully exploit the scaling behavior of the cross section, it is convenient to introduce the inelasticity parameter

$$y \equiv \frac{\nu}{E}, \qquad (7.4.53)$$

which evidently satisfies

$$0 \leq y \leq 1. \qquad (7.4.54)$$

The cross section may be expressed as

$$\frac{d^2\sigma}{dx\,dy} = \frac{2M\nu^2}{y}\frac{d^2\sigma}{dQ^2\,d\nu} = \frac{4\pi\alpha^2 s}{Q^4}\left[F_2(x)(1-y) + F_1(x)xy^2\right], \qquad (7.4.55)$$

plus a term of order $M/E$, which may be safely neglected at high energies. According to (7.4.49), the structure functions are directly related to the quark–parton distribution functions. For $ep$ scattering, we have that

$$\frac{F_2^{ep}(x)}{x} = \tfrac{4}{9}(u(x) + \bar{u}(x)) + \tfrac{1}{9}(d(x) + \bar{d}(x))$$
$$+ \tfrac{1}{9}(s(x) + \bar{s}(x)) + \cdots, \tag{7.4.56}$$

with $u(x) = f_u(x)$, and so on. The neutron structure functions may be obtained by applying an isospin rotation, which amounts to $u \leftrightarrow d$, whereupon

$$\frac{F_2^{en}(x)}{x} = \tfrac{4}{9}(d(x) + \bar{d}(x)) + \tfrac{1}{9}(u(x) + \bar{u}(x)) + \cdots, \tag{7.4.57}$$

where $u(x)$, $d(x)$, ..., refer to the quark content of the proton. The expressions (7.4.56) and (7.4.57) lead at once to the bounds

$$\tfrac{1}{4} \leq \frac{F_2^{en}(x)}{F_2^{ep}(x)} \leq 4. \tag{7.4.58}$$

Measurements respect the parton-model bounds [23], and the data have the following simple interpretation. For small values of $x$, where the sea of quark–antiquark pairs may be expected to dominate, so that

$$u(x) \simeq \bar{u}(x) \simeq d(x) \simeq \bar{d}(x), \quad x \ll 1, \tag{7.4.59}$$

then we anticipate a ratio close to unity, as is observed. At large $x$, it is the so-called valence quarks that must dominate, which is to say that

$$u(x), d(x) \gg \bar{u}(x), \bar{d}(x), \quad x \to 1. \tag{7.4.60}$$

If the up- and down-quark distributions had the same shape as $x \to 1$ (a simple but not compelling hypothesis), which would imply that

$$u(x) = 2d(x), \quad x \to 1, \tag{7.4.61}$$

then the ratio $F_2^{en}(x)/F_2^{ep}(x)$ would be expected to approach $\tfrac{2}{3}$. The prevailing interpretation that the ratio falls below $\tfrac{2}{3}$ is taken to indicate that

$$\frac{d(x)}{u(x)} \to 0, \quad x \to 1 \tag{7.4.62}$$

in most modern sets of parton distribution functions. A recent measurement by the Jefferson Lab BoNuS experiment using a spectator-tagging technique to remove the nuclear uncertainties associated with previous deuterium target measurements is shown in figure 7.11.

The parton-model expressions imply a number of useful sum rules that the quark distributions must satisfy. The electric-charge sum rule for the proton reads

$$\int_0^1 dx \left[ \tfrac{2}{3}(u(x) - \bar{u}(x)) - \tfrac{1}{3}(d(x) - \bar{d}(x)) \right] = 1, \tag{7.4.63}$$

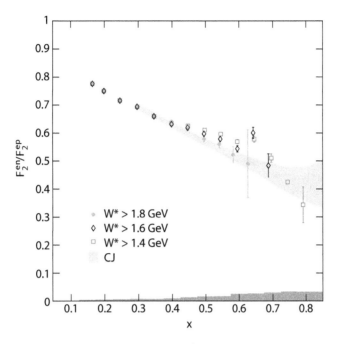

**Figure 7.11.** Dependence of the ratio $F_2^{en}(x)/F_2^{ep}(x)$ on the momentum fraction $x$, as measured by the BoNuS experiment at Jefferson Lab. The parton distribution function band (CJ) is from the CTEQ-Jefferson Lab global fitting effort. The shaded band along the $x$-axis shows the total systematic uncertainty [24].

and that for the neutron is

$$\int_0^1 dx \left[ \tfrac{2}{3}(d(x) - \bar{d}(x)) - \tfrac{1}{3}(u(x) - \bar{u}(x)) \right] = 0, \tag{7.4.64}$$

where the contributions of strange and heavier quarks vanish because the nucleon carries no net strangeness, charm, and heavier flavors. Combining these definitions, we obtain the sum rules

$$\left. \begin{matrix} \int_0^1 dx(u(x) - \bar{u}(x)) = 2 \\ \int_0^1 dx(d(x) - \bar{d}(x)) = 1 \end{matrix} \right\}, \tag{7.4.65}$$

which are simply parton-model restatements of the familiar fact that the proton is composed of two (net) up quarks and one (net) down quark. Similar expressions can be derived for other additive quantum numbers such as baryon number and strangeness.

Because the total momentum of the proton must be carried by its constituents, we may write the momentum sum rule

$$\sum_i \int_0^1 dx \, x f_i(x) = 1. \tag{7.4.66}$$

Neglecting the strange and heavier quarks, we may write

$$F_2^{ep}(x) + F_2^{en}(x) = \frac{5x}{9} \left[ u(x) + \bar{u}(x) + d(x) + \bar{d}(x) \right],$$ (7.4.67)

whereupon the fractional momentum carried by the quarks is [25]

$$\tfrac{9}{5} \int_0^1 dx \left( F_2^{ep}(x) + F_2^{en}(x) \right) \approx 0.45.$$ (7.4.68)

Unless our neglect of strange quarks was grossly in error (see problem 7.13), we are led to conclude that 55% of the proton momentum is carried by neutral partons. As we shall see in chapter 8, this role falls naturally to the gluons, the gauge bosons of quantum chromodynamics.

To proceed further in the analysis of the quark distributions without making strongly model-dependent assumptions, we must make use of information from the charged-current weak interactions. The most general form of the cross section for the inclusive reaction

$$\nu + N \rightarrow \mu + \text{anything}$$ (7.4.69)

may be derived (cf. problem 7.14) by the same methods used to derive the form (7.4.27) for deeply inelastic electron scattering. There is the important difference that the lepton tensor in this case has a $V - A$ structure, and the cross section expression is slightly complicated by the violation of parity. The general result is

$$\frac{d^2\sigma^\nu}{dQ^2\,d\nu} = \frac{G_F^2}{2\pi} E'E \left[ 2 W_1^\nu \sin^2\left(\frac{\theta}{2}\right) + W_2^\nu \cos^2\left(\frac{\theta}{2}\right) + W_3^\nu \frac{(E+E')}{M} \sin^2\left(\frac{\theta}{2}\right) \right],$$ (7.4.70)

where the final term arises from the parity-violating $\varepsilon_{\mu\nu\alpha\beta} P^\alpha q^\beta$ term in the general expansion (7.4.14) of the hadronic vertex. The cross section is conveniently recast in terms of the scaling variables $x$ and $y$ as

$$\frac{d^2\sigma^{\nu,\bar{\nu}}}{dx\,dy} = \frac{G_F^2 ME}{\pi} \left[ \mathcal{F}_1(x)xy^2 + \mathcal{F}_2(x)(1-y) \pm \mathcal{F}_3(x)xy\left(1 - \frac{y}{2}\right) \right],$$ (7.4.71)

where the dimensionless structure functions $\mathcal{F}_{1,2}(x)$ are defined in analogy with $F_{1,2}(x)$ in the electromagnetic case, and

$$\mathcal{F}_3(x) \equiv \nu W_3^\nu(x).$$ (7.4.72)

Without further analysis, we obtain the parton-model prediction for the total charged-current cross section,

$$\sigma_t(\nu N \rightarrow \ell + \text{anything}) \propto E.$$ (7.4.73)

The experimental results collected in figure 7.12 show this to be the case to excellent approximation for neutrino and antineutrino scattering at energies from

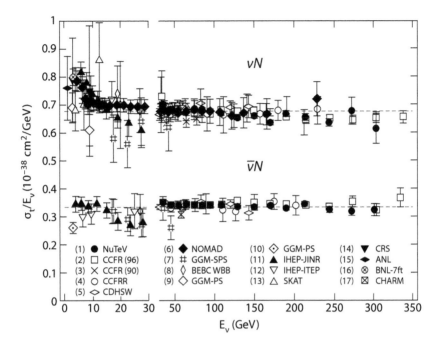

**Figure 7.12.** Slopes $\sigma_t/E$ for the muon neutrino and antineutrino charged-current total cross section as a function of incident neutrino energy. The error bars include both statistical and systematic errors. [Compiled by W. Seligman and M. H. Shaevitz for the 2010 *Review of Particle Physics* (Ref. [3]).]

30 GeV to 200 GeV, with

$$
\left.
\begin{aligned}
\frac{\sigma_t(\nu N)}{E} &= (6.77 \pm 0.14) \times 10^{-39} \text{ cm}^2/\text{GeV} \\
\frac{\sigma_t(\bar{\nu} N)}{E} &= (3.34 \pm 0.08) \times 10^{-39} \text{ cm}^2/\text{GeV}
\end{aligned}
\right\}.
\tag{7.4.74}
$$

A direct calculation (also part of problem 7.14) shows that

$$
\frac{x\mathcal{F}_3(x)}{\mathcal{F}_2(x)} =
\begin{cases}
+1, & \text{fermion target,} \\
-1, & \text{antifermion target.}
\end{cases}
\tag{7.4.75}
$$

When combined with the earlier finding [cf. (7.4.52)] that

$$
2x\mathcal{F}_1(x) = \mathcal{F}_2(x),
\tag{7.4.76}
$$

this leads to the conclusion that

$$
\frac{d^2\sigma}{dx\,dy}(\nu q) = \frac{G_F^2 ME}{\pi}\mathcal{F}_2(x)\{1\},
\tag{7.4.77}
$$

whereas

$$\frac{d^2\sigma}{dx\,dy}(\bar{\nu}q) = \frac{G_F^2 ME}{\pi}\mathcal{F}_2(x)\{(1-y)^2\}, \tag{7.4.78}$$

which reproduces the angular distributions computed in section 6.1.

As we did for the electromagnetic case, we may compute the contributions of various elementary processes to the structure functions. We find directly that, in the approximation in which $\cos\theta_C = 1$,

$$\begin{aligned}
\mathcal{F}_2^{\nu p}(x) &= 2x(d(x) + \bar{u}(x)) \\
\mathcal{F}_3^{\nu p}(x) &= 2(d(x) - \bar{u}(x))
\end{aligned} \tag{7.4.79}$$

and

$$\begin{aligned}
\mathcal{F}_2^{\bar{\nu}p}(x) &= 2x(u(x) + \bar{d}(x)) \\
\mathcal{F}_3^{\bar{\nu}p}(x) &= 2(u(x) - \bar{d}(x)).
\end{aligned} \tag{7.4.80}$$

In lieu of an extended analysis, we may content ourselves with a few elementary remarks. The combination

$$\mathcal{F}_2^{\nu p} + \mathcal{F}_2^{\nu n} = 2x(u(x) + \bar{u}(x) + d(x) + \bar{d}(x)) \tag{7.4.81}$$

is simply proportional to the analogous quantity for electromagnetic scattering,

$$\frac{F_2^{ep} + F_2^{en}}{\mathcal{F}_2^{\nu p} + \mathcal{F}_2^{\nu n}} = \frac{5}{18}, \tag{7.4.82}$$

up to neglect of strange quarks. This is nicely supported by the data for $x \gtrsim 0.1$ [26], and reinforces the conclusion that approximately half the momentum of the nucleon is carried by gluons that are inert with respect to the charged-current and electromagnetic interactions.

Let us notice finally that the structure function $\mathcal{F}_3$ measures the difference between quark and antiquark contributions, whereas $\mathcal{F}_2$ measures the sum. Thus a comparison of these two structure functions leads to an assessment of the importance of the quark–antiquark sea with respect to the valence component. More specifically, we have, for example, that

$$\mathcal{F}_3^{\nu p}(x) + \mathcal{F}_3^{\nu n}(x) = 2(u(x) - \bar{u}(x) + d(x) - \bar{d}(x)), \tag{7.4.83}$$

whence the baryon-number sum rule

$$\int_0^1 dx(\mathcal{F}_3^{\nu p}(x) + \mathcal{F}_3^{\nu n}(x)) = 6. \tag{7.4.84}$$

Look ahead to the left panel of figure 8.34 to see typical parton distributions extracted from the data on deeply inelastic lepton–nucleon scattering (and other reactions).

All this discussion has been prologue to our central interest in this section, which is a consideration of the neutral-current interactions on hadron targets.

For an isoscalar nucleon,

$$N \equiv \tfrac{1}{2}(p + n), \qquad (7.4.85)$$

we may write the charged-current cross sections as

$$\frac{d^2\sigma}{dx\,dy}(\nu N \to \mu^- X) = \frac{G_F^2 ME}{\pi} x \left[ (u(x) + d(x)) + (\bar{u}(x) + \bar{d}(x))(1 - y)^2 \right], \quad (7.4.86)$$

$$\frac{d^2\sigma}{dx\,dy}(\bar{\nu} N \to \mu^+ X) = \frac{G_F^2 ME}{\pi} x \left[ (u(x) + d(x))(1 - y)^2 + (\bar{u}(x) + \bar{d}(x)) \right]. \quad (7.4.87)$$

On comparing the Feynman rules for charged-current and neutral-current interactions with quarks (cf. figure 7.2), we may discern at once that

$$\frac{d^2\sigma}{dx\,dy}(\nu N \to \nu X) = \frac{G_F^2 ME}{4\pi} x \left\{ (L_u^2 + L_d^2) \left[ (u(x) + d(x)) + (\bar{u}(x) + \bar{d}(x))(1 - y)^2 \right] \right.$$
$$\left. + (R_u^2 + R_d^2) \left[ (u(x) + d(x))(1 - y)^2 + (\bar{u}(x) + \bar{d}(x)) \right] \right\}.$$
$$(7.4.88)$$

and

$$\frac{d^2\sigma}{dx\,dy}(\bar{\nu} N \to \bar{\nu} X) = \frac{G_F^2 ME}{4\pi} x \left\{ (L_u^2 + L_d^2) \left[ (u(x) + d(x))(1 - y)^2 + (\bar{u}(x) + \bar{d}(x)) \right] \right.$$
$$\left. + (R_u^2 + R_d^2) \left[ (u(x) + d(x)) + (\bar{u}(x) + \bar{d}(x))(1 - y)^2 \right] \right\}.$$
$$(7.4.89)$$

It is useful for orientation to make the idealization that the antiquark distributions may be neglected, whereupon compact expressions follow for the neutral-current to charged-current ratios (cf. problem 7.15)

$$R_\nu = \frac{\sigma(\nu N \to \nu X)}{\sigma(\nu N \to \mu^- X)} = \frac{1}{2} - x_W + \frac{20 x_W^2}{27}, \qquad (7.4.90)$$

$$R_{\bar{\nu}} = \frac{\sigma(\bar{\nu} N \to \bar{\nu} X)}{\sigma(\bar{\nu} N \to \mu^+ X)} = \frac{1}{2} - x_W + \frac{20 x_W^2}{9}. \qquad (7.4.91)$$

Thus, for our test value of $x_W = 0.2122$, the ratios are

$$R_\nu \approx 0.32, \qquad R_{\bar{\nu}} \approx 0.39. \qquad (7.4.92)$$

See figure 1 of Ref. [27] for a compilation of early measurements that lent support to the $SU(2)_L \otimes U(1)_Y$ electroweak theory and began to constrain the value of $\sin^2 \theta_W$.

The detailed comparison of the electroweak theory requires attention to many subtleties: the isotopic composition of heavy targets, the contributions of the quark–antiquark sea, including heavy quarks, threshold effects associated with the onset of charm production via the elementary reactions

$$\nu + d_\theta \to \mu^- + c, \qquad \bar{\nu} + \bar{d}_\theta \to \mu^+ + \bar{c}, \qquad (7.4.93)$$

and so on. Because of the extensive body of data that has been accumulated in deeply inelastic scattering, these complications can now be rather well controlled. We have, as well, the opportunity to include measurements made in hydrogen and deuterium, which separate the neutral-current cross sections on protons and neutrons, as well as data on exclusive final states. Overall, results in neutrino scattering are compatible with the standard electroweak theory and lead to a value of $\sin^2 \theta_W$ in accord with global fits to precision measurements. A notable outlier is the NuTeV determination of $\sin^2 \theta_W$ [28], based on the technique of problem 7.16.

Until now, we have discussed separately reactions mediated by $\gamma$, $W^\pm$, or $Z^0$ exchange. Deeply inelastic scattering of charged leptons provides another opportunity (as in $e^+ e^- \to f \bar{f}$) to observe $\gamma$-$Z^0$ interference effects and thus to investigate the neutral-current couplings of leptons and quarks. Within the parton model, we may write

$$
\begin{aligned}
\frac{1}{s} & \frac{d^2\sigma}{dx\,dy}(e^\mp p \to e^\mp X) \\
&= \frac{4\pi\alpha^2}{Q^4} e_q^2 (q + \bar{q}) x \left(1 - y + \frac{y^2}{2}\right) \\
&\quad - \frac{2\alpha\, G_F M_Z^2 e_q}{Q^2 \sqrt{2}(Q^2 + M_Z^2)} \left[ (L_e + R_e)(L_q + R_q)(q + \bar{q}) x \left(1 - y + \frac{y^2}{2}\right) \right. \\
&\quad \left. \pm (L_e - R_e)(L_q - R_q)(q - \bar{q}) xy \left(1 - \frac{y}{2}\right) \right] \\
&\quad + \frac{1}{2\pi} \left( \frac{G_F M_Z^2}{\sqrt{2}(Q^2 + M_Z^2)} \right)^2 \left[ (L_e^2 + R_e^2)(L_q^2 + R_q^2)(q + \bar{q}) x \left(1 - y + \frac{y^2}{2}\right) \right. \\
&\quad \left. \pm (L_e^2 - R_e^2)(L_q^2 - R_q^2)(q - \bar{q}) xy \left(1 - \frac{y}{2}\right) \right], \qquad (7.4.94)
\end{aligned}
$$

where $q(x)$ and $\bar{q}(x)$ are the distributions of quarks and antiquarks within the proton and a summation over quark species is implied. As in the analogous expression (6.4.27) for electron–positron annihilations, the first term is of electromagnetic origin, the third represents the effect of the weak neutral current, and the second is a weak–electromagnetic interference term. The weak neutral current gives rise to both parity violation and charge asymmetries, which therefore suggest particularly sensitive experimental approaches to the problem of neutral-current structure. To estimate the magnitude of these effects at low energies, it is permissible to neglect the final term because $Q^2 \ll M_Z^2$. Furthermore, if we choose $x_W = \frac{1}{4}$, not far from our leading-order test value, the electron's neutral-current coupling is purely axial, so only the first piece of the interference term need be considered. Finally, by selecting

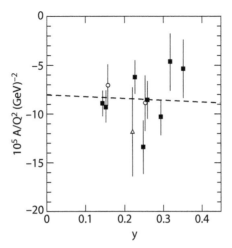

**Figure 7.13.** Measurements of the polarized-electron–deuteron asymmetry [29] at incident energies $E = 19.4$ GeV (solid rectangles), 16.2 GeV (open triangle), and 22.2 GeV (open circles). The dashed line shows the expectation of the $SU(2)_L \otimes U(1)_Y$ electroweak theory with $\sin^2 \theta_W = 0.224$.

an isoscalar target, we ensure that for the target as a whole,

$$\langle u(x) + \bar{u}(x) \rangle_{\text{target}} = \langle d(x) + \bar{d}(x) \rangle_{\text{target}}. \tag{7.4.95}$$

Neglecting the contributions of heavy quarks, we easily find that the asymmetry for polarized electron scattering from an isoscalar target is proportional to $a_e$ times a charge-weighted sum of $v_u$ and $v_d$:

$$A \equiv \frac{\sigma_R - \sigma_L}{\sigma_R + \sigma_L} \simeq \frac{-G_F Q^2 (R_e - L_e)}{4\pi \alpha \sqrt{2}} \frac{\sum_{q=u,d}(L_q + R_q)e_q}{\sum_{q=u,d} e_q^2}$$

$$= \frac{-G_F Q^2}{4\pi \alpha \sqrt{2}} \cdot \frac{9}{5} \left( 1 - \frac{20 x_W}{9} \right). \tag{7.4.96}$$

In this approximation, we expect the (parity-violating) asymmetry to be independent of $y$ and to be extremely small and negative. For $\sin^2 \theta_W = \frac{1}{4}$,

$$\frac{A}{Q^2} \simeq -7.2 \times 10^{-5}. \tag{7.4.97}$$

Precisely such an effect was measured in a meticulously executed experiment carried out by the SLAC–Yale Collaboration [29], from which the data are reproduced in figure 7.13. A less idealized analysis than the one we have just given yields, from this experiment alone, a best value for the weak mixing parameter of

$$x_W = 0.224 \pm 0.020, \tag{7.4.98}$$

consistent with other determinations. This experiment was of considerable importance in gaining acceptance for the standard model because, in addition to being the first to demonstrate directly parity violation in the neutral current, it came at a time when conflicting results from atomic physics experiments had suggested the necessity of a more involved structure for the neutral current.

At higher energies, the parity violations and charge asymmetries implied by the $SU(2)_L \otimes U(1)_Y$ electroweak theory become more dramatic. Effects of one part in $10^4$ at $Q^2 \approx 1$ GeV$^2$ rise toward a few tenths at $Q^2 = 10^4$ GeV$^2$, where the influence of the pure weak neutral-current term cannot be neglected. The asymmetries obtained from the combined H1 and ZEUS data on polarized $e^\mp p$ scattering at HERA [30] are in excellent agreement with standard-model predictions over the range 200 GeV$^2 \lesssim Q^2 \lesssim 2 \times 10^4$ GeV$^2$.

Most of our discussion of the consequences of the standard model for lepton–hadron interactions has taken place in the context of the point-coupling limit $Q^2 \ll M_{W,Z}^2$. The possibility of studying weak interactions at extremely high energies, at electron–proton colliders or in ultrahigh-energy neutrino scattering, raises the possibility of observing the damping effect of the intermediate-boson propagator upon the total cross section and other observables. It is straightforward in the parton model to calculate the total charged-current cross sections for $e^\mp p$ scattering and $(\nu, \bar{\nu})N$ scattering. If there were no intermediate bosons and if Bjorken scaling were perfect, the cross section would rise linearly with the squared c.m. energy $s$. An intermediate boson with mass $M_W \approx 80.4$ GeV causes a pronounced damping of the total cross section (roughly a 50% reduction) at $s \approx 10^5$ GeV$^2$, the HERA regime. The first evidence for the influence of the $W$ propagator was obtained in the earliest HERA measurements of the charged-current reaction $ep \to \nu + \text{anything}$ at an equivalent lab energy near 47 TeV [31]. Subsequent measurements led to quantitative determinations of $M_W$; the most precise to date is $M_W = 80.786 \pm 0.205 ^{+0.30}_{-0.098}$ GeV [32], to be compared with the current world average, $M_W = 80.399 \pm 0.023$ GeV (Ref. [3]).

The H1 and ZEUS experiments at the $e^\pm p$ collider HERA compared the momentum-transfer dependence of neutral-current ($e^\pm p \to e^\pm + \text{anything}$) and charged-current ($e^\pm p \to (\bar{\nu}_e, \nu_e) + \text{anything}$) at $\sqrt{s} = 318$ GeV. A recent summary compiled by H1 and ZEUS is given in figure 7.14. At low values of $Q^2$, the neutral-current cross section exceeds the charged-current cross section by more than two orders of magnitude, because the electromagnetic interaction is much stronger than the weak interaction at long wavelengths. There the neutral-current cross section is proportional to $1/Q^4$ due to the photon propagator, whereas the charged-current cross section is proportional to $1/(Q^2 + M_W^2)^2$. At large values of $Q^2 \gtrsim (M_W^2, M_Z^2)$, the cross sections roughly track each other. This behavior supports the notion that the intrinsic strengths of the weak and electromagnetic interactions are comparable.

Within the standard model, the charged weak current is purely left-handed, and so only left-handed electrons and right-handed positrons participate in the reactions $e^\mp p \to (\nu, \bar{\nu}) + \text{anything}$ [cf. (7.4.86) and (7.4.87)]. The HERA experiments put this prediction to the test by measuring the total cross sections for charged-current deeply inelastic scattering as functions of the lepton-beam polarization. The results, collected in figure 7.15, display the expected linear variation with polarization [36]. The $e^- p$ cross section vanishes for right-handed electrons, whereas the $e^+ p$ cross section vanishes for left-handed positrons. This behavior constrains the left-right symmetric models mentioned briefly in chapter 9.

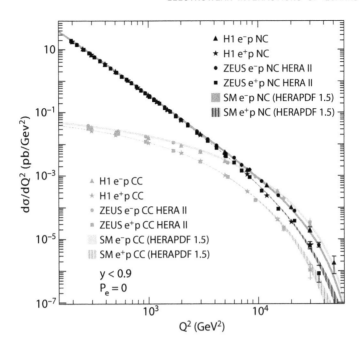

**Figure 7.14.** The $Q^2$-dependence of the neutral-current (NC) and charged-current (CC) cross sections measured by the H1 [33] and ZEUS [34] experiments at the HERA $e^{\pm}p$ collider. The curves represent the standard-model expectations derived from the HERA parton distribution functions [35].

## 7.5 HADRON–HADRON INTERACTIONS

We shall now look very briefly at manifestations of the electroweak interactions in hadron–hadron collisions. The conciseness of the presentation does not reflect a limited interest in the experimental possibilities, but only the fact that—at the level of the parton model—no new concepts are involved, beyond those we have already encountered in situations that are somewhat simpler to analyze. Indeed, the highest energies obtained in accelerator experiments have been achieved in $\bar{p}p$ collisions at the Tevatron ($\sqrt{s} \approx 2$ TeV) and the Large Hadron Collider ($\sqrt{s} = 8$ TeV, designed to reach 14 TeV). These very high energies engender a special interest in hadron–hadron collisions for the exploration of the electroweak scale. The discovery of the $W^{\pm}$ and $Z^0$ at CERN's Super Proton Synchrotron Collider was a landmark in the validation of the electroweak theory.

A high-energy hadron beam may usefully be regarded as an unseparated, broadband beam of quarks, antiquarks, and gluons (and perhaps other constituents, as yet unknown). For hard-scattering phenomena, it is the rate of encounters among energetic constituents that determines interaction rates. The spirit of the parton model is indicated in figure 7.16. In this picture, the cross section for the hadronic reaction

$$a + b \rightarrow c + \text{anything} \qquad (7.5.1)$$

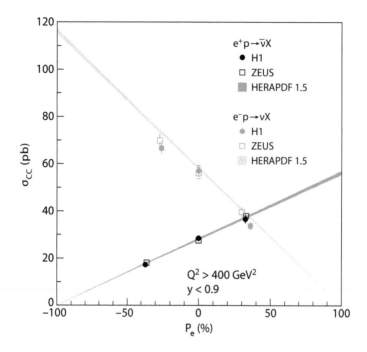

**Figure 7.15.** Total cross sections for deeply inelastic $e^{\mp} p \to (\nu, \bar{\nu}) +$ anything as a function of the longitudinal polarization of the lepton beam [33, 34, 36].

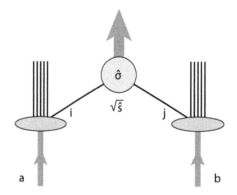

**Figure 7.16.** Parton-model representation of a hadron–hadron reaction.

is given schematically by

$$d\sigma(a + b \to c + X) = \sum_{\text{partons } i,j} f_i^{(a)}(x_a) f_j^{(b)}(x_b) \hat{\sigma}(i + j \to c + X'), \qquad (7.5.2)$$

where $f_i^{(a)}(x_a)$ is the probability of finding constituent $i$ with momentum fraction $x_a$ of hadron $a$, and $\hat{\sigma}(i + j \to c + X')$ is the cross section for the elementary process leading to the desired final state. The summation runs over all contributing parton–parton configurations. We shall have more to say about the parton-distribution functions and their evolution according to quantum chromodynamics in section 8.5.3. Our immediate interest is to have a first look at the reactions in which the electroweak gauge bosons $W^\pm$ and $Z^0$ were discovered. If we denote the invariant mass of the $i$-$j$ system as

$$\mathcal{M} = \sqrt{s\tau} \tag{7.5.3}$$

and its longitudinal momentum in the hadron–hadron c.m. frame by

$$p = \frac{x\sqrt{s}}{2}, \tag{7.5.4}$$

then the kinematic variables $x_{a,b}$ of the elementary process are related to those of the hadronic process by

$$x_{a,b} = \tfrac{1}{2}[(x^2 + 4\tau)^{1/2} \pm x]. \tag{7.5.5}$$

Within the parton-model framework, Drell and Yan [37] treated the reaction

$$a + b \rightarrow \ell^+\ell^- + \text{anything}, \tag{7.5.6}$$

in which a lepton pair of invariant mass $\mathcal{M}$ is produced with c.m. momentum fraction $x$ through the elementary reaction

$$q + \bar{q} \rightarrow \gamma^* \rightarrow \ell^+\ell^-. \tag{7.5.7}$$

The differential cross section is given by

$$\frac{d\sigma}{d\mathcal{M}^2\,dx} = \left(\frac{4\pi\alpha^2}{3\mathcal{M}^4}\right) F(\tau, x), \tag{7.5.8}$$

where the first factor is familiar as the cross section for the reaction $e^+e^- \rightarrow \mu^+\mu^-$ and the second is particular to the parton-model subprocess. It takes the form

$$F(\tau, x) = \frac{x_a x_b}{(x^2 + 4\tau)} g(x_a, x_b), \tag{7.5.9}$$

where the first factor is a Jacobian determinant connecting the variables $(x_a, x_b)$ with $(x, \mathcal{M}^2)$ and information about the quark distributions within the hadrons is contained in the function

$$g(x_a, x_b) = \tfrac{1}{3} \sum_{\text{flavors } i} e_i^2 [q_i^{(a)}(x_a)\bar{q}_i^{(b)}(x_b) + \bar{q}_i^{(a)}(x_a)q_i^{(b)}(x_b)], \tag{7.5.10}$$

where $e_i$ is the quark charge. The factor $\tfrac{1}{3}$ is a consequence of color: the quark and antiquark that annihilate into a virtual photon must have the same color as well as flavor, but we do not distinguish colors in measuring the quark distributions in deeply inelastic scattering. The parton model thus provides a link between lepton–hadron and hadron–hadron processes.

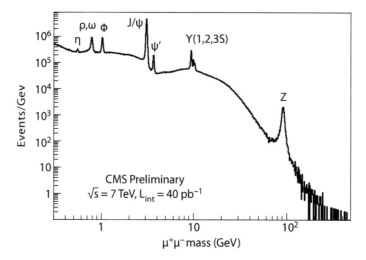

**Figure 7.17.** Opposite-sign dimuon spectrum measured by the Compact Muon Spectrometer Collaboration in $\sqrt{s} = 7$ TeV $pp$ collisions at the Large Hadron Collider [39].

This simple picture of dilepton production carries a number of significant implications. The most general of these is the scaling prediction that the combination

$$\mathcal{M}^4 \frac{d\sigma}{d\mathcal{M}^2} = \frac{4\pi\alpha^2\tau}{3} \int_\tau^1 dx \, \frac{g(x, \tau/x)}{x} \tag{7.5.11}$$

should be a function of the dimensionless variable $\tau$ alone. Although there are important strong-interaction (i.e., QCD) corrections to the parton model for this process, the scaling behavior has been established experimentally as a good first approximation [38].

The dilepton spectrum observed in high-energy collisions consists not only of the continuum described in lowest order by (7.5.11) and the preceding formulas, but also of discrete resonances that share the quantum numbers of the virtual photon or are produced by other mechanisms but decay into lepton pairs. An early look at the dimuon spectrum at the Large Hadron Collider is shown in figure 7.17.

An important example of discrete-resonance production concerns the formation of electroweak gauge bosons in quark–antiquark collisions. For the reaction

$$a + b \to W^\pm + \text{anything}, \tag{7.5.12}$$

the elementary reactions are

$$u\bar{d}_\theta \to W^+, \qquad \bar{u}d_\theta \to W^-. \tag{7.5.13}$$

The integrated cross section is

$$\sigma(a + b \to W^\pm + X) = G_F\pi\sqrt{2}\,\tau \int_\tau^1 dx \, \frac{W_{ab}^{(\pm)}(x, \tau/x)}{x}, \tag{7.5.14}$$

where $\tau = M_W^2/s$ and

$$W_{ab}^{(+)} = \tfrac{1}{3} \left\{ [u^{(a)}(x_a)\bar{d}^{(b)}(x_b) + \bar{d}^{(a)}(x_a)u^{(b)}(x_b)] \cos^2 \theta_C \right.$$

$$\left. + [u^{(a)}(x_a)\bar{s}^{(b)}(x_b) + \bar{s}^{(a)}(x_a)u^{(b)}(x_b)] \sin^2 \theta_C \right\}. \qquad (7.5.15)$$

At CERN's $S\bar{p}pS$ Collider, which operated at 540 and 630 GeV, the production cross section times leptonic branching ratio of approximately 0.5 nb enabled the discovery [40] of the charged intermediate boson by the UA1 and UA2 Collaborations. For the Tevatron experiments CDF and D0, studying 1.96-GeV proton–antiproton collisions, $\sigma(\bar{p}p \to W)B(W \to \ell\nu) \approx 2.8$ nb. Their samples of 10 fb$^{-1}$ will yield $O(10^7)$ $W$-bosons per experiment. With its large event samples, the Tevatron has made highly precise measurements of the $W$-boson mass, $M_W = 80.387 \pm 0.016$ GeV. In initial running of the Large Hadron Collider at $\sqrt{s} = 8$ TeV, the ATLAS and CMS experiments benefit from a larger $\sigma(pp \to W)B(W \to \ell\nu) \approx 12.5$ nb. At a luminosity of $5 \times 10^{33}$ cm$^{-2}$ s$^{-1}$, the LHC experiments expect to observe approximately 60 $W$-bosons per second.

The production and leptonic decay of the charged intermediate boson owes an interesting experimental signature to the $V - A$ structure of the charged current. If we idealize the proton as composed entirely of quarks and the antiproton as composed entirely of antiquarks, the intermediate bosons will be produced with spins aligned along the direction of the incident antiproton, because only left-handed fermions and right-handed antifermions participate in the formation reactions (7.5.13). Then, for the same reason (as we discussed in §6.2), the decay products that are fermions will be emitted opposite to the direction of $W$ polarization. As a consequence, the negative leptons from the decay $W^- \to \ell^- \bar{\nu}_\ell$ will be preferentially emitted in the direction defined by the incident protons, whereas positive leptons (from $W^+$ decay) will tend to follow the antiproton direction. The predicted decay angular distribution [cf. (6.2.24)] is proportional to $(1 + \cos\theta^*)^2$. These idealized conditions held, to high approximation, in the pioneering experiments at CERN's Super Proton Synchrotron Collider. At $\sqrt{s} = 540$ and 630 GeV, the elementary collisions that form $W$ (and $Z$) bosons are overwhelmingly valence quarks from the proton on valence antiquarks from the antiproton. An early decay angular distribution, recorded by the UA1 Collaboration, is shown in figure 7.18. The observed charge asymmetry, which is opposite to what might be expected for the background of strong-interaction processes, provided important evidence that the new object observed as an isolated lepton plus missing energy had the characteristics anticipated for the $W$-boson of the SU(2)$_L \otimes$ U(1)$_Y$ theory. The $V - A$ structure of the charged current enables a direct way to study the spin structure of the proton in polarized $\bar{p}p \to W^\pm +$ anything [42].

Corresponding calculations for hypothetical additional $W'$-bosons calibrate searches and make it possible to set limits. For the artificial but informative construct of a sequential gauge boson with standard-model couplings, (7.5.14) and (7.5.15) apply, now with $\tau = M_{W'}^2/s$. [Other assumptions for the couplings of $W'$ to fermions would modify (7.5.15).] Early limits from experiments at the Large Hadron Collider surpass $M_{W'} > 2$ TeV [43].

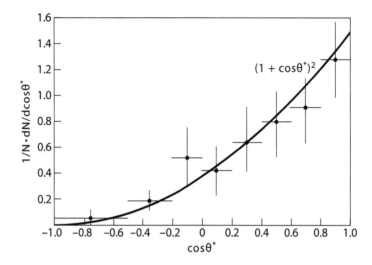

**Figure 7.18.** Decay angular distribution of the emission angle $\theta^*$ of the electron (positron) with respect to the proton (antiproton) direction in the $W$ rest frame, for the reaction $\bar{p}p \to W^\pm + $ anything at $\sqrt{s} = 546$ and $630$ GeV [41]. The curve shows the standard-model expectation, $(1 + \cos\theta^*)^2$, for $W$-bosons produced with helicity $= -1$.

The cross section for the production of neutral intermediate bosons may be estimated in precisely the same manner from the calculable rates for the reactions

$$u\bar{u} \to Z^0, \qquad d\bar{d} \to Z^0, \qquad s\bar{s} \to Z^0. \tag{7.5.16}$$

The cross section may be written as

$$\sigma(a + b \to Z^0 + X) = \frac{G_F\pi}{\sqrt{2}}\tau \int_\tau^1 dx \, \frac{Z_{ab}(x, \tau/x)}{x}, \tag{7.5.17}$$

where $\tau = M_Z^2/s$ and

$$Z_{ab}(x_a, x_b) = \tfrac{1}{3}\sum_q [q^{(a)}(x_a)\bar{q}^{(b)}(x_b) + \bar{q}^{(a)}(x_a)q^{(b)}(x_b)](L_q^2 + R_q^2), \tag{7.5.18}$$

where $L_q$ and $R_q$ are the familiar neutral-current chiral couplings (cf. figure 7.2). A fuller account of dilepton production in $p^\pm p$ collisions must incorporate the contributions from both (virtual) $\gamma$ and $Z^0$ in the intermediate state, in analogy to the description of the reaction $e^+e^- \to q\bar{q}$ given in (7.3.3).

The rate $\sigma(\bar{p}p \to Z^0)B(Z^0 \to \ell^+\ell^-) \approx 0.07$ nb at the $S\bar{p}pS$ Collider sufficed for the discovery of the neutral intermediate boson [44], a key event in the history of the electroweak theory. At the Tevatron, where $\sigma(\bar{p}p \to Z^0)B(Z^0 \to \ell^+\ell^-) \approx 0.25$ nb, the final samples in CDF and D0 should run to 400 thousand in each leptonic channel. Among the detailed measurements carried out at the Tevatron, the D0 Collaboration has reported the most precise determinations of the chiral couplings of $Z^0$ to the light quarks $u$ and $d$ [45]. In 8-TeV running at the LHC, the rate grows to $\sigma(pp \to Z^0)B(Z^0 \to \ell^+\ell^-) \approx 1.1$ nb, for a yield of 5 $Z^0$ per second at $\mathcal{L} = 5 \times 10^{33}$ cm$^{-2}$ s$^{-1}$.

As we saw for the charged intermediate bosons, corresponding calculations for hypothetical additional $Z'$ bosons calibrate searches and make it possible to set limits. The useful benchmark of a sequential $Z'$ with standard model couplings is governed by (7.5.17) and (7.5.18), with $\tau = M_{Z'}^{2}/s$. [Other assumptions for the couplings of $Z'$ to fermions would modify (7.5.18).] The most telling direct searches, from the LHC experiments, give limits approaching $M_{Z'} > 2$ TeV [46]. Complementary searches look for evidence of $Z' \to W^{+}W^{-}$. Global fits to electroweak parameters and neutral-current studies away from the $Z^{0}$ pole are sensitive to a $Z'$. Problem 7.11 explores the luminosity required for a future lepton collider to exploit the discovery of a heavy $Z'$.

## 7.6 FURTHER TESTS OF THE ELECTROWEAK THEORY

Stringent bounds on strangeness-changing neutral currents motivated the Glashow–Iliopoulos–Maiani mechanism (Ref. [5]) and its six-quark generalization developed in section 7.1. The bounds on flavor-changing neutral currents involving heavier flavors are less restrictive but provide significant additional tests of the electroweak theory, as well as opportunities for new physics to reveal itself [47].

Within the standard model, the rate anticipated for the decay $D^{0} \to \mu^{+}\mu^{-}$ is very small: $B(D^{0} \to \mu^{+}\mu^{-}) \gtrsim 4 \times 10^{-13}$ [48]. The Belle Collaboration bounds $B(D^{0} \to \mu^{+}\mu^{-}) < 1.4 \times 10^{-7}$ at 90% CL [49]. The observation of $D^{0}$–$\bar{D}^{0}$ mixing [50] has intensified interest in the search for new physics in charmed meson decays [51]. The LHC$b$ and CDF collaborations have given tantalizing evidence for CP-violating effects in $D^{0}$ decays into pairs of charged hadrons [52].

The dilepton decays of mesons containing $b$ quarks have the most accessible standard-model rates [53], and this is the sector in which experiments have come nearest to standard-model sensitivity. The current experimental upper bound on leptonic $B_{s}$ decays [54] is 1.2× the standard-model expectation [55], $B(B_{s} \to \mu^{+}\mu^{-}) < (3.5 \pm 0.2) \times 10^{-9}$. The corresponding upper bound for $B^{0}$ is approximately one order of magnitude above the standard-model expectation, $B(B^{0} \to \mu^{+}\mu^{-}) = (0.10 \pm 0.01) \times 10^{-9}$. The sensitivity of the LHC experiments is growing rapidly.

The world sample of top quarks remains modest, so the study of rare top decays is less advanced than for $K$, $D$, and $B$ mesons. The branching fractions anticipated in the standard model for the decays $t \to (u, c) + (\gamma, Z, g)$ are exceedingly small, less than $10^{-12}$. Representative current limits are $B(t \to (u, c) + \gamma) < 5.9 \times 10^{-3}$ [56], $B(t \to (u, c) + Z) < 0.73\%$ [57], $B(t \to ug) < 5.7 \times 10^{-5}$ and $B(t \to cg) < 2.7 \times 10^{-4}$ [58], all at 95% CL. Copious top-quark production at the Large Hadron Collider will advance the study of rare top decays and also make possible precise studies of single-top production through the electroweak interaction [59].

To the extent that flavor-changing neutral currents are highly suppressed in nature, it would seem that any physics beyond the standard electroweak theory should be governed by some new symmetry or dynamical scheme to cap new sources of flavor change. According to the "minimal flavor violation" hypothesis, the quark-mixing matrix is the sole source of flavor change [60]. For the moment, this is more a name for an observed regularity than an established principle.

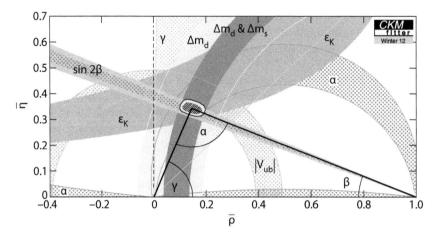

**Figure 7.19.** Constraints in the $(\bar{\rho}, \bar{\eta})$ plane as of winter 2012, as determined by the CKMfitter Group [64], in which the experimental inputs are detailed. The hashed region near the vertex $\alpha$ shows the global combination at 68% CL.

A straightforward test for the completeness of the CKM picture is to ask whether the magnitudes $|V_{ij}|$ are consistent with the hypothesis that the quark-mixing matrix is unitary. Particular attention has been accorded to the first row of the CKM matrix, looking for deviations from the unitarity requirement

$$S_u \equiv |V_{ud}|^2 + |V_{us}|^2 + |V_{ub}|^2 = 1, \qquad (7.6.1)$$

which would signal new physics. (Because $|V_{ub}|^2 \ll 1$, this is essentially a test of the Cabibbo picture.) A careful analysis of current experimental information yields $S_u = 0.9999 \pm 0.0006$ [61], which constrains new physics effects both at tree level and in quantum corrections. Ongoing studies of neutron decays should resolve a persistent lifetime puzzle [62], and may lead to an improved determination of $|V_{ud}|$.

Sustained experimental effort has produced a rich library of information about decay rates, neutral-meson mixings, and CP violation in $K$- and $B$-meson decays [63]. One application of that body of knowledge has been to probe in depth the unitarity of the CKM matrix $W^\dagger = I$, where $I$ is the $3 \times 3$ identity, by examining $\sum_i V_{ij} V_{ik}^* = \delta_{jk}$ and $\sum_j V_{ij} V_{kj}^* = \delta_{ik}$. The six vanishing conditions may be represented as triangles in the complex plane, each with an area proportional to $\text{Im}[V_{ij} V_{k\ell} V_{i\ell}^* V_{kj}^*]$, a parametrization-independent measure of CP violation (cf. problem 7.5).

The most commonly displayed unitarity triangle (cf. problem 7.4) is constructed from the constraint

$$V_{ud} V_{ub}^* + V_{cd} V_{cb}^* + V_{td} V_{tb}^* = 0. \qquad (7.6.2)$$

It is conventional to normalize the triangle, dividing the complex vector for each leg by the well-determined $V_{cd} V_{cb}^*$. The vertices of the triangle are then $(0, 0)$, $(1, 0)$, and $(\bar{\rho}, \bar{\eta})$. Among the tests available in this formalism are whether the triangle closes and whether different data sets yield a common vertex, $(\bar{\rho}, \bar{\eta})$. The plot in figure 7.19, which is representative of recent work, shows consistency among many experimental constraints.** That the imaginary coordinate $\bar{\eta}$ differs from zero shows that the Kobayashi–Maskawa mechanism is at work. A crucial prediction, that CP violation in $K$ physics is small because of flavor suppression but CP violation should be appreciable in $B$ physics, is fulfilled. More-detailed analysis shows that the quark-mixing matrix is the dominant source of CP violation in meson decays. On current evidence, including the null searches for flavor-changing neutral currents just reviewed, new-physics contributions are extremely small in $s \leftrightarrow d$, $b \leftrightarrow d$, $s \leftrightarrow b$, and $c \leftrightarrow u$ transitions.

The consistency of the CKM picture does not yet exclude a fourth generation of quarks. Direct constraints on $|V_{tb}|$ are consistent with a value near unity but are not yet terribly restrictive [65]. Global fits to the precision electroweak data would allow mixing between the third and fourth families at the level seen between the first and second families [66].

Finally, the robustness of the CKM unitarity triangle does not mean that there is no new physics to be found. The unitarity triangle analysis is sensitive mainly to processes that change flavor by two units. Even in the well-studied rare $K$ and $B$ decays (flavor change by one unit), many examples of new physics that could have passed the unitarity-triangle screen—supersymmetry, little Higgs models with $T$-parity, and warped extra dimensions—could give large departures [67]. New sources of CP violation and flavor-changing neutral currents occur in models that do not enforce minimal flavor violation. Today's experiments leave ample space between current bounds and standard-model expectations for departures from the CMK paradigm in many rare decays. Because the unitarity triangle is described well by the standard model, it will pay to examine CP violation in $b \to s$ transitions and rare decays, where standard-model contributions are small. Indeed, some argue that cracks have begun to appear in the CKM edifice [68].

The ability of the electroweak theory incorporating Cabibbo–Kobayashi–Maskawa mixing to account for—and predict—a vast number of observables in flavor physics is highly impressive. We must remember, however, that experiments have validated a framework, not an explanation. Just as the standard model makes no predictions for quark and lepton masses, it has nothing to say about the mixing angles and the Kobayashi–Maskawa phase. These can arise in the electroweak theory, but we do not know what determines their values.

## 7.7 A BRIEF LOOK AT QUANTUM CORRECTIONS

In a fully defined theory such as quantum electrodynamics, the study of higher-order corrections to well-measured quantities serves two purposes: to elaborate and test the predictions of the quantum field theory as precisely as possible and to prospect for evidence of unknown degrees of freedom at the high scales probed by loop integrals.

Successful calculations of lepton anomalous magnetic moments are among the great quantitative triumphs of modern physics. Measurements with a one-electron quantum cyclotron determine the electron magnetic moment, given in terms of the

---

** A persistent tension among determinations of $|V_{ub}|$, not indicated in the plot, is the subject of problem 7.23.

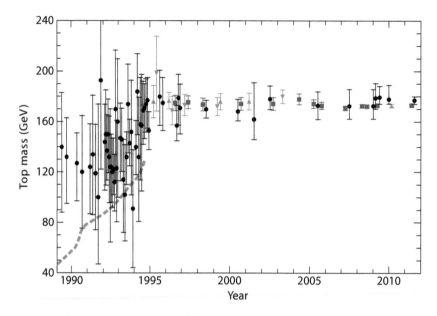

**Figure 7.20.** Indirect determinations of the top-quark mass from fits to electroweak observables (circles) and 95% confidence-level lower bounds on the top-quark mass inferred from direct searches in $\bar{p}p$ collisions, assuming that standard decay modes dominate (broken line). Direct measurements of $m_t$ by the CDF (triangles) and D0 (inverted triangles) Collaborations are shown at the time of initial evidence, discovery claim, and thereafter. The world average from direct observations is shown as boxes. (Updated from [71].)

gyromagnetic ratio as $g/2 = 1.001\,159\,652\,180\,73(28)$ [69], which agrees, within 5 parts per trillion, with QED calculations through order $(\alpha/\pi)^4$ [70].

In the electroweak theory, we encounter another situation, in which new degrees of freedom are anticipated, but crucial parameters such as particle masses are unknown. Here we may ask what value of an unknown parameter would bring calculations into best agreement with experiment. From the early 1990s, measurements of phenomena involving the $b$-quark demonstrated (cf. problem 7.12) that $b$ was the $I_3 = -\frac{1}{2}$ member of a weak-isospin doublet. The top quark had to be there, as the $I_3 = \frac{1}{2}$ partner, but it had not yet been observed. Could quantum corrections anticipate $m_t$? As we will see presently by means of an example, higher-order processes involving virtual top quarks are an important element in quantum corrections to the predictions the electroweak theory makes for many observables. Typical loop corrections are quadratically sensitive to $m_t$, albeit with small coefficients, so the situation was promising—as long as the electroweak theory could be counted upon.

The top mass favored by simultaneous fits to many electroweak observables is shown as a function of time in figure 7.20. By the end of 1994, the indirect determinations favored $m_t \approx 175 \pm 25$ GeV, successfully anticipating the masses reported in the discovery papers: $176 \pm 8 \pm 10$ GeV for CDF and $199^{+19}_{-21} \pm 22$ GeV for D0. Today, direct measurements at the Tevatron determine the top-quark mass to a precision of 0.5%, $m_t = 173.2 \pm 0.9$ GeV [72], far more precise than the best indirect determinations.

The parameter

$$\rho \equiv \frac{M_W^2}{M_Z^2 \cos^2 \theta_W}, \tag{7.7.1}$$

which measures the relative strength of neutral-current and charged-current amplitudes at low energies, reflects the weak-isospin and weak-hypercharge content of the auxiliary scalar fields that hide the electroweak symmetry, as we have seen in problem 6.16. The cross sections for $(\nu_\mu, \bar{\nu}_\mu)e$ scattering derived in (6.4.11) and (6.4.12) generalize to

$$\sigma(\nu_\mu e \to \nu_\mu e) = \frac{G_F^2 mE}{2\pi} \rho^2 \left[ (2x_W - 1)^2 + \frac{4x_W^2}{3} \right] \tag{7.7.2}$$

and

$$\sigma(\bar{\nu}_\mu e \to \bar{\nu}_\mu e) = \frac{G_F^2 mE}{2\pi} \rho^2 \left[ \frac{(2x_W - 1)^2}{3} + 4x_W^2 \right], \tag{7.7.3}$$

whereas the low-energy limit of the charged-current cross section (6.2.5),

$$\sigma(\nu_\mu e \to \mu \nu_e) = \frac{2G_F^2 mE}{\pi} \left[ 1 - (\mu^2 - m^2)/2mE \right]^2, \tag{7.7.4}$$

remains unchanged. (Here $m$ and $\mu$ refer to the electron mass and muon mass.) The ratio $\sigma(\nu_\mu e \to \nu_\mu e)/\sigma(\bar{\nu}_\mu e \to \bar{\nu}_\mu e)$ determines the weak mixing parameter $x_W$, and the comparison of neutral-current and charged-current cross sections yields $\rho$. For example, the CHARM-II Collaboration reported [73]

$$\sin^2 \theta_W(\nu_\mu e) = 0.237 \pm 0.007 \text{ (stat.)} \pm 0.007 \text{ (sys.)}, \tag{7.7.5}$$

$$\rho(\nu_\mu e) = 1.006 \pm 0.014 \text{ (stat.)} \pm 0.033 \text{ (sys.)}. \tag{7.7.6}$$

In the standard model with a single Higgs doublet, $\rho = 1$ at tree level.

The $\rho$-parameter is also influenced by quantum corrections and, in particular, by weak-isospin breaking induced by large mass splittings within an SU(2)$_L$ doublet of fermions. Both W$^\pm$ and Z$^0$ masses are influenced by a variety of loop corrections [74], but the most important correction to $\rho$ is due to the heavy-quark–loop contributions shown in figure 7.21. The full calculation is lengthy, but we can obtain the contribution relevant here by evaluating the heavy-quark–loop diagrams and using dimensional regularization (cf. §8.2.2 and appendix B.4) to eliminate all but the finite parts. Then, defining $\rho = 1 + \Delta\rho$, we find

$$\Delta\rho = \frac{3G_F}{8\pi^2 \sqrt{2}} \left[ m_t^2 + m_b^2 + \frac{2m_t^2 m_b^2}{(m_t^2 - m_b^2)} \ln\left(\frac{m_b^2}{m_t^2}\right) \right]. \tag{7.7.7}$$

For heavy-lepton contributions, the color factor 3 is replaced by 1.

Note first that in the limit $m_b \to m_t$, $\Delta\rho \to 0$ independent of the value of $m_t$ (expand the logarithm in the neighborhood of argument unity to see this). The custodial SU(2) symmetry mentioned in the chapter 6 reading list is respected, and $\rho = 1$. Weak-isospin breaking manifested by $m_b \neq m_t$ is a prerequisite to

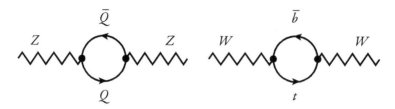

**Figure 7.21.** Contributions of heavy-quark loops to masses of the electroweak gauge bosons. The $Z^0$ mass receives $b\bar{b}$ and $t\bar{t}$ contributions; the $W^+$ mass is influenced by $t\bar{b}$ loops.

nonvanishing $\Delta\rho$ from fermion loops. In the limit $m_b \ll m_t$, which approximates the real world, we find the celebrated behavior,

$$\Delta\rho \to \frac{3G_F m_t^2}{8\pi^2\sqrt{2}}, \tag{7.7.8}$$

that made electroweak radiative corrections a telling probe of the top-quark mass.

Measurements on and near the $Z^0$ pole by the LEP experiments ALEPH, DELPHI, L3, and OPAL and by the SLD experiment at the Stanford Linear Collider were decisive in testing and refining the electroweak theory. What has been achieved overall is a comprehensive test of the electroweak theory as a quantum field theory at a precision of one part in a thousand for several observables. An important asset of global fits to many observables is their sensitivity to virtual effects and thus to parameters that have not been measured directly. The successful inference of the range of top-quark masses is a prime example. Now that $m_t$ is measured at high precision, it becomes a fixed parameter in the global fits, which may probe for the next unknown quantity. That is the Higgs-boson mass, for which the sensitivity in radiative corrections is subtle, $\propto \ln M_H$.

Figure 7.22 shows how the quality of the LEP Electroweak Working Group Summer 2011 global fit [75] varies with $M_H$. The fit is evidently improved by the inclusion of quantum corrections involving a Higgs boson that has standard-model interactions with the electroweak gauge bosons $W^\pm$ and Z. A satisfactory fit does not prove that the standard-model Higgs boson exists, but it offers guidance for the search and sets up a consistency check when a putative Higgs boson is observed. It is important to note that, although the global fits give evidence for the effect of the Higgs boson in the vacuum, they do not have any sensitivity to couplings of the Higgs boson to fermions free of the *assumption* that Higgs–Yukawa couplings set the fermion masses. The best-fit value is $M_H = 92^{+34}_{-26}$ GeV, largely in the range excluded by LEP, and the upper bound at 95% CL (without taking into account the LEP search) is 161 GeV. We will look more closely at inferences about $M_H$ when we consider the search for the standard-model Higgs boson in section 7.9.

The extraordinary precision of measurements on the $Z^0$ pole has given them a decisive weight in our assessment of the electroweak theory. They are, however, blind to new physics that does not directly modify the $Z^0$ properties. A heavy $Z'$ that does not mix appreciably with $Z^0$ is an important example. For this reason, experiments off the $Z^0$ pole, even of lower precision, command our attention—particularly in the search for physics beyond the standard model.

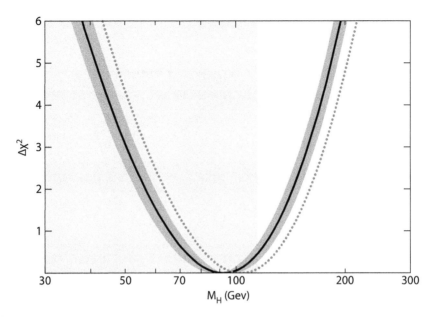

**Figure 7.22.** $\Delta\chi^2 = \chi^2 - \chi^2_{\min}$ from a fit to an ensemble of electroweak measurements as a function of the standard-model Higgs-boson mass. The solid line is the result of the fit, and the surrounding band represents an estimate of the theoretical uncertainty due to missing higher-order corrections. The shaded rectangle denotes the 95% CL lower bound on $M_H > 114.4$ GeV from direct searches at LEP [76]. The dotted curve includes additional low-$Q^2$ data. (Adapted from the LEP Electroweak Working Group; see Ref. [75].)

The weak mixing parameter [cf. (6.3.50)] is defined in terms of (running) couplings,

$$\sin^2\theta_{\rm W}(Q^2) = \frac{\alpha(Q^2)}{\alpha_2(Q^2)} = \frac{1/\alpha_2(Q^2)}{1/\alpha_Y(Q^2) + 1/\alpha_2(Q^2)}, \qquad (7.7.9)$$

so its value depends on the scale at which it is measured. We will develop this scale dependence at some length in the context of unified theories of the strong, weak, and electromagnetic interactions in section 9.3 and in problem 9.17. We will examine the evolution of the fine structure constant in section 8.2 and study the evolution of the coupling constant in non-Abelian gauge theories in section 8.3. In the range of scales currently accessible to experiment, the evolution of the weak mixing parameter is predicted within the electroweak theory itself. Theoretical expectations are compared with a digest of the relevant data in figure 7.23. A detailed narrative of data treatment and the comparison with theory is given in section 10.3 of Ref. [78]. Overall, the agreement is quite acceptable. In particular, the parity-violating left-right asymmetry observed [79] in polarized Møller scattering, $e^-e^- \rightarrow e^-e^-$, establishes the trend of low-scale running at more than six standard deviations and is in reasonable agreement with the prediction at $Q^2 = 0.026$ GeV$^2$.

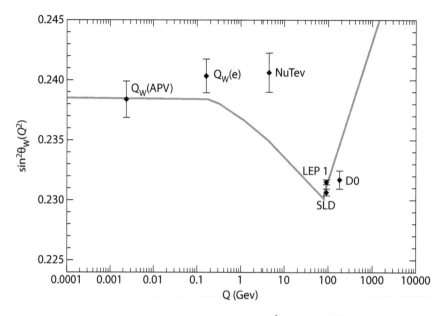

**Figure 7.23.** Evolution of the weak mixing parameter $\sin^2 \theta_W$ in the $\overline{\text{MS}}$ scheme, according to Ref. [77] (solid curve). The minimum occurs at $Q = M_W$, where the derivative of the weak mixing parameter changes sign as the influence of weak-boson loops drops out. The experimental determinations are from weak charges in atomic parity violation [$Q_W$(APV)] and Møller scattering [$Q_W(e)$], deeply inelastic $\nu N$ scattering, and studies at the $Z^0$ pole [D0 point (Ref. [45]) displaced for clarity; figure adapted from Ref. [78].]

## 7.8 THE SCALE OF FERMION MASSES

Within the standard electroweak theory, the overall scale of the fermion masses is set by the vacuum expectation value $v/\sqrt{2} \approx 174$ GeV of the Higgs field, but each fermion mass $m_i = \zeta_i v/\sqrt{2}$ involves a distinct Yukawa coupling $\zeta_i$ [cf. (6.3.42) and (7.1.32)]. The Yukawa couplings that reproduce the observed quark and charged-lepton masses range over many orders of magnitude, from $\zeta_e \approx 3 \times 10^{-6}$ for the electron to $\zeta_t \approx 1$ for the top quark. The origin of these Yukawa couplings is unknown and, at present, mysterious. Through its Yukawa couplings, the Higgs scalar is the only element of the standard model that distinguishes quark and lepton generations. In Veltman's phrase [80], it knows something that we do not know. In an important sense, therefore, *all fermion masses involve physics beyond the standard model.* The approach to fermion masses in unified theories is treated briefly in section 9.6.

Although the $SU(2)_L \otimes U(1)_Y$ electroweak theory shows how fermion masses might arise, we cannot be sure that finding the agent of electroweak symmetry breaking will bring a full understanding of fermion masses—even if the answer is an elementary Higgs boson. This is because we do not know that fermion masses are set on the electroweak scale. This point merits closer examination.

The observation of a nonzero fermion mass implies that the electroweak gauge symmetry $SU(2)_L \otimes U(1)_Y$ is broken, but electroweak symmetry breaking is only a necessary, not a sufficient, condition for the generation of fermion mass. In the standard-model framework, new physics at an unknown scale must give rise to the Yukawa couplings. We shall see in section 8.9 that if the strong interactions hide the electroweak symmetry, a separate mechanism—perhaps at a different scale—is required to generate fermion masses. Where might that different scale lie?

Partial-wave unitarity sets a model-independent upper bound on the energy scale of fermion mass generation [81]. The strategy, exhibited in problem 6.12, is simply to add explicit fermion mass terms to the electroweak Lagrangian in place of the Yukawa terms of (6.3.42) and (7.1.32). Explicit Dirac mass terms link the left-handed and right-handed fermions and thus violate the $SU(2)_L \otimes U(1)_Y$ gauge symmetry of the electroweak theory. If they persisted to arbitrarily high energies, such hard masses would destroy the renormalizability of the theory. That is why we ruled out explicit fermion masses when we formulated the electroweak theory. We might instead think it overly ambitious to demand that a theory make sense at all energies and to consider explicit fermion masses in the framework of an *effective field theory* valid over a limited range of energies, to be supplanted at higher energies by a theory that entails a different set of degrees of freedom [82].

Because the gauge symmetry is broken in a theory with explicit fermion masses $m_i$, at lowest order in perturbation theory, scattering amplitudes for the production of pairs of longitudinally polarized gauge bosons in fermion-antifermion annihilations grow with c.m. energy roughly as $G_F m_i E_{cm}$. (In the standard electroweak theory, this behavior is canceled by the contribution of direct-channel Higgs-boson exchange.) The resulting partial-wave amplitudes saturate partial-wave unitarity for the standard model with a Higgs mechanism at a critical c.m. energy (Ref. [81])

$$\sqrt{s_i} \simeq \frac{4\pi\sqrt{2}}{\sqrt{3\eta_i}\, G_F m_i} = \frac{8\pi v^2}{\sqrt{3\eta_i}\, m_i}, \tag{7.8.1}$$

where $\eta_i = 1(3)$ for leptons (quarks). As usual, the parameter $v$ sets the scale of electroweak symmetry breaking. If the electron mass were hard, the critical energy would be $\sqrt{s_e} \approx 1.7 \times 10^9$ GeV; the corresponding energy for the top quark is $\sqrt{s_t} \approx 3$ TeV. The fact that a hard electron mass would imply a saturation of partial-wave unitarity only at a prodigiously high energy means that although the behavior of $\sigma(e^+e^- \to W^+W^-)$ shown in figure 6.18 validates the gauge symmetry of the electroweak theory, it does not establish that the theory is renormalizable. Our current measurements do not establish the $He\bar{e}$ coupling of the standard model. Quantum corrections test the Higgs boson interactions with gauge bosons but do not probe the Higgs boson couplings to fermions. Until the Higgs boson is discovered and shown to couple to fermion pairs, the idea that a single agent hides the electroweak symmetry and also generates fermion mass is only a hypothesis.

## 7.9 SEARCH FOR THE HIGGS BOSON

We saw in our discussion of evidence for the virtual influence of the Higgs boson in section 7.7 that global fits, made within the framework of the standard

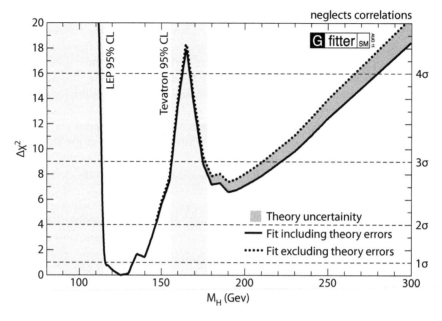

**Figure 7.24.** $\Delta\chi^2$ as a function of the Higgs-boson mass for the August 2011 Gfitter complete fit, taking account of direct searches at LEP and the Tevatron. The solid (dashed) line gives the results when including (ignoring) theoretical errors (from Ref. [87]).

electroweak theory, favor a light Higgs boson and exhibit some tension with the direct searches that preceded the LHC. The LEP experiments, which focused on the $e^+e^- \to HZ^0$ channel, set a lower bound on the standard-model Higgs-boson mass of $M_H > 114.4$ GeV at 95% CL [76, 83, 84]. The Tevatron experiments CDF and D0 also search for the standard-model Higgs boson, examining a variety of production channels and decay modes appropriate to different Higgs-boson masses. Their most recent combined result excludes the ranges 100 GeV $< M_H <$ 103 GeV and 147 GeV $< M_H <$ 180 GeV at 95% CL [85]. The Tevatron experiments report a broad excess, particularly in the $b\bar{b}$ channel, over the range 115 GeV $< M_H <$ 140 GeV. See Ref. [86] for an overview of past searches.

The disjoint exclusion regions from LEP and the Tevatron made it somewhat complicated to specify the remaining mass ranges favored for the standard-model Higgs boson. A useful example from the Gfitter team is shown in figure 7.24. In the Gfitter analysis [87], at $2\sigma$-significance ($\approx$ 95% CL), the standard-model Higgs-boson mass must lie in the interval 114 GeV $\lesssim M_H \lesssim$ 143 GeV. This reflects the state of knowledge before the first significant results from the ATLAS and CMS experiments at the Large Hadron Collider. It is worth noting that the best-fit value, omitting information from the direct searches, is $M_H = 96^{+31}_{-24}$ GeV. Similar studies have been carried out over several years by the LEP Electroweak Working Group (Ref. [75]; cf. figure 7.22) and by the Particle Data Group (Ref. [78]).

The standard electroweak theory gives an excellent account of many pieces of data over a wide range of energies, and its main elements can be stated compactly.

Nevertheless, it leaves too many gaps in our understanding for it to be considered a complete theory (cf. §7.10–7.12). We therefore have reason to consider extensions to the standard model, for which the standard-model fits to the electroweak measurements may not apply. Accordingly, healthy skepticism dictates that we regard the inferred constraints on $M_H$ as a potential test of the standard model, not as rigid boundaries on where the agent of electroweak symmetry breaking must show itself. It is prudent that we search for the agent of electroweak symmetry breaking over the entire mass range allowed by general arguments, and this is what the LHC experiments set out to do.

In analyses of approximately $5 \, \text{fb}^{-1}$ of 7-TeV $pp$ collisions at the end of 2011, ATLAS and CMS gave 95% CL exclusions of the *standard-model* Higgs boson in the range $127 \, \text{GeV} < M_H < 600 \, \text{GeV}$ and extended the LEP lower bound to $M_H > 115.5 \, \text{GeV}$ [88]. The two collaborations also presented provocative evidence for excesses over expected background, particularly in the $\gamma\gamma$ and $\ell^+\ell^-\ell^+\ell^-$ ($ZZ^*$) channels, in the neighborhood of 125 GeV. By July 2012, the addition of approximately $5 \, \text{fb}^{-1}$ of 8-TeV data enabled ATLAS [89] and CMS [90] to report the discovery at $5\sigma$ significance of a neutral, even-integer-spin particle in the neighborhood of 125–126 GeV. The product of production cross section and branching fraction for the $\gamma\gamma$, $ZZ^*$, and $WW^*$ channels is consistent, within limited statistics, with expectations for the standard-model Higgs boson.

Cross sections for the principal reactions under study at the LHC are shown in figure 7.25, for initial running at $\sqrt{s} = 7 \, \text{TeV}$ and at the design energy of $\sqrt{s} = 14 \, \text{TeV}$. The largest cross section for Higgs production at both the LHC and the Tevatron occurs in the reaction $p^\pm p \to H + \text{anything}$, which proceeds by gluon fusion through heavy-quark loops (cf. problem 8.25). The shoulder in that cross section near $M_H = 400 \, \text{GeV}$ reflects the behavior of the top-quark loop. A fourth generation of heavy quarks would raise the $gg \to H$ rate significantly, increasing the sensitivity of searches at the Tevatron and LHC. Vector–boson fusion is an important secondary production mechanism and would be of particular importance for a Higgs boson with reduced (or vanishing) couplings to fermions [92].

For small Higgs-boson masses, figure 6.26 shows that the dominant decay is into $b\bar{b}$ pairs. However, the reaction $p^\pm p \to H + \text{anything}$ followed by the decay $H \to b\bar{b}$ is swamped by QCD production of $b\bar{b}$ pairs. Consequently, experiments must rely on rare decay modes ($\tau^+\tau^-$ or $\gamma\gamma$, for example) with lower backgrounds or resort to different production mechanisms for which specific reaction topologies reduce backgrounds. Accordingly, the production of Higgs bosons in association with electroweak gauge bosons has received close scrutiny at the Tevatron. The rare $\gamma\gamma$ channel has been an important target for LHC experiments for light Higgs-boson masses. [Problem 7.27 explores the $H \to \gamma\gamma$ decay rate in the standard electroweak theory.] At higher masses, the hadron collider experiments have exploited good sensitivity to the $gg \to H \to (W^+W^-$ or $ZZ)$ reaction chain to set their exclusion limits.

Now that a Higgs-boson candidate is found, it will be of great interest to map its decay pattern in order to characterize the mechanism of electroweak symmetry breaking. The Higgs mechanism shows how electroweak symmetry could be broken to reproduce the low-energy features of the real world, but it is not the only possibility. It is by no means guaranteed that the same agent hides electroweak symmetry and generates fermion mass [93]. In section 8.9,

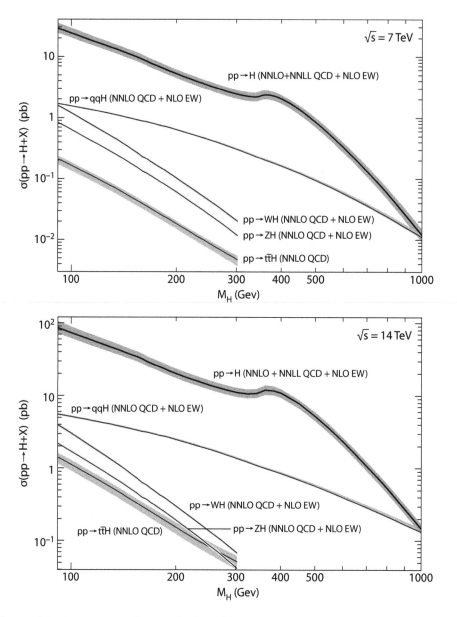

**Figure 7.25.** Cross sections for standard-model Higgs-boson production in *pp* collisions at 7 TeV (upper panel) and 14 TeV (lower panel). From high to low cross sections, the reactions are gluon–gluon fusion, vector-boson fusion, and *WH*, *ZH*, and *t̄tH* associated production (from Ref. [91]).

we shall see that spontaneous chiral symmetry breaking in QCD could hide the electroweak symmetry without generating fermion masses. Indeed, many extensions to the standard model significantly alter the decay pattern of the Higgs boson. In supersymmetric models, five Higgs bosons are expected, and the branching fractions of the lightest one may be very different from those presented in figure 6.26.

Supersymmetric extensions of the electroweak theory entail considerable model dependence but yield high-quality fits to the precision data [94, 95]. The tension between fits that prefer light masses and direct searches that disfavor a light Higgs boson is not present in the supersymmetric world. On the other hand, in its simplest form, the minimal supersymmetric standard model would be challenged if the mass of the lightest CP-even Higgs boson, $h$, exceeded about 135 GeV [96]. A recent 25-parameter fit to the "phenomenological minimal supersymmetric standard model" concludes that 117 GeV $\lesssim M_h \lesssim$ 129 GeV (Ref. [95]).

If new strong dynamics—rather than a perturbatively coupled elementary scalar—hides the electroweak symmetry, then the mass of the composite stand-in for the Higgs boson can range up to several hundred GeV. (If the electroweak symmetry is dynamically broken, the Higgs model can be seen as the analogue of the sigma model (cf. §5.4) in quantum chromodynamics.) Standard-model fits that allow an extra generation of quarks and leptons also permit Higgs-boson masses as large as a few hundred GeV [97].

A precise determination of Higgs-boson couplings is one of the strengths of an $e^+e^- \rightarrow HZ$ "Higgs factory," which could also allow access to so-called invisible decay modes. The LHC itself will supply crucial clues to the origin of fermion masses. With the LHC's large data sets, it is plausible that Higgs-boson couplings can eventually be measured at levels that test the electroweak theory and provide interesting constraints on extensions to the standard model.

The discovery of a standard-model-like Higgs boson carries implications for the range of validity of the electroweak theory, to which we now turn.

## 7.10 INCOMPLETENESS OF THE ELECTROWEAK THEORY

By demanding that the electroweak theory be complete and consistent as a quantum field theory, we can establish bounds on the Higgs boson mass and uncover evidence that discoveries will not end with the Higgs boson. Scalar field theories make sense on all energy scales only if they are noninteracting, or "trivial" [98]. The vacuum of quantum field theory is a generalized dielectric medium that screens (or antiscreens) charge. Accordingly, the effective charge is a function of the distance or, equivalently, of the energy scale. This is the famous phenomenon of the running coupling constant, about which we will have more to say in sections 8.2 and 8.3.

In $\lambda\phi^4$ theory (compare the interaction term in the Higgs potential), it is easy to calculate the variation of the coupling constant $\lambda$ in perturbation theory by summing quantum corrections given by bubble graphs. The coupling constant $\lambda(\kappa)$ on a physical scale $\kappa$ is related to the coupling constant on a higher scale $\Lambda$ by

$$\frac{1}{\lambda(\kappa)} = \frac{1}{\lambda(\Lambda)} + \frac{3}{2\pi^2} \ln\left(\frac{\Lambda}{\kappa}\right). \tag{7.10.1}$$

In order for the Higgs potential to be stable (i.e., for the energy of the vacuum state not to race off to $-\infty$), $\lambda(\Lambda)$ must not be negative. Applied to (7.10.1), this condition leads to an inequality,

$$\frac{1}{\lambda(\kappa)} \geq \frac{3}{2\pi^2} \ln\left(\frac{\Lambda}{\kappa}\right), \tag{7.10.2}$$

that implies an *upper bound*,

$$\lambda(\kappa) \le \frac{2\pi^2}{3\ln(\Lambda/\kappa)}, \tag{7.10.3}$$

on the coupling strength at the physical scale $\kappa$. If the theory is to make sense to arbitrarily high energies—or short distances—we must consider the limit $\Lambda \to \infty$ while holding $\kappa$ fixed at some reasonable physical scale; the bound (7.10.3) then forces $\lambda(\kappa)$ to zero. The scalar field theory has become free field theory; in theorist's jargon, it is trivial.

Rearranging and exponentiating both sides of (7.10.3) gives the condition

$$\Lambda \le \kappa \exp\left(\frac{2\pi^2}{3\lambda(\kappa)}\right), \tag{7.10.4}$$

from which we can infer a limit on the Higgs-boson mass. Choosing the physical scale as $\kappa = M_H$, using the definition $M_H^2 = 2\lambda(M_H)v^2$, we find that

$$\Lambda \le \Lambda^\star \equiv M_H \exp\left(\frac{4\pi^2 v^2}{3M_H^2}\right). \tag{7.10.5}$$

For any given Higgs-boson mass, we can identify a maximum energy scale $\Lambda^\star$ at which the theory ceases to make sense, so that new physics must enter. From this perspective, the electroweak theory is at best an effective theory, valid over a finite range of energies. This perturbative analysis leading to (7.10.5) must be supplemented by lattice-field-theory analyses as the Higgs-boson mass approaches 1 TeV and the interactions become strong [99]. A two-loop analysis leads to the bounds [100]

$$M_H|_{\Lambda^*=M_{\text{Pl}}} \lesssim 180 \text{ GeV}; \qquad M_H|_{\Lambda^*=1\text{ TeV}} \lesssim 700 \text{ GeV}. \tag{7.10.6}$$

The correlation between $M_H$ and $\Lambda^*$ is indicated by the upper curve in figure 7.26. If the elementary Higgs boson takes on the largest mass allowed by perturbative unitarity arguments, the electroweak theory lives on the brink of instability.

We remarked in section 5.6 that the analysis of spontaneous symmetry breaking presented there was entirely classical in that quantum corrections to the Higgs potential were neglected and that we should verify in specific applications that these do not alter essential conclusions. Such considerations lead to an interesting lower bound on the mass of the Higgs boson in the electroweak theory. When the one-loop corrections to the Higgs potential (6.3.19) are calculated, the effective potential can be written in the form

$$V(\varphi^\dagger\varphi) = -\mu^2\varphi^\dagger\varphi + B(\varphi^\dagger\varphi)^2 \ln(\varphi^\dagger\varphi/M^2), \tag{7.10.7}$$

where the parameter $M$ has been chosen to absorb the erstwhile $(\varphi^\dagger\varphi)^2$ term. The coefficient $B$ includes contributions of gauge boson, Higgs boson, and fermion loops.

The effective potential has a local minimum at a value of $\langle\varphi^\dagger\phi\rangle_0$ defined by

$$\langle\varphi^\dagger\phi\rangle_0 \left[\ln\left(\frac{\langle\varphi^\dagger\phi\rangle_0}{M^2}\right) + \frac{1}{2}\right] = \frac{\mu^2}{2B}, \tag{7.10.8}$$

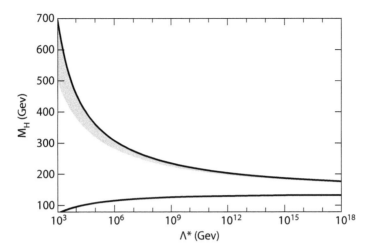

**Figure 7.26.** Bounds on the standard-model Higgs-boson mass that follow from requirements that the electroweak theory be consistent up to the energy $\Lambda^*$. The upper bound follows from triviality conditions; the lower bound follows from the requirement that $V(v) < V(0)$. The shaded band below the triviality bound reflects a range of criteria for perturbativity. Values of $(\Lambda^*, M_H)$ for which the electroweak theory can be complete and consistent lie between the two boundaries (adapted from Ref. [100]).

at which the value of the potential is

$$V(\langle \varphi^\dagger \phi \rangle_0) = -B \langle \varphi^\dagger \phi \rangle_0^2 \left[ \ln \left( \frac{\langle \varphi^\dagger \phi \rangle_0}{M^2} \right) + 1 \right]. \qquad (7.10.9)$$

This local minimum will be an absolute minimum provided that $V(\langle \varphi^\dagger \phi \rangle_0) < V(0)$, which is ensured if

$$\ln \left( \frac{\langle \varphi^\dagger \phi \rangle_0}{M^2} \right) > -1. \qquad (7.10.10)$$

Inserting this constraint in the expression for the mass of the physical Higgs boson,

$$M_H^2 \equiv V''|_{\langle \varphi^\dagger \phi \rangle_0} = 4B \langle \varphi^\dagger \phi \rangle_0 \left[ \ln \left( \frac{\langle \varphi^\dagger \phi \rangle_0}{M^2} \right) + \frac{3}{2} \right], \qquad (7.10.11)$$

yields a lower bound on the Higgs-boson mass [101],

$$M_H^2 > 2B \langle \varphi^\dagger \phi \rangle_0. \qquad (7.10.12)$$

Extending this analysis to the case in which the electroweak theory is complete and consistent up to scale $\Lambda^*$ leads to the lower bound on the Higgs-boson mass shown as the lower curve in figure 7.26.

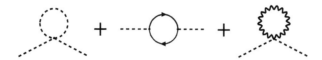

**Figure 7.27.** One-loop radiative corrections to the Higgs-boson mass.

Evidently the theory can be self-consistent up to very high energies, provided that the Higgs boson is relatively light. For the electroweak theory to make sense all the way up to a unification scale $\Lambda^\star = u \approx 10^{16}$ GeV, the Higgs-boson mass must lie in the interval 134 GeV $\lesssim M_H \lesssim$ 177 GeV. If $M_H$ is not within this chimney, the electroweak theory is incomplete; it is an effective theory that will be subsumed in a more comprehensive description. The most favored mass interval cited in section 7.9 lies below the chimney. Problem 7.28 investigates the implications of the triviality and stability bounds in light of current information about a standard-model Higgs boson.

The condition of absolute vacuum stability that leads to the lower bound on $M_H$ displayed in figure 7.26 is more stringent than is required by observational evidence. It is consistent with observations for the ground state of the electroweak theory to be a *false* (metastable) vacuum that has survived quantum fluctuations until now, so that the mean time to tunnel from our vacuum to a deeper vacuum exceeds the age of the universe. Renormalization-group-improved one-loop calculations of the tunneling probability at zero temperature characterize the stability, metastability, or instability of the electroweak theory in terms of the Higgs-boson mass and the top-quark mass [102]. Should $(M_H, m_t)$ settle in the metastable zone, we would have a provocative hint for new physics below the Planck scale [103].

## 7.11 THE HIERARCHY PROBLEM

Beyond the classical approximation, scalar mass parameters receive quantum corrections from loops that contain particles of spins $J = 0, \frac{1}{2}$, and 1, as shown symbolically in figure 7.27. The dashed lines represent the Higgs boson, solid lines with arrows represent fermions and antifermions, and wavy lines stand for gauge bosons. The quantum corrections that determine the running mass lead potentially to divergences,

$$\Delta M_H^2 = C_H \lambda \int_{p^2}^{\Lambda_H^2} dk^2 + C_q \zeta_q^2 \int_{p^2}^{\Lambda_q^2} dk^2 + C_g g^2 \int_{p^2}^{\Lambda_g^2} dk^2 + \cdots , \qquad (7.11.1)$$

where $\lambda$, $\zeta_q$, and $g$ are coupling constants and the coefficients $C_i$ are calculable in any particular theory. The loop integrals appear to be quadratically divergent, $\propto \Lambda_i^2$. In a world limited to the electroweak theory, the quadratic dependences on cutoff parameters may be seen as artifacts of the regularization procedure [104]. Taking a more complete view, we expect physically meaningful reference points and not merely cutoff parameters. In the absence of nearby new physics, the reference scale $\Lambda$ would naturally be large. If the fundamental interactions are described by quantum chromodynamics and the electroweak theory, then a natural reference scale is the Planck mass, $\Lambda \sim M_{\text{Pl}} = (\hbar c / G_{\text{N}})^{1/2} \approx 1.2 \times 10^{19}$ GeV. As we shall see in chapter 9, in a unified theory of the strong, weak, and electromagnetic interactions, a natural scale is the unification scale, $\Lambda \sim u \approx 10^{15} - 10^{16}$ GeV. Both estimates are very large compared to the electroweak scale and so imply a very long range of integration.

In order for the mass shifts induced by quantum corrections to remain modest, either something must limit the range of integration, or new physics must damp the integrand. The challenge of preserving widely separated electroweak and reference scales in the presence of quantum corrections is known as the *hierarchy problem*. Unless we suppose that $M_H^2(\Lambda^2)$ and the quantum corrections are finely tuned to yield $M_H^2(p^2) \lesssim (1 \text{ TeV})^2$, some new physics—a new symmetry or new dynamics—must intervene at an energy of approximately 1 TeV to tame the integral in (7.11.1). (See problem 7.21 for illustrations of the fine-tuning problem.)

Let us review the argument for the hierarchy problem: The unitarity argument of section 6.6 showed that new physics must be present on the 1-TeV scale, either in the form of a Higgs boson or other new phenomena. Within the electroweak theory (cf. §7.9), a light Higgs boson is favored by precision measurements. But a low-mass Higgs boson is imperiled by quantum corrections. New physics not far above the 1-TeV scale could bring the reference scale $\Lambda$ low enough to mitigate the threat. That is what happens in models of large [105] or warped [106] extra dimensions, in which $M_{\text{Pl}}$ is seen as a mirage, based on a mistaken extrapolation of Newton's law of gravitation to very short distances, or a new cutoff emerges, set by the scale of the extra dimension.

If the reference scale is indeed very large, then either various contributions to the Higgs-boson mass must be precariously balanced or new physics must control the contribution of the integral in (7.11.1). It is important to keep in mind that fine-tuning, perhaps guided by environmental selection, might be the way of the world [107]. However, experience teaches us to be alert for symmetries or dynamics behind precise cancellations.

A new symmetry, not present in the standard model, could resolve the hierarchy problem. Exploiting the fact that fermion loops contribute with an overall minus sign relative to boson loops (because of Fermi statistics), *supersymmetry* [108] balances the contributions of fermion and boson loops. In unbroken supersymmetry, the masses of bosons are degenerate with those of their fermion counterparts, so the cancellation is exact. If supersymmetry is present in our world, it must be broken. The contribution of the integrals may still be acceptably small if the fermion-boson mass splittings $\Delta M$ are not too large. The condition that $g^2 \Delta M^2$ be "small enough" leads to the requirement that superpartner masses be less than about 1 TeV. It is provocative to note that, with superpartners at $\mathcal{O}(1 \text{ TeV})$, the $\text{SU}(3)_{\text{c}} \otimes \text{SU}(2)_{\text{L}} \otimes \text{U}(1)_{\text{Y}}$ coupling constants run to a common value at a unification scale of about $10^{16}$ GeV (cf. problem 9.11).

Theories of dynamical symmetry breaking (cf. §8.9) offer a second solution to the problem of the enormous range of integration in (7.11.1). In technicolor models, the Higgs boson is composite, and its internal structure comes into play on the scale of its binding, $\Lambda_{\text{TC}} \simeq \mathcal{O}(1 \text{ TeV})$. The integrand is damped, the effective range of integration is cut off, and mass shifts are under control.

A recurring hope among theorists has been the notion that the Higgs boson might be naturally light because it is the pseudo-Nambu–Goldstone boson of some approximate global symmetry. "Little Higgs" models [109] introduce additional gauge bosons, vectorlike quarks, and scalars on the TeV scale. These conspire, thanks to a global symmetry, to cancel the quadratic divergences in (7.11.1) that result from loops of standard-model particles and defer the hierarchy problem to about 10 TeV. In contrast to supersymmetry, the cancellations arise from loops containing particles of the same spin. In "twin Higgs" models [110], the new states do not carry standard-model charges. The new physics at $\sim 10$ TeV raises impediments to conventional hopes for perturbative unification of the strong, weak, and electromagnetic interactions, as discussed in section 9.3.

## 7.12  THE VACUUM ENERGY PROBLEM

The cosmological constant problem—why empty space is so nearly massless—is one of the great mysteries of science [111]. It is the reason why gravity has weighed on the minds of electroweak theorists, despite the utterly negligible role that gravity normally plays in particle reactions. Recall that the gravitational attraction between an electron and proton is 41 orders of magnitude smaller than the electrostatic attraction at the same separation!

At the vacuum expectation value $\langle\phi\rangle_0$ of the Higgs field, the (position-independent) value of the Higgs potential is

$$V(\langle\phi^\dagger\phi\rangle_0) = \frac{\mu^2 v^2}{4} = -\frac{|\lambda|\, v^4}{4} < 0. \tag{7.12.1}$$

Identifying $M_H^2 = -2\mu^2$, we see that the Higgs potential contributes a uniform vacuum energy density,

$$\varrho_H \equiv \frac{M_H^2 v^2}{8}. \tag{7.12.2}$$

From the perspective of general relativity, this amounts to adding to Einstein's equation a cosmological constant, $\Lambda = (8\pi\, G_N/c^4)\varrho_H$, where $G_N$ is Newton's gravitational constant [112].

The discovery [113] that the expansion of the universe is accelerating raises the intriguing possibility that the cosmological constant may be different from zero, but the essential fact is that the observed vacuum energy density must be very small indeed [114],

$$\varrho_{vac} \lesssim 10^{-46} \text{ GeV}^4 \approx \text{(a few meV)}^4. \tag{7.12.3}$$

Therein lies the puzzle: if we take $v = (G_F\sqrt{2})^{-1/2} \approx 246$ GeV and insert the mass of the LHC boson, $M_H \approx 126$ GeV (Refs. [89,90]), into (7.12.2), we find that the Higgs field's contribution to the vacuum energy density is

$$\varrho_H \approx 1.2 \times 10^8 \text{ GeV}^4, \tag{7.12.4}$$

some 54 orders of magnitude larger than the upper bound inferred from the cosmological constant. This mismatch has been a source of dull headaches for more than three decades.

The problem is still more serious in a unified theory of the strong, weak, and electromagnetic interactions, in which other (heavy!) Higgs fields have nonzero vacuum expectation values that may give rise to still larger vacuum energies (cf. §9.2). At a fundamental level, we can, therefore, conclude that a spontaneously broken gauge theory of the strong, weak, and electromagnetic interactions—or merely of the electroweak interactions—cannot be complete. The vacuum energy problem must be an important clue. But to what?

The tentative evidence for a nonzero cosmological constant recasts the problem in two important ways. First, instead of looking for a principle that would forbid a cosmological constant, perhaps a symmetry principle that would set it exactly to zero, we may be called upon to explain a tiny cosmological constant. Second, if the interpretation of the accelerating expansion in terms of dark energy is correct, we may gain observational access to some new stuff whose equation of state and other properties we can try to measure. Perhaps that will give us the clues that we need to solve this old problem, and to understand how it relates to the electroweak theory.

# 7.13 REFLECTIONS

The electroweak theory is a remarkable achievement. It gives a deeper understanding of two of the fundamental forces of nature—electromagnetism and the charged-current weak interaction—and introduces a previously unobserved neutral-current weak interaction. It meets the most important criteria for a good theory: we get more out than we put in, and it raises new and significant questions. Diverse experiments have tested the gauge sector and the flavor sector of the electroweak theory rigorously over the past two decades, so that we may rightly regard the electroweak theory as a new law of nature—subject, as always, to revision in light of future experience.

Our urgent task is to consolidate our understanding by identifying the agent of electroweak symmetry breaking. The clues in hand suggest that it represents a novel fundamental interaction operating on the 1-TeV scale. *We do not know the nature of the mysterious new force.*

A leading possibility is that the agent of electroweak symmetry breaking is an elementary scalar, the Higgs boson of the $SU(2)_L \otimes U(1)_Y$ standard model. An essential step toward coming to terms with the new force that shapes our world is to search for the Higgs boson and explore its properties. Global fits to electroweak measurements indicate that a standard-model Higgs boson should be found with a mass not much more than 200 GeV. At the beginning of 2012, direct searches highlighted the range $115 \text{ GeV} \lesssim M_H \lesssim 130 \text{ GeV}$. We now have a Higgs-boson candidate at $M_H \approx 126 \text{ GeV}$.

Should the agent of electroweak symmetry breaking turn out to be a Higgs boson, will there be one, or several, and will it—or they—be elementary or composite? Is the Higgs boson indeed relatively light, as anticipated by the global fits to electroweak precision measurements? Does the Higgs boson give mass only to the electroweak gauge bosons, or does it also endow the fermions with mass? Proceeding step by step, does the "$H$" couple to fermions? Are the branching fractions for decays into fermion pairs in accord with the standard model? A difficult follow-up question: if the Higgs boson *is* responsible for fermion mass, is what determines the masses and mixings of the fermions?

Higgs bosons could couple to particles beyond those known in the standard model. Will the pattern of Higgs-boson decays imply novel interactions? Will unexpected or rare decays of the Higgs boson reveal new kinds of matter? If more than one, apparently elementary, Higgs boson is found, will that be a sign for a supersymmetric generalization of the standard model, or for a different sort of two-Higgs-doublet model? What stabilizes the Higgs-boson mass below 1 TeV? How can a light Higgs boson coexist with the absence of signals for new phenomena?

If, on the contrary, electroweak symmetry is broken by some form of new strong dynamics, how can we diagnose the origins and properties of the new dynamics? What takes the place of a Higgs boson? How do the fermions acquire masses and flavor patterns? Might electroweak symmetry breaking be an emergent phenomenon? Or could electroweak symmetry breaking be related through extra spacetime dimensions to gravity?

Whatever the outcome, we can be confident that the origin of gauge-boson masses will be understood through explorations of the 1-TeV scale. *We have not identified the energy scale at which we can expect to decode the pattern of fermion masses.*

For all its successes, the electroweak theory leaves many questions unanswered. It does not explain the negative coefficient $\mu^2 < 0$ of the quadratic term that shapes the Higgs potential (6.3.19) required to hide the electroweak symmetry. It shows how the masses of the quarks and leptons might arise but does not predict their values. It does not even give a qualitative understanding of why the quark-mixing parameters are small and hierarchical, nor why the pattern of neutrino mixing should be so different. (See §9.6 for a brief discussion of fermion masses within unified theories.) The Cabibbo–Kobayashi–Maskawa framework describes what we know of CP violation but does not explain its origin. The CP violation observed in the quark sector, in accord with the CKM paradigm, seems far too small to account for the excess of matter over antimatter in the universe (cf. §9.5). The discovery of neutrino flavor mixing, with its implication that neutrinos have mass, calls for an extension of the electroweak theory.

Moreover, the standard-model Higgs sector is unstable against large radiative corrections. Although fine tuning is a logical possibility, our experience argues that a solution to this gauge hierarchy problem will come in the form of new dynamics or a new symmetry. Neutrinos are the only dark-matter candidates within the (extended) standard model. They appear to contribute only a small share of the inferred dark-matter energy density and as relativistic (hot) dark matter, not the cold dark matter required for structure formation in the early universe. However, candidate solutions to the hierarchy problem entail new physics on the TeV scale, and the weakly-interacting-massive-particle response to the dark-matter question suggests a mass in the few-hundred–GeV range [115]. These hints suggest that, in addition to the electroweak-symmetry-breaking physics that we confidently expect to see at the LHC, there is every likelihood of more new phenomena. A season of change is upon us!

# PROBLEMS

7.1. Consider the generation of fermion masses and mixings in the four-quark version of the electroweak theory. (a) Verify that the interactions between the scalar fields and quarks may be cast in the form

$$\mathcal{L}_{\text{Yukawa}} = -\zeta_1[(\bar{L}_u\tilde{\phi})u_R + \bar{u}_R(\tilde{\phi}^\dagger L_u)] - \zeta_2[(\bar{L}_u\phi)d_R + \bar{d}_R(\phi^\dagger L_u)]$$
$$- \zeta_3[(\bar{L}_u\phi)s_R + \bar{s}_R(\phi^\dagger L_u)] - \zeta_4[(\bar{L}_c\tilde{\phi})c_R + \bar{c}_R(\tilde{\phi}^\dagger L_c)]$$
$$- \zeta_5[(\bar{L}_c\phi)d_R + \bar{d}_R(\phi^\dagger L_c)] - \zeta_6[(\bar{L}_c\phi)s_R + \bar{s}_R(\phi^\dagger L_c)].$$

(b) Replace the scalar field $\phi$ by its vacuum expectation value

$$\langle\phi\rangle_0 = \begin{pmatrix} 0 \\ v/\sqrt{2} \end{pmatrix}, \tag{6.3.21}$$

to obtain a series of mass terms as in (6.3.41). Solve for the Yukawa couplings $\zeta_1, \zeta_2, \ldots, \zeta_6$ that yield the correct quark masses for $u$, $d$, $s$, $c$.
(c) How could symmetry relations among the Yukawa couplings $\zeta_i$ imply connections between the quark masses and the Cabibbo mixing angle?

7.2. Show that the $n \times n$ unitary matrix $V$ defined in (7.1.23)–(7.1.25) to describe the mixing of quarks of different flavors can be parametrized in terms of $n(n-1)/2$ real mixing angles and $(n-1)(n-2)/2$ complex phases, after the freedom to redefine the phases of quark fields has been taken into account. For the special case of $n = 3$ quark doublets, show that $V$ may be written in the form

$$V = \begin{bmatrix} 1 & 0 & 0 \\ 0 & c_2 & s_2 \\ 0 & -s_2 & c_2 \end{bmatrix} \begin{bmatrix} c_1 & s_1 & 0 \\ -s_1 & c_1 & 0 \\ 0 & 0 & 1 \end{bmatrix} \begin{bmatrix} 1 & 0 & 0 \\ 0 & 1 & 0 \\ 0 & 0 & e^{i\phi} \end{bmatrix} \begin{bmatrix} 1 & 0 & 0 \\ 0 & c_3 & s_3 \\ 0 & -s_3 & c_3 \end{bmatrix},$$

where $s_i = \sin\theta_i$ and $c_i = \cos\theta_i$. Discuss the implications of the phase for CP invariance. [M. Kobayashi and T. Maskawa, *Prog. Theoret. Phys. (Kyoto)* **49**, 652 (1973).]

7.3. A standard convention for the $3 \times 3$ quark-mixing matrix has become

$$V = \begin{pmatrix} c_{12}c_{13} & s_{12}c_{13} & s_{13}e^{-i\delta} \\ -s_{12}c_{23} - c_{12}s_{23}s_{13}e^{i\delta} & c_{12}c_{23} - s_{12}s_{23}s_{13}e^{i\delta} & s_{23}c_{13} \\ s_{12}s_{23} - c_{12}c_{23}s_{13}e^{i\delta} & -c_{12}s_{23} - s_{12}c_{23}s_{13}e^{i\delta} & c_{23}c_{13} \end{pmatrix},$$

with the angles $\theta_{ij}$ in the first quadrant and $\delta$ in $(-\pi, \pi]$. [L.-L. Chau and W.-Y. Keung, *Phys. Rev. Lett.* **53**, 1802 (1984).] The hierarchy $s_{13} \ll s_{23} \ll s_{12} \ll 1$ is a useful idealization of experimental information. A compact parametrization, informed by the rough magnitudes of the elements, aids in anticipating and integrating experimental results.
(a) Note that the quark-mixing matrix is in first approximation diagonal, symmetrical, and with a hierarchical structure for the mixing between generations.

Now take $V_{us} \equiv \lambda$ as an expansion parameter for deviations from the identity matrix. Express $V_{ub} = A\lambda^2$, where [cf. (7.1.28)] $A \approx 0.808$ is of order unity. Set $V_{ub} = 0$, and use the requirement of unitarity, $V^\dagger V = I$, to solve for the diagonal elements. Show that, to $O(\lambda^2)$,

$$V = \begin{pmatrix} 1 - \frac{1}{2}\lambda^2 & \lambda & 0 \\ -\lambda & 1 - \frac{1}{2}\lambda^2 & A\lambda^2 \\ 0 & -A\lambda^2 & 1 \end{pmatrix} + O(\lambda^3).$$

(b) Now extend the parametrization to $O(\lambda^3)$. Assign $s_{12} = \lambda$, $s_{23} = A\lambda^2$, and $s_{13}e^{-i\delta} = A\lambda^3(\rho - i\eta)$. By imposing the unitarity condition, show that

$$V = \begin{pmatrix} 1 - \frac{1}{2}\lambda^2 & \lambda & A\lambda^3(\rho - i\eta) \\ -\lambda & 1 - \frac{1}{2}\lambda^2 & A\lambda^2 \\ A\lambda^3(1 - \rho - i\eta) & -A\lambda^2 & 1 \end{pmatrix} + O(\lambda^4).$$

[L. Wolfenstein, *Phys. Rev. Lett.* **51**, 1945 (1983).]

7.4. The unitarity condition $V^\dagger V = I$ may be expressed through the equations

$$\sum_i V_{ij} V_{ik}^* = \delta_{jk} \quad \text{and} \quad \sum_j V_{ij} V_{kj}^* = \delta_{ik}.$$

It can be informative to express the orthogonality conditions (for $j \neq k$ and $i \neq j$, respectively) as triangles in a complex plane.

(a) Starting from the Wolfenstein parametrization at $O(\lambda^3)$, compute the equations for the six unitarity triangles. Note that the four conditions involving adjacent rows or columns lead to nearly degenerate triangles. Identify, in generic notation, the orthogonality conditions corresponding to the two nondegenerate triangles.

(b) Divide out an overall factor of $-A\lambda^3$ from your equation expressing the condition

$$V_{ud} V_{ub}^* + V_{cd} V_{cb}^* + V_{td} V_{tb}^* = 0, \qquad (\star)$$

and plot the resulting triangle in terms of the Wolfenstein parameters. [A. Ceccucci, Z. Ligeti, and Y. Sakai, "The CKM Quark Mixing Matrix," §11 of the 2010 *Review of Particle Properties*, Ref. [3].]

(c) Returning to the generic notation for the quark-mixing-matrix elements, show that the unitarity condition $(\star)$ can be brought to the normalized form

$$R_t e^{-i\beta} + R_u e^{i\gamma} = 1,$$

where the nontrivial lengths of the sides are

$$R_t = \left| \frac{V_{td} V_{tb}^*}{V_{cd} V_{cb}^*} \right| \quad \text{and} \quad R_u = \left| \frac{V_{ud} V_{ub}^*}{V_{cd} V_{cb}^*} \right|.$$

and the corresponding angles are

$$\beta = \arg\left(-\frac{V_{cd}V_{cb}^*}{V_{td}V_{tb}^*}\right) \quad \text{and} \quad \gamma = \arg\left(-\frac{V_{ud}V_{ub}^*}{V_{cd}V_{cb}^*}\right).$$

(d) Sketch the normalized unitarity triangle; label the sides and angles, including $\alpha \equiv \pi - \beta - \gamma$.

7.5. (a) Express the area $\bar{\mathbb{A}}$ of the normalized unitarity triangle in terms of the Wolfenstein parameters.
(b) Compute the area $\mathbb{A}$ of the unnormalized unitarity triangle ($\star$) in terms of the sines and cosines of the mixing angles and the phase $\delta$.
(c) Express $\mathbb{A}$ in terms of the Wolfenstein parameters.
The area $\mathbb{A}$, which is the same for all six unitarity triangles, is given by half of the Jarlskog invariant $\mathcal{J}$ that controls the strength of CP violation. [C. Jarlskog, *Phys. Rev. Lett.* **55**, 1039 (1985).]

7.6. Extensive measurements of the properties of the neutral weak boson $Z^0$ have been summarized by the LEP Electroweak Working Group (Ref. [10]), under the (well-tested) assumption of lepton universality, $\Gamma(Z \to e^+e^-) = \Gamma(Z \to \mu^+\mu^-) = \Gamma(Z \to \tau^+\tau^-)$:

| | |
|---|---|
| $M_Z$ | $91.1875 \pm 0.0021$ GeV/$c^2$ |
| $\Gamma_Z$ | $2.4952 \pm 0.0023$ GeV |
| $\Gamma(Z \to \text{hadrons})$ | $1744.4 \pm 2.0$ MeV |
| $\sigma_{\text{had}}(\text{pole})$ | $41.540 \pm 0.037$ nb |
| $\Gamma_{\text{leptonic}} = \Gamma(Z \to \ell^+\ell^-)$ | $83.984 \pm 0.086$ MeV |
| $\Gamma_{\text{invisible}}$ | $499.0 \pm 1.5$ MeV |
| $a_\ell$ | $-0.50123 \pm 0.00026$ |
| $v_\ell$ | $-0.03783 \pm 0.00041$ |

Let us confront these results with the lowest-order predictions of the electroweak theory.
(a) Use the value of the $Z^0$ mass to determine the weak mixing parameter $x_W = \sin^2\theta_W$.
(b) Using your value of $x_W$, predict the $W^\pm$ mass and compare it with the current world average.
(c) Use the measured value of the $Z^0$ mass to predict the partial width, $\Gamma(Z \to \nu\bar{\nu})$. By comparing with $\Gamma_{\text{invisible}}$, determine the number of light neutrino species.
(d) Use your value of the weak mixing parameter $x_W$ to predict $\Gamma(Z \to \ell^+\ell^-)$, $\Gamma(Z \to \text{hadrons})$, $a_\ell$, and $v_\ell$. Neglect the masses of all fermions but the top quark.

7.7. Compute the cross section for the reaction $e^+e^- \to \nu_e\bar{\nu}_e$ in the electroweak theory. What, if anything, is the importance of the $W$-exchange diagram in the neighborhood of the $Z^0$ peak?

7.8. Verify the general expression (7.3.3) for the parton-level reaction, $e^+e^- \to q\bar{q}$.

7.9. (a) Plot the variation of the asymmetries $A_{\text{peak}}^{(\ell^+\ell^-)}$, $A_{\text{peak}}^{(c\bar{c})}$, and $A_{\text{peak}}^{(b\bar{b})}$ in the interval $0.2 \le x_{\text{W}} \le 0.25$.
(b) Add to your plot the measurements given in (7.3.20a)–(7.3.20c) and infer a "best value" of $x_{\text{W}}$.

7.10. (a) Starting from the general expression (6.4.27) for the $e^+e^- \to \tau^+\tau^-$ cross section, derive an expression for the polarization of the produced fermion as a function of the production angle, $z = \cos\theta$, at the $Z^0$ mass.
(b) Show that $P(z = -1) = 0$.
(c) Give an expression for $P(z = +1)$, and show that the value is highly sensitive to the magnitude of the weak mixing parameter, for $x_{\text{W}} \approx \frac{1}{4}$.
(d) Plot your prediction as a function of $\cos\theta$ and compare with the LEP measurements collected in figure 4.7 of Ref. [10]. [For a description of the experimental techniques, see, for example, P. Abreu et al. [DELPHI Collaboration], *Eur. Phys. J.* C **14**, 585 (2000).]]

7.11. The search for new forces of nature through the observation of new gauge bosons presents an important opportunity for high-energy colliders. Although the coupling to fermions of a new neutral gauge boson $Z'$ will be dictated by the new gauge symmetry, it is a useful exercise to examine the requirements necessary to observe a "sequential" $Z'$-boson with fermion couplings identical to those of the standard-model $Z^0$ and no interactions with the known $W^\pm$ and $Z^0$. The sequential $Z'_{\text{seq}}$ is an artificial, but nevertheless informative, construct.
(a) Calculate the partial widths $\Gamma(Z'_{\text{seq}} \to e^+e^-)$ and $\Gamma(Z'_{\text{seq}} \to \nu\bar{\nu})$ [cf. (6.3.65) and (6.3.64)] and plot the values for $Z'_{\text{seq}}$ masses between 1 and 5 TeV.
(b) Express the partial width for the decay $Z'_{\text{seq}} \to$ hadrons, including decays into all kinematically allowed fermion flavors, in a form analogous to (7.2.11). Add the hadronic width to your plot.
(c) Construct a formula for the $Z'_{\text{seq}}$ total width analogous to (7.2.13), and add the total width to your plot.
(d) Now compute the cross section for the reaction $\mu^+\mu^- \to Z'_{\text{seq}} \to$ hadrons in the neighborhood of the $Z'_{\text{seq}}$ peak. Estimate the integrated luminosity needed to produce 1000 $Z'_{\text{seq}} \to$ hadrons events. What instantaneous luminosity would be required to produce this sample in 1 year ($\equiv 10^7$ s) of running?

7.12. Three observables concerning the $b$ quark are sensitive to different combinations of the chiral couplings: $\Gamma(Z^0 \to b\bar{b})$ is determined by $(L_b^2 + R_b^2)$ [(7.2.10)], $A_{\text{peak}}^{(b\bar{b})}$ is sensitive to $(L_b^2 - R_b^2)/(L_b^2 + R_b^2)$ [(7.3.18)], and the low-energy forward-backward asymmetry $A(b\bar{b})$ is proportional to $(R_b - L_b)$ [(7.3.13)]. Generalize the standard $SU(2)_L \otimes U(1)_Y$ electroweak theory to include right-handed charged-current interactions of $b$, so that $L_b = \tau_{3L} - 2Q_b x_{\text{W}}$ and $R_b = \tau_{3R} - 2Q_b x_{\text{W}}$. Working to leading order, display allowed regions in the $I_{3L}$-$I_{3R}$ plane and determine the weak-isospin quantum numbers of $b$. [For a thorough analysis and useful compendium of data, see D. Schaile and P. M. Zerwas, *Phys. Rev.* D **45**, 3262 (1992).]

7.13. Reconsider the momentum sum rule (7.4.66), taking into account the contribution of strange quarks. How large a strange-quark sea would be required to account for the "missing" momentum? What implications would such a sea have for the neutrino-induced production of charm?

7.14. Derive the general result (7.4.70) of the cross section for the inclusive reaction $\nu_\mu N \to \mu^- +$ anything, and place the result in the scaling form (7.4.71). For scattering on a structureless target, show that $x\mathcal{F}_3(x)/\mathcal{F}_2(x) = \pm 1$ for a (fermion, antifermion) target.

7.15. Neglecting the antiquarks within a nucleon, compute the ratios of neutral-current to charged-current cross sections given in (7.4.90) and (7.4.91).

7.16. (a) Generalize the differential cross sections for (anti)neutrino–nucleon charged-current [(7.4.86) and (7.4.87)] and neutral-current [(7.4.88) and (7.4.89)] scattering to include the contributions of strange quarks and anti-quarks.
(b) Compute the combination

$$R^- \equiv \frac{\sigma_{NC}^\nu - \sigma_{NC}^{\bar{\nu}}}{\sigma_{CC}^\nu - \sigma_{CC}^{\bar{\nu}}}.$$

(c) Under what condition is $R^-$ insensitive to the strange sea? In that case, show that $R^- = \frac{1}{2} + \sin^2\theta_W$. [E. A. Paschos and L. Wolfenstein, *Phys. Rev. D* **7**, 91 (1973).]

7.17. Working in the framework of the parton model, derive the general form (7.4.94) of the cross section for the reaction $e^\mp p \to e^\mp +$ anything.

7.18. Because the neutrino–nucleon cross section increases with energy, Earth is opaque to ultrahigh-energy neutrinos. From what you know of the measured cross section for charged-current $\nu N$ interactions [cf. figure 7.12 and (7.4.74)], compute the "water-equivalent" interaction length, $\mathcal{L}_{int} = 1/[\sigma_{\nu N}(E_\nu)N_A]$, where $N_A$ is Avogadro's number. If the diameter of Earth is 11 kilotonnes/cm$^2$, that is, $1.1 \times 10^{10}$ cmwe (centimeters of water equivalent), at what energy does the charged-current interaction length become smaller than one Earth diameter? How does the W-boson modify your conclusions? [R. Gandhi, C. Quigg, M. H. Reno, and I. Sarcevic, *Astropart. Phys.* **5**, 81 (1996).]

7.19. (a) Compute the peak cross section for the reaction $\mu^+\mu^- \to H \to$ hadrons, assuming that the only decay products are $e^+e^-$ $\mu^+\mu^-$, $\tau^+\tau^-$, $\bar{u}u$, $\bar{d}d$, $\bar{s}s$, $\bar{c}c$, and $\bar{b}b$, for a Higgs-boson mass $M_H = 120$ GeV. How does your result compare with the peak cross section observed for the reaction $e^+e^- \to Z \to$ hadrons, as quoted in problem 7.6? (b) Under the same assumptions, what integrated luminosity would be required to observe a 120-GeV Higgs boson in the reaction $\mu^+\mu^- \to H$? Assume that the uncertainty in c.m. energy is 0.1%. (b) How finely (in terms of $\delta p/p$) must the circulating muon momentum be controlled so that $\delta\sqrt{s} = \Gamma(H)$?

7.20. Show that the total cross section for associated production of a Higgs boson and $Z^0$ boson in quark–antiquark annihilations, averaged over initial quark colors, is

$$\sigma(q_i\bar{q}_i \to HZ) = \frac{\pi\alpha^2}{72}\left(\frac{2K}{\sqrt{\hat{s}}}\right)\frac{(K^2 + 3M_Z^2)}{(\hat{s} - M_Z^2)^2 + M_Z^2\Gamma_Z^2}\frac{(L_i^2 + R_i^2)}{x_W^2(1 - x_W)^2},$$

where $\sqrt{\hat{s}}$ is the c.m. energy of the $q\bar{q}$ pair, $K$ is the c.m. momentum of the emerging particles and as usual $L_i$ and $R_i$ are the chiral couplings of the quarks to the $Z^0$ and $x_W = \sin^2 \theta_W$. [See §IV.E of E. Eichten, I. Hinchliffe, K. D. Lane, and C. Quigg, *Rev. Mod. Phys.* **56**, 579 (1984); Addendum: *Rev. Mod. Phys.* **58**, 1065 (1986).]

7.21. A fine-tuning problem may be seen to arise even when the scale $\Lambda$ in (7.11.1) is not extremely large. What has been called the "LEP paradox" refers to a tension within the precise measurements of electroweak observables carried out at LEP and elsewhere. On one hand, the global fits of Refs. [87] and [75] point to a light standard-model Higgs boson. On the other hand, a straightforward effective-operator analysis of possible beyond-the-standard-model contributions to the same observables gives no hint of any new physics—of the kind needed to resolve the hierarchy problem—below about 5 TeV. The one-loop quantum corrections to the Higgs-boson mass shown schematically in figure 7.27 contribute

$$\delta M_H^2 = \frac{G_F \Lambda^2}{4\pi^2 \sqrt{2}} (6M_W^2 + 3M_Z^2 + M_H^2 - 12m_t^2),$$

where $\Lambda$ represents the cutoff scale at which new physics might be presumed to intervene.

(a) For the choice $\Lambda = 5$ TeV, make a bar graph of the individual contributions and estimate the degree of fine tuning required to preserve $M_H \lesssim 1$ TeV.

(b) What level of fine tuning would be required for $\Lambda = 100$ TeV? [R. Barbieri and A. Strumia, "The 'LEP paradox'," hep-ph/0007265; G. Burdman et al., *JHEP* **0702**, 009 (2007).]

(c) What degree of fine tuning would be required at $\Lambda = 5$ TeV and 100 TeV for the case of a bosogamous Higgs boson that couples only to bosons, not to fermions?

7.22. Compute the rate for the dominant decay of the top quark, $t \to bW^+$, neglecting the $b$-quark mass. You will need to sum over the possible polarizations of the $W$-boson. Show that the rate takes the form

$$\Gamma(t \to bW^+) = \frac{G_F m_t^3}{8\pi \sqrt{2}} |V_{tb}|^2 \left( 1 - \frac{M_W^2}{m_t^2} \right)^2 \left( 1 + \frac{2M_W^2}{m_t^2} \right),$$

where $V_{tb}$ measures the strength of the $t \to bW$ coupling. What fraction of the decays produce $W$ bosons with longitudinal polarization? [I. I. Y. Bigi et al., *Phys. Lett.* **B181**, 157 (1986).]

7.23. Generalize the low-energy charged-current effective Lagrangian to include the possibility of a small right-handed–current interaction among the quarks,

$$\mathcal{L}_{\text{eff}}^{\text{cc}} = -\frac{4G_F}{\sqrt{2}} \bar{u}\gamma^\mu \left[ \mathsf{V}\frac{(1-\gamma_5)}{2} + \varepsilon\mathsf{X}\frac{(1+\gamma_5)}{2} \right] d\, \bar{\ell}_L \gamma_\mu \nu_L,$$

where $u$ and $d$ are composite spinors describing uplike and downlike quarks, $\mathsf{V}$ is the standard Cabibbo–Kobayashi–Maskawa (left-handed) quark-mixing matrix given by (7.1.27), X is its right-handed analogue, and the factor $\varepsilon$ measures the relative strength of right-handed *versus* left-handed interactions in the low-energy limit. Now consider three independent determinations, within the standard model, of the element $|V_{ub}|$: The inclusive rate for the transition $b \to u\ell\nu$ is proportional to

$$\left| V_{ub}^{incl} \right|^2 = |V_{ub}|^2 + |\varepsilon|^2 \, |X_{ub}|^2 \,.$$

The exclusive rate for the transition $B \to \pi\ell\nu$ is governed by the vector current and so is proportional to

$$\left| V_{ub}^{B\to\pi} \right|^2 = |V_{ub} + \varepsilon X_{ub}|^2 \,.$$

The annihilation decay, $B \to \tau\nu_\tau$, is mediated by the axial current, so that

$$\left| V_{ub}^{B\to\tau\nu} \right|^2 = |V_{ub} - \varepsilon X_{ub}|^2 \,.$$

To leading order in the small parameter $\varepsilon$, the three rates depend on $|V_{ub}|$ and $\varepsilon \, \mathrm{Re}\,(X_{ub}/V_{ub})$.

(a) Using current values for the three experimental determinations, plot the allowed regions for the three determinations (at one standard deviation) in the $|V_{ub}|$ - $\varepsilon \, \mathrm{Re}\,(X_{ub}/V_{ub})$ plane.

(b) Find a common solution, if possible, and discuss the statistical significance of a constraint on, or evidence for, a contribution from right-handed currents. [[cf. Figure 1 of A. J. Buras, K. Gemmler, and G. Isidori, *Nucl. Phys. B* **843**, 107 (2011), which uses as inputs $\left| V_{ub}^{incl} \right| = (4.11 \pm 0.28) \times 10^{-3}$, $\left| V_{ub}^{B\to\pi} \right| = (3.38 \pm 0.36) \times 10^{-3}$, and $\left| V_{ub}^{B\to\tau\nu} \right| = (5.14 \pm 0.57) \times 10^{-3}$. See also figure 1 of A. Crivellin, *Phys. Rev. D* **81**, 031301 (2010).]]

7.24. Consider the effect of a right-handed $u \to b$ transition on the $y$ distribution of $\bar{\nu}N \to \mu^+ +$ anything, where $b$ has charge $-\frac{1}{3}$ and mass $m_b = 5$ GeV.

(a) Extend the parton-model discussion of section 7.4 to the case of heavy-quark production, for which a minimum momentum transfer is required to excite the heavy quark. Show that in place of the usual scaling variable $x \equiv -q^2/2q \cdot P$, the appropriate scaling variable is $z = x + m_b^2/2MEy$.

(b) Express the cross section in terms of scaling variables as

$$\frac{d^2\sigma}{dxdy} = \frac{G_F^2 ME}{\pi} \mathcal{F}_2(z) \left\{ 1 - y + \frac{x}{z}\left[ \frac{y^2}{2} \pm \left( y - \frac{y^2}{2} \right) \right] \right\} \theta(1-z),$$

where the $\pm$ sign corresponds to $(\nu, \bar{\nu})$ scattering for left-handed interactions with quarks and to $(\bar{\nu}, \nu)$ for right-handed interactions. The $\theta$ function is a step-function approximation of the kinematic threshold.

(c) Verify that in the limit $m_b \to 0$ or $E \to \infty$, you recover the familiar forms (7.4.77) and (7.4.78).

(d) Now suppose that $(u \to d)_L$ and $(u \to b)_R$ transitions both occur at full strength, that is, without Cabibbo suppression. Plot the two contributions to $d\sigma(\bar{\nu}_\mu p \to \mu^+ + X)/dy$ along with their sum. For this exercise, suppose that the up-quark distribution in the proton is given by $u(x) = (1.78/\sqrt{x})(1 - x^{1.51})^{3.5}$, and assume a monochromatic $\bar{\nu}$ beam with $E = 50$ GeV. Neglect sea-quark contributions [R. M. Barnett, *Phys. Rev. Lett.* **36**, 1163 (1976)].

7.25. Assuming the availability of precise determinations of $\Gamma(Z)$ and the $Z \to \ell^+\ell^-$ branching fraction from $e^+e^-$ Z-factory experiments, devise a strategy to infer the W-boson width from the ratio

$$R_\ell = \frac{\sigma(\bar{p}p \to W + \text{anything}) \cdot B(W \to \ell\nu)}{\sigma(\bar{p}p \to Z + \text{anything}) \cdot B(Z \to \ell^+\ell^-)}.$$

If the decay $Z \to t\bar{t}$ is not observed, what implications would $\Gamma(W)$ have for the mass of the top quark? [For an early application, see F. Abe et al. [CDF Collaboration], *Phys. Rev. Lett.* **64**, 152 (1990). The current state of the art (now superseded by direct measurements) is exhibited in A. Abulencia et al. [CDF Collaboration], *J. Phys. G* **34**, 2457 (2007).]

7.26. Examine the angular distribution of photons emitted in the reaction $d\bar{u} \to W^-\gamma$, which is governed by the following Feynman diagrams.

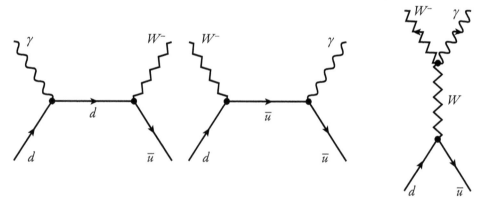

(a) Show that, averaged over quark colors, the differential cross section is

$$\frac{d\sigma}{d\hat{t}} = \frac{\pi\alpha^2}{6\hat{s}^2} \frac{|V_{ud}|^2}{\sin^2\theta_W} \left[ \left( \frac{1}{1 + \hat{t}/\hat{u}} + Q_d \right)^2 \frac{\hat{t}^2 + \hat{u}^2 + 2M_W^2\hat{s}}{\hat{t}\hat{u}} \right],$$

where $V_{ud}$ is an element of the quark mixing matrix, $Q_d$ is the electric charge of the down quark (in units of the proton charge), and $\hat{t}$ measures the momentum transfer between $d$ and $W^-$ [K. O. Mikaelian, M. A. Samuel, and D. Sahdev, *Phys. Rev. Lett.* **43**, 746 (1979)].

(b) Express the Mandelstam invariants in terms of the c.m. scattering angle variable $\cos\theta^*$ and show that the vanishing of the differential cross section at $\hat{t}/\hat{u} = 2$ corresponds to $\cos\theta^* = Q_d = -\frac{1}{3}$. The extinction is understood in terms of classical radiation zeros [S. J. Brodsky and R. W. Brown, *Phys. Rev. Lett.* **49**, 966 (1982)]. This phenomenon has been reported in the reaction $\bar{p}p \to W^\pm + \gamma + \text{anything}$ by the D0 Collaboration, D0 Note 6172-CONF (March 11, 2011), http://j.mp/g2Dwnb.

7.27. Within the standard $SU(2)_L \otimes U(1)_Y$ electroweak theory, the decay of the Higgs boson into two photons proceeds through fermion loops or $W$-boson loops:

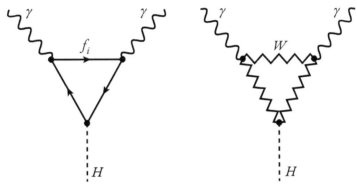

(a) Show that decay rate may be written as

$$\Gamma(H \to \gamma\gamma) = \frac{G_F \alpha^2 M_H^3}{8\pi^3 \sqrt{2}} \left| \sum_i N_c^{(i)} Q_i^2 \mathcal{F}(\epsilon_i) + \mathcal{W}(\epsilon_W) \right|^2,$$

where $N_c^{(i)}$ is the number of colors of fermion $i$ and $Q_i$ is its charge. The scaling variables are $\epsilon_i = 4m_i^2/M_H^2$ and $\epsilon_W = 4M_W^2/M_H^2$, and the loop contributions are

$$\mathcal{F}(\epsilon_i) = \frac{\epsilon_i}{2}\left[1 + (\epsilon_i - 1)\varphi(\epsilon_i)\right],$$

$$\mathcal{W}(\epsilon_W) = -\tfrac{1}{4}\left[2 + 3\epsilon_W + 3(\epsilon_W^2 - 2\epsilon_W)\varphi(\epsilon_W)\right],$$

where

$$\varphi(\epsilon) = \begin{cases} -\left[\sin^{-1}(1/\sqrt{\epsilon})\right]^2, & \epsilon > 1, \\ \tfrac{1}{4}\left[\ln(\zeta_+/\zeta_-) + i\pi\right]^2, & \epsilon < 1, \end{cases}$$

and $\zeta_\pm = 1 \pm \sqrt{1-\epsilon}$.

(b) For a light Higgs boson, with $M_H \approx 120$ GeV, which fermion makes the dominant contribution?

(c) How would $\Gamma(H \to \gamma\gamma)$ change if the Higgs boson did not couple to fermions?

[A convenient reference is M. Spira et al., *Nucl. Phys. B* **453**, 17 (1995). See also W. J. Marciano, C. Zhang, and S. Willenbrock, *Phys. Rev. D* **85**, 013002 (2012).]

7.28. Add to figure 7.26 current knowledge of exclusion limits or observations for the standard-model Higgs boson. What can you conclude about the necessity for new physics to render the electroweak theory consistent?

## ⸾⸾⸾⸾⸾⸾⸾⸾⸾⸾⸾⸾ FOR FURTHER READING ⸾⸾⸾⸾⸾⸾⸾⸾⸾⸾⸾⸾⸾⸾

**Standard model and beyond resource letter.** For a well-chosen annotated bibliography, see

J. L. Rosner, *Am. J. Phys.* **71**, 302 (2003).

**Charmed particles.** An influential general survey appeared in the prospectus

M. K. Gaillard, B. W. Lee, and J. L. Rosner, *Rev. Mod. Phys.* **47**, 277 (1975).

The progression from the discovery of the $J/\psi$ to the discovery of open charm is recounted in

R. Cahn and G. Goldhaber, *The Experimental Foundations of Particle Physics,* 2nd ed., Cambridge University Press, Cambridge, 2009, chap. 9.

For a recent survey of charm flavor physics, see

M. Artuso, B. Meadows, and A. A. Petrov, *Ann. Rev. Nucl. Part. Sci.* **58**, 249 (2008).

A primer on physics opportunities at the next generation of $e^+e^-$ flavor factories is given by

M. Ciuchini and A. Stocchi, *Ann. Rev. Nucl. Part. Sci.* **61**, 491 (2011).

**Alternatives to gauge theories.** Means of reproducing the low-energy phenomenology of the standard model without appealing to local gauge invariance were investigated in

J. D. Bjorken, *Phys. Rev. D* **19**, 335 (1979),

P. Q. Hung and J. J. Sakurai, *Nucl. Phys. B* **143**, 81 (1978) [*Erratum: Nucl. Phys. B* **148**, 538 (1979)].

Now that the electroweak theory has been validated through quantum corrections as a quantum field theory, these are of mainly historical interest, though similar approaches might be useful in the setting of other interactions, as yet unknown.

**CP violation.** The discovery of CP nonconservation in the neutral kaon system and the search for its origin are described in the Nobel lectures by

V. L. Fitch, *Rev. Mod. Phys.* **53**, 367 (1981),

J. W. Cronin, *Rev. Mod. Phys.* **53**, 373 (1981),

Manifestations and implications of CP nonconservation are discussed in depth in

G. C. Branco, L. Lavoura, and J. P. Silva, *CP Violation,* Oxford University Press, Oxford and New York, 1999.

I. I. Bigi and A. I. Sanda, *CP Violation,* 2nd ed., Cambridge University Press, Cambridge, 2009.

The possibility that CP violation arises from complex elements of the quark mixing matrix was raised by

M. Kobayashi and T. Maskawa, *Prog. Theoret. Phys. (Kyoto)* **49**, 652 (1973).

See also the Nobel lectures

M. Kobayashi, *Rev. Mod. Phys.* **81**, 1019 (2009),

T. Maskawa, *Rev. Mod. Phys.* **81**, 1027 (2009).

For an excellent primer on CP violation, emphasizing the CKM paradigm, see

*The* BABAR *Physics Book,* ed. P. F. Harrison and H. R. Quinn, SLAC Report 504, October, 1998, chap. 1.

**Invariants that quantify CP violation.** For a compact historical perspective, see

C. Jarlskog, "On Invariants of Quark and Lepton Mass Matrices in the Standard Model," arXiv:1102.2823.

See also

J. D. Bjorken and I. Dunietz, *Phys. Rev. D* **36**, 2109 (1987).

An interesting geometrical realization is given in

C. Jarlskog and R. Stora, *Phys. Lett.* **B208**, 268 (1988).

**Quark-mixing matrix elements.** Tests of the Cabibbo universality hypothesis relating the strengths of $u \leftrightarrow d$, $u \leftrightarrow s$, and $v \leftrightarrow e$ transitions are reviewed in

N. Cabibbo, E. C. Swallow, and R. Winston, *Ann. Rev. Nucl. Part. Sci.* **53**, 39 (2003).

Our knowledge of the (Cabibbo–Kobayashi–Maskawa) quark-mixing matrix is reviewed in

A. Ceccucci, Z. Ligeti, and Y. Sakai, "The CKM Quark-Mixing Matrix," in Ref. [3], p. 146.

Consult the online CKMfitter and UTfit documents cited in the next entry for additional determinations of the mixing parameters.

**The unitarity triangle.** Comprehensive tests of the Cabibbo–Kobayashi–Maskawa paradigm, building on the elements established in problems 7.3 and 7.4, have been carried out by

J. Charles et al [CKMfitter Group], *Eur. Phys. J.* **C41**, 1 (2005); updated results and plots available at http://ckmfitter.in2p3.fr;

M. Ciuchini et al [UTfit Collaboration], *JHEP* **0107**, 013 (2001); updated results and plots available at http://www.utfit.org.

**Global symmetries.** For an overview of the exact and approximate continuous global symmetries of the standard model, emphasizing the Higgs sector of the electroweak theory, see

S. Willenbrock, "Symmetries of the Standard Model," in *Physics in D $\geq$ 4: TASI 2004*, ed. J. Terning, C.E.M. Wagner, and D. Zeppenfeld, World Scientific, Singapore, 2006, p. 3, hep-ph/0410370.

**Nonleptonic weak decays.** Within the framework of quantum chromodynamics, a partial understanding of the relative enhancement and suppression of certain operators that occur in nonleptonic decays has been achieved by

M. K. Gaillard and B. W. Lee, *Phys. Rev. Lett.* **33**, 108 (1974),

G. Altarelli and L. Maiani, *Phys. Lett.* **52B**, 351 (1974).

The strong interactions also induce new classes of diagrams, with important implications for nonleptonic enhancement, as shown in

M. A. Shifman, A. I. Vainshtein, and V. I. Zakharov, *Nucl. Phys. B* **120**, 316 (1977).

**Intermediate Bosons.** The evolution of the intermediate boson hypothesis—before the discovery experiments—is reviewed in

P. Q. Hung and C. Quigg, *Science* **210**, 1205 (1980).

For an account of the search for the intermediate bosons, see the Nobel lectures of

S. van der Meer, *Rev. Mod. Phys.* **57**, 689 (1985),

C. Rubbia, *Rev. Mod. Phys.* **57**, 699 (1985).

For related history, see

> H. Schopper, *LEP — The Lord of the Collider Rings at CERN 1980–2000*, Springer, Heidelberg and Berlin, 2009,
>
> R. Cashmore, L. Maiani, and J.-P. Revol (ed.), *Prestigious Discoveries at CERN, Eur. Phy. J. C* **34**, No. 1 (2004); reprinted by Springer, Heidelberg and Berlin, 2010.

An extensive account of theoretical and experimental undertakings is given in

> R. Tenchini and C. Verzegnassi, *The Physics of the Z and W Bosons*, World Scientific, Singapore, 2007.

Motivations and explorations of properties of hypothetical additional electroweak gauge bosons may be found in the suggested reading for chapter 9. For minireviews of $W'$ and $Z'$ searches, see

> M.-C. Chen and B. A. Dobrescu, "$W'$ Searches," in Ref. [3], p. 480; "$Z'$ searches," in Ref. [3], p. 483.

**Flavor-changing neutral currents.** Current bounds are summarized by

> L. Wolfenstein, T. G. Trippe, and C.-J. Lin, "Tests of Conservation Laws," in Ref. [3], p. 89.

**Validation of the electroweak theory.** For a compact authoritative rendering of the role of precision measurements in establishing the electroweak theory, consult

> P. Langacker, *J. Phys. G* **29**, 35 (2003).

Lepton–nucleon scattering experiments are emphasized in

> E. Derman and W. J. Marciano, *Ann. Phys.* **121**, 147 (1979),
>
> M. J. Musolf et al. *Phys. Rept.* **239**, 1 (1994).

**Parton model.** An important early paper is

> J. D. Bjorken and E. A. Paschos, *Phys. Rev.* **185**, 1975 (1969).

Thorough developments of the quark–parton model are given by

> F. E. Close, *Introduction to Quarks and Partons*, Academic, New York, 1979,
>
> R. P. Feynman, *Photon–Hadron Interactions*, Benjamin, Reading, MA, 1972.

A broad modern perspective is provided by

> R. Devenish and A. Cooper-Sarkar, *Deep Inelastic Scattering*, Oxford University Press, Oxford and New York, 2004.

Many applications of the parton-model philosophy were treated in

> S. M. Berman, J. D. Bjorken, and J. B. Kogut, *Phys. Rev. D* **4**, 3388 (1971).

**Operator product expansion.** This fruitful alternative to the heuristic parton-model formulation was propounded by

> K. G. Wilson, *Phys. Rev.* **179**, 1499 (1969).

**Diboson production.** Pair-production of electroweak gauge bosons is reviewed in

> M. S. Neubauer, *Ann. Rev. Nucl. Part. Sci.* **61**, 223 (2011).

**Single top-quark production.** The first observations of electroweak production of single top quarks are described in

> A. Heinson and T. R. Junk, *Ann. Rev. Nucl. Part. Sci.* **61**, 171 (2011).

**Feynman rules.** The rules given in chapter 6 are complete only for diagrams without loops. A convenient source for the rules in a general gauge of the type described in section 6.8 is

K. Fujikawa, B. W. Lee, and A. I. Sanda, *Phys. Rev. D* **6**, 2923 (1972).

For a complete set of Feynman rules suitable for calculating higher-order effects, see

K. I. Aoki et al., *Prog. Theor. Phys. Suppl.* **73**, 1 (1982),
M. Veltman, *Diagrammatica: The Path to Feynman Diagrams,* Cambridge University Press, Cambridge, 1994.

Pay attention to differing conventions.

**Quantum corrections to electroweak observables.** A compact textbook introduction appears in chapter 7 of

P. Ramond, *Journeys beyond the Standard Model,* Westview Press, Boulder, CO, 2003.

For an approachable orientation, see

J. L. Rosner, *Phys. Rev. D* **42**, 3107 (1990).

A simple renormalization framework for evaluating radiative corrections is constructed in

A. Sirlin, *Phys. Rev. D* **22**, 971 (1980).

Applications to $e^+e^-$ annihilations in the neighborhood of the $Z^0$ resonance are emphasized in

B. W. Lynn, M. E. Peskin, and R. G. Stuart, "Radiative Corrections in $SU(2)_L \otimes U(1)_Y$: LEP / SLC," in *Physics at LEP,* ed. J. Ellis and R. D. Peccei, CERN Report 90-152, vol. 1, p. 90, and SLAC-PUB-3725, http://j.mp/y8EMTB.

Important tools include the codes

ZFITTER: D. Y. Bardin et al., *Comput. Phys. Commun.* **133**, 229 (2001); A. B. Arbuzov et al., *Comput. Phys. Commun.* **174**, 728 (2006); http://zfitter.com,
TOPAZ0: G. Montagna et al., *Comput. Phys. Commun.* **76**, 328 (1993); G. Montagna, O. Nicrosini, F. Piccinini, and G. Passarino, *Comput. Phys. Commun.* **117**, 278 (1999); http://j.mp/w9OAh0.

For an early review, see

M. I. Vysotsky, V. A. Novikov, L. B. Okun, and A. N. Rozanov, *Phys. Usp.* **39**. 503 (1996) [*Usp. Fiz. Nauk* **166**, 539 (1996)].

The definitive treatment of calculational progress in the LEP era is

D. Bardin and G. Passarino, *The Standard Model in the Making,* Oxford University Press, Oxford, 1999.

The potential to uncover physics beyond the standard model through virtual effects is developed in

D. C. Kennedy and B. W. Lynn, *Nucl. Phys.* **B322**, 1 (1989),
M. E. Peskin and T. Takeuchi, *Phys. Rev. D* **46**, 381 (1992).

A digest of oblique parameters (those that affect only the vacuum-polarization amplitudes for electroweak gauge bosons) determined by the Gfitter group may be found at http://j.mp/xW2nLl.

**Electroweak physics on the 1-TeV scale.** For recent examinations of key questions to be addressed by experiments at the large hadron collider, see

C. Quigg, *Ann. Rev. Nucl. Part. Sci.* **59**, 505 (2009),
J. D. Hobbs, M. S. Neubauer, and S. Willenbrock, *Rev. Mod. Phys.* **84**, 1477 (2012).

**Misgivings about elementary scalars.** A classic treatment is
K. G. Wilson, *Phys. Rev. D* **3**, 1818 (1971), particularly §V.

**Quantum corrections to the Higgs potential.** Among early investigations, see
S. Coleman and E. Weinberg, *Phys. Rev. D* **7**, 1888 (1973),
S. Weinberg, *Phys. Rev. D* **7**, 2887 (1973),
E. Gildener and S. Weinberg, *Phys. Rev. D* **13**, 3333 (1976).

**Validity of the electroweak theory.** For a thoughtful discussion of stability and triviality bounds on the Higgs-boson mass, and their implications, see
A. Djouadi, *Phys. Rept.* **457**, 1 (2008), §1.4.3.

**Two Higgs-doublet models.** For a recent review of the theory and phenomenology, consult
G. C. Branco et al., *Phys. Rept.* **516**, 1 (2012).

**Dynamical symmetry breaking.** Discussions directly relevant to gauge theories include
R. Jackiw, "Dynamical Symmetry Breaking," in *Laws of Hadronic Matter*, 1973 Erice School, ed. A. Zichichi, Academic Press, New York, 1975, p. 225,
S. Weinberg, *Phys. Rev. D* **13**, 974 (1976); *D* **19**, 1277 (1978),
L. Susskind, *Phys. Rev. D* **20**, 2619 (1979).

A convenient summary and reprint collection is
E. Farhi and R. Jackiw (ed.), *Dynamical Gauge Symmetry Breaking*, World Scientific, Singapore, 1982.

Detailed reviews of the technicolor approach to dynamical symmetry breaking have been given by
K. D. Lane, "Two Lectures on Technicolor," arXiv:hep-ph/0202255,
C. T. Hill and E. H. Simmons, *Phys. Rept.* **381**, 235 (2003) [Erratum: *Phys. Rept.* **390**, 553 (2004)].

**Extra spacetime dimensions.** An excellent introduction is given in
C. Csaki, "TASI lectures on extra dimensions and branes" in *Particle Physics and Cosmology: The Quest for Physics Beyond the Standard Model(s)*, ed. H. E. Haber and A. E. Nelson, World Scientific, Singapore, 2004, p. 605, arXiv:hep-ph/0404096.

For a pedagogical review of little Higgs models, gauge–Higgs unification, and warped extra dimension approaches to electroweak symmetry breaking, see
G. Bhattacharyya, *Rept. Prog. Phys.* **74**, 026201 (2011).

**The vacuum energy problem.** An introduction is given by
L. Abbott, "The Mystery of the Cosmological Constant," *Sci. Am.* **258**, 106 (May, 1988).

For historical perspective, see
N. Straumann, "The History of the Cosmological Constant Problem," arXiv:gr-qc/0208027,
Ya. B. Zel'dovich, *Sov. Phys. Usp.* **11**, 381 (1968); a new translation, with commentary, appears in *Gen. Rel. Grav.* **40**, 1557 (2008) [http://j.mp/nrgLSE].

||||||||||||||||| REFERENCES |||||||||||||||||||||||

1. For an assessment of the state of the art and the demands of experimental analyses, see J. Laiho, E. Lunghi, and R. S. Van de Water, *Phys. Rev. D* **81**, 034503 (2010); "Flavor Physics in the LHC era: The Role of the Lattice," arXiv:1204.0791; http://latticeaverages.org.

2. N. Cabibbo, *Phys. Rev. Lett.* **10**, 531 (1963),

3. K. Nakamura et al. [Particle Data Group], *J. Phys. G* **37**, 075021 (2010). Consult J. Beringer et al. [Particle Data Group], *Phys. Rev. D* **86**, 010001 (2012) for updates.

4. A. V. Artamonov et al. [E949 Collaboration], *Phys. Rev. Lett.* **101**, 191802 (2008).

5. S. L. Glashow, J. Iliopoulos, and L. Maiani, *Phys. Rev. D* **2**, 1285 (1970). For additional context, see J. Iliopoulos, *Scholarpedia*, **5** (5), 7125 (2010) [http://j.mp/pSQYSg].

6. See, for example, J. L. Rosner, "The Arrival of Charm," *AIP Conf. Proc.* **459**, 9 (1999).

7. M. Kobayashi and T. Maskawa, *Prog. Theor. Phys.* **49**, 652 (1973).

8. A comprehensive survey of how the Kobayashi–Maskawa theory of CP violation was created, how it was confirmed experimentally, and how theory and experiment may advance in the future can be found in the Special Issue Commemorating the Nobel Prize Awarded to M. Kobayashi and T. Maskawa, ed. K. Higashijima et al., *Prog. Theor. Phys.* **122**, No. 1 (2009).

9. E. A. Kuraev and V. S. Fadin, *Yad. Fiz.* **41**, 733 (1985) [*Sov. J. Nucl. Phys.* **41**, 466 (1985)] is an influential early analysis. Much detailed work is reported in G. Altarelli, R. Kleiss, and C. Verzegnassi (ed.), "Z Physics At LEP-1, Vol. 1: Standard Physics," CERN Report 89-08 (1989). For convenient analytical approximations to the dominant terms, see R. N. Cahn, *Phys. Rev. D* **36**, 2666 (1987) [Erratum: *Phys. Rev. D* **40**, 922 (1989)].

10. The ALEPH, DELPHI, L3, OPAL, SLD Collaborations, the LEP Electroweak Working Group, the SLD Electroweak and Heavy Flavour Groups, *Phys. Rept.* **427**, 257 (2006); update, arXiv:1012.2367.

11. B. Holdom et al., *PMC Phys. A* **3**, 4 (2009), arXiv:0904.4698.

12. See the ALEPH Collaboration Web site at http://aleph.web.cern.ch/aleph. See also C. Grupen, I. Hughes, J. Lynch, and R. Settles (editors), *The ALEPH 'Experience,'* 2nd ed. (2006), http://j.mp/xEWUrs.

13. See the DELPHI Collaboration Web site at http://delphiwww.cern.ch.

14. See the L3 Collaboration Web site at http://l3.web.cern.ch/l3.

15. See the OPAL Collaboration Web site at http://j.mp/zpqUQD.

16. See the SLD Collaboration Web site at http://j.mp/A3SdPd. For highlights of the SLD physics program at the SLAC Linear Collider, see P. C. Rowson, D. Su, and S. Willocq, *Ann. Rev. Nucl. Part. Sci.* **51**, 345 (2001).

17. J. Alcaraz et al, [ALEPH, DELPHI, L3, and OPAL Collaborations and LEP Electroweak Working Group], "A Combination of preliminary electroweak measurements and constraints on the standard model," arXiv:hep-ex/0612034.

18. The LEP Collaborations ALEPH, DELPHI, L3, and OPAL and the LEP Triple Gauge Couplings Working Group, "A Combination of Results on Charged Triple-Gauge-Boson Couplings Measured by the LEP Experiments," LEPEWWG/TGC/2005-01, http://j.mp/zbJ74c.

19. Earlier we denoted charges by $Q_i$, to avoid confusion with the electron spinor $e$; here we change notation to $e_i$, to avoid confusion with the four-momentum–transfer variable $Q^2$.

20. J. D. Bjorken, *Phys. Rev.* **179**, 1547 (1969).

21. A convenient summary of the early SLAC–MIT experiments appears in J. I. Friedman and H. W. Kendall, *Ann. Rev. Nucl. Sci.* **22**, 203 (1972).

22. M. D. Mestayer et al. [SLAC–MIT Collaboration], *Phys. Rev. D* **27**, 285 (1983). For $ep$ scattering at $Q^2 = 10$ GeV$^2$, the result is $\sigma_S/\sigma_T = 0.22 \pm 0.10$. For further analysis of SLAC $ep$ and $ed$ scattering data, see L. W. Whitlow et al., *Phys. Lett. B* **250**, 193 (1990). The corresponding measurement for neutrino–nucleon scattering yields $\sigma_S/\sigma_T = 0.10 \pm 0.07$ for $\nu = 50$ GeV: H. Abramowitz et al., *Phys. Lett.* **107B**, 141 (1981). Measurements in high-energy muon–nucleon scattering are reported in J. J. Aubert, *Phys. Lett.* **121B**, 87 (1983); A. C. Benvenuti et al. [BCDMS Collaboration], *Phys. Lett. B* **223**, 485 (1989); M. Arneodo et al. [New Muon Collaboration], *Nucl. Phys. B* **483**, 3 (1997). A convenient summary of measurements at the $ep$ collider HERA is given by B. Reisert [H1 and ZEUS Collaborations], *PoS* **EPS-HEP2009**, 309 (2009).

23. U. K. Yang and A. Bodek, *Phys. Rev. Lett.* **82**, 2467 (1998). A. Accardi et al., *Phys. Rev. D* **84**, 014008 (2011) critically assess the theoretical uncertainties surrounding the extraction of $F_2^{en}$ from measurements on deuterium and other nuclear targets, a topic of persistent controversy. See especially figure 7.

24. N. Baillie et al., [CLAS Collaboration] *Phys. Rev. Lett.* **108**, 142001 (2012).

25. See, for example, table 2.2 of T. Sloan, G. Smadja, and R. Voss, *Phys. Rep.* **162**, 26 (1988).

26. See figure 16.9 of Ref. [3], which includes a refinement for the contribution of strange quarks and antiquarks.

27. M. Jonker et al. [CHARM Collaboration], *Phys. Lett.* **99B**, 265 (1981).

28. G. P. Zeller et al. [NuTeV Collaboration], *Phys. Rev. Lett.* **88**, 091802 (2002); [Erratum: *Phys. Rev. Lett.* **90**, 239902 (2003)].

29. C. Prescott et al., *Phys. Lett.* **77B**, 347 (1978); *Phys. Lett.* **84B**, 524 (1979). For general theoretical analyses of polarized electron–nucleon scattering, see R. N. Cahn and F. J. Gilman, *Phys. Rev. D* **17**, 1313 (1978); E. Derman and W. J. Marciano, *Ann. Phys.* **121**, 147 (1979).

30. H1 and ZEUS Collaborations, "Electroweak Neutral Currents at HERA," H1prelim-06-142 / ZEUS-prel-06-022, http://j.mp/qrRFtp.

31. T. Ahmed et al. (H1 Collaboration), *Phys. Lett. B* **324**, 241 (1994); M. Derrick et al. (ZEUS Collaboration), *Phys. Rev. Lett.* **75**, 1006 (1995).

32. A. Aktas et al. (H1 Collaboration), *Phys. Lett. B* **632**, 35 (2006). For a useful digest of such measurements, see Z. Zhang, *Nucl. Phys. Proc. Suppl.* **191**, 271 (2009) [arXiv:0812.4662].

33. F. D. Aaron et al. [H1 Collaboration], *JHEP* **1209**, 061 (2012).

34. S. Chekanov et al. [ZEUS Collaboration], *Eur. Phys. J. C* **61**, 223 (2009), *Eur. Phys. J. C* **62**, 625 (2009); H. Abramowicz et al. [ZEUS Collaboration], *Eur. Phys. J. C* **70**, 945 (2010), "Measurement of High-$Q^2$ Neutral Current Deep Inelastic $e^+p$ Scattering Cross Sections with a Longitudinally Polarised Position Beam at HERA," arXiv: 1208.6138.

35. F. D. Aaron et al. [H1 and ZEUS Collaborations], *JHEP* **1001**, 109 (2010); figure from http://j.mp/pRuq4N.

36. The figure is from http://j.mp/pZlows.

37. S. D. Drell and T.-M. Yan, *Phys. Rev. Lett.* **25**, 316 (1970); *Annals Phys.* **66**, 578 (1971).

38. For an informative early review of the dilepton production, with tests of scaling, see I. R. Kenyon, *Rept. Prog. Phys.* **45**, 1261 (1982).

39. CMS Collaboration, "Performance of CMS muon reconstruction in $pp$ collision events at $\sqrt{s} = 7$ TeV," arXiv:1206.4071.

40. The first observations of the $W^\pm$ intermediate bosons were made in two experiments studying $\sqrt{s} = 540$ GeV proton–antiproton interactions at the CERN S$\bar{p}p$S Collider. The initial results of the UA1 detector are reported in G. Arnison et al., *Phys. Lett.*

**122B,** 103 (1983); *Phys. Lett.* **129B,** 273 (1983). Those of the UA2 detector appear in M. Banner et al. *Phys. Lett.* **122B,** 476 (1983). The mass, production rates, and other properties were in very good general agreement with the expectations of the $SU(2)_L \otimes U(1)_Y$ electroweak theory.

41. G. Arnison et al. (UA1 Collaboration), *Phys. Lett.* **B166,** 484 (1986).
42. A. Adare et al. [PHENIX Collaboration], *Phys. Rev. Lett.* **106,** 062001 (2011); M. M. Aggarwal et al. [STAR Collaboration], *ibid.* 062002 (2011).
43. G. Aad et al. [ATLAS Collaboration], *Phys. Lett. B* **705,** 28 (2011); The CMS Collaboration, "Search for $W'$ in the leptonic channels in $pp$ collisions at $\sqrt{s} = 7$ TeV," CMS Physics Analysis Summary EXO-11-024.
44. The first observations of the $Z^0$ intermediate boson were reported by G. Arnison et al. [UA1 Collaboration], *Phys. Lett.* **126B,** 398 (1983), and by P. Bagnaia et al [UA2 Collaboration], *Phys. Lett.* **129B,** 130 (1983).
45. V. M. Abazov et al. [D0 Collaboration], *Phys. Rev. D* **84,** 012007 (2011).
46. G. Aad et al. [ATLAS Collaboration], "Search for dilepton resonances in $pp$ collisions at $\sqrt{s} = 7$ TeV with the ATLAS detector," arXiv:1108.1582; The CMS Collaboration, "Search for Resonances in the Dilepton Mass Distribution in $pp$ Collisions at $\sqrt{s} = 7$ TeV," CMS Physics Analysis Summary EXO-11-019.
47. A useful resource is D. Asner et al. [Heavy Flavor Averaging Group], "Averages of $b$-Hadron, $c$-Hadron, and $\tau$-Lepton Properties," arXiv:1010.1589. For updates, see http://j.mp/zu6LRb.
48. G. Burdman, E. Golowich, J. L. Hewett, and S. Pakvasa, *Phys. Rev. D* **66,** 014009 (2002); E. Golowich, J. Hewett, S. Pakvasa, and A. A. Petrov, *Phys. Rev. D* **79,** 114030 (2009).
49. M. Petric et al [Belle Collaboration], *Phys. Rev. D* **81,** 091102 (2010).
50. B. Aubert et al. [BABAR Collaboration], *Phys. Rev. Lett.* **98,** 211802 (2007); M. Staric, et al. [Belle Collaboration], *Phys. Rev. Lett.,* 211803 (2007).
51. The Heavy Flavor Averaging Group maintains an evolving set of references on charmed-particle properties: http://j.mp/yFhZJN.
52. R. Aaij et al [LHCb Collaboration], *Phys. Rev. Lett.* **108,** 111602 (2012). T. Aaltonen et al. [CDF Collaboration], *Phys. Rev. D* **85,** 012009 (2012).
53. An informative orientation is given in chapter 9 of *The BABAR Physics Book,* ed. P. F. Harrison and H. R. Quinn, SLAC Report 504, October, 1998.
54. R. Aaij et al [LHCb Collaboration], *Phys. Rev. Lett.* 108, **231801** (2012). S. Chatrchyan et al [CMS Collaboration], *JHEP* **1204,** 033 (2012). G. Aad et al. [ATLAS Collaboration], *Phys. Lett. B* **713,** 387 (2012). T. Aaltonen et al. [CDF Collaboration], *Phys. Rev. Lett.* **107,** 191801 (2011) [Erratum: *Phys. Rev. Lett.* 239903 (2011)]. S. Stone, "New Physics from Flavour," plenary lecture at 36th International Conference on High Energy Physics, http://j.mp/PSqC2z.
55. A. J. Buras, M. V. Carlucci, S. Gori, and G. Isidori, *JHEP* **1010,** 009 (2010); A. J. Buras, *Acta Phys. Polon.* **B41,** 2487 (2010). The latest estimate is from K. de Bruyn et al., "A New Window for New Physics in $B_s^0 \to \mu^+\mu^-$," arXiv:1204.1737.
56. S. Chekanov et al. [ZEUS Collaboration], *Phys. Lett. B* **559,** 153 (2003).
57. G. Aad. et al. [ATLAS Collaboration], "A Search for Flavour Changing Neutral Currents in Top-Quark Decays in $pp$ Collision Data Collected with the ATLAS Detector at $\sqrt{s} = 7$ TeV," arXiv:1206.0257.
58. G. Aad et al. [ATLAS Collaboration], *Phys. Lett. B* **712,** 351 (2012).
59. For a review of single-top studies at the Tevatron, plus a digest of the scientific interest, see A. Heinson and T. R. Junk, *Ann. Rev. Nucl. Part. Sci.* **61,** 171 (2011).
60. G. D'Ambrosio, G. F. Giudice, G. Isidori, and A. Strumia, *Nucl. Phys. B* **645,** 155 (2002); A. J. Buras, *Acta Phys. Polon. B* **34,** 5615 (2003). For the extension of these notions to the lepton sector, see V. Cirigliano, B. Grinstein, G. Isidori, and M. B. Wise, *Nucl. Phys. B* **728,** 121 (2005).

61. E. Blucher and W. J. Marciano, "$V_{ud}$, $V_{us}$, the Cabibbo Angle and CKM Unitarity," in Ref. [3], p. 771. See also T. Gershon, "Overview of the CKM Matrix," arXiv:1112.1984, to appear in *Proceedings of the XXV International Symposium on Lepton Photon Interactions at High Energies*; and the series of workshops on the CKM unitarity triangle, http://j.mp/zMuZHt.

62. S. Paul, *Nucl. Instrum. Meth. A* **611**, 157 (2009).

63. D. Kirkby and Y. Nir, "CP Violation in Meson Decays," in Ref. [3], §12; T. E. Browder and R. Faccini, *Annu. Rev. Nucl. Part. Sci.* **53**, 353 (2003); A. Höcker and Z. Ligeti, *Annu. Rev. Nucl. Part. Sci.* **56**, 501 (2006).

64. CKMfitter Group, http://ckmfitter.in2p3.fr. For corresponding UT*fit* plots, see http://j.mp/wO1iEf.

65. T. Aaltonen et al. [CDF Collaboration], *Phys. Rev. D* **82**, 112005 (2010), reports $|V_{tb}| > 0.71$ at 95% CL.

66. M. S. Chanowitz, *Phys. Rev. D* **79**, 113008 (2009).

67. A. J. Buras, *PoS* **BEAUTY 2011**, 008 (2011).

68. E. Lunghi and A. Soni, "Demise of CKM and Its Aftermath," arXiv:1104.2117.

69. D. Hanneke, S. Fogwell Hoogerheide, and G. Gabrielse, *Phys. Rev. A* **83**, 052122 (2011).

70. G. Gabrielse et al., *Phys. Rev. Lett.* **97**, 030802 (2006).

71. C. Quigg, *Phys. Today* **50**, 20 (May, 1997); extended version circulated as arXiv:hep-ph/9704332.

72. Tevatron Electroweak Working Group and the CDF and D0 Collaborations, "Combination of CDF and D0 Results on the Mass of the Top Quark Using Up to 5.8 fb$^{-1}$ of Data," arXiv:1107.5255.

73. P. Vilain et al. [CHARM-II Collaboration], *Phys. Lett. B* **281**, 159 (1992).

74. All the bits are displayed, for example, in F. Antonelli, M. Consoli, and G. Corbò, *Phys. Lett. B* **91**, 90 (1980). See also A. Sirlin, *Phys. Rev. D* **22**, 971 (1980). The possibility of using the $\rho$-parameter to bound heavy fermion masses was emphasized by M.J.G. Veltman, *Nucl. Phys. B* **123**, 89 (1977).

75. The LEP Electroweak Working Group's constraints on the mass of the standard-model Higgs boson are updated regularly at http://lepewwg.web.cern.ch/LEPEWWG/.

76. R. Barate et al. [LEP Working Group for Higgs boson searches], *Phys. Lett.* **B565**, 61 (2003), http://lephiggs.web.cern.ch.

77. J. Erler and M. J. Ramsey-Musolf, *Phys. Rev. D* **72**, 073003 (2005).

78. J. Erler and P. Langacker, "Electroweak Model and Constraints on New Physics," in Ref. [3], §10.

79. P. L. Anthony et al. [SLAC E158 Collaboration], *Phys. Rev. Lett.* **95**, 081601 (2005),

80. M.J.G. Veltman, "Reflections on the Higgs System," CERN academic training lectures Yellow Report CERN-97-05, http://j.mp/w5yQti (2007).

81. T. Appelquist and M. S. Chanowitz, *Phys. Rev. Lett.* **59**, 2405 (1987) [Erratum: *Phys. Rev. Lett.* **60**, 1589 (1988)]. See also F. Maltoni, J. M. Niczyporuk, and S. Willenbrock, *Phys. Rev. D* **65**, 033004 (2002); D. A. Dicus and H.-J. He, *Phys. Rev. D* **71**, 093009 (2005).

82. H. Georgi, *Ann. Rev. Nucl. Part. Sci.* **43**, 209 (1993).

83. M. M. Kado and C. G. Tully, *Ann. Rev. Nucl. Part. Sci.* **52**, 65 (2002).

84. W. D. Schlatter and P. M. Zerwas, *Eur. Phys. J. H* **36**, 579 (2012).

85. Tevatron New Phenomena and Higgs Working Group and CDF and D0 Collaborations, "Updated Combination of CDF and D0 Searches for Standard Model Higgs Boson Production with up to 10.0 fb$^{-1}$ of Data," arXiv:1207.0449.

86. G. Bernardi, M. Carena, and T. Junk, "Higgs bosons: Theory and searches," in Ref. [3], p. 448.

87. For constraints on the mass of the standard-model Higgs boson and other parameters from the Gfitter Collaboration, see H. Flächer et al., *Eur. Phys. J. C* **60**, 543 (2009) [Erratum: *Eur. Phys. J. C* **71**, 1718 (2011)]; M. Baak et al., "Updated Status of the Global Electroweak Fit and Constraints on New Physics," arXiv:1107.0975; http://gfitter.desy.de.

88. S. Chatrchyan et al. [CMS Collaboration], *Phys. Lett. B* **710**, 26 (2012); G. Aad et al. [ATLAS Collaboration], *Phys. Rev. D* **86**, 032003 (2012).

89. G. Aad et al. [ATLAS Collaboration], *Phys. Lett. B* **718**, 1 (2012).

90. S. Chatrchyan et al. [CMS Collaboration], *Phys. Lett. B* **718**, 30 (2012).

91. S. Dittmaier et al. [LHC Higgs Cross Section Working Group], "Handbook of LHC Higgs Cross Sections: 1. Inclusive Observables," arXiv:1101.0593; http://j.mp/z2HV9g.

92. The possible importance of this reaction for hadron colliders was noted by R. N. Cahn and S. Dawson, *Phys. Lett. B* **136**, 196 (1984) [Erratum: *Phys. Lett. B* **138**, 464 (1984)]; and earlier for $e^+e^-$ collisions by D.R.T. Jones and S. T. Petcov, *Phys. Lett. B* **84**, 440 (1979).

93. For branching fractions calculated in a model with standard-model $HWW$ and $HZZ$ couplings, but Yukawa couplings set to zero, see the LHC Higgs Cross Section Working Group page, http://j.mp/xE8MrZ.

94. S. Heinemeyer, W. Hollik, A. M. Weber, and G. Weiglein, *JHEP* **0804**, 039 (2008).

95. S. S. AbdusSalam et al., *Phys. Rev. D* **81**, 095012 (2010).

96. A thorough discussion appears in section 8.1 of S. P. Martin, "A Supersymmetry Primer," version 6, arXiv:hep-ph/9709356 (September 2011).

97. G. D. Kribs, T. Plehn, M. Spannowsky, and T. M. P. Tait, *Phys. Rev. D* **76**, 075016 (2007); V. A. Novikov, A. N. Rozanov, and M. I. Vysotsky, *Phys. Atom. Nucl.* **73**, 636 (2010), arXiv:0904.4570.

98. K. G. Wilson and J. B. Kogut, *Phys. Rept.* **12**, 75 (1974).

99. Z. Fodor, K. Holland, J. Kuti, D. Nogradi, and C. Schroeder, *PoS* **LAT2007**, 056 (2007); P. Gerhold and K. Jansen, *JHEP* **0907**, 025 (2009).

100. A. Wingerter, *Phys. Rev. D* **84**, 095012 (2011). See important earlier work in T. Hambye and K. Riesselmann, *Phys. Rev. D* **55**, 7255 (1997); J. A. Casas, J. R. Espinosa, and M. Quirós, *Phys. Lett. B* **342**, 171 (1995).

101. A. D. Linde, *Pis'ma Zh. Eksp. Teor. Fiz.* **23**, 73 (1976) [English translation: *Sov. Phys.—JETP Letters* **23**, 64 (1976)]; S. Weinberg, *Phys. Rev. Lett.* **36**, 294 (1973).

102. G. Isidori, G. Ridolfi, and A. Strumia, *Nucl. Phys. B* **609**, 387 (2001); C. D. Froggatt, H. B. Nielsen, and Y. Takanishi, *Phys. Rev. D* **64**, 113014 (2001); M. B. Einhorn and D.R.T. Jones, *JHEP* **0704**, 051 (2007).

103. J. Elias-Miro et al., *Phys. Lett. B* **709**, 222 (2012). F. Bezrukov, M. Y. Kalmykov, B. A. Kniehl and M. Shaposhnikov, "Higgs boson mass and new physics," arXiv:1205.2893.

104. W. A. Bardeen, "On Naturalness in the Standard Model," FERMILAB-CONF-95-391-T, http://j.mp/A8zouc.

105. N. Arkani-Hamed, S. Dimopoulos, and G. R. Dvali, *Phys. Lett. B* **429**, 263 (1998); I. Antoniadis et al., *Phys. Lett. B* **436**, 257 (1998).

106. L. Randall and R. Sundrum, *Phys. Rev. Lett.* **83**, 3370 (1999); *Phys. Rev. Lett.* **83**, 4690 (1999).

107. S. Weinberg, "Living in the Multiverse," in *Universe or Multiverse?*, ed. B. Carr, Cambridge University Press, Cambridge & New York, 2007, chapter 2 [hep-th/0511037]. For thoughtful presentations of the landscape point of view regarding fermion masses and mixings and other parameters, see A. N. Schellekens, physics/0604134; arXiv:0807.3249.

108. A. Djouadi, *Phys. Rept.* **459**, 1 (2008); S. Weinberg, *The Quantum Theory of Fields: vol. 3, Supersymmetry,* Cambridge University Press, Cambridge, 2005. See also Ref. [96].

109. M. Schmaltz and D. Tucker-Smith, *Ann. Rev. Nucl. Part. Sci.* **55**, 229 (2005); M. Perelstein, *Prog. Part. Nucl. Phys.* **58**, 247 (2007); H.-C. Cheng and I. Low, *JHEP* **0408**, 061 (2004).

110. Z. Chacko, H.-S. Goh, and R. Harnik, *Phys. Rev. Lett.* **96**, 231802 (2006).

111. S. Weinberg, *Rev. Mod. Phys.* **61**, 1 (1989); J. Frieman, M. Turner, and D. Huterer, *Ann. Rev. Astron. Astrophys.* **46**, 385 (2008). Also see S. Weinberg, "The Cosmological constant problems," in *Sources and Detection of Dark Matter and Dark Energy in the Universe,* edited by D. B. Cline, Springer-Verlag, Berlin, 2001, p. 18 [astro-ph/0005265].

112. A. D. Linde, *Pisma Zh. Eksp. Teor. Fiz* **19**, 320 (1974); [English transl.: *JETP Lett.* **19** 183 (1974)]; M.J.G. Veltman, *Phys. Rev. Lett.* **34**, 777 (1975); P.J.E. Peebles and B. Ratra, *Rev. Mod. Phys.* **75**, 559 (2003).

113. Powerful evidence for the accelerated expansion of the universe was presented by A. G. Riess et al. (High-$z$ Supernova Team), *Astron. J.* **116**, 1009 (1998), and by S. Perlmutter et al. (Supernova Cosmology Project), *Astrophys. J.* **517**, 565 (1999). For a recent assessment, see M. Sullivan et al., *Astrophys. J.* **737**, 102 (2011). See also the Nobel lectures by S. Perlmutter, *Rev. Mod. Phys.* **84**, 1127 (2012); B. P. Schmidt, *Rev. Mod. Phys.* **84**, 1151 (2012) ; A. G. Riess, *Rev. Mod. Phys.* **84**, 1165 (2012).

114. E. Komatsu et al. (WMAP Collaboration), *Astrophys. J. Suppl.* **192**, 18 (2011).

115. G. Bertone, D. Hooper, and J. Silk, *Phys. Rept.* **405**, 279 (2005).

# Eight ||||||||||||||||||||||||||||||||||||||||||||||||||||||||||||||||||||||||||||||||||||||||||||||

## Strong Interactions among Quarks

Having learned that local gauge invariance provides the key to understanding the weak and electromagnetic interactions, we turn our attention once again to the strong interactions. The work of many people, from Yang and Mills [1] through Sakurai [2], Ne'eman [3], Englert and Brout [4], to 't Hooft [5], among others, showed that it is unlikely that a flavor symmetry such as isospin or flavor-SU(3) could be the basis of a successful theory of the hadronic interactions. Furthermore, at what we currently perceive to be the constituent level of quarks and leptons, flavor has been seen to be an attribute tied more directly to the weak interactions than to the strong. The property that distinguishes quarks from leptons is *color,* so it is natural to construct a theory of the strong interactions among quarks based upon a local color gauge symmetry. The resulting theory has acquired the name quantum chromodynamics, or QCD.

QCD is a remarkably simple, successful, and rich theory of the strong interactions. The theory provides a dynamical basis for the *quark-model* description of the *hadrons*, the strongly interacting particles such as protons and pions that are accessible for direct laboratory study. Interactions among the quarks are mediated by vector force particles called *gluons*, which themselves experience strong interactions. The nuclear force that binds protons and neutrons together in atomic nuclei emerges from the interactions among quarks and gluons.

For reasons connected with the nonobservation of free quarks, it is appealing to have *long-range forces* among the quarks, as would be mediated by massless gauge bosons in an unbroken gauge theory. With no spontaneous symmetry breaking, the formulation of QCD is technically simpler than that of the $SU(2)_L \otimes U(1)_Y$ electroweak theory. The analysis of the theory is considerably complicated by the fact that the strong interactions are strong. There is, thus, at first sight no guarantee that perturbation theory—our most highly developed tool—will be at all useful. In a general field theory, the effective coupling constant is not a constant but depends on a momentum or distance scale because of renormalization effects. It is a particular property of non-Abelian gauge theories that the effective coupling constant decreases at short distances or, equivalently, at high momenta. Such theories are said to be asymptotically free. This raises the hope that, in the regime of short-distance phenomena, perturbative methods may indeed be reliable. Furthermore, it suggests a reconciliation of the quasi-freedom of quarks, embodied in the parton model, with the nonobservation of isolated quarks.

QCD underlies a wealth of physical phenomena, from the structure of nuclei to the inner workings of neutron stars and the cross sections for the highest-energy elementary-particle collisions. We shall investigate only a small sample of these, in the interest of restricting the book to finite size, because of the absence of complete or graceful solutions to certain problems and because the limited techniques we have used are ill suited to some others. Nevertheless, we shall be able to expose the essential features of QCD, to see why it appears to be the fundamental theory of the strong interactions, and to prepare the reader to confront specialized monographs and the research literature. The emphasis will be not so much on the shortcomings of elementary methods as on the considerable amount that may be done with so little formal apparatus.

We shall begin by reviewing the motivation for hadronic color and for the choice of $SU(3)_c$ as the strong-interaction gauge group. Next, we construct the QCD Lagrangian and extract the Feynman rules for tree diagrams. An analysis of the spectroscopic consequences of the theory in the crude approximation of lowest-order perturbation theory reveals no obvious contradictions with experiment. Thereafter, we turn to a study of the coupling constant evolution in the context of charge renormalization in electrodynamics. This provides the occasion to introduce two useful techniques for the evaluation of the integrals over loop momenta, which occur in the computation of Feynman diagrams. Turning to QCD we derive the celebrated property of asymptotic freedom. Having found this signal for the existence of a regime in which perturbation theory may be expected to apply, we consider four concrete applications: electron–positron annihilation into hadrons, deeply inelastic lepton–nucleon scattering, jet production in hadron collisions, and two-photon processes. To deal with the $Q^2$-evolution of the nucleon and photon structure functions, we develop the intuitive method applied to this problem by Altarelli and Parisi [6]. At the end of the chapter, we discuss some applications for which perturbative methods do not suffice and assess the current status and open problems in QCD.

## 8.1 A COLOR GAUGE THEORY

Several pieces of evidence have been arrayed in favor of the hypothesis that the quarks $u, d, s, c, b, t$ are color triplets. These include the resolution of the spin-statistics problem for baryons, the magnitude of the cross section of electron–positron annihilation into hadrons, the branching ratios for $\tau$-decay, the $\pi^0$ lifetime,* and the requirement of anomaly cancellation in the standard model of weak and electromagnetic interactions. The known leptons, in contrast, are all

---

* For a critical look at the circumstances under which the number of colors can be determined in $\pi^0 \to \gamma\gamma$ decay, see O. Bär and U. J. Wiese, *Nucl. Phys. B* **609**, 225 (2001).

colorless states. The distinction suggests the possibility that color may play the part of a charge of the strong interactions. We attempt, therefore, to form a dynamical theory based on color gauge symmetry. Early steps in this direction were taken by Nambu [7]. It was soon recognized [8] that such a theory might provide a basis for the simple rules that mesons are quark–antiquark states and baryons are three-quark states. Let us see why $SU(3)_c$ is a promising choice for the color symmetry group.

When we entertain the possibility that the color quantum number reflects a continuous symmetry of the strong-interaction Lagrangian, rather than merely a discrete degree of freedom, three candidates for the symmetry come immediately to mind: $SO(3)$, $SU(3)$, and $U(3)$. The choice of a gauge group is to be governed by the two empirical facts: the quarks are color triplets, but all the known hadrons are color singlets. Simple arguments discourage the adoption of $SO(3)$ and $U(3)$, as we shall now see.

The $SO(3)$ symmetry makes no distinction between color and anticolor, so in the computation of forces there will be no distinction between quarks and antiquarks. The existence of $(q\bar{q})$ mesons thus implies the existence of $(qq)$ diquark states, which would be fractionally charged. Such states are not observed as free particles. (We noted a similar shortcoming of the Yang–Mills theory for nucleon–(anti)nucleon interactions in section 4.3 and problem 4.4.) We shall also see in problem 8.14 that an $SO(3)$ gauge theory with six quark flavors would not be asymptotically free.

In $U(3)$ color gauge theory, the color-singlet gauge boson that occurs in the product

$$\mathbf{3} \otimes \mathbf{3}^* = \mathbf{1} \oplus \mathbf{8} \tag{8.1.1}$$

(in $SU(3)$ notation) cannot be dispensed with. It would mediate long-range strong interactions between color-singlet hadrons and is thus ruled out by the same experimental facts that eliminate the Yang–Mills theory. Thus, $U(3)$ is excluded, and we are left with $SU(3)$ as the candidate gauge group.

### 8.1.1 QCD: FIRST STEPS

To construct the $SU(3)$ color gauge theory for the interactions of color-triplet quarks, we follow the procedure used in section 4.2 for the Yang–Mills theory. The color-octet gauge bosons that will emerge in the theory are called gluons because of their role in binding quarks together within hadrons. The Lagrangian for the theory will have the standard form

$$\mathcal{L} = \bar{\psi}(i\gamma^\mu \mathcal{D}_\mu - m)\psi - \tfrac{1}{2}\mathrm{tr}\,G_{\mu\nu}G^{\mu\nu}, \tag{8.1.2}$$

where the composite spinor of color-triplet quarks is

$$\psi = \begin{pmatrix} q_{\text{red}} \\ q_{\text{green}} \\ q_{\text{blue}} \end{pmatrix}, \tag{8.1.3}$$

and the gauge-covariant derivative is

$$\mathcal{D}_\mu = \partial_\mu + ig B_\mu, \tag{8.1.4}$$

where the object $B_\mu$ is a three-by-three matrix in color space formed from the eight color gauge fields $b_\mu^l$ and the generators $\lambda^l/2$ of the SU(3) gauge group as

$$B_\mu = \tfrac{1}{2}\boldsymbol{\lambda} \cdot \mathbf{b}_\mu = \tfrac{1}{2}\lambda^l b_\mu^l. \tag{8.1.5}$$

The gluon field-strength tensor is

$$\begin{aligned} G_{\mu\nu} &= \tfrac{1}{2}\mathbf{G}_{\mu\nu} \cdot \boldsymbol{\lambda} = \tfrac{1}{2}G_{\mu\nu}^l \lambda^l \\ &= (ig)^{-1}\left[\mathcal{D}_\nu, \mathcal{D}_\mu\right] = \partial_\nu B_\mu - \partial_\mu B_\nu + ig\left[B_\nu, B_\mu\right]. \end{aligned} \tag{8.1.6}$$

The $\lambda$-matrices are familiar [9] from the study of flavor-SU(3) symmetry. They have a number of simple properties, including

$$\text{tr}(\lambda^l) = 0, \tag{8.1.7}$$

$$\text{tr}(\lambda^k \lambda^l) = 4 T_R \delta^{kl} = 2\delta^{kl}, \tag{8.1.8}$$

so that $T_R$ (also called $T_F$) $= \tfrac{1}{2}$, and

$$[\lambda^j, \lambda^k] = 2i f^{jkl}\lambda^l, \tag{8.1.9}$$

which parallel those of the Pauli isospin matrices given in (4.2.18) and (4.2.25). Indeed, in the canonical $RGB$ basis

$$\lambda_1 = \begin{pmatrix} 0 & 1 & 0 \\ 1 & 0 & 0 \\ 0 & 0 & 0 \end{pmatrix}, \qquad \lambda_2 = \begin{pmatrix} 0 & -i & 0 \\ i & 0 & 0 \\ 0 & 0 & 0 \end{pmatrix},$$

$$\lambda_3 = \begin{pmatrix} 1 & 0 & 0 \\ 0 & -1 & 0 \\ 0 & 0 & 0 \end{pmatrix}, \qquad \lambda_4 = \begin{pmatrix} 0 & 0 & 1 \\ 0 & 0 & 0 \\ 1 & 0 & 0 \end{pmatrix},$$

$$\lambda_5 = \begin{pmatrix} 0 & 0 & -i \\ 0 & 0 & 0 \\ i & 0 & 0 \end{pmatrix}, \qquad \lambda_6 = \begin{pmatrix} 0 & 0 & 0 \\ 0 & 0 & 1 \\ 0 & 1 & 0 \end{pmatrix}, \tag{8.1.10}$$

$$\lambda_7 = \begin{pmatrix} 0 & 0 & 0 \\ 0 & 0 & -i \\ 0 & i & 0 \end{pmatrix}, \qquad \lambda_8 = \frac{1}{\sqrt{3}}\begin{pmatrix} 1 & 0 & 0 \\ 0 & 1 & 0 \\ 0 & 0 & -2 \end{pmatrix},$$

the upper-left $2 \times 2$ blocks of $\lambda_1$, $\lambda_2$, $\lambda_3$ are simply the Pauli isospin matrices $\tau_1$, $\tau_2$, $\tau_3$. Using (8.1.8) and (8.1.9), it is easy to compute the antisymmetric structure constants of the Lie group as

$$f^{jkl} = (4i)^{-1}\text{tr}(\lambda^l[\lambda^j, \lambda^k]). \tag{8.1.11}$$

**TABLE 8.1**
Antisymmetric Structure Constants of SU(3)

| |
|---|
| $f_{123} = 1$ |
| $f_{147} = f_{246} = f_{257} = f_{345} = f_{516} = f_{637} = \frac{1}{2}$ |
| $f_{458} = f_{678} = \sqrt{3}/2$ |

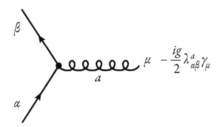

**Figure 8.1.** Feynman rule for the quark–antiquark–gluon vertex in QCD.

The nonzero elements are given by the entries in table 8.1 and their permutations. The field-strength tensor may be rewritten in component form as

$$G^l_{\mu\nu} = \partial_\nu b^l_\mu - \partial_\mu b^l_\nu + g f^{jkl} b^j_\mu b^k_\nu. \tag{8.1.12}$$

The non-Abelian gauge symmetry reflected in the final term entails three-gluon and four-gluon interactions, along with the $gq\bar{q}$ vertex (cf. (8.1.2) and figure 8.15).

Knowing the QCD Lagrangian, we may now study the interactions between quarks in a very simplified fashion, just as we did the interactions among nucleons in the Yang–Mills theory in section 4.3. The point of the exercise will be to verify that color singlets enjoy a preferred status. This encourages the hope—now realized—that the spectrum of QCD, once computed, will display the systematics that inspired the invention of the theory (cf. section 8.8).

The quark–gluon interaction term in the QCD Lagrangian is

$$\mathcal{L}_{\text{int}} = -\frac{g}{2} b^a_\mu \bar{\psi} \gamma^\mu \lambda^a \psi, \tag{8.1.13}$$

which leads at once to the Feynman rule for the quark–antiquark–gluon vertex. For the transition depicted in figure 8.1 of a quark with color index $\alpha$ ($= R$, $G$, $B$ or $1, 2, 3$) into a quark with color index $\beta$ and a gluon with Lorentz index $\mu$ and color label $a$ ($= 1, 2, \ldots, 8$), the vertex factor is simply

$$-\frac{ig}{2} \lambda^a_{\alpha\beta} \gamma_\mu. \tag{8.1.14}$$

**TABLE 8.2**
Value of the Color Casimir Operator in Small Representations of SU(3).

| Representation | $\langle \mathbf{T}^2 \rangle$ |
|---|---|
| **1** | 0 |
| **3** or **3*** | 4/3 |
| **6** or **6*** | 10/3 |
| **8** | 3 |
| **10** or **10*** | 6 |
| **27** | 8 |

Thus the one-gluon-exchange force between quarks is proportional to

$$\mathcal{E} = \frac{g^2}{4} \sum_a \lambda^a_{\alpha\beta} \lambda^a_{\gamma\delta} \tag{8.1.15}$$

for the transition $\alpha + \gamma \to \beta + \delta$. We shall take the quantity (8.1.15) as representative of the interaction energy between quarks and develop the consequences of QCD for the hadron spectrum, according to that measure.

To compute the interaction energy it is necessary to evaluate the expection value of products such as $\frac{1}{4}\boldsymbol{\lambda}^{(1)} \cdot \boldsymbol{\lambda}^{(2)}$, where the superscripts label the interacting quarks and the $\boldsymbol{\lambda}^{(i)}$ are 8-vectors in color space. The SU($N$) techniques are quite standard [10]. It will save writing to define the SU(3)$_c$ generators

$$\mathbf{T} \equiv \tfrac{1}{2}\boldsymbol{\lambda} \tag{8.1.16}$$

and to evaluate the expectation value $\langle \mathbf{T}^2 \rangle$ in various representations of interest.

In SU($N$), it is equivalent to average the square of any single generator over the representation, or to perform the sum over all the generators. The former tactic is simpler, and it is particularly convenient to choose $I_3$, the third component of isospin in the flavor analogy, as the designated generator. Consequently, the expectation value in a representation of dimension $d$ is

$$\langle \mathbf{T}^2 \rangle_d = (N^2 - 1) \sum_{\text{rep. } d} \frac{I_3^2}{d}, \tag{8.1.17}$$

where $N^2 - 1$ is the number of generators of SU($N$). Results for the low-dimensional representations of SU(3) are given in table 8.2. To evaluate $\langle \mathbf{T}^{(1)} \cdot \mathbf{T}^{(2)} \rangle$, we use the familiar identity

$$\langle \mathbf{T}^{(1)} \cdot \mathbf{T}^{(2)} \rangle = \frac{\langle \mathbf{T}^2 \rangle - \langle \mathbf{T}^{(1)2} \rangle - \langle \mathbf{T}^{(2)2} \rangle}{2}. \tag{8.1.18}$$

The *interaction energies* for two-body systems composed of quark–quark and quark–antiquark are given in table 8.3. For the $(q\bar{q})$ systems, the one-gluon-exchange contribution is attractive for the color singlet but repulsive for the color octet. Similarly for the diquark systems, the color triplet is attracted but the color

TABLE 8.3

Interaction energies for few-quark systems

| Configuration | $\sum_{i<j}\langle \mathbf{T}^{(i)} \cdot \mathbf{T}^{(j)}\rangle$ |
|---|---|
| $(q\bar{q})_1$ | $-\frac{4}{3}$ |
| $(q\bar{q})_8$ | $+\frac{1}{6}$ |
| $(qq)_{3^*}$ | $-\frac{2}{3}$ |
| $(qq)_6$ | $\frac{1}{3}$ |
| $(qqq)_1$ | $-2$ |
| $(qqq)_8$ | $-\frac{1}{2}$ |
| $(qqq)_{10}$ | $+1$ |
| $(qqqq)_3$ | $-2$ |

sextet is repelled. Of all the two-body channels, the color-singlet $q\bar{q}$ is the most attractive. On the basis of this analysis, we may choose to believe that colored mesons should not exist, whereas color-singlets should be found [11].

To anticipate the behavior of three-body systems and beyond, let us posit that the interaction is merely the sum of two-body forces, so that

$$\mathcal{E} = \sum_{i<j} \langle \mathbf{T}^{(i)} \cdot \mathbf{T}^{(j)} \rangle, \tag{8.1.19}$$

which is easily computed as

$$\sum_{i<j} \langle \mathbf{T}^{(i)} \cdot \mathbf{T}^{(j)} \rangle = \frac{\langle \mathbf{T}^2 \rangle - \sum_i \langle \mathbf{T}^{(i)2} \rangle}{2}. \tag{8.1.20}$$

For three-quark systems, the results in table 8.3 show that the color singlet is again the most attractive channel. This is the desired result.

Several potentially important effects have been neglected in these calculations: (1) Multiple gluon exchanges between quarks have been ignored, and we have put forward no arguments for trust in lowest-order perturbation theory. (2) Configurations including the three-gluon vertex, such as the $(qqq)$ state shown in figure 8.2, have not been taken into account. (3) What may be the most serious limitation of the toy calculation is the neglect of the energetics associated with the creation of an isolated colored state. In section 8.8 we shall review a plausibility argument that implies an infinite cost in energy to isolate a colored system. If that is so, the attraction provided by one-gluon exchange will be insufficient to bind colored states.

In spite of these shortcomings, the elementary "maximally attractive channel" calculation we have just completed does make it plausible that color singlets are energetically favored states. In addition, it is easy to see that there is no long-range interaction with color singlets. As an example, consider whether a quark is bound to a baryon. The final entry in table 8.3 shows that the interaction energy of the

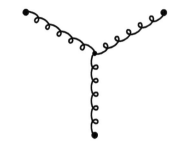

**Figure 8.2.** A baryon configuration that is not considered in the sum over two-body forces.

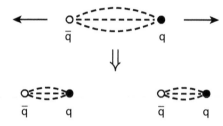

**Figure 8.3.** Attempting to separate a quark and antiquark results in the creation of a quark–antiquark pair from the vacuum, so that color is always neutralized locally.

quark-plus-baryon system is precisely that which binds the baryon, with no additional attraction.

We lack an analytical proof that QCD is a confining theory [12], and that colored objects cannot be liberated, but lattice gauge theory provides powerful evidence in support of color confinement (cf. section 8.8). A host of searches for fractionally charged matter have established that free quarks are exquisitely rare [13]. The definitive observation of free quarks would require a serious revision of our thinking.

### 8.1.2 THE STRING PICTURE OF HADRONS

Suppose that the interaction among quarks is so strong at large distances that a $(q\bar{q})$ pair is always created when the quarks are widely separated, as depicted in figure 8.3. Studies of multiple production in high-energy collisions suggest that the $(q\bar{q})$ pair is created as a typical hadron of mass $\sim 1$ GeV at a separation of $\sim 1$ fm. That would imply that between a separating quark and antiquark, there is a linear energy density of order

$$k = \frac{\Delta E}{\Delta r} = 1\frac{\text{GeV}}{\text{fm}} = 0.2 \text{ GeV}^2 \approx \frac{5}{\text{fm}^2}. \qquad (8.1.21)$$

This picture is consonant with the evidence for linear Regge trajectories of the light hadrons. For the families of hadrons composed entirely of light quarks, the

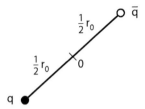

**Figure 8.4.** A massless quark and antiquark connected by a linear string.

spins and masses are correlated through

$$J(M^2) = \alpha_0 + \alpha' M^2, \tag{8.1.22}$$

with slope

$$\alpha' \approx 0.8 \text{ to } 0.9 \text{ GeV}^{-2}. \tag{8.1.23}$$

The connection between linear energy density and linear Regge trajectories is provided by the string model formulated by Nambu [14].

Consider a massless quark and antiquark connected by a string of length $r_0$, which is characterized by an energy density per unit length $k$. The setup is sketched in figure 8.4. For a given value of the length $r_0$, the largest achievable angular momentum occurs when the ends of the string move with the speed of light. In this circumstance, the speed of any point along the string will be

$$\beta(r) = \frac{2r}{r_0}. \tag{8.1.24}$$

The total mass of the system is then

$$M = 2 \int_0^{r_0/2} dr \, k \left[1 - \beta(r)^2\right]^{-1/2} = \frac{k r_0 \pi}{2}, \tag{8.1.25}$$

whereas the orbital angular momentum of the string is

$$L = 2 \int_0^{r_0/2} dr \, k r \beta(r) \left[1 - \beta(r)^2\right]^{-1/2} = \frac{k r_0^2 \pi}{8}, \tag{8.1.26}$$

Inferring from (8.1.25) that $r_0^2 = 4M^2/k^2\pi^2$, we find that

$$L = \frac{M^2}{2\pi k}, \tag{8.1.27}$$

which corresponds to a linear Regge trajectory, with slope

$$\alpha' = \frac{1}{2\pi k}. \tag{8.1.28}$$

This connection yields

$$k = \begin{cases} 0.18 \text{ GeV}^2, \\ 0.20 \text{ GeV}^2, \end{cases} \quad \text{for} \quad \alpha' = \begin{cases} 0.9 \text{ GeV}^{-2}, \\ 0.8 \text{ GeV}^{-2}, \end{cases} \tag{8.1.29}$$

consistent with our heuristic estimate of the energy density. Thus we see that a linear energy density implies linearly rising Regge trajectories and that the connection makes quantitative sense.

At least for high angular momenta, it is plausible that the light baryons could be pictured as quark and diquark connected by a string [15]. The $3$—$3^*$ arrangement is the same as for a quark–antiquark meson, so we expect that baryons and mesons on the leading trajectory should be characterized by the same energy density and, hence, the same Regge slope.

We shall see in section 8.8.2 that a linear potential between color sources at large separations indeed arises in a fully nonperturbative (lattice) treatment of QCD.

### 8.1.3 THE PROTON VIEWED IN QCD: A FIRST LOOK

Before proceeding to a specific study of the predictions of QCD, it will be useful to present a qualitative discussion of the implications of an interacting field theory of quarks and gluons. A convenient setting for this discussion is the deeply inelastic scattering of leptons from a proton target, in which an electroweak gauge boson of $(\text{mass})^2 = -Q^2$ probes the target structure on a length scale characterized by $1/\sqrt{Q^2}$.

Viewed at very long wavelengths, the proton appears structureless, but as $Q^2$ increases and the resolution becomes fine, the proton is revealed as a composite object characterized, for example, by rapidly falling elastic form factors that decrease as $1/Q^4$. According to the parton model, which ignores interactions among the constituents of the proton, the picture for deeply inelastic scattering then becomes exceedingly simple, as sketched in figure 8.5. Once $Q^2$ has become large enough for the quark constituents to be resolved, no finer structure is seen. The quarks are structureless, have no size, and thus introduce no length scale, When $Q^2$ exceeds a few GeV$^2$, all fixed mass scales become irrelevant, so the $Q^2$-dependence of structure functions can be determined by dimensional analysis. We saw in section 7.4 that this description is not only simple, but extraordinarily successful in reproducing the essential features of the data. It is also a description that appears difficult to reconcile with the idea of an underlying quantum field theory, for reasons that will immediately become apparent.

In an interacting field theory, a more complex picture of hadron structure emerges. As $Q^2$ increases beyond the magnitude required to resolve quarks, the quarks themselves are found to have an apparent structure, which arises from the interactions mediated by the boson fields. This is indicated in figure 8.6. The quantum fluctuations shown there lead to deviations from Bjorken scaling in deeply inelastic scattering. As we saw in section 7.4, the structure functions $F_i(x, Q^2)$ measure the distribution of quarks in a fast-moving proton as a function of the

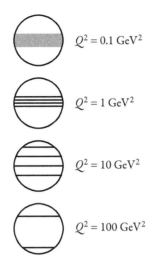

**Figure 8.5.** Schematic parton-model view of the proton.

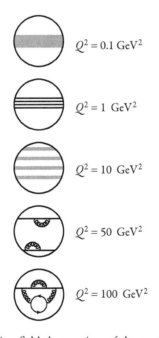

**Figure 8.6.** Schematic interacting-field-theory view of the proton.

momentum fraction

$$x = \frac{p_{\text{quark}}}{P_{\text{proton}}}. \tag{8.1.30}$$

As $Q^2$ grows, the structure functions undergo a characteristic evolution. At large values of $x$ ($0.3 \lesssim x < 1$), uncertainty-principle considerations make it increasingly

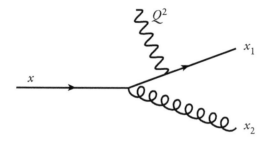

**Figure 8.7.** Virtual dissociation of a quark into a quark and a gluon. A high-$Q^2$ probe may catch the system in midfluctuation.

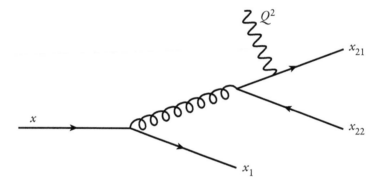

**Figure 8.8.** Virtual dissociation of a gluon into a quark and antiquark, which enhances the population of low-$x$ quarks and antiquarks seen by a high-$Q^2$ probe.

likely with increasing $Q^2$ that a quark with momentum fraction $x$ will be caught in middissociation into components with momentum fractions $x_1 + x_2 = x$, as shown in figure 8.7. For small values of $x$, the population of both quarks and antiquarks will be enhanced by the virtual dissociation of the boson fields, such as the second-order fluctuation illustrated in figure 8.8.

It is, therefore, plausible to expect, in any interacting field theory, that as $Q^2$ increases, the structure function will fall at large values of $x$ and rise at small values of $x$, as sketched in figure 8.9. It remains for a quantitative analysis to show whether these effects are reliably calculable in a particular field theory—specifically in QCD—using perturbation-theory techniques. In the general case, we have no reason to anticipate that the effects will be subtle, as they are in nature. Indeed, the straightforward inference is that structure functions should decrease as inverse powers of $Q^2$. It would, therefore, appear doubtful that approximate parton-model behavior could be accomodated in an interacting field theory. In the next two sections, we shall begin to see why non-Abelian gauge theories—QCD among them—constitute an important, and indeed unique, exceptional case.

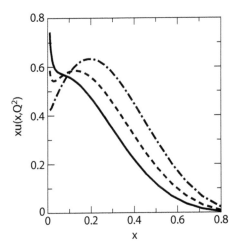

**Figure 8.9.** Schematic evolution of the up-quark parton distribution at low (dot-dash), medium (dash), and high (solid) values of $Q^2$.

## 8.2 CHARGE RENORMALIZATION IN ELECTRODYNAMICS

In quantum field theories, observables such as scattering amplitudes may be sensitive to higher-order corrections, in addition to Born diagrams. The modifications to lowest-order contributions are, in general, dependent upon kinematic variables. A convenient way of expressing an important class of these modifications is to introduce a so-called running coupling constant, which is to say an effective coupling strength that depends upon the kinematical circumstances. A phenomenon of this sort is, in fact, familiar in the classical electrodynamics of a polarizable substance. In the absence of a test charge, molecules within the medium will orient themselves to ensure that the substance is locally neutral, as shown in figure 8.10(a). A test charge placed in the medium will polarize the substance, attracting opposite charges and repelling like charges, as shown in figure 8.10(b). At any radius $r$ (larger than the molecular scale) from the test charge, the total charge enclosed within a sphere centered on the test charge will be smaller than the test charge itself. By Gauss's law, the effective charge will be smaller in magnitude than the test charge,

$$Q_{\text{eff}} = \frac{Q_{\text{test}}}{\epsilon},\tag{8.2.1}$$

where $\epsilon \geq 1$ is the dielectric constant of the medium. By this we mean that the electric field due to the test charge in the medium is related to what the field would have been in vacuo by

$$\mathbf{E}_{\text{medium}} = \frac{\mathbf{E}_{\text{vacuo}}}{\epsilon}.\tag{8.2.2}$$

At very short distances, closer to the test charge than the molecular size, screening cannot occur, and the effective charge is equal to the full magnitude of the test

(a)                                      (b)

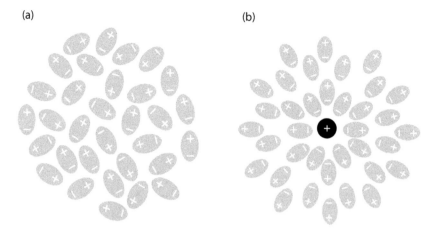

**Figure 8.10.** Polarization of a dielectric medium by a test charge. (a) Unpolarized medium. (b) Medium polarized by a positive test charge.

charge. Thus we may say that the effective electric charge increases at short distances. In quantum electrodynamics, the vacuum itself behaves as a polarizable medium in which virtual electron–positron pairs are polarized by the test charge. Consequently, the effective charge shows a similar dependence on the distance scale, which we shall now derive.

### 8.2.1 CHARGE RENORMALIZATION IN QUANTUM ELECTRODYNAMICS

The lowest-order manifestation of the vacuum polarization phenomenon occurs in the one-loop corrections to Coulomb scattering. For definiteness, we consider the scattering of an electron from an infinitely massive "proton," for which the Born diagram is shown in figure 8.11(a). The invariant amplitude is (see the Feynman rules in appendix B)

$$\mathcal{M} = -ie^2 \bar{u}(p')\gamma_\mu u(p)\frac{g^{\mu\nu}}{q^2}\bar{U}(P')\gamma_\nu U(P),\qquad(8.2.3)$$

for which the momentum transfer is

$$q = p - p'\qquad(8.2.4)$$

and $U$ denotes the proton spinors. In the limit of very large proton mass, these become

$$\left.\begin{array}{l}U_\lambda(P) = \sqrt{2M}\left(\begin{array}{c}\chi_\lambda\\0\end{array}\right)\\[12pt]\bar{U}_{\lambda'}(P') = \sqrt{2M}(\overline{\chi}_{\lambda'}, 0)\end{array}\right\},\qquad(8.2.5)$$

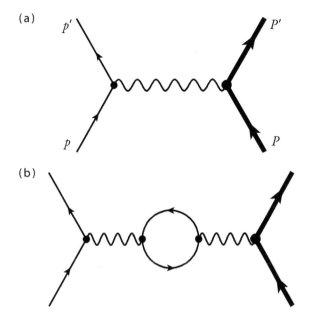

**Figure 8.11.** Feynman diagrams for Coulomb scattering. (a) Born term; (b) one-loop correction to the photon propagator.

so that

$$\bar{U}_{\lambda'}(P')\gamma^{\nu}U_{\lambda}(P) = 2M\delta_{\lambda\lambda'}\delta^{\nu 0}. \tag{8.2.6}$$

The last result follows because the $\gamma$-matrices mix large and small components. In this limit, the invariant amplitude takes the form

$$\mathcal{M} = -\frac{ie^2}{q^2}\bar{u}(p')\gamma_0 u(p) \cdot 2M, \tag{8.2.7}$$

which is precisely the form of a Coulomb interaction, namely, the interaction of the lepton current with a purely timelike potential.

Next we examine the contribution of a one-loop vacuum-polarization graph shown in figure 8.11(b). The invariant amplitude is given by

$$\mathcal{M}' = -\text{tr} \int \frac{d^4k}{(2\pi)^4}\left[(+ie)\bar{u}(p')\gamma_{\mu}u(p)\frac{-ig^{\mu\rho}}{q^2}(ie)\gamma_{\rho}\frac{i(\not{k}+m)}{k^2-m^2+i\varepsilon}\right.$$
$$\left. \times (ie)\gamma_{\sigma}\frac{i(\not{k}-\not{q}+m)}{(k-q)^2-m^2+i\varepsilon}\frac{-ig^{\nu\sigma}}{q^2}(-ie)\bar{U}(P')\gamma_{\nu}U(P)\right], \tag{8.2.8}$$

corresponding to the replacement in (8.2.3) of $-ig^{\mu\nu}/q^2$ by

$$\left(\frac{-i}{q^2}\right)I^{\mu\nu}\left(\frac{-i}{q^2}\right), \tag{8.2.9}$$

where

$$
\begin{aligned}
I^{\mu\nu} &= -\mathrm{tr} \int \frac{d^4k}{(2\pi)^4} (ie)^2 \gamma^\mu \frac{i(\slashed{k}+m)}{k^2 - m^2 + i\varepsilon} \gamma^\nu \frac{i(\slashed{k}-\slashed{q}+m)}{(k-q)^2 - m^2 + i\varepsilon} \\
&= e^2 \mathrm{tr} \int \frac{d^4k}{(2\pi)^4} \gamma^\mu \frac{i(\slashed{k}+m)}{k^2 - m^2 + i\varepsilon} \gamma^\nu \frac{i(\slashed{k}-\slashed{q}+m)}{(k-q)^2 - m^2 + i\varepsilon} .
\end{aligned} \tag{8.2.10}
$$

The closed-loop integral diverges for $k \to \infty$, and its evaluation thus will require some cutoff procedure. To proceed further, it is convenient to bring the integrand to an exponential form, in order that gaussian methods may eventually be used. With the aid of the identity

$$
\frac{i(\slashed{k}+m)}{k^2 - m^2 + i\varepsilon} = (\slashed{k}+m) \int_0^\infty dz\, e^{iz(k^2 - m^2 + i\varepsilon)} , \tag{8.2.11}
$$

we may write

$$
\begin{aligned}
I^{\mu\nu} &= e^2 \int \frac{d^4k}{(2\pi)^4} \int_0^\infty dz_1 \int_0^\infty dz_2 \, (\mathrm{tr}[\gamma^\mu(\slashed{k}+m)\gamma^\nu(\slashed{k}-\slashed{q}+m)] \\
&\quad \times \exp\{iz_1(k^2 - m^2 + i\varepsilon) + iz_2[(k-q)^2 - m^2 + i\varepsilon]\}) .
\end{aligned} \tag{8.2.12}
$$

A convenient, and gauge-invariant, cutoff procedure is the method of regulators due to Pauli and Villars [16], which consists in adding to the Lagrangian unphysical fields with adjustable masses that are taken to infinity at the end of the calculation. The procedure succeeds if observables are finite and independent of the cutoff masses. In cases such as the one before us, this amounts to the replacement

$$
\begin{aligned}
I^{\mu\nu}(q^2, m^2) \to \bar{I}^{\mu\nu}(q^2) &= I^{\mu\nu}(q^2, m^2) + \sum_{i=1}^N a_i I^{\mu\nu}(q^2, M_i^2) \\
&= \sum_{i=0}^N a_i I^{\mu\nu}(q^2, M_i^2),
\end{aligned} \tag{8.2.13}
$$

where the number and coefficients of the regulator terms are chosen to render the loop integral convergent and, evidently, $a_0 = 1$, $M_0^2 = m^2$. Evaluation of integrals of the general form (8.2.12) will be facilitated by completing the square in the exponential. Because the final expression will be convergent, we are free to introduce the new integration variable

$$
l = \frac{kz_1 + (k-q)z_2}{z_1 + z_2} , \tag{8.2.14}
$$

in terms of which

$$
\begin{aligned}
k &= l + qz_2/(z_1 + z_2), \\
k - q &= l - qz_1/(z_1 + z_2).
\end{aligned} \tag{8.2.15}
$$

Carrying out the trace, completing the square in the exponential, and inverting the order of integration, we have

$$
I^{\mu\nu}(q^2, m^2) = 4e^2 \int_0^\infty dz_1 \int_0^\infty dz_2 \exp\left\{ i \left[ \frac{q^2 z_1 z_2}{z_1 + z_2} - (m^2 - i\varepsilon)(z_1 + z_2) \right] \right\}
$$
$$
\times \int \frac{d^4 l}{(2\pi)^4} e^{il^2(z_1+z_2)} \left( 2l^\mu l^\nu - (l^\mu q^\nu + l^\nu q^\mu) \frac{z_1 - z_2}{z_1 + z_2} - 2q^\mu q^\nu \frac{z_1 z_2}{(z_1 + z_2)^2} \right.
$$
$$
\left. + g^{\mu\nu} \left\{ m^2 - \left[ l^2 - l \cdot q \frac{z_1 - z_2}{z_1 + z_2} - q^2 \frac{z_1 z_2}{(z_1 + z_2)^2} \right] \right\} \right). \tag{8.2.16}
$$

The momentum integrals are conveniently organized according to powers of $l$ as

$$
\mathcal{I}_2^{\mu\nu} = \int \frac{d^4 l}{(2\pi)^4} e^{il^2(z_1+z_2)} (2l^\mu l^\nu - g^{\mu\nu} l^2), \tag{8.2.17}
$$

$$
\mathcal{I}_1^{\mu\nu} = \int \frac{d^4 l}{(2\pi)^4} e^{il^2(z_1+z_2)} (g^{\mu\nu} l \cdot q - l^\mu q^\nu - l^\nu q^\mu) \frac{z_1 - z_2}{z_1 + z_2}, \tag{8.2.18}
$$

$$
\mathcal{I}_0^{\mu\nu} = \int \frac{d^4 l}{(2\pi)^4} e^{il^2(z_1+z_2)} \left[ (g^{\mu\nu} q^2 - 2q^\mu q^\nu) \frac{z_1 z_2}{(z_1 + z_2)^2} + g^{\mu\nu} m^2 \right]. \tag{8.2.19}
$$

By symmetric integration [cf. (B.3.11a), (B.3.17)], $\mathcal{I}_1^{\mu\nu}$ vanishes and the other pieces become

$$
\mathcal{I}_2^{\mu\nu} = \frac{1}{16\pi^2 i (z_1 + z_2)^2} \frac{ig^{\mu\nu}}{2(z_1 + z_2)} (2 - 4)
$$
$$
= \frac{-g^{\mu\nu}}{16\pi^2 (z_1 + z_2)^3} \tag{8.2.20}
$$

and

$$
\mathcal{I}_0^{\mu\nu} = \frac{1}{16\pi^2 i (z_1 + z_2)^2} \left\{ 2(g^{\mu\nu} q^2 - q^\mu q^\nu) \frac{z_1 z_2}{(z_1 + z_2)^2} + g^{\mu\nu} \left[ m^2 - \frac{q^2 z_1 z_2}{(z_1 + z_2)^2} \right] \right\}. \tag{8.2.21}
$$

We therefore obtain

$$
I^{\mu\nu}(q^2, m^2) = \frac{-i\alpha}{\pi} \int_0^\infty dz_1 \int_0^\infty dz_2 \frac{1}{(z_1+z_2)^2} \exp\left\{ i \left[ \frac{q^2 z_1 z_2}{z_1+z_2} - (m^2 - i\varepsilon)(z_1+z_2) \right] \right\}
$$
$$
\times \left\{ 2(g^{\mu\nu} q^2 - q^\mu q^\nu) \frac{z_1 z_2}{(z_1 + z_2)^2} + g^{\mu\nu} \left[ m^2 - \frac{q^2 z_1 z_2}{(z_1 + z_2)^2} - \frac{i}{z_1 + z_2} \right] \right\}. \tag{8.2.22}
$$

The first term in braces has been constructed to satisfy the current-conservation requirement

$$
q_\mu I^{\mu\nu} = 0 = q_\nu I^{\mu\nu}. \tag{8.2.23}
$$

The term proportional to $g^{\mu\nu}$ does not obey this condition but can be shown to vanish identically, if we note that its contribution to $I^{\mu\nu}(q^2)$ may be rewritten as

$$\Delta^{\mu\nu} = -\frac{i\alpha}{\pi} g^{\mu\nu} \int_0^\infty dz_1 \int_0^\infty dz_2 \frac{1}{(z_1 + z_2)^3} i\sigma \frac{\partial}{\partial\sigma} \frac{1}{\sigma}$$
$$\times \sum_{i=0}^N a_i \exp\left\{ i\sigma \left[ \frac{q^2 z_1 z_2}{z_1 + z_2} - (M_i^2 - i\varepsilon)(z_1 + z_2) \right] \right\}_{\sigma=1}. \qquad (8.2.24)$$

Because the integrals are convergent, we may interchange the order of the integrations and the differentiation $\partial/\partial\sigma$, so that

$$\Delta^{\mu\nu} = \frac{\alpha}{\pi} g^{\mu\nu} \sigma \frac{\partial}{\partial\sigma} \left( \int_0^\infty dz_1 \int_0^\infty dz_2 \frac{1}{(z_1 + z_2)^3 \sigma} \right.$$
$$\left. \times \sum_{i=0}^N a_i \exp\left\{ i\sigma \left[ \frac{q^2 z_1 z_2}{z_1 + z_2} - (M_i^2 - i\varepsilon)(z_1 + z_2) \right] \right\} \right)_{\sigma=1}. \qquad (8.2.25)$$

If we now rescale the integration variables as

$$z_i \to \frac{z_i}{\sigma}, \qquad (8.2.26)$$

the expression in parentheses is seen to be independent of $\sigma$, so the contribution $\Delta^{\mu\nu}$ vanishes.

We are left with the task of evaluating

$$\bar{I}^{\mu\nu}(q^2) = -\frac{2i\alpha}{\pi} (g^{\mu\nu} q^2 - q^\mu q^\nu) \int_0^\infty dz_1 \int_0^\infty dz_2 \frac{z_1 z_2}{(z_1 + z_2)^4}$$
$$\times \sum_{i=0}^N a_i \exp\left\{ i\sigma \left[ \frac{q^2 z_1 z_2}{z_1 + z_2} - (M_i^2 - i\varepsilon)(z_1 + z_2) \right] \right\}. \qquad (8.2.27)$$

This is conveniently accomplished by rescaling the integration variables. Introduce the factor

$$1 = \int_0^\infty d\sigma \, \delta(\sigma - x_1 - x_2) \qquad (8.2.28)$$

under the integral signs and let $z_i \to \sigma z_i$, so that

$$\bar{I}^{\mu\nu}(q^2) = \frac{2i\alpha}{\pi} (q^\mu q^\nu - g^{\mu\nu} q^2) \int_0^\infty dz_1 \int_0^\infty dz_2 \, z_1 z_2 \delta(1 - z_1 - z_2)$$
$$\times \int_0^\infty \frac{d\sigma}{\sigma} \sum_{i=0}^N a_i \exp[i\sigma(q^2 z_1 z_2 - M_i^2 + i\varepsilon)], \qquad (8.2.29)$$

which appears logarithmically divergent at the point $\sigma = 0$. This mild divergence can be removed with a single regulator, by the choice

$$a_1 = -1, \qquad M_1^2 = \Lambda^2 \gg m^2, \, q^2. \qquad (8.2.30)$$

Now, for $q^2 < 4m^2$, we find

$$
\begin{aligned}
\bar{I}^{\mu\nu}(q^2) &= I^{\mu\nu}(q^2, m^2) - I^{\mu\nu}(q^2, \Lambda^2) \\
&= \frac{2i\alpha}{\pi}(q^\mu q^\nu - g^{\mu\nu}q^2) \int_0^1 dz\, z(1-z) \ln\left[\frac{\Lambda^2}{m^2 - q^2 z(1-z)}\right] \\
&= \frac{i\alpha}{3\pi}(q^\mu q^\nu - g^{\mu\nu}q^2)\left\{ \ln\left(\frac{\Lambda^2}{m^2}\right) \right. \\
&\qquad \left. - 6\int_0^1 dz\, z(1-z)\ln\left[1 - \frac{q^2 z(1-z)}{m^2}\right] \right\}.
\end{aligned}
\tag{8.2.31}
$$

Returning to the problem of original interest, the scattering of an electron in a Coulomb field, we note that the $q^\mu q^\nu$ factor does not contribute to the scattering amplitude by virtue of current conservation at the electron vertex (or, more operationally, because of the Dirac equation). Hence, the modified photon propagator is

$$
\begin{aligned}
\frac{-ig^{\mu\nu}}{q^2} + \left(\frac{-i}{q^2}\right)\bar{I}^{\mu\nu}\left(\frac{-i}{q^2}\right) &= \frac{-ig^{\mu\nu}}{q^2}\left\{ 1 - \frac{\alpha}{3\pi}\ln\left(\frac{\Lambda^2}{m^2}\right) \right. \\
&\qquad \left. + \frac{2\alpha}{\pi}\int_0^1 dz\, z(1-z)\ln\left[1 - \frac{q^2 z(1-z)}{m^2}\right] \right\}.
\end{aligned}
\tag{8.2.32}
$$

A number of limiting cases are worth comment. For $q^2 \to 0$, the long-wavelength limit, the propagator is modified by the factor

$$
1 - \frac{\alpha}{3\pi}\ln\left(\frac{\Lambda^2}{m^2}\right) \equiv Z_3,
\tag{8.2.33}
$$

so that the Coulomb scattering amplitude is changed by the same factor. In other words, the electric charge measured in Coulomb scattering is not the "bare" charge $e$ that appears in the Lagrangian, but a renormalized quantity $e_R$ that differs by a factor $\sqrt{Z_3}$. The term $(-\alpha/3\pi)\ln(\Lambda^2/m^2)$ is but the first term in the renormalization of the electric charge, which can be carried out systematically to all orders in perturbation theory. We accommodate this by replacing the bare charge $e$ by the renormalized charge $e_R$ in all our expressions for observables, although it will usually be convenient to omit the subscript. It is thus the renormalized charge that is measured in experiments as $e_R^2 \approx 4\pi/137$.

Of greater interest for us in what follows is the modification to the Coulomb scattering amplitude for large spacelike values of $q^2$, such that $-q^2 \gg m^2$. In this circumstance, the remaining integral in (8.2.32) may be approximated by

$$
\frac{2\alpha}{\pi}\int_0^1 dz\, z(1-z)\ln\left(\frac{-q^2}{m^2}\right) = \frac{\alpha}{3\pi}\ln\left(\frac{-q^2}{m^2}\right).
\tag{8.2.34}
$$

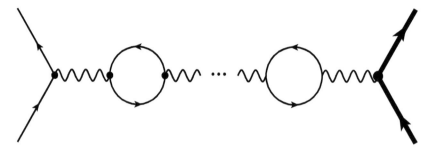

**Figure 8.12.** Leading-logarithmic corrections to the photon propagator in spinor electrodynamics.

As a result, the photon propagator is given by

$$\frac{-ig^{\mu\nu}}{q^2}\left[1 - \frac{\alpha}{3\pi}\ln\left(\frac{\Lambda^2}{m^2}\right) + \frac{\alpha}{3\pi}\ln\left(\frac{-q^2}{m^2}\right)\right],\qquad(8.2.35)$$

which amounts to the replacement of $\alpha_R$ in the scattering amplitude by the "running coupling constant"

$$\alpha_R(q^2) = \alpha_R(m^2)\left[1 + \frac{\alpha_R(m^2)}{3\pi}\ln\left(\frac{-q^2}{m^2}\right) + O(\alpha_R^2)\right].\qquad(8.2.36)$$

We see that to this approximation the effective electric charges increases logarithmically with $-q^2$, which is to say with improving resolution. This is the quantum-mechanical analogue of the macroscopic charge screening discussed at the beginning of this section.

For small values of $-q^2 \ll m^2$, it is straightforward (see problem 8.9) to derive the modification to Coulomb's law in position space that results from vacuum polarization. The additional term in the force law, first calculated by Uehling [17] in 1935, induces a shift in the $s$-wave energy levels of atoms that is especially significant for high-$Z$ muonic atoms.

It is possible to continue the study of modifications to Coulomb's law in higher orders of perturbation theory. For this purpose, the most important class of Feynman diagrams is the "sum of bubbles" indicated in figure 8.12, which corresponds to the most divergent set of logarithms. It does not require detailed calculation to see that for $-q^2 \gg m^2$, the running coupling constant will be given by

$$\alpha_R(q^2) = \alpha_R(m^2)\left\{1 + \frac{\alpha_R(m^2)}{3\pi}\ln\left(\frac{-q^2}{m^2}\right) + \left[\frac{\alpha_R(m^2)}{3\pi}\ln\left(\frac{-q^2}{m^2}\right)\right]^2 + \cdots\right\}$$

$$\rightarrow \frac{\alpha_R(m^2)}{1 - \frac{\alpha_R(m^2)}{3\pi}\ln\left(\frac{-q^2}{m^2}\right)},\qquad(8.2.37)$$

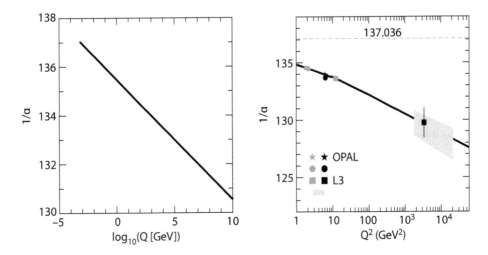

**Figure 8.13.** Evolution of the fine structure constant. Left panel: evolution of $1/\alpha$ computed in pure QED, the theory of photons and electrons alone, given by (8.2.39). Right panel: Evolution computed taking into account all charged particles (solid line), compared with LEP measurements (from Ref. [18]).

where we have identified

$$\sum_{n=0}^{\infty} x^n = 1/(1-x),\tag{8.2.38}$$

without worrying about convergence issues. To this approximation, it is convenient to write

$$\frac{1}{\alpha_R(q^2)} = \frac{1}{\alpha_R(m^2)} - \left(\frac{1}{3\pi}\right)\ln\left(-\frac{q^2}{m^2}\right)\tag{8.2.39}$$

and to represent the $q^2$-evolution of the coupling constant by the graph shown in the left panel of figure 8.13.

Pure QED, the theory of photons and electrons, describes nature only over a finite (but highly important) range of energies. For momentum transfers greater than a few hundred MeV, the evolution of $1/\alpha$ will be influenced by other virtual particles, beginning with the muon. The solid line in the right panel of figure 8.13 shows the result of such a calculation [19]. Values of $\alpha(Q^2)$ extracted from measurements of the reaction $e^+e^- \to e^+e^-$ display the predicted behavior.

This brief calculation has shown how a running coupling constant may arise in an interacting field theory. We have not, in this limited discussion, shown how the renormalization program may be carried through in detail, nor have we justified the implicit claim of the foregoing discussion that the effects of vacuum polarization can be accounted for systematically by replacing $\alpha$ by $\alpha(q^2)$ as computed from the bubble sum of the photon propagator. Those issues are well treated in the standard textbooks on QED, and the reader who has not yet experienced the renormalization

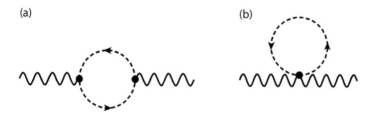

(a)    (b)

**Figure 8.14.** One-loop corrections to the photon propagator in scalar electrodynamics.

program is counseled to do so, pen in hand, although it will not be required for what follows.

## 8.2.2 CHARGE RENORMALIZATION IN SCALAR ELECTRODYNAMICS

Although the principal interest in this chapter is QCD and, among other things, the comportment of the running coupling constant in that theory, it will be useful to spend a little more time on electrodynamics. We consider the evolution of the running coupling constant in scalar electrodynamics, both because the result is interesting and instructive and also in order to introduce a second regularization technique that is better suited to non-Abelian gauge theories than the method of Pauli and Villars.

Applying the Feynman rules for scalar electrodynamics given in appendix B.5, we find that the lowest-order corrections to the photon propagator are the two diagrams shown in figure 8.14. In the same notation employed previously in this section, we wish to evaluate

$$I^{\mu\nu}(q^2, m^2) = e^2 \int \frac{d^4k}{(2\pi)^4} \left\{ \frac{(2k-q)^\mu (2k-q)^\nu}{(k^2 - m^2 + i\varepsilon)[(k-q)^2 - m^2 + i\varepsilon]} - \frac{2g^{\mu\nu}}{k^2 - m^2 + i\varepsilon} \right\},$$

$$(8.2.40)$$

where the first term corresponds to the bubble of figure 8.14(a) and the second, to the contact term of figure 8.14(b). We combine the terms over a common denominator as

$$I^{\mu\nu}(q^2, m^2) = e^2 \int \frac{d^4k}{(2\pi)^4} \frac{(2k-q)^\mu (2k-q)^\nu - 2g^{\mu\nu}[(k-q)^2 - m^2 + i\varepsilon]}{(k^2 - m^2 + i\varepsilon)[(k-q)^2 - m^2 + i\varepsilon]} \quad (8.2.41)$$

and promote the denominators to an exponent by means of the identity

$$i(k^2 - m^2 + i\varepsilon)^{-1} = \int_0^\infty dz \, e^{iz(k^2 - m^2 + i\varepsilon)}. \quad (8.2.42)$$

The result is

$$I^{\mu\nu}(q^2, m^2) = -e^2 \int \frac{d^4k}{(2\pi)^4} \int_0^\infty dz_1 \int_0^\infty dz_2 \Big\{ (2k-q)^\mu (2k-q)^\nu$$
$$-2g^{\mu\nu}[(k-q)^2 - m^2 + i\varepsilon] \Big\} \tag{8.2.43}$$
$$\times \exp\{iz_1(k^2 - m^2 + i\varepsilon) + iz_2[(k-q)^2 - m^2 + i\varepsilon]\}.$$

We complete the square in the exponential on introducing the new variable

$$l = \frac{kz_1 + (k-q)z_2}{z_1 + z_2} \tag{8.2.14}$$

and regularize the integral by changing the dimensionality of spacetime to $n$. We are left with

$$I^{\mu\nu}(q^2, m^2) = -e^2 \int_0^\infty dz_1 \int_0^\infty dz_2 \exp\left\{ i \left[ \frac{q^2 z_1 z_2}{z_1 + z_2} - (m^2 - i\varepsilon)(z_1 + z_2) \right] \right\}$$
$$\times \int \frac{d^n l}{(2\pi)^n} e^{il^2(z_1 + z_2)} \left\{ 4l^\mu l^\nu + \frac{(z_1 - z_2)^2}{(z_1 + z_2)^2} q^\mu q^\nu \right.$$
$$\left. - 2g^{\mu\nu} \left[ l^2 - m^2 + \frac{q^2 z_1^2}{(z_1 + z_2)^2} \right] \right\}, \tag{8.2.44}$$

where odd powers of $l$ have been omitted in anticipation of symmetric integration, and the order of integrations has been interchanged. As we did for spinor QED, we may rearrange the momentum integral as

$$\int \frac{d^n l}{(2\pi)^n} e^{il^2(z_1 + z_2)} \left( \frac{(z_1 - z_2)^2}{(z_1 + z_2)^2} (q^\mu q^\nu - g^{\mu\nu} q^2) \right.$$
$$\left. + 4l^\mu l^\nu - 2g^{\mu\nu} \left\{ l^2 - m^2 + q^2 \frac{[z_1^2 - (z_1 - z_2)^2/2]}{(z_1 + z_2)^2} \right\} \right). \tag{8.2.45}$$

The first term in large parentheses satisfies the requirements of current conservation, and we shall evaluate its contribution presently. First, we show by the 't Hooft–Veltman method of dimensional regularization [20] that the remaining terms do not, in fact, contribute. Using the expressions for $n$-dimensional gaussian integrals given in appendix B.4, we may write

$$\int \frac{d^n l}{(2\pi)^n} e^{il^2(z_1 + z_2)} \left( 4l^\mu l^\nu - 2l^2 g^{\mu\nu} + g^{\mu\nu} \left\{ 2m^2 - 2\frac{[z_1^2 - (z_1 - z_2)^2/2]}{(z_1 + z_2)^2} q^2 \right\} \right)$$
$$= \frac{ie^{-i\pi n/4} g^{\mu\nu}}{[4\pi(z_1 + z_2)]^{n/2}} \left\{ \frac{in(4/n - 2)}{2(z_1 + z_2)} + 2m^2 - \frac{2[z_1^2 - (z_1 - z_2)^2/2]}{(z_1 + z_2)^2} q^2 \right\}. \tag{8.2.46}$$

The contribution to $I^{\mu\nu}$ is, therefore.

$$
\Delta^{\mu\nu} = \frac{ie^2 e^{-i\pi n/4}}{(4\pi)^{n/2}} g^{\mu\nu} \int_0^\infty dz_1 \int_0^\infty dz_2 \exp\left\{i\left[\frac{q^2 z_1 z_2}{z_1 + z_2} - (m^2 - i\varepsilon)(z_1 + z_2)\right]\right\}
$$
$$
\times \frac{2}{(z_1 + z_2)^{1+n/2}} \left\{i\left(1 - \frac{n}{2}\right) + m^2(z_1 + z_2) - \frac{[z_1^2 - (z_1 - z_2)^2/2]q^2}{z_1 + z_2}\right\}.
$$
$$(8.2.47)$$

Introducing a scaling parameter $\sigma$ as before, we have

$$
\Delta^{\mu\nu} = \frac{-2e^2 e^{-i\pi n/4}}{(4\pi)^{n/2}} g^{\mu\nu} \int_0^\infty dz_1 \int_0^\infty dz_2\, \delta(1 - z_1 - z_2) \int_0^\infty \frac{d\sigma}{\sigma^{n/2-1}}
$$
$$
\times \exp[i\sigma(q^2 z_1 z_2 - m^2 + i\varepsilon)] \times \left[\frac{(1 - n/2)}{\sigma} + i(z_1 z_2 q^2 - m^2)\right],
$$
$$(8.2.48)$$

in which, in anticipation of the restrictions on the $z_1$ and $z_2$ integrations, we have replaced $z_1 + z_2$ by 1 and dropped terms that are odd in $(1 - 2z_i)$. The integral over $\sigma$ may be expressed as

$$
\int_0^\infty d\sigma\, \frac{\partial}{\partial\sigma}\{\sigma^{1-n/2} \exp[i\sigma(q^2 z_1 z_2 - m^2 + i\varepsilon)]\},
$$
$$(8.2.49)$$

which is convergent and equal to zero for sufficiently small values of $n$, namely, $n < 2$. Thus, by dimensional regularization, we obtain the anticipated gauge-invariant result,

$$
\Delta^{\mu\nu} = 0.
$$
$$(8.2.50)$$

Let us return to the gauge-invariant contribution. We have

$$
I^{\mu\nu} = \frac{ie^2 e^{-i\pi n/4}}{(4\pi)^{n/2}} (q^\mu q^\nu - q^2 g^{\mu\nu}) \int_0^\infty dz_1 \int_0^\infty dz_2
$$
$$
\times \exp\left\{i\left[\frac{q^2 z_1 z_2}{z_1 + z_2} - (m^2 - i\varepsilon)(z_1 + z_2)\right]\right\} \frac{(z_1 - z_2)^2}{(z_1 + z_2)^{2+n/2}}
$$
$$
= \frac{ie^2 e^{-i\pi n/4}}{(4\pi)^{n/2}} (q^\mu q^\nu - q^2 g^{\mu\nu}) \int_0^\infty dz_1 \int_0^\infty dz_2 \delta(1 - z_1 - z_2) \int_0^\infty \frac{d\sigma}{\sigma^{n/2-1}}
$$
$$
\times \exp[i\sigma(q^2 z_1 z_2 - m^2 + i\varepsilon)](z_1 - z_2)^2
$$
$$
= \frac{ie^2 e^{-i\pi n/4}}{(4\pi)^{n/2}} (q^\mu q^\nu - q^2 g^{\mu\nu}) \int_0^1 dz(1 - 2z)^2
$$
$$
\times \int_0^\infty \frac{d\sigma}{\sigma^{n/2-1}} \exp\{i\sigma[q^2 z(1 - z) - m^2 + i\varepsilon]\}.
$$
$$(8.2.51)$$

The $\sigma$-integration produces a gamma function, so that, finally, we have

$$I^{\mu\nu} = \frac{ie^2 \, e^{-i\pi n/4}}{(4\pi)^{n/2}} (q^\mu q^\nu - q^2 g^{\mu\nu}) \Gamma\left(2 - \frac{n}{2}\right) \int_0^1 \frac{dz \, (1 - 2z)^2}{[q^2 z(1 - z) - m^2 + i\varepsilon]^{2-n/2}},$$

(8.2.52)

which has simple poles at $n = 4, 6, 8, \ldots$. To regularize the integral, we subtract from (8.2.52) the pole at $n = 4$ with its residue, given by

$$I^{\mu\nu}_{\text{pole}} = \frac{-ie^2}{16\pi^2} (q^\mu q^\nu - q^2 g^{\mu\nu}) \frac{2}{4 - n} \int_0^1 dz \, (1 - 2z)^2.$$

(8.2.53)

Subtracting the pole term and taking the limit $n \to 4$ in the regularized expression, we obtain

$$\bar{I}^{\mu\nu} = \lim_{n \to 4} (I^{\mu\nu} - I^{\mu\nu}_{\text{pole}}) = \frac{-ie^2}{16\pi^2} (q^\mu q^\nu - q^2 g^{\mu\nu})$$
$$\times \left\{ \int_0^1 dz \, (1 - 2z)^2 \ln\left[1 - \frac{q^2 z(1 - z)}{m^2}\right] + K \right\},$$

(8.2.54)

where the constant $K$ arises from the $n$-dependence other than in the denominator of the integrand in (8.2.52). It is arbitrary in the same way as the Pauli–Villars cutoff is arbitrary and contributes to the overall charge renormalization.

Let us compare the $q^2$-evolution of the coupling constant with the result we found in spinor electrodynamics. For $-q^2 \gg m^2$, we have that

$$\bar{I}^{\mu\nu} = \frac{-i\alpha}{4\pi} (q^\mu q^\nu - q^2 g^{\mu\nu}) \left[ K + \ln\left(\frac{-q^2}{m^2}\right) \int_0^1 dz \, (1 - 2z)^2 \right]$$

$$= \frac{-i\alpha}{4\pi} (q^\mu q^\nu - q^2 g^{\mu\nu}) \left[ K + \ln\left(\frac{-q^2}{m^2}\right) \left(\frac{1}{3}\right) \right].$$

(8.2.55)

Forming the photon propagator as in (8.2.32) and using current conservation at the scalar vertex, we find, for the running renormalized coupling constant in scalar electrodynamics,

$$\alpha_R(q^2) = \alpha_R(m^2) \left[ 1 + \frac{\alpha_R(m^2)}{12\pi} \ln\left(\frac{-q^2}{m^2}\right) + \cdots \right],$$

(8.2.56)

which is to be compared with the expression (8.2.36) obtained in spinor electrodynamics. The rate of change of the coupling constant in scalar electrodynamics is one-fourth of that in spinor electrodynamics. This is precisely the ratio of the cross section for the reactions $e^+ e^- \to \sigma^+ \sigma^-$ and $e^+ e^- \to \mu^+ \mu^-$ evaluated in problems 1.4 and 1.5, which are directly related to the same polarization tensor.

It is of some importance that, whereas the variation in the effective coupling constant is logarithmically divergent, the quantity $\partial \alpha(q^2)/\partial(\ln q^2)$ is finite. This tame

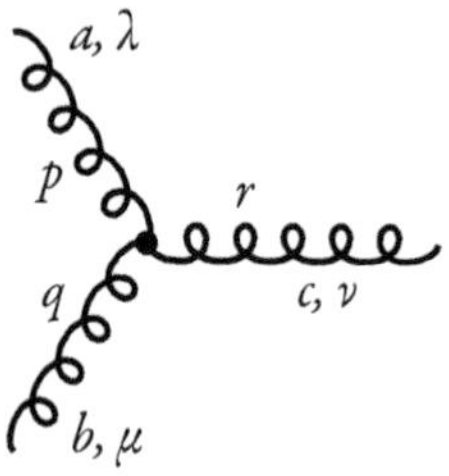

$$-gf^{abc}\left[(p-q)_{\nu}g_{\lambda\mu}+(q-r)_{\lambda}g_{\mu\nu}+(r-p)_{\mu}g_{\lambda\nu}\right]$$

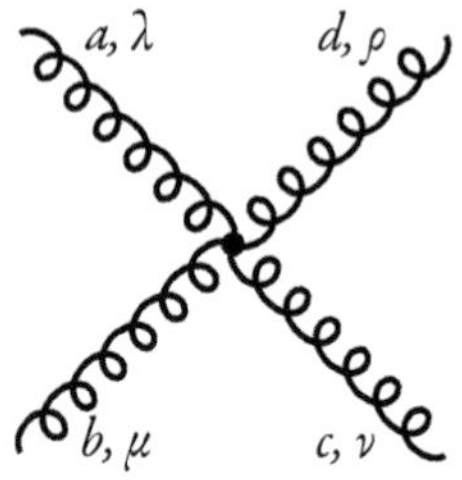

$$-ig^2\left[f^{abe}f^{cde}(g_{\lambda\nu}g_{\mu\rho}-g_{\lambda\rho}g_{\mu\nu})\right.$$
$$+f^{ace}f^{bde}(g_{\lambda\mu}g_{\nu\rho}-g_{\lambda\rho}g_{\mu\nu})$$
$$\left.+f^{ade}f^{bce}(g_{\lambda\mu}g_{\nu\rho}-g_{\lambda\nu}g_{\mu\rho})\right]$$

**Figure 8.15.** Feynman rules for multigluon interactions in non-Abelian gauge theories. All gluon momenta are incoming in the three-gluon vertex, namely, $p+q+r=0$.

behavior, when combined with the observation that the effects of renormalization are multiplicative in QED, permits the introduction of the powerful general techniques of the renormalization group.

We now investigate the running coupling constant in QCD. This will lead us to the notion of asymptotic freedom and the prospect of reliable perturbative calculations for the strong interactions.

## 8.3 THE RUNNING COUPLING CONSTANT IN QCD

In order to discuss the effects of higher-order corrections in QCD, we must first develop the Feynman rules for the theory in a way that leads to consistent results for diagrams containing closed gluon loops. There is no particular difficulty in evaluating the multigluon vertex factors shown [21] in figure 8.15, but some technical matters must be confronted. These arise from the familiar problem in gauge theories that the gauge-invariant Lagrangian does not uniquely determine the gauge field in terms of a source. In our treatment of electrodynamics in section 3.6, we resolved this ambiguity in the definition of the photon propagator by imposing the covariant gauge condition

$$\partial_{\mu}A^{\mu}=0. \tag{8.3.1}$$

The resulting form of the photon propagator,

$$\frac{-ig^{\mu\nu}}{q^2+i\varepsilon}, \tag{8.3.2}$$

and associated Feynman rules are convenient for many, but by no means all, computations in electrodynamics. Equivalently, in the case of QED, we may modify the source-free Lagrangian by adding to it a "gauge-fixing" term $(-1/2\xi)(\partial_\mu A^\mu)^2$, so that

$$\mathcal{L}_{\text{photon}} = -\frac{1}{4} F_{\mu\nu} F^{\mu\nu} - \left(\frac{1}{2\xi}\right) (\partial_\mu A^\mu)^2. \qquad (8.3.3)$$

The Euler-Lagrange equations then lead to the equations of motion

$$\Box A^\nu - (1 - \xi^{-1}) \partial^\nu (\partial_\mu A^\mu) = J^\nu, \qquad (8.3.4)$$

from which the photon propagator is

$$\frac{-i \left[ g^{\mu\nu} + (\xi - 1) q^\mu q^\nu / (q^2 + i\varepsilon) \right]}{q^2 + i\varepsilon}. \qquad (8.3.5)$$

This is entirely parallel to the gauge-fixing procedure for spontaneously broken gauge theories such as the Abelian Higgs model mentioned in section 5.3. Again, we recognize the well-known special cases of Feynman (or Lorenz) gauge for $\xi = 1$ and the Landau gauge for $\xi = 0$. The parameter $\xi$ is a reminder of the arbitrariness of the gauge-fixing condition. It must be absent from any physical quantity we calculate.

It is important to remark that the Lagrangian (8.3.3) remains gauge invariant provided that the local phase rotation that generates the gauge transformation

$$A_\mu(x) \to A_\mu(x) - \partial_\mu \alpha(x) \qquad (3.5.14)$$

satisfies the condition

$$\Box \alpha(x) = 0. \qquad (8.3.6)$$

Maintaining the gauge invariance is important in order that the physical requirements of local current conservation and transversality of the gauge fields remain in force. The covariant gauges are not the only possible choices, nor are they the most convenient for every sort of calculation. They will prove adequate to our limited needs, however, and will lead us to acknowledge an important technical issue, as we shall now see.

The preceding discussion would seem to apply, mutatis mutandis, to the propagator for any massless gauge field. In particular for the gluon propagator, we need only add a Kronecker delta in color space, so that the gluon propagator is

$$\frac{-i\delta_{jk} \left[ g^{\mu\nu} + (\xi - 1) q^\mu q^\nu / (q^2 + i\varepsilon) \right]}{q^2 + i\varepsilon}. \qquad (8.3.7)$$

The use of this form indeed leads to consistent results for tree diagrams. Because of the non-Abelian nature of the color-gauge transformations

$$b_\mu^l \to b_\mu'^l = b_\mu^l - \left(\frac{1}{g}\right) \partial_\mu \alpha^l - f_{jkl} \alpha^j b^k \qquad (8.3.8)$$

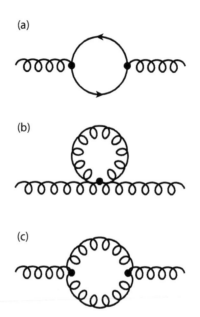

(a)

(b)

(c)

**Figure 8.16.** One-loop corrections to the gluon propagator in QCD.

upon the gluon field, it is no longer possible to find a condition analogous to (8.3.6) that preserves the gauge-invariant appearance of the Lagrangian. For diagrams involving closed gluon loops, this leads to problems that, at the technical level of this volume, are best encountered by direct computation. In this way we shall also see how the general cure solves our specific problems.

Consider the one-loop modifications to the gluon propagator, the analogue to the calculation discussed in section 8.2 for the photon. In addition to the quark-loop diagram of figure 8.16(a), there is the tadpole term of figure 8.16(b), which arises from the four-gluon vertex and the gluon loop shown in figure 8.16(c). We work in the dimensional regularization scheme to maintain gauge invariance. The quark propagator with momentum $q$, mass $m$, and incoming and outgoing color indices $\alpha$ and $\beta$ is simply transcribed from QED as

$$\frac{i\delta_{\alpha\beta}(\slashed{q}+m)}{q^2-m^2+i\varepsilon}. \tag{8.3.9}$$

Now, the quark-loop diagram may be evaluated directly from the corresponding QED expressions (8.2.31) and (8.2.34) simply by replacing

$$e^2 \to g^2 \operatorname{tr}\left(\frac{\lambda^a}{2}\frac{\lambda^b}{2}\right) = g^2\frac{\delta^{ab}}{2} \tag{8.3.10}$$

for each quark flavor to account for the difference in Feynman rules. In the short-wavelength limit, we find, for the logarithmically divergent part,

$$\bar{I}^{\mu\nu}_{\text{quarks}}(q^2) = \frac{-ig^2}{16\pi^2}(q^\mu q^\nu - q^2 g^{\mu\nu}) \cdot \frac{4}{3}\ln\left(\frac{-q^2}{\mu^2}\right)n_{\text{f}}\frac{\delta^{ab}}{2}, \tag{8.3.11}$$

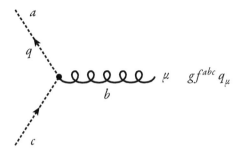

**Figure 8.17.** Feynman rule for the gluon–ghost–ghost interaction in QCD.

where $n_f$ is the number of quark flavors appearing in the loops and $-\mu^2$ is a spacelike renormalization point.

Within the dimensional regularization scheme, massless tadpole graphs do not contribute [20]. Hence, we have

$$\bar{I}^{\mu\nu}_{\text{tadpole}}(q^2) = 0. \tag{8.3.12}$$

Evaluation of the gluon loop is algebraically tedious but entirely parallel to the methods used for scalar electrodynamics in the previous section. The result is

$$\bar{I}^{\mu\nu}_{\text{gluons}}(q^2) = \frac{ig^2}{16\pi^2} f^{acd} f^{bcd} \left[ \frac{11}{6} q^\mu q^\nu - \frac{19}{12} q^2 g^{\mu\nu} \right.$$
$$\left. + \frac{(1-\xi)}{2} (q^\mu q^\nu - q^2 g^{\mu\nu}) \right] \ln\left( \frac{-q^2}{\mu^2} \right), \tag{8.3.13}$$

which is evidently not purely transverse—that is, not proportional to $(q^\mu q^\nu - q^2 g^{\mu\nu})$. This nontransversity is a reflection of the fact that the gauge-fixing term has broken the gauge invariance of the theory. A precursor of this effect is to be found in the results of problem 8.6, where the reaction $q\bar{q} \to gg$ is considered. There it will be seen that, in contrast to electrodynamics, the current-conservation conditions $k_{1\nu} T^{\mu\nu} = 0 = k_{2\mu} T^{\mu\nu}$, where $\epsilon^*_{1\nu} \epsilon^*_{2\mu} T^{\mu\nu}$ is the total amplitude corresponding to the tree diagrams, are satisfied only if the gluons are themselves transverse—that is, on-mass-shell. This is a much weaker condition than exists in electrodynamics and suggests the delicacy of the comportment of diagrams with loops.

A systematic resolution of this problem is beyond the elementary means that we have chosen to employ in this book. We shall simply state that such a resolution exists and that it may be characterized by the introduction of a color octet of fictitious scalar particles commonly known as Faddeev–Popov ghosts, which appear only in closed loops. These particles are called ghosts because, although spinless, they obey Fermi statistics; thus each closed ghost loop appears multiplied by an explicit minus sign, like a fermion loop. A ghost that carries momentum $q$ and incoming and outgoing color indices $\alpha$ and $\beta$ has the usual propagator for massless scalars, supplemented by a color factor,

$$\frac{\delta_{\alpha\beta}}{q^2 + i\varepsilon}. \tag{8.3.14}$$

Its interactions with gluons are given by the Feynman rule shown in figure 8.17.

**Figure 8.18.** Ghost-loop correction to the gluon propagator in QCD.

It is now straightforward (problem 8.10) to evaluate the modification to the gluon propagator given by the ghost loop shown in figure 8.18 using the method of dimensional regularization. It is

$$\bar{I}^{\mu\nu}_{\text{ghosts}}(q^2) = \frac{-ig^2}{16\pi^2} f^{acd} f^{bcd} \left( \frac{1}{6} q^\mu q^\nu + \frac{1}{12} q^2 g^{\mu\nu} \right) \ln \left( \frac{-q^2}{\mu^2} \right), \tag{8.3.15}$$

which is not transverse. The sum of the gluon-loop and ghost-loop contributions,

$$\bar{I}^{\mu\nu}_{\text{gluons}}(q^2) + \bar{I}^{\mu\nu}_{\text{ghosts}}(q^2) = \frac{ig^2}{16\pi^2} f^{acd} f^{bcd} \left[ \frac{5}{3} + \frac{(1-\xi)}{2} \right]$$
$$\times (q^\mu q^\nu - q^2 g^{\mu\nu}) \ln \left( \frac{-q^2}{\mu^2} \right), \tag{8.3.16}$$

is, however, transverse, as required. Thus we have verified that in this simplest case the Faddeev–Popov ghosts have performed the task for which they were introduced: the removal of unphysical polarizations from the contributions of gluon loops.

The modification to the gluon propagator in one-loop order is now given by

$$\bar{I}^{\mu\nu}(q^2) = \bar{I}^{\mu\nu}_{\text{quarks}}(q^2) + \bar{I}^{\mu\nu}_{\text{gluons}}(q^2) + \bar{I}^{\mu\nu}_{\text{ghosts}}(q^2) \tag{8.3.17}$$

$$= \frac{ig^2}{16\pi^2} \delta^{ab} \ln \left( \frac{-q^2}{\mu^2} \right) (q^\mu q^\nu - q^2 g^{\mu\nu}) \left\{ N \left[ \frac{5}{3} + \frac{(1-\xi)}{2} \right] - \frac{2n_f}{3} \right\},$$

which is explicitly gauge dependent and, thus, cannot alone describe the $q^2$-dependent evolution of observable quantities. In passing to (8.3.17) we have simplified the color factor** for the boson loops using

$$f^{acd} f^{bcd} \equiv C_A \delta^{ab} = N \delta^{ab}, \tag{8.3.18}$$

where $N$ is the number of colors—that is, the dimension of the fundamental representation of the $SU(N)_c$ gauge group. It remains to evaluate the one-loop corrections to the quark propagator, shown in figure 8.19(a), and the one-loop vertex correction, drawn in figure 8.19(b). The modified quark propagator may be taken over directly from quantum electrodynamics upon inserting the required color factor [cf. (8.3.10)]. It is

$$G(p) = \frac{i\not{p}}{p^2} \left[ \delta_{\alpha\beta} - \xi \frac{g^2}{16\pi^2} \ln \left( \frac{-p^2}{\mu^2} \right) \sum_a \frac{\lambda^a_{\alpha\gamma} \lambda^a_{\gamma\beta}}{4} \right], \tag{8.3.19}$$

** See appendix A.5 for a diagrammatic approach to the color algebra.

(a)

(b)

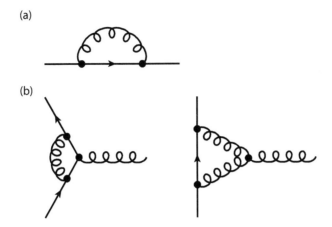

**Figure 8.19.** One-loop corrections to (a) the fermion propagator; (b) the quark–antiquark–gluon vertex in QCD.

where $\alpha$, $\beta$, $\gamma$ are color indices of the quarks and $a$ is the color index of the gluon. We may replace

$$\frac{1}{4}\sum_a \lambda^a_{\alpha\gamma}\lambda^a_{\gamma\beta} \equiv C_F \delta_{\alpha\beta} = \frac{N^2-1}{2N}\delta_{\alpha\beta} \qquad (8.3.20)$$

to obtain

$$G(p) = \frac{i\not{p}\,\delta_{\alpha\beta}}{p^2}\left[1 - \xi\frac{(N^2-1)}{2N}\frac{g^2}{16\pi^2}\ln\left(\frac{-p^2}{\mu^2}\right)\right]. \qquad (8.3.21)$$

The vertex modification involves a piece familiar from electrodynamics plus one involving the three-gluon vertex. The one-loop-corrected vertex is

$$\Gamma^{\alpha\beta;a}_\mu(q^2) = -ig\gamma_\mu\left(\frac{\lambda^a_{\alpha\beta}}{2} - \frac{g^2}{16\pi^2}\ln\left(\frac{-q^2}{\mu^2}\right)\right.$$
$$\left.\times\left\{\xi\frac{(N^2-1)}{2N}\frac{\lambda^a_{\alpha\beta}}{2} - \frac{i}{2}\left[1 - \frac{(1-\xi)}{4}\right]f^{abc}\lambda^b_{\alpha\gamma}\lambda^c_{\gamma\beta}\right\}\right), \qquad (8.3.22)$$

which may be simplified using the identity

$$f^{abc}\lambda^b\lambda^c = iN\lambda^a, \qquad (8.3.23)$$

whereupon

$$\Gamma^{\alpha\beta;a}_\mu(q^2) = -ig\frac{\lambda^a_{\alpha\beta}}{2}\gamma_\mu\left(1 - \frac{g^2}{16\pi^2}\ln\left(\frac{-q^2}{\mu^2}\right)\left\{\xi\frac{(N^2-1)}{2N} + \left[1 - \frac{(1-\xi)}{4}\right]N\right\}\right). \qquad (8.3.24)$$

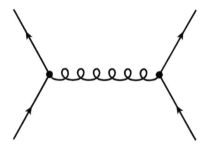

**Figure 8.20.** Lowest-order contribution to quark–quark scattering in QCD.

Note that the QED results may be recovered by setting

$$\left.\begin{array}{r}\dfrac{N^2-1}{2N} \to 1 \\ N \to 0 \\ \dfrac{g^2}{4\pi} \to \alpha \end{array}\right\}. \qquad (8.3.25)$$

To assemble the pieces for a physical process, it is convenient to work in Landau gauge ($\xi = 0$), for which there is no wave function renormalization to this order. It is instructive to consider the reaction

$$ud \to ud, \qquad (8.3.26)$$

for which the one-gluon-exchange Born diagram is shown in figure 8.20. Denoting the leading-order amplitude (cf. problem 8.19) as $\mathcal{M}_\mathrm{B}$, we may write the amplitude modified by one-loop corrections as

$$\mathcal{M} = \mathcal{M}_\mathrm{B}\left[1 + \frac{g^2}{16\pi^2}\ln\left(\frac{-q^2}{\mu^2}\right)\left(\frac{2n_\mathrm{f}}{3} - \frac{13N}{6} - 2\cdot\frac{3N}{4}\right)\right], \qquad (8.3.27)$$

where the terms multiplying the logarithm correspond to the contribution of fermion and boson loops to the gluon propagator and to the correction at both vertices arising from the three-gluon interaction [22]. Combining terms and defining $\alpha_\mathrm{s} \equiv g^2/4\pi$ as the strong-interaction coupling constant, we may evidently define, for this process, a running coupling constant

$$\alpha_\mathrm{s}(q^2) = \alpha_\mathrm{s}(\mu^2)\left[1 + \frac{\alpha_\mathrm{s}(\mu^2)}{12\pi}\ln\left(\frac{-q^2}{\mu^2}\right)(2n_\mathrm{f} - 11N) + O(\alpha_\mathrm{s}^2)\right], \qquad (8.3.28)$$

which is similar in form to the running coupling constant (8.2.36) of QED. Although it has been camouflaged in our abbreviated discussion, the expression, of course, refers to the renormalized coupling constant.

Consider now the implications of the form (8.3.28). The fermion-loop contribution has qualitatively the same effect as in QED: it is a vacuum polarization term that tends to enhance the coupling constant at short distances or large values of $q^2$.

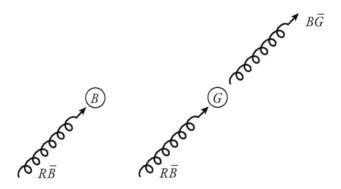

**Figure 8.21.** $R\bar{B}$ gluon incident on a blue quark (left panel) may find the blue charge dispersed as a result of quantum fluctuations (right panel).

The contribution due to the three-gluon vertex is of the *opposite sign* and tends to decrease the interaction strength at short distances. It corresponds not to screening, but to antiscreening.

The possibility of antiscreening can readily be understood in qualitative terms using the sketch in figure 8.21. Suppose that our "test charge," in the language of the earlier electrodynamics discussion, is a blue quark and that the probe we employ to measure its charge is a red-antiblue gluon. It may happen that, while the probe is en route to its target, the blue quark radiates a virtual blue-antigreen gluon and thus fluctuates into a green quark—to which the probe is blind. Rather than being concentrated at the quark's location, the net color charge will thus be dispersed throughout the surrounding gluon cloud. Therefore, only by inspecting the test charge from large distances will we be able to measure its full effect.

The combined effect of quark loops and gluon loops is thus the result of a competition between screening and antiscreening. As long as $11N$ exceeds $2n_f$, the antiscreening combination will dominate and the running coupling constant will become small at short distances or at large values of $q^2$. Our discussion has been lacking both in generality and in systematic rigor, but the result is indeed generally true. The summation of leading logarithms may be carried out, as in the case of electrodynamics, most conveniently by using renormalization group techniques. As our one-loop expression (8.3.28) makes plausible, the running coupling constant of QCD ($N = 3$) takes the form

$$\frac{1}{\alpha_s(q^2)} = \frac{1}{\alpha_s(\mu^2)} + \frac{(33 - 2n_f)}{12\pi} \ln\left(\frac{-q^2}{\mu^2}\right). \tag{8.3.29}$$

As long as the number of color-triplet quarks does not exceed 16, antiscreening prevails. In a world of six active quarks, the coefficient of the logarithm in (8.3.29) is $+7/4\pi$. The profound significance of this behavior for a calculable theory of the strong interactions was recognized [23] by Gross and Wilczek and by Politzer in 1973. The existence of a regime in which $\alpha_s(q^2) \ll 1$ implies a realm in which QCD perturbation theory should be valid. This property of non-Abelian gauge theories is known as *asymptotic freedom*. It makes plausible the parton-model assertion that at

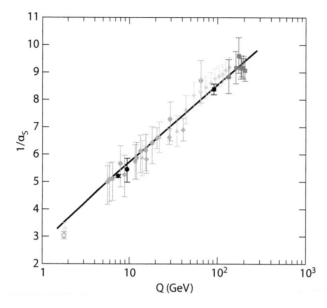

**Figure 8.22.** Measurements of the strong coupling $1/\alpha_s(Q^2)$ as a function of the energy scale ln $Q$, compared with the evolution predicted in QCD [27].

very short distances (i.e, when examined by very high $Q^2$ probes) quarks may behave nearly as free particles within hadrons. By contrast, the growth of the coupling constant at large distances implies the existence of a domain in which the strong interactions become formidable. This strong-coupling regime is of key importance for quark—or color—confinement.

The decrease of $\alpha_s$ with $Q$ has been demonstrated by measurements in many experimental settings [24] . Over the past decade, the precision of $\alpha_s$ determinations has improved dramatically, thanks to a plethora of results from various processes aided by improved calculations at higher orders in perturbation theory [25]. The scale dependence of $\alpha_s$ itself has been computed [26] to order $\alpha_s^5$. A representative selection of experimental determinations is shown in figure 8.22. The trend toward asymptotic freedom is clear, and the agreement with the predicted evolution is excellent, within the uncertainties in the measurements. An interesting challenge for the future will be to measure $\alpha_s(Q^2)$ with precision sufficient to detect the expected change of slope at the top-quark threshold. Problems 9.15 and 9.16 offer additional insights into the effect of the spectrum on the evolution of $\alpha_s$.

It is conventional, and enlightening, to rewrite the evolution equation (8.3.29) in the form

$$\frac{1}{\alpha_s(Q^2)} = \frac{33 - 2n_f}{12\pi} \ln\left(\frac{Q^2}{\Lambda_{QCD}^2}\right), \qquad (8.3.30)$$

where $\Lambda_{QCD}$ is the QCD scale parameter, with dimensions of energy. Several subtleties attend this simple and useful parametrization. First, if we enforce the requirement that $\alpha_s(Q^2)$ be continuous at flavor thresholds, then $\Lambda_{QCD}$ must depend

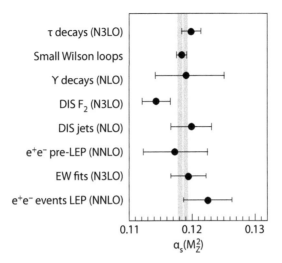

**Figure 8.23.** Determinations of $\alpha_s(M_Z^2)$ from several processes. In most cases, the value measured at a scale $\mu^2$ has been evolved to $\mu^2 = M_Z^2$. Error bars include the theoretical uncertainties. Perturbative calculations are carried out at next-to-leading order (NLO) and beyond (adapted from Ref. [25]).

on the number of active quark flavors. Second, the value of $\Lambda_{QCD}$ depends on the renormalization scheme; the canonical choice is the modified minimal subtraction ($\overline{MS}$) scheme [28]. The $n_f$ and scheme dependence is given via labels on $\Lambda$, such as $\Lambda_{\overline{MS}}^{(n_f)}$. Representative estimates of the QCD scale are [25] $\Lambda_{\overline{MS}}^{(5)} = 213$ MeV, $\Lambda_{\overline{MS}}^{(4)} = 296$ MeV, and $\Lambda_{\overline{MS}}^{(3)} = 338$ MeV. The appearance of a dimensional quantity to parametrize the running coupling is sometimes called *dimensional transmutation*.

When evolved to a common scale $Q = M_Z$, the various determinations of $\alpha_s$ lead to consistent values, as shown in figure 8.23. A representative mean value is (Ref. [25])

$$\alpha_s(M_Z^2) = 0.1184 \pm 0.0007. \tag{8.3.31}$$

The agreement of the determination of $\alpha_s$ from the hadron spectrum, via lattice QCD and from high-energy scattering processes, via perturbative QCD (and factorization for deeply inelastic scattering), indicates that QCD describes both hadron and partons. In other words, this single theory may account for all facets of the strong interactions.

## 8.4 PERTURBATIVE QCD: A FIRST EXAMPLE

We now begin to consider specific applications of QCD perturbation theory. The general spirit of these calculations is that by carrying out a computation to some modest order of perturbation theory, but replacing the strong coupling constant $\alpha_s$ by the running coupling constant $\alpha_s(q^2)$, we can construct a systematic approximation to the observable. Such a program can be given a precise definition

(a)                     (b)                                 (c)

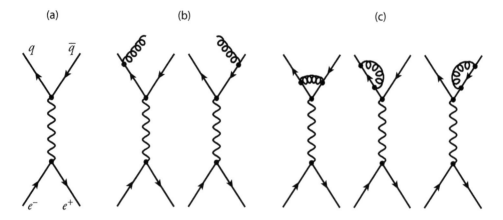

**Figure 8.24.** Feynman diagrams that enter the computation of $\sigma(e^+e^- \to q\bar{q})$.

within the framework of the renormalization group approach, as long as a perturbation expansion makes sense, in that the running coupling must be, by an appropriate standard, small. The values of $\alpha_s(q^2)$ inferred from experiment (cf. figure 8.22) show that this requirement is satisfied for many processes of interest. The overall strategy is frequently called renormalization-group-improved perturbation theory. Quark-hadron duality connects the parton-level results with observations.

The simplest illustration of QCD perturbation theory is the calculation of the cross section for hadron production in electron–positron annihilations. As we have discussed many times, this process is represented in the quark-parton model by the elementary transition illustrated in figure 8.24(a), which yields

$$\sigma_{\text{qpm}}(e^+e^- \to \text{hadrons}; s) = \frac{4\pi\alpha^2}{3s}\left[3\sum_q e_q^2 \theta(s - 4m_q^2)\right], \qquad (8.4.1)$$

where $e_q$ and $m_q$ are the charge and mass of quark flavor $q$ and the step function $\theta$ is a crude representation of kinematic threshold effects. The factor 3 preceding the sum over active flavors is a consequence of quark color. The rough agreement between measurements of the ratio of hadron production to muon-pair production at modest energies and the prediction (8.4.1), shown as the dashed line in figure 8.25, is powerful evidence that quarks are color triplets.

The parton-level prediction is modified by real and virtual emission of gluons, much as the quantum electrodynamics prediction for $\sigma(e^+e^- \to \mu^+\mu^-) = 4\pi\alpha^2/3s$ is changed by real and virtual emission of photons. To order $\alpha_s$, this entails the calculation of the two gluon-radiation graphs shown in figure 8.24(b) and of the three virtual-gluon graphs of figure 8.24(c). In a schematic but self-evident notation, the cross section is given by

$$\sigma(e^+e^- \to \text{hadrons}) = \underbrace{|A|^2}_{\alpha_s^0} + \underbrace{|B|^2 + |A \otimes C|}_{\alpha_s^1} + O(\alpha_s^2). \qquad (8.4.2)$$

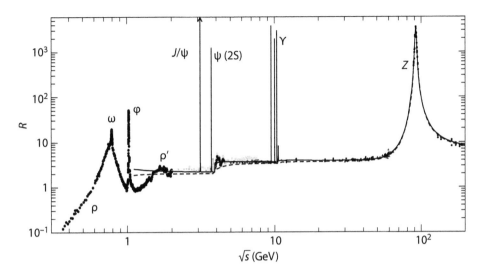

**Figure 8.25.** World data on the ratio $R \equiv \sigma(e^+e^- \to \text{hadrons})/\sigma(e^+e^- \to \mu^+\mu^-)$, compared with predictions of the quark-parton model (dashed curve) and perturbative QCD at three loops (solid line) from Ref. [24]). The low-energy expression (8.4.1) is augmented by the contribution of the $Z^0$ pole.

The ultraviolet divergences that appear separately in each loop diagram are eliminated in the usual renormalization program and introduce no particular problems. This cancellation is a consequence of the gauge invariance of the theory—or, equivalently of the Ward–Takahashi identities—and may be seen to follow from the explicit results of the preceding section. The cancellation of infrared divergences is more subtle and has interesting consequences for the definition of experimental observables [29].

Consider the gluon-radiation graphs of figure 8.24(b). If we denote by $q_i$ the final momentum of a quark that has radiated a gluon of momentum $k$, the virtual-quark propagator is proportional to

$$\frac{1}{(q_i + k)^2 - m^2} = \frac{1}{k^2 + 2q_i \cdot k}, \tag{8.4.3}$$

which vanishes in the soft-gluon limit as $k_\mu \to 0$ and, for the emission of massless gluons with $k^2 = 0$, also vanishes if

$$k \cdot q_i = 0. \tag{8.4.4}$$

The latter condition occurs, for massless quarks, when the quark and gluon momenta are collinear. These infrared singularities are canceled by similar terms in the vertex and quark-propagator corrections, which will occur in the $|A \otimes C|$ interference terms. This means that, whereas the total cross section given in (8.4.2) is infrared finite, the cross sections for specific final states

$$\sigma(e^+e^- \to q\bar{q}) = |A|^2 + |A \otimes C| + \cdots \tag{8.4.5}$$

and

$$\sigma(e^+e^- \to q\bar{q}g) = |B|^2 + \cdots \tag{8.4.6}$$

have infrared divergences and cannot meaningfully be calculated, for general values of the kinematic variables [30].

Once this is recognized, it is straightforward to evaluate the cross section. A practical approach is to first use dimensional regularization to characterize the divergences of the real and virtual contributions separately. The soft and collinear singularities of the real ($|B|^2$) and virtual ($|A \otimes C|$) pieces coincide and cancel, so that a finite physical cross section ensues. The inclusive cross section has the same form, modulo color factors, as that calculated in QED by Jost and Luttinger [31] . To leading order in the running coupling $\alpha_s(s)$, the result is [32]

$$\sigma_{\text{QCD}}(e^+e^- \to \text{hadrons}) = \sigma_{\text{qpm}} \left[ 1 + \frac{\alpha_s(s)}{\pi} + \mathcal{O}(\alpha_s^2) \right]. \tag{8.4.7}$$

The strong-interactions corrections to the total cross section for hadron production are thus seen to be calculable and free of infrared problems; positive, and decreasing with increasing $s$; small, in the asymptotically free region, with higher-order corrections [33] that are not enormous. The QCD prediction for

$$R \equiv \frac{\sigma(e^+e^- \to \text{hadrons})}{\sigma(e^+e^- \to \mu^+\mu^-)}, \tag{8.4.8}$$

now known through order $\alpha_s^3$, is shown as the solid line in figure 8.25, which also includes contributions of diagrams in which the photon is replaced by $Z^0$.

The good behavior of the inclusive cross section is an example of the Kinoshita–Lee–Nauenberg theorem [34]. Infrared divergences cancel in sufficiently inclusive observables, yielding finite results to all orders in perturbation theory.

Although the definition of a three-jet cross section corresponding to the $q\bar{q}g$ final state is plagued by infrared difficulties, it is nevertheless apparent that three-jet events are to be expected as a consequence of the gluon-radiation diagrams of figure 8.24. At modest energies, it is a reasonable approximation to omit the contribution from $Z^0$-exchange and consider only the photon-exchange diagrams. Neglecting quark masses, we denote the fractional energies carried by the outgoing quark, antiquark, and gluon by

$$x_i = \frac{2E_i}{\sqrt{s}} = \frac{2q_i \cdot Q}{Q^2}, \tag{8.4.9a}$$

$$x_g = \frac{2E_g}{\sqrt{s}} = \frac{2k \cdot Q}{Q^2}, \tag{8.4.9b}$$

so that $x_1 + x_2 + x_g = 2$, with $0 \leq x \leq 1$. The gluon + (quark or antiquark) subenergies are

$$s_i = (q_i + k)^2 = 2q_i \cdot k = Q^2(1 - x_i), \quad i \neq j. \tag{8.4.10}$$

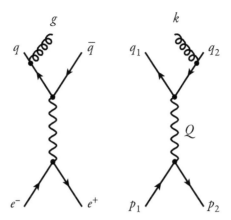

**Figure 8.26.** Lowest-order contributions to the reaction $e^+e^- \to q\bar{q}g$: species labeled on left, momenta on right.

After integration over angles, the differential cross section for $q\bar{q}g$ production is [35]

$$\frac{d\sigma}{dx_1 dx_2} = \frac{2\alpha_s}{3\pi} \sigma_{\text{qpm}} \frac{x_1^2 + x_2^2}{(1-x_1)(1-x_2)}, \tag{8.4.11}$$

where $\sigma_{\text{qpm}}$, the quark-parton-model cross section for the reaction $e^+e^- \to q\bar{q}$, is given by (8.4.1). As our general discussion anticipated, the divergences in (8.4.11) occur for

$$x_1 \to 1 \quad \Leftrightarrow \quad \theta_{2g} \to 0, \text{ collinear emission;}$$
$$x_2 \to 1 \quad \Leftrightarrow \quad \theta_{1g} \to 0, \text{ collinear emission;} \tag{8.4.12}$$
$$x_1 \to 1, \ x_2 \to 1 \quad \Leftrightarrow \quad E_g \to 0, \text{ soft emission.}$$

It follows that the divergences are unphysical. They arise in kinematical regimes in which parton energies or parton-pair invariant masses are comparable to hadron masses, so the concept of an individual parton is ambiguous. As long as we exclude configurations in which kinematical variables (subenergies) are too small, (8.4.11) should give a faithful representation of three-jet cross differential cross section. The observation in high-energy $e^+e^-$ annihilations at DESY's PETRA ring of three-jet topologies that displayed the anticipated features established the existence of the gluon [36]. The path toward a systematic confrontation of QCD with experiment was indicated by Sterman and Weinberg [37], who showed how to define infrared-finite energy-weighted cross sections that are calculable within QCD.

At still higher orders in $\alpha_s$, the hadron-production cross section receives contributions from diagrams that have an obvious (though infrared dangerous) interpretation in terms of four (or more!) jet events. It has been of considerable interest to devise means of confronting these expectations with experiment. Beyond the problem of infrared safety, we must deal with the hadronization of partons, which enters in the experimental definition of jet cross sections and in the determination of experimental efficiencies. This is an active and nontrivial area of research.

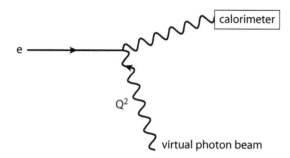

**Figure 8.27.** Conceptual experiment to measure the momentum spectrum of electrons in a beam prepared as monochromatic. The scattered electron is not shown.

Modern definitions of jets—taking account of infrared safety, calculability, ease of measurement, and the extension to hadronic collisions—are surveyed in Ref. [38].

## 8.5 QCD CORRECTIONS TO DEEPLY INELASTIC SCATTERING

According to the parton model, a hadron is a collection of quasifree quarks, antiquarks, and gluons. In the infinite-momentum frame, in which the longitudinal momentum of the hadron is very large, each parton carries a fraction $x$ of the hadron's momentum. A parton distribution function $f_i(x_i)$ specifies the probability of finding a parton of species $i$ with momentum fraction $x_i$. In an interacting field theory such as QCD that entails vacuum fluctuations, the partonic content of a hadron depends on the resolution of the probe. Next, we shall develop and apply a highly intuitive formalism that generalizes the parton distributions to $f_i(x_i, Q^2)$ and stipulates the evolution of parton distributions with momentum transfer $Q^2$.

### 8.5.1 AN ELECTRODYNAMIC PROTOTYPE

The evolution with $Q^2$ of the structure functions measured in deeply inelastic scattering may be analyzed in QCD perturbation theory. To make clear the logical structure, it is helpful to revert to electrodynamics and to consider a *Gedankenexperiment* to measure the momentum spectrum of electrons in a "monochromatic" beam by observing the scattering of virtual photons, as shown in figure 8.27. A perfect calorimeter measures the energy of the backscattered photon and thus determines the incident energy of the electron from which the photon was scattered. If the momentum of the prepared electron beam is defined to be 1, then in zeroth order the momentum distribution of electrons in the beam is

$$\frac{d\mathcal{N}}{dz} = \mathcal{N}\delta(z-1),\qquad (8.5.1)$$

where

$$z \equiv \frac{\text{measured momentum}}{\text{prepared momentum}}.\qquad (8.5.2)$$

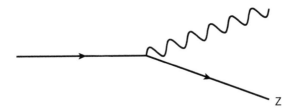

**Figure 8.28.** Fluctuation of an electron into an electron-carrying momentum fraction $z$ and a photon.

The virtual dissociation of an electron into an electron plus a photon, depicted in figure 8.28, induces in the beam a component with momentum fraction $z < 1$. The sensitivity of the apparatus (figure 8.27) to these fluctuations is a function of the invariant mass of the virtual photon, as uncertainty-principle reasoning makes clear. The fluctuations are calculable in QED.

To see this explicitly, let us define the parameter

$$\tau \equiv \ln \left( \frac{Q^2}{q_0^2} \right) \tag{8.5.3}$$

and let the quantity

$$\frac{\alpha}{2\pi} P_{e \leftarrow e}(z)\, d\tau \tag{8.5.4}$$

represent the probability of observing an electron carrying a fraction $z$ of the parent electron's momentum. Then, if $e(z, \tau)$ is the number density of electrons that a probe with resolving power characterized by $\tau$ would observe with fractional momenta between $z$ and $z + dz$, it follows at once that

$$\frac{de}{d\tau}(x, \tau) = \frac{\alpha(\tau)}{2\pi} \int_0^1 dy \int_0^1 dz\, \delta(zy - x) e(y, \tau) P_{e \leftarrow e}(z)$$
$$= \frac{\alpha(\tau)}{2\pi} \int_x^1 \frac{dy}{y}\, e(y, \tau) P_{e \leftarrow e}\left( \frac{x}{y} \right). \tag{8.5.5}$$

For an initial distribution $e(y, \tau) = \mathcal{N}\delta(y - 1)$ corresponding to the beam prepared in (8.27), we recover

$$\frac{de}{d\tau}(x, \tau) = \mathcal{N}\frac{\alpha(\tau)}{2\pi} P_{e \leftarrow e}(x). \tag{8.5.6}$$

By virtue of the same fluctuations, there are now photons in the beam as well. We let the probability of finding a photon carrying a fraction $z$ of the parent electron's momentum be represented by

$$\frac{\alpha}{2\pi} P_{\gamma \leftarrow e}(z)\, d\tau \tag{8.5.7}$$

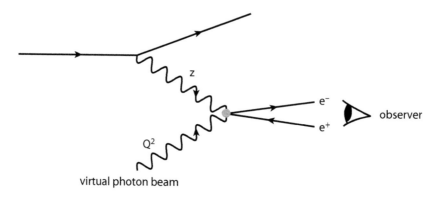

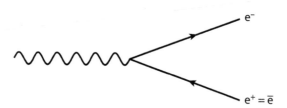

**Figure 8.29.** Conceptual experiment to measure the momentum spectrum of photons in an electron beam prepared as monochromatic.

**Figure 8.30.** Fluctuation of a photon into an electron–positron pair.

and let $\gamma(z, \tau)$ be the number density of photons observed with fractional momenta between $z$ and $z + dz$ by a probe with resolving power characterized by $\tau$. A schematic apparatus to observe these photons is shown in figure 8.29. If the source of the virtual-photon probe is (for example) a nitrogen nucleus, the *Gedankenapparat* is recognized as a surrogate for the development of electromagnetic showers in the atmosphere. Indeed, the theory of cascade showers presented by Rossi [39] has much in common with the present discussion. The evolution of the photon distribution is given by

$$\frac{d\gamma}{d\tau}(x, \tau) = \frac{\alpha(\tau)}{2\pi} \int_x^1 \frac{dy}{y} \, e(y, \tau) P_{\gamma \leftarrow e}\left(\frac{x}{y}\right). \tag{8.5.8}$$

Of course, photons themselves may fluctuate into electron–positron pairs, as indicated in figure 8.30, so we are led to define

$$\frac{\alpha}{2\pi} P_{e \leftarrow \gamma}(z) \tag{8.5.9}$$

as the probability for finding an electron (or a positron) carrying a fraction $z$ of the momentum of the parent photon.

The evolution of the electron distribution is now given by

$$\frac{de}{d\tau}(x, t) = \frac{\alpha(\tau)}{2\pi} \int_x^1 \frac{dy}{y} \left[ e(y, \tau) P_{e \leftarrow e}\left(\frac{x}{y}\right) + \gamma(y, \tau) P_{e \leftarrow \gamma}\left(\frac{x}{y}\right) \right], \qquad (8.5.10)$$

and the induced positron component will obey an equation identical in form,

$$\frac{d\bar{e}}{d\tau}(x, t) = \frac{\alpha(\tau)}{2\pi} \int_x^1 \frac{dy}{y} \left[ \bar{e}(y, \tau) P_{e \leftarrow e}\left(\frac{x}{y}\right) + \gamma(y, \tau) P_{e \leftarrow \gamma}\left(\frac{x}{y}\right) \right]. \qquad (8.5.11)$$

The photon, in turn, evolves according to

$$\frac{d\gamma}{d\tau}(x, \tau) = \frac{\alpha(\tau)}{2\pi} \int_x^1 \frac{dy}{y} \left[ e(y, \tau) + \bar{e}(y, \tau) \right] P_{\gamma \leftarrow e}\left(\frac{x}{y}\right). \qquad (8.5.12)$$

To solve equations of this kind, it is convenient to define moments of the distribution functions through Mellin transformations as

$$M_n(\tau) = \int_0^1 dx \, x^{n-1} e(x, \tau), \qquad (8.5.13a)$$

$$\overline{M}_n(\tau) = \int_0^1 dx \, x^{n-1} \bar{e}(x, \tau), \qquad (8.5.13b)$$

and similarly for the photon. The evolution of the moments is then easily computed. It takes a particularly simple form for the combination $M_n(\tau) - \overline{M}_n(\tau)$, for which

$$\frac{d}{d\tau}\left[ M_n(\tau) - \overline{M}_n(\tau) \right] = \int_0^1 dx \, x^{n-1} \left( \frac{de}{d\tau}(x, \tau) - \frac{d\bar{e}}{d\tau}(x, \tau) \right)$$

$$= \frac{\alpha(\tau)}{2\pi} \int_0^1 dx \, x^{n-1} \int_0^1 dy$$

$$\times \int_0^1 dz \, \delta(zy - x) P_{e \leftarrow e}(z) \left[ e(y, \tau) - \bar{e}(y, \tau) \right]$$

$$= \frac{\alpha(\tau)}{2\pi} \int_0^1 dx \, x^{n-1} \int_0^1 \frac{dy}{y} P_{e \leftarrow e}\left(\frac{x}{y}\right) \left[ e(y, \tau) - \bar{e}(y, \tau) \right]$$

$$= \frac{\alpha(\tau)}{2\pi} \int_0^1 dy \, y^{n-1} \left[ e(y, \tau) - \bar{e}(y, \tau) \right]$$

$$\times \int_0^1 d\left(\frac{x}{y}\right) \left(\frac{x}{y}\right)^{n-1} P_{e \leftarrow e}\left(\frac{x}{y}\right)$$

$$= \frac{\alpha(\tau)}{2\pi} \left[ M_n(\tau) - \overline{M}_n(\tau) \right] A_n, \qquad (8.5.14)$$

where

$$A_n \equiv \int_0^1 dz \, z^{n-1} P_{e \leftarrow e}(z) \qquad (8.5.15)$$

is the moment of the so-called splitting function, $P_{e \leftarrow e}(z)$. Now writing

$$\Delta_n \equiv \left[ M_n(\tau) - \overline{M}_n(\tau) \right], \tag{8.5.16}$$

we have that

$$\frac{d(\ln \Delta_n(\tau))}{d\tau} = \frac{\alpha(\tau)}{2\pi} A_n. \tag{8.5.17}$$

If the running coupling itself evolves as

$$\alpha(\tau) = \frac{\alpha(0)}{1 + b\alpha(0)\tau}, \tag{8.5.18}$$

the general leading-logarithmic form appropriate for QED and QCD, then the differential equation (8.5.17) is easily integrated to

$$\begin{aligned}
\ln\left(\frac{\Delta_n(\tau)}{\Delta_n(0)}\right) &= \frac{\alpha(0)A_n}{2\pi} \int_0^\tau \frac{d\tau}{1 + b\alpha(0)\tau} \\
&= \frac{A_n}{2\pi b} \ln\left[1 + b\alpha(0)\tau\right] \\
&= \frac{A_n}{2\pi b} \ln\left(\frac{\alpha(0)}{\alpha(\tau)}\right).
\end{aligned} \tag{8.5.19}$$

We therefore obtain the simple prediction for the evolution of moments that

$$\frac{\Delta_n(\tau)}{\Delta_n(0)} = \left[\frac{\alpha(\tau)}{\alpha(0)}\right]^{-A_n/2\pi b}, \tag{8.5.20}$$

as well as an especially simple prediction for

$$\frac{\ln\left[\Delta_n(\tau)/\Delta_n(0)\right]}{\ln\left[\Delta_k(\tau)/\Delta_k(0)\right]} = \frac{A_n}{A_k}. \tag{8.5.21}$$

These specific forms are valid at leading order in renormalization-group-improved perturbation theory. The evolution of the moments is completely specified by the exponents $A_n$, which may be calculated without reference to the electron and photon distribution functions. To describe the evolution of individual moments, rather than moment-by-moment ratios, it is necessary to know or determine the running coupling $\alpha(0)$.

Experiments at electron–positron colliders have determined the electromagnetic structure function of the photon through the reaction

$$e^+e^- \to e^+e^-\mu^+\mu^-. \tag{8.5.22}$$

The most extensive measurements, carried out at LEP [40], are in good agreement with QED predictions made more explicit in (8.7.26) and serve to validate the techniques used to extract the hadronic structure function of the photon (cf. section 8.7).

## 8.5.2 SPLITTING FUNCTIONS IN QCD

In fact, nothing of the procedure we have followed is specific to QED. The same method can be adapted to quantum chromodynamics, as was done by Altarelli and Parisi [6], by identifying the electron, positron, and photon distributions as quark, antiquark, and gluon distributions and allowing for the possibility of a gluon fluctuating into two gluons.

The splitting functions $P_{B \leftarrow A}(z)$, which are to be computed in perturbation theory, satisfy some obvious sum rules. Fermion number conservation,

$$\int dx \left( \frac{dq^i}{d\tau}(x, \tau) - \frac{d\bar{q}^i}{d\tau}(x, \tau) \right) = 0, \tag{8.5.23}$$

implies that

$$\int_0^1 dz \, P_{q \leftarrow q}(z) = 0. \tag{8.5.24}$$

Here the superscript $i$ has been introduced as a flavor index for the quarks. Momentum conservation,

$$\int_0^1 dx \, x \left( \sum_i \frac{dq^i}{d\tau}(x, \tau) + \sum_i \frac{d\bar{q}^i}{d\tau}(x, \tau) + \frac{dG}{d\tau}(x, \tau) \right) = 0, \tag{8.5.25}$$

imposes two constraints:

$$\int_0^1 dz \, z \left( P_{q \leftarrow q}(z) + P_{g \leftarrow q}(z) \right) = 0 \tag{8.5.26}$$

and

$$\int_0^1 dz \, z \left( 2n_f P_{q \leftarrow g}(z) + P_{g \leftarrow g}(z) \right) = 0, \tag{8.5.27}$$

where $n_f$ denotes the number of quark flavors. In addition, momentum conservation at the elementary vertices requires a number of symmetry properties to hold for $z \neq 1$:

$$P_{q \leftarrow q}(z) = P_{g \leftarrow q}(1 - z), \tag{8.5.28}$$

$$P_{q \leftarrow g}(z) = P_{q \leftarrow g}(1 - z), \tag{8.5.29}$$

$$P_{g \leftarrow g}(z) = P_{g \leftarrow g}(1 - z). \tag{8.5.30}$$

Two methods are in use for the actual evaluation of the splitting functions. We shall sketch the "old-fashioned," which is to say time-ordered, perturbation theory approach taken by Altarelli and Parisi. This has the merit of making plain the generality of the results and making contact with the classic techniques of Weizsäcker and Williams [41]. Completely equivalent calculations may be carried out by using the covariant techniques of ordinary Feynman graphs [42], which are perhaps more transparent but have the appearance of being process dependent.

(a)                                    (b)

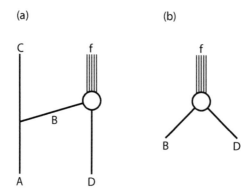

**Figure 8.31.** (a) The compound process $A + D \to C + f$ used in the calculation of the splitting function $P_{B \leftarrow A}$. (b) The subprocess $B + D \to f$.

The procedure is, in any case, straightforward: for $z \neq 1$, the splitting function $P_{B \leftarrow A}(z)$ can be related to the square of a matrix element.

To see how the connection is made, let us evaluate in the infinite-momentum frame the probability $\mathcal{P}_{BA}(z)$ of finding a parton of type $B$ with momentum fraction $z$ in a beam of partons of type $A$. According to our previous definitions, we may write simply

$$d\mathcal{P}_{BA}(z)\, dz = \frac{\alpha_s}{2\pi} P_{B \leftarrow A}(z)\, dz\, d\tau. \tag{8.5.31}$$

Consider the reaction

$$A + D \to C + f \tag{8.5.32}$$

sketched in figure 8.31(a), in which $ABC$ is the calculable elementary vertex of interest and particle $D$ merely provides the kinematical crutch of the process

$$B + D \to f \tag{8.5.33}$$

indicated in figure 8.31(b). The differential cross section for the compound reaction may be written as

$$d\sigma(A + D \to C + f) = d\mathcal{P}_{BA}(z)\, d\sigma(B + D \to f)\, dz. \tag{8.5.34}$$

For massless partons, it is then a straightforward exercise in kinematics [43] to show that with four-momenta defined as

$$k_A = (P; \vec{0}, P) \tag{8.5.35a}$$

$$k_B = \left\{ [(zP)^2 + p_\perp^2]^{1/2} ; \vec{p}_\perp, zP \right\} \tag{8.5.35b}$$

$$\simeq (zP + p_\perp^2/2zP; \vec{p}_\perp, zP)$$

$$k_C = \left\{ [(1-z)^2 P^2 + p_\perp^2]^{1/2} ; -\vec{p}_\perp, (1-z)P \right\} \tag{8.5.35c}$$

$$\simeq [(1-z)P + p_\perp^2/2(1-z)P; -\vec{p}_\perp, (1-z)P],$$

the differential probability is given by

$$dP_{BA}(z) = \frac{\alpha_s}{2\pi} \frac{z(1-z)}{2p_\perp^2} \overline{|\mathcal{M}(A \to BC)|^2} \, d(\ln p_\perp^2), \qquad (8.5.36)$$

where $\overline{|\mathcal{M}|^2}$ is the spin-averaged square of the matrix element. With the (asymptotic) identification $d(\ln p_\perp^2) \to d\tau$, this implies that the splitting function is

$$P_{B \leftarrow A}(z) = \frac{z(1-z)}{2p_\perp^2} \overline{|\mathcal{M}(A \to BC)|^2}, \qquad z \neq 1. \qquad (8.5.37)$$

We therefore find for the quark–quark splitting function,

$$P_{q \leftarrow q}(z < 1) = \left( \frac{N^2 - 1}{2N} \right) \left( \frac{1 + z^2}{1 - z} \right), \qquad (8.5.38)$$

where the color factor has been written for SU($N$). It remains to fix the behavior of $P_{q \leftarrow q}$ at $z = 1$. This is done in two steps. First, we interpret the denominator $1/(1-z)$ in the sense of a distribution $1/(1-z)_+$, defined by the property

$$\int_0^1 \frac{dz\, f(z)}{(1-z)_+} = \int_0^1 dz \frac{f(z) - f(1)}{(1-z)} = \int_0^1 dz \ln(1-z) \frac{df}{dz}(z) \qquad (8.5.39)$$

for a function $f$ that is regular at $z = 1$. Second, we add a term proportional to $\delta(1-z)$ with a coefficient chosen to ensure compliance with the fermion conservation sum rule (8.5.24). This determines the final result for SU(3)$_c$ as

$$P_{q \leftarrow q} = \frac{4}{3} \left[ \frac{1 + z^2}{(1-z)_+} + \frac{3}{2}\delta(1-z) \right]. \qquad (8.5.40)$$

The other splitting functions, which are computed in similar fashion, are

$$P_{g \leftarrow q} = \frac{4}{3} \left[ \frac{1 + (1-z)^2}{z} \right], \qquad (8.5.41)$$

$$P_{q \leftarrow g}(z) = \frac{1}{2} \left[ z^2 + (1-z)^2 \right], \qquad (8.5.42)$$

for each quark flavor, and

$$P_{g \leftarrow g}(z) = 6 \left[ \frac{z}{(1-z)_+} + \frac{(1-z)}{z} + z(1-z) + \left( \frac{11}{12} - \frac{n_f}{18} \right)\delta(1-z) \right]. \qquad (8.5.43)$$

What is required to compute the evolution of the moments of quark distributions is the moments of the splitting functions. As an example, let us evaluate

$$A_n(q \leftarrow q) \equiv \int_0^1 dz\, z^{n-1} P_{q \leftarrow q}(z)$$

$$= \frac{4}{3} \left[ \frac{3}{2} + \int_0^1 dz \frac{(z^{n-1} + z^{n+1} - 2)}{1-z} \right]. \qquad (8.5.44)$$

The remaining integral may be computed easily by noting that

$$\int_0^1 dz \, \frac{z^{n-1} - 1}{1 - z} = - \sum_{j=0}^{n-2} \int_0^1 dz \, z^j = - \sum_{j=1}^{n-1} \frac{1}{j}. \tag{8.5.45}$$

Thus we have

$$A_n(q \leftarrow q) = \frac{4}{3} \left( \frac{3}{2} - \sum_{j=1}^{n-1} \frac{1}{j} - \sum_{k=1}^{n+1} \frac{1}{k} \right)$$

$$= \frac{4}{3} \left[ -\frac{1}{2} + \frac{1}{n(n+1)} - 2 \sum_{j=2}^{n} \frac{1}{j} \right], \tag{8.5.46}$$

and after similar arithmetic,

$$A_n(g \leftarrow q) = \frac{4}{3} \left[ \frac{n^2 + n + 2}{n(n^2 - 1)} \right], \tag{8.5.47}$$

$$A_n(q \leftarrow g) = \frac{1}{2} \left[ \frac{n^2 + n + 2}{n(n+1)(n+2)} \right], \tag{8.5.48}$$

$$A_n(g \leftarrow g) = 6 \left[ -\frac{1}{12} + \frac{1}{n(n-1)} + \frac{1}{n(n+1)(n+2)} - \sum_{j=2}^{n} \frac{1}{j} - \frac{n_f}{18} \right]. \tag{8.5.49}$$

It is natural for a weak-coupling theory such as QED to give reliable results for the splitting functions and, hence, the Mellin moments at low orders in a perturbation expansion. For the strong interactions, the asymptotic freedom of QCD (cf. figure 8.22) presents the hope that even leading-order results might be reliable, for large enough values of $Q^2$.

### 8.5.3 EXPERIMENTAL VALIDATION

The evolution of what is called the nonsinglet moment, which corresponds to the difference between the quark and antiquark distribution functions, is particularly simple to evaluate, as the discussion of electromagnetism has shown. It is also accessible experimentally in rather direct fashion because, as noted in section 7.4, the structure function $\mathcal{F}_3$ measures the difference between quark and antiquark distributions. For an isoscalar-nucleon target, the difference of neutrino and antineutrino charged-current cross sections is seen from (7.4.71), and (7.4.86), and (7.4.87) to yield

$$xF_3 = u(x) - \bar{u}(x) + d(x) - \bar{d}(x)$$

$$= \left[ \frac{d^2\sigma}{dx\,dy}(\nu N \to \mu^- X) - \frac{d^2\sigma}{dx\,dy}(\bar{\nu} N \to \mu^+ X) \right]$$

$$\times \frac{\pi}{G_F^2 ME} \cdot \frac{1}{1 - (1 - y)^2}. \tag{8.5.50}$$

Transcribing the results of our QED analysis, we have, for the evolution of the nonsinglet moments, defined as

$$\Delta_n(\tau) \equiv \int_0^1 dx\, x^n \mathcal{F}_3(x, \tau) = \int_0^1 dx\, x^{n-1} \left[ q(x, \tau) - \bar{q}(x, \tau) \right], \tag{8.5.51}$$

the simple and characteristic predictions

$$\frac{\Delta_n(\tau)}{\Delta_n(0)} = \left( \frac{\alpha_s(\tau)}{\alpha_s(0)} \right)^{-A_n(q \leftarrow q)/2\pi b} \tag{8.5.52}$$

and

$$\frac{\ln\left(\Delta_n(\tau)/\Delta_n(0)\right)}{\ln\left(\Delta_k(\tau)/\Delta_n(0)\right)} = \frac{A_n(q \leftarrow q)}{A_k(q \leftarrow q)}, \tag{8.5.53}$$

where the numerical values of the nonsinglet moments are $A_1(q \leftarrow q) = 0$ (by fermion number conservation), $A_2(q \leftarrow q) = -1.78$, $A_3(q \leftarrow q) = -2.78$, $A_4(q \leftarrow q) = -3.49$, $A_5(q \leftarrow q) = -4.04$, $A_6(q \leftarrow q) = -4.50$; and $A_7(q \leftarrow q) = -4.89$. For QCD [cf. (8.3.29)], the coefficient of the logarithmic running is

$$b = \frac{33 - 2n_f}{12\pi}. \tag{8.5.54}$$

The evolution of moments of the structure function in QCD is, thus, logarithmic, so that Bjorken scaling may plausibly hold to good approximation over a wide range in $Q^2$. One indication this is a special property may be seen simply as follows. Let us imagine a theory of the strong interactions in which the coupling $\bar{\alpha}_s$ is small, so that perturbation theory makes sense, but constant. The differential equation (8.5.17) for the evolution of the nonsinglet moment now takes the form

$$\frac{d(\ln \Delta_n(\tau))}{d\tau} = \frac{\bar{\alpha}_s}{2\pi} A_n, \tag{8.5.55}$$

which has the elementary solution

$$\Delta_n(\tau) = \Delta_n(0) e^{\bar{\alpha}_s A_n \tau / 2\pi}$$

$$= \Delta_n(0) \left( \frac{Q^2}{Q_0^2} \right)^{\bar{\alpha}_s A_n / 2\pi}, \tag{8.5.56}$$

corresponding to power-law violations of Bjorken scaling [44].

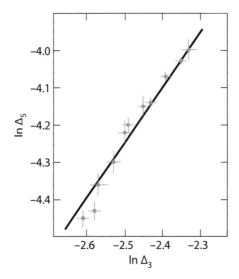

**Figure 8.32.** Scatter plot of logarithms of the fifth versus third nonsinglet moments determined in the CDHS neutrino experiment (See Ref. [45].). The indicated errors include the correlations between the two measurements.

The logarithmic ratios of the nonsinglet moments have been studied in many experiments. An early and influential analysis, based on the high-statistics sample of deeply inelastic neutrino and antineutrino interactions in iron measured in the CERN–Dortmund–Heidelberg–Saclay detector [45], is represented in figure 8.32. The figure shows the slope plot $\ln \Delta_5(\tau)$ versus $\ln \Delta_3(\tau)$ obtained from their measurements of the structure function $\mathcal{F}_3$ [46]. According to the first-order QCD analysis we have just carried out, the slope should be

$$r_{53} = \frac{A_5(q \leftarrow q)}{A_3(q \leftarrow q)} = 1.46, \tag{8.5.57}$$

which is in reasonable agreement with the experimental best fit,

$$r_{53} = 1.68 \pm 0.11. \tag{8.5.58}$$

The agreement between the experimental result and the QCD prediction is improved by about one (experimental) standard deviation if next-to-leading order corrections are included (solid line in figure 8.32).

The moment-by-moment test has the virtue of great simplicity and historically discriminated against candidate descriptions other than QCD, but it does not confront the predictions of QCD in a particularly differential or incisive fashion. The next step is to consider the prediction (8.5.52) for the evolution of the moments themselves. The behavior of the nonsinglet moments is rather simple:

$$\Delta_n(\tau) = \Delta_n(0) \left( \frac{\alpha_s(0)}{\alpha_s(\tau)} \right)^{6A_n/(33-2n_f)} . \tag{8.5.59}$$

As long as $33 - 2n_f > 0$, which is to say that the theory is asymptotically free, we have the general expectation that

$$\lim_{Q^2 \to \infty} \Delta_n \to 0, \quad n > 1, \tag{8.5.60}$$

because $A_{n>1}(q \leftarrow q) < 0$. [As we have already remarked following (8.5.53), the $\tau$-independence of the first nonsinglet moment, $\Delta_1$, implied by $A_1 = 0$, expresses fermion-number conservation.] This means, in particular, that the momentum fraction carried by valence quarks, which is characterized by the nonsinglet moment $\Delta_2$, will vanish as $Q^2 \to \infty$:

$$\int_0^1 dx \, x q_v(x, Q^2 \to \infty) = 0. \tag{8.5.61}$$

We shall look further into the partition of the hadron momentum among the constituents in section 8.5.4.

Increasingly comprehensive data sets gathered in fixed-target $eN$, $\mu N$, and $\nu N$ scattering experiments and in $e^{\pm}p$ collisions at the electron–proton collider, HERA, have deepened the dialogue between theory and experiment. It is now standard practice to fit QCD expressions, computed beyond leading order, to a vast universe of data, including information from hadron collider experiments. Two examples of the accord between QCD and experiment are shown in figure 8.33. In the left panel, the reduced cross section for deeply inelastic neutral-current $ep$ scattering,

$$\sigma_r^{NC}(x, Q^2) = \frac{d^2\sigma}{dx dQ^2} \cdot \frac{Q^4 x}{2\pi\alpha^2[1 + (1 - y)^2]}, \tag{8.5.62}$$

measured at HERA, is compared with next-to-leading-order predictions. In (8.5.62) $\alpha$ denotes the fine structure constant, $x$ is the Bjorken scaling variable, and $y$ is the inelasticity of the scattering process related to $Q^2$ and $x$ by $y = Q^2/xs$, where $s$ is the squared c.m. energy of the incoming electron and proton. The right panel shows the nonsinglet structure function $xF_3(x, Q^2)$ measured in the NuTeV experiment at Fermilab.

The data display a very weak dependence on $Q^2$ at moderate values of $x$; this extends the early evidence for Bjorken scaling [50]. At smaller values of $x$, we observe a growth with $Q^2$, reflecting the growing population of low-$x$ partons anticipated schematically in figure 8.9. The decrease with $Q^2$ seen at larger values of $x$ follows from the degrading of the valence-quark density.

Fits to the structure functions measured in deeply inelastic scattering and to other observables allow the systematic extraction of parton distribution functions, the strong coupling $\alpha_s$, and other parameters. It is now possible to attach uncertainties to parton distribution functions. Typical of the current state of the art are

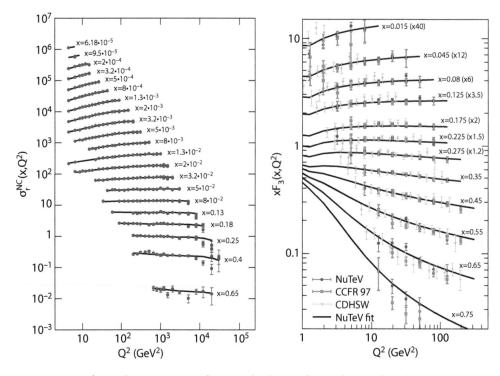

**Figure 8.33.** Left panel: Comparison of next-to-leading-order predictions based on the CT10 parton distributions for the reduced neutral-current cross section in $e^+ p$ deeply inelastic scattering measured at HERA (Ref. [47], adapted from Ref. [48]). Correlated systematic shifts are included, and each data set is scaled from the previous one by a factor of 2. Right panel: Nonsinglet structure function $xF_3$ measured in the NuTeV $\nu$ Fe-scattering experiment at Fermilab (Ref. [49]) compared with a leading-order QCD fit, modified to include nonleading effects for $\sigma_L/\sigma_T$ (cf. (8.5.63) and following), nonperturbative effects, and the charmed-quark mass.

the 2008 MSTW parton distributions $xf_i(x, Q^2)$ shown in figure 8.34 at low and moderately high values of $Q^2$. The schematic evolution sketched in figure 8.9 is here made quantitative.

It is conventional to separate quark (and antiquark) distributions into *valence* components that account for a hadron's net quantum numbers and *sea* contributions, in which quarks balance antiquarks overall. Neither a symmetry nor QCD dynamics demand that $\bar{u}(x) = \bar{d}(x)$ locally, and experiment has now revealed a flavor asymmetry in the light-quark sea of the proton [52]. Problem 8.17 examines how this might come about.

Our knowledge of the spin structure of the proton at the constituent level is drawn from polarized deeply inelastic scattering experiments, in which polarized leptons or photons probe the structure of a polarized proton and polarized proton–proton collisions [53].

We close this brief survey with a remark on the longitudinal cross section in deeply inelastic scattering, which provides a test of parton spin, as noted in

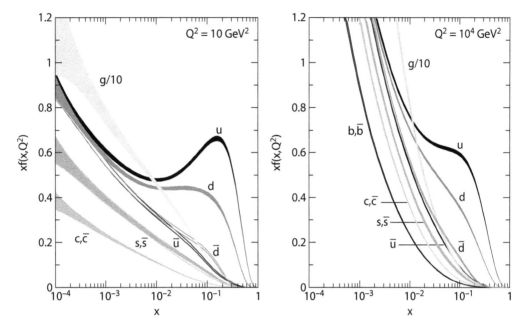

**Figure 8.34.** MSTW 2008 next-to-leading-order parton distributions $xf_i(x, Q^2)$, with one-standard-deviation uncertainties, at $Q^2 = 10$ GeV$^2$ (left panel) and $Q^2 = 10^4$ GeV$^2$ (right panel), from Ref. [51].

section 7.4. We showed in problem 1.3 that, for the collinear reaction $\gamma + q \to q$, the cross section for absorption of longitudinally polarized photons vanishes. We remarked, however, that an intrinsic transverse momentum for the partons would give rise to a nonvanishing longitudinal-to-transverse-cross-section ratio characterized by

$$\frac{\sigma_S}{\sigma_T} \propto \frac{\langle p_\perp^2 \rangle}{Q^2}. \tag{8.5.63}$$

Evidently quarks (and gluons) that arise from the virtual dissociation of gluons or quarks need not be collinear with the incident hadron, and thus these processes may also induce $\sigma_S/\sigma_T \neq 0$, even if the partons that carry electroweak charges are all spin-$\frac{1}{2}$ quarks. By measuring the longitudinal cross section, we not only test the Callan–Gross relation [54] $F_2 = 2xF_1$ for spin-$\frac{1}{2}$ partons, we also peer into the proton structure beyond the naive parton model. The sensitivity of $\sigma_S/\sigma_T$ to parton splitting means that the effect should be most prominent at small values of $x$, where it is sensitive to the gluon distribution function.

We recast the reduced cross section (8.5.62) for deeply inelastic neutral-current $ep$ scattering as

$$\sigma_r^{NC}(x, Q^2) = F_2(x, Q^2) - \frac{y^2 \, F_L(x, Q^2)}{1 + (1 - y)^2}, \tag{8.5.64}$$

where

$$F_L(x, Q^2) = F_2(x, Q^2) - 2xF_1(x, Q^2) \tag{8.5.65}$$

is the longitudinal structure function. Neglecting the transverse momentum of the partons inside the nucleon, the elementary processes at order $\alpha_s$ that give rise to quarks and antiquarks with large transverse momentum are

$$(q_i, \bar{q}_i) + \gamma^* \to (q_f, \bar{q}_f) + g, \tag{8.5.66a}$$

$$g + \gamma^* \to q + \bar{q}, \tag{8.5.66b}$$

where $\gamma^*$ is the virtual photon of virtuality $Q^2$. As we did in problem 1.3, it is convenient to analyze the problem in the frame in which the virtual photon carries no energy. The calculation entails keeping track of the initial parton's longitudinal momentum fraction and the struck (anti)quark's transverse momentum. The longitudinal structure function is [55]

$$F_L(x, Q^2) = \frac{\alpha_s(Q^2)}{2\pi} x^2 \left\{ \int_x^1 \frac{du}{u^3} \left[ \frac{8}{3} F_2(x, Q^2) + 4 \sum_i e_i^2\, u G(u, Q^2) \left(1 - \frac{x}{u}\right) \right] \right\}, \tag{8.5.67}$$

where the sum runs over active quark flavors and the two contributions correspond to processes (8.5.66a) and (8.5.66b), respectively. The relative importance of the two contributions is determined by the relative weights of the $q + \bar{q}$ and gluon densities. At small values of $x$, the gluon density predominates, and so a measurement of $F_L(x, Q^2)$ should provide direct information about the small-$x$ gluon distribution. Measurements by the H1 and ZEUS Collaborations at HERA agree well with prior knowledge of the gluon distributions and the expectations of perturbative QCD [56].

### 8.5.4 ASYMPTOTIC LIMIT OF HADRON STRUCTURE

The evolution of the remaining moments of parton distributions is somewhat more complicated because of mixing between the gluon distribution $G(x, \tau)$ and the "singlet" quark distribution,

$$\Sigma(x, \tau) \equiv \sum_{i=\text{flavors}} (q_i(x, \tau) + \bar{q}_i(x, \tau)). \tag{8.5.68}$$

The evolution of the moments is described by coupled Altarelli–Parisi equations, which may be represented in matrix form by

$$\frac{d}{d\tau} \begin{pmatrix} \Sigma_n \\ G_n \end{pmatrix} = \frac{\alpha_s}{2\pi} \begin{pmatrix} A_n(q \leftarrow q) & 2n_f A_n(q \leftarrow g) \\ A_n(g \leftarrow q) & A_n(g \leftarrow g) \end{pmatrix} \begin{pmatrix} \Sigma_n \\ G_n \end{pmatrix}. \tag{8.5.69}$$

Of special interest is the behavior of the $n = 2$ moments as $Q^2 \to \infty$, which describe the momentum fractions carried by quarks and gluons in that asymptotic regime. The matrix of moments of the splitting functions for $n = 2$ is

$$A_2 = \begin{pmatrix} -16/9 & n_{\mathrm{f}}/3 \\ 16/9 & -n_{\mathrm{f}}/3 \end{pmatrix}, \tag{8.5.70}$$

which manifestly satisfies the requirements of momentum conservation.

To consider the $Q^2 \to \infty$ limit, we diagonalize the two-by-two matrix $A_2$. It has two eigenvalues,

$$\left. \begin{aligned} \lambda_+ &= -\frac{(16 + 3n_{\mathrm{f}})}{9} \\ \lambda_- &= 0 \end{aligned} \right\}, \tag{8.5.71}$$

the weights of which will evolve with $Q^2$ as

$$\left( \frac{\alpha_{\mathrm{s}}(0)}{\alpha_{\mathrm{s}}(\tau)} \right)^{6\lambda_\pm/(33 - 2n_{\mathrm{f}})}, \tag{8.5.72}$$

and so the eigenvector corresponding to $\lambda_+$ disappears in the limit $Q^2 \to \infty$. The eigenvector corresponding to the zero eigenvalue $\lambda_-$ is easily found to be

$$\begin{pmatrix} \Sigma_2^\infty \\ G_2^\infty \end{pmatrix} = \frac{1}{16 + 3n_{\mathrm{f}}} \begin{pmatrix} 3n_{\mathrm{f}} \\ 16 \end{pmatrix}. \tag{8.5.73}$$

Thus for six active flavors of quarks, the momentum fraction in gluons is

$$\int_0^1 dx\, x G(x, Q^2 \to \infty) = G_2^\infty = \frac{8}{17} \approx 0.47, \tag{8.5.74}$$

and each species of quark or antiquark carries

$$\int_0^1 dx\, x q_{\mathrm{s}}(x, Q^2 \to \infty) = \tfrac{1}{2}\Sigma_2^\infty = \frac{3}{68} \approx 0.044. \tag{8.5.75}$$

The equilibrium partition reflects both the relative strengths of the quark–antiquark–gluon and three-gluon couplings and the number of active fermion species. It is easy to verify that the momentum sum rule

$$\sum_{\substack{i = \text{parton} \\ \text{species}}} \int_0^1 dx\, x f_i(x) = 1 \tag{8.5.76}$$

is satisfied:

$$0 \text{ (valence)} + \tfrac{8}{17} \text{ (gluons)} + 6 \text{ flavors} \cdot 2 \text{ (quarks + antiquarks)} \cdot \tfrac{3}{68} = 1. \tag{8.5.77}$$

Because the valence-quark share of the momentum vanishes in the $Q^2 \to \infty$ limit, the result (8.5.77) is universal, for any hadron.

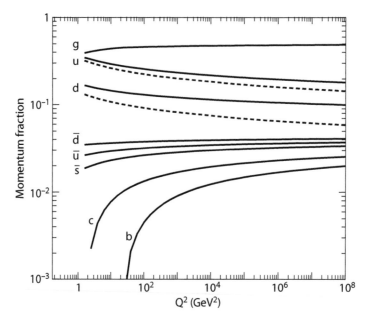

**Figure 8.35.** $Q^2$-evolution of the momentum fractions carried by various parton species in the proton, according to the CTEQ6.6M parton distributions (Ref [57].). For up and down quarks, the solid lines show the total (valence plus sea) momentum fractions, and the dashed lines show the valence contribution alone.

To gain more insight into the question What is a proton? let us look at the flavor content of the proton at finite values of $Q^2$, as measured by the momentum fraction

$$\int_0^1 dx\, x f_i(x, Q^2) \tag{8.5.78}$$

carried by each parton species. This is shown in figure 8.35 for the CTEQ6.6M parton distributions [57]. As $Q^2$ increases, momentum is partitioned more and more equally among the quark and antiquark flavors, reflecting the trend toward the asymptotic values (8.5.61), (8.5.74), and (8.5.75), expected in QCD with six quark flavors and no light, colored superpartners. For the foreseeable future, our experiments probe the proton structure very far from asymptopia!

## 8.5.5 BEYOND THE PARTON-MODEL PARADIGM

The great majority of phenomenological studies hew to the parton-model idealization that partons are independent and have negligible transverse momentum. Accordingly, the standard sets of parton distribution functions provide detailed information about how longitudinal momentum is partitioned among quarks, antiquarks, and gluons, but no information about transverse degrees of freedom or correlations among partons. It is reasonable to imagine that, in the limit as $x \to 1$, the information that a proton is a spin-$\frac{1}{2}$ particle (for example) should manifest itself in correlations among the partons. Generalized parton distributions

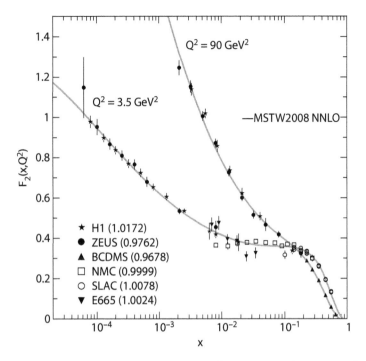

**Figure 8.36.** The proton structure function $F_2(x, Q^2)$ given at $Q^2 = 3.5 \, \text{GeV}^2$ and $Q^2 = 90 \, \text{GeV}^2$ (from Ref. [24]). The individual data sets have been renormalized by the factors shown in brackets in the legend, as determined in the MSTW2008 global analysis (Ref. [51]..)

inferred from exclusive scattering processes are a tool for probing such features of hadron structure [58]. Spectroscopic studies make it plausible that baryon properties are consistent with an important quark–diquark component of the baryon wave function (Ref. [15]). If so, we should expect to find some evidence in scattering data for the presence of diquarks.

Another challenge to the idealization that partons are independent and non-interacting arises at small values of $x$ and large values of $Q^2$. In this regime, the number densities $q(x, Q^2)$, $\bar{q}(x, Q^2)$, and $G(x, Q^2)$ calculated using the evolution equations (Ref. [6]) may become so large that the partons overlap and recombination may occur [59]. The H1 and ZEUS experiments at the $e^\mp p$ collider HERA, which operated at c.m. energies up to $\sqrt{s} = 320 \, \text{GeV}$, probed the small-$x$ domain and established a rapid rise in the parton densities as $x \to 0$, as shown in the plot of the structure function $F_2(x, Q^2)$ in figure 8.36. However, the anticipated recombination phenomena [60] have not yet been demonstrated.

## 8.6 JETS IN HADRON–HADRON COLLISIONS

Not every aspect of the strong interactions among hadrons, even at high energies or at large momentum transfer, can be calculated within perturbative QCD. Because colored quarks and gluons are confined, a full calculation necessarily entails intrinsically nonperturbative elements: the hadron structure for the beam and target, and how partons materialize into hadrons. In the parton model, the problem separates into the perturbative computation of short-distance matrix elements specific to the reaction in question plus long-distance components such as parton distribution functions and fragmentation functions that are hoped to be universal. The latter must for now be taken from experiment; how they evolve with scale can be calculated in perturbative QCD. The distinction between short and long distances (or short and long time scales) is reminiscent of the Born–Oppenheimer approximation in molecular physics, which separates electronic and nuclear motion on the basis of different intrinsic time scales. In a number of important settings, QCD *factorization theorems* justify the separation into independent short-distance and long-distance pieces [61].

The reactions that may occur at lowest order, $O(\alpha_s^2)$, in QCD are two-body to two-body processes leading to final states that consist of two outgoing partons observed as two jets of hadrons, with equal and opposite transverse momenta. It is convenient to express the cross section in terms of the rapidities $y_1$ and $y_2$ of the two jets and their common transverse momentum, $p_\perp$, with respect to the beam direction chosen as the z-axis.[†] Because of its transformation properties under Lorentz boosts, the rapidity,

$$y \equiv \frac{1}{2} \ln \left| \frac{E + p_z}{E - p_z} \right|, \tag{8.6.1}$$

is a highly convenient longitudinal variable for an individual particle or a jet. The pseudorapidity,

$$\eta \equiv - \ln \tan \frac{\theta}{2}, \tag{8.6.2}$$

is a close approximation to the rapidity in the setting of collider detectors and can be measured, even when the mass of the outgoing object is unknown (cf. problem 8.18).

Using the parton-model picture of hadron–hadron collisions introduced in Figure 7.16 and (7.5.2), we may write the differential cross section as

$$\frac{d\sigma}{dy_1 dy_2 dp_\perp} = \sum_{i,j} \frac{2\pi p_\perp}{(1 + \delta_{ij})s} \left[ f_i^{(a)}(x_a, \hat{s}) f_j^{(b)}(x_b, \hat{s}) \hat{\sigma}_{ij}(\hat{s}, \hat{t}, \hat{u}) + (i \leftrightarrow j) \right], \tag{8.6.3}$$

where for simplicity we have chosen $\hat{s}$ as the scale at which to evaluate the parton distribution functions. The summation runs over all contributing parton–parton configurations. It is implicit in this formula that, when necessary for comparison with experiment, the transition from outgoing partons to detected objects will be parametrized or modeled by a parton-shower Monte Carlo.

Expectations derived from perturbative QCD have been tested extensively in studies of jets produced in $\bar{p}p$ collisions at the Tevatron and are being tested further in $pp$ collisions at the Large Hadron Collider. A representative example, shown in figure 8.37, is the doubly differential dijet cross section measured by the D0

---

[†] We neglect the intrinsic transverse momentum carried by the partons.

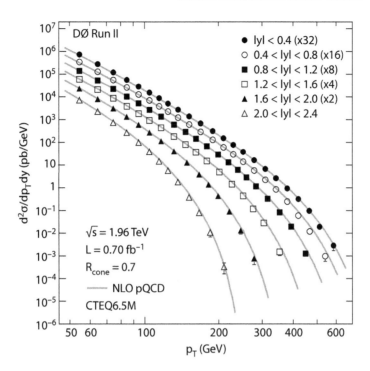

**Figure 8.37.** The inclusive jet cross section measured in $\bar{p}p$ collisions at $\sqrt{s} = 1.96$ TeV by the D0 Collaboration (Ref. [62]) as a function of transverse momentum in 6 rapidity bins. The data points are multiplied by 2, 4, 8, 16, and 32 for the bins $1.6 < |y| < 2.0$, $1.2 < |y| < 1.6$, $0.8 < |y| < 1.2$, $0.4 < |y| < 0.8$, and $|y| < 0.4$, respectively. Solid curves show next-to-leading-order (one-loop) perturbative QCD predictions. A 6.1% uncertainty on the integrated luminosity is not included. Theoretical predictions carry an uncertainty of approximately 10%.

Collaboration at Fermilab. Over 10 decades, the observed cross section follows the calculated values. Studies of rapidity-transverse momentum distributions, angular distributions, and dijet invariant-mass distributions probe the completeness of QCD and test the notion that the quarks are structureless (cf. problem 8.20). As the perturbative-QCD description of dijet production is validated, it becomes an ever-more-powerful tool for characterizing the structure of the proton through parton distribution functions. Many examples of new physics hypothesized for the Large Hadron Collider involve phenomena at large values of $\hat{s}$. To anticipate signals and backgrounds at high scales, we require secure knowledge of the parton distribution functions at large values of $x$, because $\hat{s} = x_a x_b s$. We can explore the parton distribution functions in that regime by studying low-mass dijet production at large values of $y_{boost}$ (cf. problem 8.21).

The signals and backgrounds that we wish to study at the Large Hadron Collider entail multiparton final states far greater in complexity than the dijet events mentioned here. Sixty years ago, Feynman diagrams entered as an astoundingly efficient calculus for higher-order effects in quantum electrodynamics. So they remain, and remarkable advances in methodology and automated calculation have

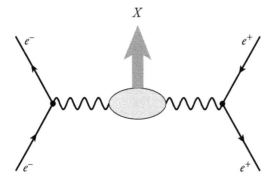

**Figure 8.38.** The two-photon process $e^+e^- \to e^+e^- X$.

brought into reach calculations that only a short time ago would have been prohibitively time consuming. But as our concerns in quantum chromodynamics have moved toward processes involving many external particles, we have found the method of Feynman diagrams to be highly inefficient. A new wave of inventions is vastly expanding the realm of conceivable calculations in perturbative QCD, extending to next-to-leading order amplitudes and beyond [63].

## 8.7 TWO-PHOTON PROCESSES AND THE PHOTON-STRUCTURE FUNCTION

The idea of the equivalent-photon (or Weizsäcker–Williams) approximation (Ref. [41]), which was prominent in our derivation of the parton splitting functions, leads naturally to the possibility of studying $\gamma\gamma$ collisions in electron–positron or electron–electron colliding beams experiments, as indicated in figure 8.38. In the double-equivalent-photon approximation, the cross section for the reaction

$$e^+e^- \to e^+e^- X \qquad (8.7.1)$$

at symmetric beam energies $E$ is given by [64]

$$\sigma_{ee \to eeX}(E) \simeq 2 \left(\frac{\alpha}{\pi}\right)^2 \ln^2 \left(\frac{E}{m_e}\right) \int_0^{4E^2} \frac{d\hat{s}}{\hat{s}} \, f\left(\frac{\sqrt{\hat{s}}}{2E}\right) \sigma_{\gamma\gamma \to X}(\hat{s}), \qquad (8.7.2)$$

where $\sqrt{\hat{s}}$ is the invariant mass of the produced system, the flux factor is

$$f(x) = (2 + x^2)^2 \ln \left(\frac{1}{x}\right) - (1 - x^2)(3 + x^2), \qquad (8.7.3)$$

and $m_e$ is the electron mass. Indeed, QED reactions such as

$$e^+e^- \to e^+e^-\ell^+\ell^- \qquad (8.7.4)$$

were already considered more than 75 years ago by Landau and Lifshitz and Williams [65]. In modern notation and in the extreme relativistic limit, their result can be expressed as

$$\sigma \simeq \frac{224\alpha^4}{27\pi m_\ell^2} \ln^2\left(\frac{E}{m_e}\right) \ln\left(\frac{E}{m_\ell}\right). \tag{8.7.5}$$

This cross section becomes increasingly important at high energies and competes, with growing success, with the annihilation channel. For example,

$$\frac{\sigma(e^+e^- \to e^+e^-\mu^+\mu^-)}{\sigma(e^+e^- \to \mu^+\mu^-)} \simeq 1 \tag{8.7.6}$$

for $E \sim 1$ GeV and grows to $\sim 10^3$ at beam energies of 15–20 GeV. Whatever its physics interest (cf. the final paragraph of section 8.5.1) [66], this two-photon process is a significant source of experimental background.

Two-photon production of discrete resonances is also of practical importance. The cross section for the production of a spin-$J$ particle $h^0$ in the reaction

$$e^+e^- \to e^+e^-h^0 \tag{8.7.7}$$

is given by

$$\sigma(E) = 16\alpha^2 \frac{\Gamma(h^0 \to \gamma\gamma)}{M_h^3}(2J+1)\ln^2\left(\frac{E}{m_e}\right) f\left(\frac{M_h}{2E}\right). \tag{8.7.8}$$

This implies, for example, a ratio

$$\frac{\sigma(e^+e^- \to e^+e^-\pi^0)}{\sigma(e^+e^- \to \mu^+\mu^-)} \tag{8.7.9}$$

that is of order unity for $E \simeq 5$ GeV [67]. This method has been used to excellent effect to determine (for example) the decay rate $\Gamma(\chi_{c2}(1P) \to \gamma\gamma)$ [68], and to discover the $\chi_{c2}(2P)(3930)$ radial excitation [69].

From discrete resonances, we pass on to the inclusive production of hadrons, which is represented in the parton model by the reaction

$$e^+e^- \to e^+e^-q\bar{q} \tag{8.7.10}$$

depicted in figure 8.39. This process yields a sizable cross section at high energies,

$$\frac{\sigma(e^+e^- \to e^+e^- + \text{hadrons})}{\sigma(e^+e^- \to e^+e^-\mu^+\mu^-)} \simeq 3 \sum_{\text{flavors}} e_q^4 = \frac{35}{27}, \tag{8.7.11}$$

in the standard model of fractionally charged $u$-, $d$-, $s$-, $c$-, and $b$-quarks. As problem 8.27 will show, the ratio (8.7.11) is specific to the model of fractionally charged quarks and, therefore, should permit a decisive test of that hypothesis.

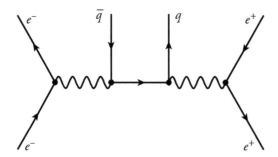

**Figure 8.39.** The two-photon process $e^+e^- \to e^+e^- +$ hadrons.

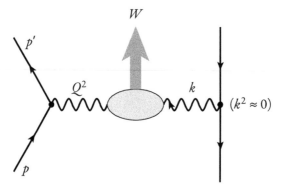

**Figure 8.40.** Kinematics for the measurement of the photon structure function in $e^+e^-$ collisions.

Experiments have established that the predicted class of events exists and that the fractional charge assignment is favored over an integer-charge quark model. Measurements of jet production and inclusive charged-hadron production in $\gamma\gamma$ collisions are in accord with QCD predictions, now at next-to-leading order [70].

In all these applications, the dominant contribution to the cross section is from two nearly real photons, because of the enhancement in flux due to the photon propagator. Consider instead the kinematic configuration sketched in figure 8.40, in which a highly virtual photon is used to probe the structure of the second photon, which we shall take to be nearly real. The relevant class of events can be selected by observing the final-state electron and positron. Write the four-momenta of the incident and scattered electrons as usual as

$$p = (E; 0, 0, E),$$
$$p' = (E'; E' \sin\theta, 0, E' \cos\theta), \tag{8.7.12}$$

so the probe carries momentum $q = p - p'$ and the "target" photon has four-momentum

$$k \approx (E_\gamma; 0, 0, -E_\gamma). \tag{8.7.13}$$

As usual for deeply inelastic scattering processes, we define [cf. (7.4.7) and (7.4.4)]

$$Q^2 = -q^2 = 4EE' \sin^2\left(\frac{\theta}{2}\right), \tag{8.7.14}$$

$$\nu = k \cdot q = 2E_\gamma \left[E - E' \cos^2\left(\frac{\theta}{2}\right)\right]. \tag{8.7.15}$$

It is also useful to define the scaling variables

$$x \equiv \frac{Q^2}{2\nu} = \frac{Q^2}{Q^2 + W^2}, \tag{8.7.16}$$

where

$$W^2 = (k+q)^2 = 2\nu - Q^2 \tag{8.7.17}$$

and

$$y \equiv \frac{\nu}{p \cdot k} = \frac{\nu}{2EE_\gamma} = 1 - \left(\frac{E'}{E}\right)\cos^2\left(\frac{\theta}{2}\right). \tag{8.7.18}$$

As in the case of lepton–nucleon scattering, we write the target tensor in the form

$$W_{\mu\nu} = W_1\left(-g_{\mu\nu} + \frac{q_\mu q_\nu}{q^2}\right) + W_2\left(k_\mu - \frac{k \cdot q}{q^2}q_\mu\right)\left(k_\nu - \frac{k \cdot q}{q^2}q_\nu\right), \tag{8.7.19}$$

so that

$$\frac{d^2\sigma}{dx\,dy} = \frac{16\pi\alpha^2 EE_\gamma}{Q^4}[(1-y)F_2(x, Q^2) + xy^2 F_1(x, Q^2)], \tag{8.7.20}$$

where

$$F_2 = \nu W_2, \qquad F_1 = W_1. \tag{8.7.21}$$

The photon is a particularly interesting target particle for deeply inelastic scattering because of its two-component nature. In many circumstances, the photon displays a hadronic character that may largely be understood in the context of the vector–meson-dominance model, the conjecture that the total (hadronic) photoabsorption cross section may be calculated from diffractive photoproduction of vector mesons. However, as our preceding discussion has reminded us, the photon also has a pointlike component because of its elementary coupling to charged particles. This latter component should be observable in hard-scattering processes. What is more, as was first recognized by Witten [71], the contributions of this pointlike component to the structure functions may be calculable a priori even in the presence of strong interactions.

The structure functions of a photon target may be computed in the parton-model approximation—that is, neglecting strong-interaction corrections—directly

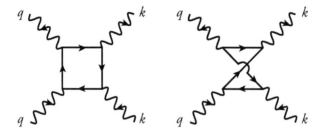

**Figure 8.41.** Feynman diagrams for the photon structure function in the parton model.

from the box diagrams shown in figure 8.41 [72]. However, we may also proceed in parallel with our earlier discussion (cf. section 8.5) simply by computing to leading order the *electromagnetic* evolution of the quark distribution within the photon [73]. We write, in analogy with (8.5.10),

$$\frac{dq}{d\tau}(x, \tau) = \frac{\alpha(\tau)}{2\pi} \int_x^1 \frac{dy}{y} P_{q \leftarrow \gamma}\left(\frac{x}{y}\right) \gamma(y, \tau), \tag{8.7.22}$$

where we begin from a monochromatic photon beam characterized by

$$\gamma(y, \tau) = \delta(y - 1). \tag{8.7.23}$$

The essential form of the splitting function was encountered earlier in (8.5.42). It is nothing but

$$P_{q \leftarrow \gamma}(z) = 3e_q^2[z^2 + (1 - z)^2], \tag{8.7.24}$$

where 3 is a color factor. Consequently, to this approximation, we find

$$\frac{dq}{d\tau}(x, \tau) = \frac{3\alpha e_q^2}{2\pi}[x^2 + (1 - x)^2], \tag{8.7.25}$$

which can be integrated at once to yield

$$q(x, \tau) = \frac{3\alpha e_q^2}{2\pi}[x^2 + (1 - x)^2] \ln\left(\frac{Q^2}{\Lambda^2}\right) + \cdots \tag{8.7.26}$$

plus nonlogarithmic terms that are asymptotically negligible. In anticipation of the coming discussion on strong-interaction corrections, we have exercised our freedom to write the scale of $Q^2$ in the logarithm as $\Lambda^2$. The $\ln(Q^2)$ dependence, which reflects the intuitive picture of increased resolution of the virtual fluctuations as $Q^2$ increases, may be traced in the diagram language to the pointlike $\gamma q \bar{q}$ coupling and the integration over transverse momentum in the reaction $\gamma \gamma_{\text{virtual}} \to q\bar{q}$. The same integration over transverse momentum gives rise to a nonvanishing longitudinal cross section, just as happened under the influence of QCD evolution in deeply inelastic lepton scattering.

Now consider the strong-interaction evolution as well, according to

$$\frac{dq_i}{d\tau}(x, \tau) = \frac{1}{2\pi} \int_x^1 \frac{dy}{y} \left\{ \alpha(\tau) P_{q \leftarrow \gamma} \left( \frac{x}{y} \right) \gamma(y, \tau) e_i^2 \right. \tag{8.7.27}$$

$$\left. + \alpha_s(\tau) \left[ P_{q \leftarrow q} \left( \frac{x}{y} \right) q_i(y, \tau) + P_{q \leftarrow g} \left( \frac{x}{y} \right) G(y, \tau) \right] \right\}$$

$$\frac{dG}{d\tau}(x, \tau) = \frac{\alpha_s(\tau)}{2\pi} \int_x^1 \frac{dy}{y} \left[ P_{g \leftarrow q} \left( \frac{x}{y} \right) \sum_{\text{flavors}} (q_i(y, \tau) + \bar{q}_i(y, \tau)) \right.$$

$$\left. + P_{g \leftarrow g} \left( \frac{x}{y} \right) G(y, \tau) \right], \tag{8.7.28}$$

and for the photon itself an evolution equation analogous to (8.5.12). It is convenient to rewrite these equations in terms of singlet and nonsinglet quark distributions as

$$\Sigma(x, \tau) = \sum_{i=\text{flavors}} [q_i(x, \tau) + \bar{q}_i(x, \tau)] \tag{8.7.29}$$

and

$$\Delta^{(i)}(x, \tau) = q_i(x, \tau) - \left( \frac{1}{2n_f} \right) \Sigma(x, \tau), \tag{8.7.30}$$

so that

$$\frac{d\Delta^{(i)}}{d\tau}(x, \tau) = \frac{1}{2\pi} \int_x^1 \frac{dy}{y} \left[ \alpha(\tau)(e_i^2 - \langle e^2 \rangle) P_{q \leftarrow \gamma} \left( \frac{x}{y} \right) \gamma(y, \tau) \right.$$

$$\left. + \alpha_s(\tau) P_{q \leftarrow q} \left( \frac{x}{y} \right) \Delta^{(i)}(x, \tau) \right], \tag{8.7.31}$$

where

$$\langle e^2 \rangle \equiv \left( \frac{1}{2n_f} \right) \sum_{i=\text{flavors}} e_i^2, \tag{8.7.32}$$

and

$$\frac{d\Sigma}{d\tau}(x, \tau) = \frac{1}{2\pi} \int_x^1 \frac{dy}{y} \left\{ 2n_f \alpha(\tau) \langle e^2 \rangle P_{q \leftarrow \gamma} \left( \frac{x}{y} \right) \gamma(x, \tau) \right. \tag{8.7.33}$$

$$\left. + \alpha_s(\tau) \left[ P_{q \leftarrow q} \left( \frac{x}{y} \right) \Sigma(x, \tau) + 2n_f P_{q \leftarrow g} \left( \frac{x}{y} \right) G(y, \tau) \right] \right\}.$$

Again, the solution is straightforward in terms of moments. The important features are illustrated by the computation of the nonsinglet distribution. Taking moments of (8.7.30), we find that

$$\frac{d\Delta_n^{(i)}}{d\tau} = \frac{\alpha_s(\tau)}{2\pi} A_n(q \leftarrow q)\Delta_n^{(i)} + \frac{\alpha}{2\pi}(e_i^2 - \langle e^2 \rangle) A_n(q \leftarrow \gamma),$$ (8.7.34)

where

$$A_n(q \leftarrow \gamma) = 3 \left[ \frac{n^2 + n + 2}{n(n+1)(n+2)} \right].$$ (8.7.35)

Upon using the expression (8.3.30) for $\alpha_s(\tau)$, (8.7.34) becomes

$$\frac{d\Delta_n^{(i)}}{d\tau} = \frac{A_n(q \leftarrow q)\Delta_n^{(i)}}{2\pi b\tau} + \frac{\alpha}{2\pi}(e_i^2 - \langle e^2 \rangle) A_n(q \leftarrow \gamma),$$ (8.7.36)

where, as usual for QCD,

$$b = \frac{33 - 2n_f}{12\pi}.$$ (8.7.37)

The inhomogeneous differential equation has the elementary solution,

$$\Delta_n^{(i)}(\tau) = \Delta_n^{(i)}(0)[b\tau]^{A_n(q \leftarrow q)/2\pi b} + \frac{\alpha(e_i^2 - \langle e^2 \rangle) A_n(q \leftarrow \gamma)\tau}{2\pi[1 - A_n(q \leftarrow q)/2\pi b]},$$ (8.7.38)

as may be verified by direct computation. The second term has the form of the parton-model result [cf. the second term of (8.7.36) and problem 8.26], renormalized by the factor $[1 - A_n(q \leftarrow q)/2\pi b]^{-1}$. The rescaling factor is less than unity for $n > 1$ because the corresponding nonsinglet moments are negative. The first term in (8.7.38), which by virtue of the presence of $\Delta_n^{(i)}(0)$ can be said to contain all the details of the hadronic structure of the photon, behaves as a constant (for $n = 1$) or as a negative power of $\tau = \ln(Q^2/\Lambda^2)$ (for $n > 1$). Thus, it is the pointlike (second) term that dominates as $Q^2 \to \infty$, and the asymptotic behavior of the nonsinglet quark distribution is an absolute prediction, to leading order in renormalization-group-improved perturbation theory. The dominant effect of the strong interactions is to soften the quark distribution by gluon radiation and, thus, to diminish the structure functions near $x = 1$.

The coupled differential equations for the quark singlet and gluon moments may be solved in similar fashion, using the method of section 8.5, and, thus, the observable structure functions may be evaluated. The effect upon $F_2^\gamma$ of the QCD corrections that we have just explored is indicated in figure 8.42. There we see as well that higher-order corrections tend to diminish the photon structure function further in the neighborhood of $x = 1$.

Figure 8.43 confronts a next-to-leading-order calculation [75] with values of $F_2^\gamma(x, Q^2)$ measured by the OPAL Collaboration [76]. These data, and the much more extensive collection in the *Review of Particle Physics* [77], clearly show the expected growth with $\ln Q^2$ and indicate an increase with $x$ at small and intermediate values of $x$. The photon structure function exhibits positive scaling violations for all

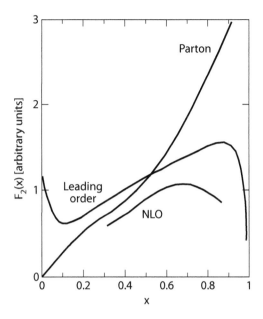

**Figure 8.42.** Sketch of the photon structure function behavior in the parton model, at leading order in QCD and incorporating higher-order corrections (after Ref. [74]).

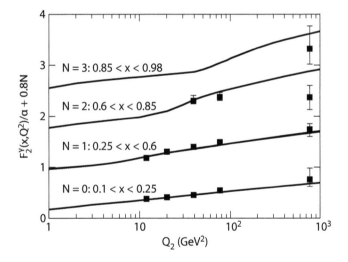

**Figure 8.43.** Comparison of a next-to-leading-order calculation of $F_2^\gamma(x, Q^2)$ with measurements by the OPAL Collaboration at LEP (after Ref. [75]).

values of $x$, as apposed to the proton structure function (cf. figures 8.33, 8.34, and 8.36), for which the scaling violations are positive at small values of $x$ but negative at large values of $x$. The pointlike character of the photon means that the direct production of $q\bar{q}$ pairs at large $x$ is more efficient than the degradation of the quark population due to gluon radiation. There is a suggestion in the full data set of the predicted turnover as $x \to 1$, but the result is not definitive. Overall, $F_2^\gamma(x, Q^2)$ and the quark distributions $q^\gamma(x, Q^2)$ are quite well known, but we have only schematic knowledge of the gluon distribution $G^\gamma(x, Q^2)$.

From perturbative applications of QCD, we now pass on to three topics related to the realm in which the strong interactions are strong: confinement of colored objects, QCD as a model for dynamical electroweak symmetry breaking, and an idealized analogue of QCD with a very large number of colors.

## 8.8 COLOR CONFINEMENT

The applications of QCD that we have considered to this point all are made possible by asymptotic freedom and the aptness of perturbation theory. The other face of QCD, the low-energy realm in which the strong interactions are too strong to treat in perturbation theory, governs the structure of hadrons, the low-$Q^2$ parton distributions that are the starting point for perturbative-QCD evolution, and emergent phenomena, including chiral symmetry breaking. We shall first consider a very idealized argument that makes plausible the possibility of color confinement; then we shall cite some of the outstanding achievements of lattice QCD.

### 8.8.1 A DIELECTRIC ANALOGY

We have seen that it is typical in field theories for the coupling constant to depend on the distance scale. This dependence can be expressed in terms of a dielectric constant $\epsilon$. We define

$$\epsilon(r_0) \equiv 1 \tag{8.8.1}$$

and write

$$g^2(r) = \frac{g^2(r_0)}{\epsilon(r)}. \tag{8.8.2}$$

We assert that the implication of asymptotic freedom is that in QCD, the effective color charge decreases at short distances and increases at large distances. In other words, the dielectric "constant" will obey

$$\epsilon(r) > 1, \quad \text{for } r < r_0, \tag{8.8.3a}$$

$$\epsilon(r) < 1, \quad \text{for } r > r_0. \tag{8.8.3b}$$

Indeed, to second order in the strong coupling we may write

$$\epsilon(r) = \left[1 + \frac{1}{2\pi}\frac{g^2(r_0)}{4\pi}\left(11 - \frac{2n_f}{3}\right)\ln\left(\frac{r}{r_0}\right) + O(g^4)\right]^{-1} \tag{8.8.4}$$

in QCD, where $n_f$ is the number of active quark flavors.

Let us now consider an idealization based upon electrodynamics. In quantum electrodynamics, we choose

$$\epsilon_{\text{vacuum}} = 1 \tag{8.8.5}$$

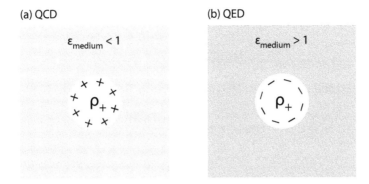

**Figure 8.44.** Charge induced by a positive test charge placed at the center of a hole in a dielectric medium. (a) Dia-electric case $\epsilon_{medium} < 1$ hoped to resemble QCD. (b) Dielectric case $\epsilon_{medium} > 1$ of normal electrodynamics.

and can show [78] that physical media have $\epsilon > 1$. The displacement field is

$$\mathbf{D} = \mathbf{E} + \mathbf{P}, \tag{8.8.6}$$

and atoms are polarizable with $\mathbf{P}$ parallel to the applied field $\mathbf{E}$, so that $|\mathbf{D}| > |\mathbf{E}|$. Because the dielectric constant is defined through

$$\mathbf{D} = \epsilon \mathbf{E} \tag{8.8.7}$$

in these simple circumstances, we conclude that $\epsilon > 1$.

Now let us consider, in contrast to the familiar situation, the possibility of a dielectric medium with

$$\epsilon_{medium} = 0, \tag{8.8.8}$$

a perfect dia-electric medium, or at least

$$\epsilon_{medium} \ll 1, \tag{8.8.9}$$

a very effective dia-electric medium. We can easily show that if a test charge is placed within the medium, a bubble will develop around it. To see this, consider the arrangement depicted in figure 8.44(a), a (spherically symmetric) positive charge distribution $\rho_+$ placed in the medium. Suppose that a bubble is formed. Then, because the dielectric constant of the medium is less than unity, the induced charge on the inner surface of the bubble will also be positive. The test charge and the induced charge thus repel, and the bubble is stable against collapse. In normal QED, the induced charge will be negative, as indicated in figure 8.44(b), and will attract the test charge. In a normal dielectric medium, a bubble is unstable against collapse.

By minimizing the energy, we can estimate the size of a stable bubble in the dia-electric medium. Within the bubble, the electrical energy $W_{in}$ is finite and independent of the dielectric constant of the medium. The displacement field is radial—and, hence, continuous—across the spherical boundary at radius $R$. Thus it is given outside the bubble by

$$\mathbf{D}_{\text{out}}(r > R) = \frac{\hat{\mathbf{r}} Q}{r^2}, \tag{8.8.10}$$

where $Q$ is the total test charge. The induced charge density on the bubble surface is

$$\begin{aligned}
\sigma_{\text{induced}} &= \frac{(1 - \epsilon)\,|\mathbf{D}(R)|}{4\pi\epsilon} \\
&= \frac{(1 - \epsilon)Q}{4\pi\epsilon R^2},
\end{aligned} \tag{8.8.11}$$

which has the same sign as $Q$, as earlier asserted. Outside the bubble, the electric field is determined by the total interior charge

$$Q + \frac{(1 - \epsilon)Q}{\epsilon} = \frac{Q}{\epsilon}, \tag{8.8.12}$$

so that

$$\mathbf{E}_{\text{out}}(r > R) = \frac{\hat{\mathbf{r}} Q}{\epsilon r^2}. \tag{8.8.13}$$

The energy stored in electric fields outside the bubble is then

$$\begin{aligned}
W_{\text{out}} &= \frac{1}{8\pi} \int d^3\mathbf{r}\, \mathbf{D}_{\text{out}}(r) \cdot \mathbf{E}_{\text{out}}(r) \\
&= \frac{1}{2} \int_R^\infty r^2\, dr\, \frac{Q^2}{\epsilon r^4} = \frac{Q^2}{2\epsilon R}.
\end{aligned} \tag{8.8.14}$$

As the dielectric constant of the medium approaches zero, $W_{\text{out}}$ becomes large compared with $W_{\text{in}}$, so that the total energy

$$W_{\text{el}} \equiv W_{\text{in}} + W_{\text{out}} \to W_{\text{out}}, \qquad \text{as } \epsilon \to 0. \tag{8.8.15}$$

We must consider as well the energy required to hew such a bubble out of the medium. For a bubble of macroscopic size, it is reasonable to expect that

$$W_{\text{bubble}} = \frac{4\pi R^3}{3} v + 4\pi R^2 s + \cdots, \tag{8.8.16}$$

where $v$ and $s$ are nonnegative constants. The total energy of the system,

$$W = W_{\text{el}} + W_{\text{bubble}}, \tag{8.8.17}$$

can now be minimized with respect to $R$. In the regime where the volume term dominates $W_{\text{bubble}}$, the minimum occurs at

$$R = \left( \frac{Q^2}{2\epsilon} \times \frac{1}{4\pi v} \right)^{1/4} \neq 0, \tag{8.8.18}$$

for which

$$W_{el} \approx \left( \frac{Q^2}{2\epsilon} \right)^{3/4} (4\pi v)^{1/4} \qquad (8.8.19)$$

and

$$W_{bubble} \approx \frac{1}{3} \left( \frac{Q^2}{2\epsilon} \right)^{3/4} (4\pi v)^{1/4}, \qquad (8.8.20)$$

so that

$$W \approx \frac{4}{3} \left( \frac{Q^2}{2\epsilon} \right)^{3/4} (4\pi v)^{1/4}. \qquad (8.8.21)$$

Thus, in a very effective dia-electric medium, a test charge will induce a bubble or hole of finite radius. Notice, however, that in the limit of a perfect dia-electric medium,

$$W \to \infty \qquad \text{as } \epsilon \to 0. \qquad (8.8.22)$$

An isolated charge in a perfect dia-electric medium thus has infinite energy. This is the promised counterpart of the argument used in section 8.1 to wish away colored objects.

If, instead of an isolated charge, we place a test dipole within the putative bubble in the dia-electric medium, we can again show that the minimum-energy configuration occurs for a hole of finite radius about the test dipole. In this case, however, the field lines need not extend to infinity, so the hole radius remains finite as $\epsilon \to 0$, and so does the total energy of the system. The analogy between the exclusion of chromoelectric flux from the QCD vacuum and the exclusion of magnetic flux from a superconductor is now suggestive. To separate the dipole charges to $\pm\infty$ requires an infinite amount of work, as the previous example showed. This is the would-be analogue of quark confinement.

## 8.8.2 INSIGHTS FROM LATTICE QCD

The development of lattice gauge theory has made possible a quantitative description of phenomena that emerge at the low-energy scale associated with infrared slavery and confinement. By recasting QCD on a discrete spacetime lattice, this approach gives an explicit, regulated definition of the theory's behavior in the ultraviolet. Accordingly, lattice QCD lends itself to computational methods that integrate the functional integral of QCD numerically.

An important piece of evidence in favor of color confinement is the computation of the potential energy between static sources of color. The potential, illustrated in figure 8.45, looks Coulomb-like at short distances, in accord with asymptotic freedom, and linear at long distances, as suggested by the arguments of section 8.1.2.

Lattice QCD also provides a way to compute the hadron mass spectrum directly from the QCD Lagrangian. An ab-initio computation [81] of light-hadron masses,

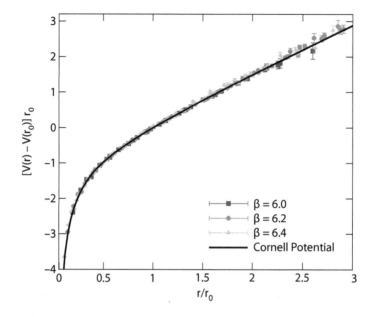

**Figure 8.45.** The potential energy $V(r)$ between static sources of color (in an approximation without sea quarks). The zero of energy and the units are set by a conventional distance $r_0$, defined by $r_0^2\, dV/dr = 1.65$. The data points are from lattice QCD, generated at several values of $\beta = 6/g^2$, which—via dimensional transmutation—corresponds to varying the spacing between lattice sites. The black curve is a fit of these data to the Coulomb + linear potential model applied to quarkonium in Ref. [79] (from Ref. [80].).

incorporating the effects of $2+1$ flavors of dynamical fermions $(u, d; s)$, is shown in figure 8.46. The agreement between theory and experiment is highly satisfying. Such calculations yield as by-products refined estimates [82] of the light-quark masses, for example,

$$\tfrac{1}{2}(m_u + m_d) = 3.54^{+0.64}_{-0.35} \text{ MeV}, \qquad m_s = 91.1^{+14.6}_{-6.2} \text{ MeV}, \qquad (8.8.23)$$

evaluated in the $\overline{\text{MS}}$ scheme at 2 GeV. These results make precise the crucial insight that most of the mass of hadrons such as the proton and neutron arises not from the masses of their quark constituents, but from the quarks' kinetic energy and the energy stored in the gluon field [83].

Lattice calculations at finite temperature aim to predict the phase diagram of QCD in the regime explored by heavy-ion collisions. Calculations with $2+1$ flavors of dynamical quarks have shown that hot QCD contains a phase in which (quasiparticle guises of) quarks and gluons are no longer confined and the chiral symmetry of quarks is restored [84]. As shown in the top and middle panels of figure 8.47, the deconfinement and chiral-symmetry-restoring transitions are smooth rather than discontinuous, but the two order parameters change at essentially the same temperature, roughly set by $\Lambda_{\text{QCD}}$. Another aspect of chiral symmetry breaking is illustrated in the bottom panel of Figure 8.47 [85]. As its momentum decreases,

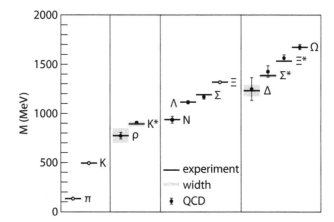

**Figure 8.46.** Meson ($\pi$, $K$, $\rho$, $K^*$) and baryon masses (others) from lattice QCD, using the $\pi$, $K$, and $\Xi$ masses to fix the three free parameters $m_u = m_d$, $m_s$, and $\Lambda_{QCD}$ (from Ref. [81]).

the current quark of perturbative QCD evolves into a constituent quark. The constituent-quark mass arises from the cloud of low-momentum gluons generated by quantum fluctuations around the current quark. Dynamical chiral symmetry breaking is an inherently nonperturbative effect that spontaneously generates a quark mass, even in the chiral limit $m \to 0$. This evidence gives a quantitative basis for the highly useful chiral quark model mentioned in section 8.11, in which constituent quarks and Goldstone bosons are taken as the apt degrees of freedom at low scales.

## 8.9 QCD-INDUCED ELECTROWEAK SYMMETRY BREAKING

The spontaneous symmetry breaking in the standard electroweak theory is modeled on the Ginzburg–Landau description of superconductivity [86]. In the Ginzburg–Landau phenomenology (cf. problem 5.7), the macroscopic order parameter, which corresponds to the wave function of superconducting charge carriers, acquires a nonzero vacuum expectation value in the superconducting state. Within a superconductor, the photon picks up a mass $M_\gamma = 1/\lambda$, where $\lambda$ is the London penetration depth that characterizes the exclusion of magnetic flux by the Meissner effect. In like manner, the auxiliary scalars introduced to hide the electroweak symmetry acquire a nonzero vacuum expectation value. The weak gauge bosons become massive in the process.

In the case of superconductivity, we have an example of a gauge-symmetry-breaking mechanism that does not rely on introducing an ad hoc order parameter. The microscopic Bardeen–Cooper–Schrieffer theory [87] reveals a dynamical origin of the order parameter through the formation of correlated states of elementary fermions, the Cooper pairs of electrons. Could the electroweak symmetry also be broken dynamically, without the need to introduce scalar fields? As we shall now see, quantum chromodynamics is, in principle, a source of electroweak symmetry breaking.

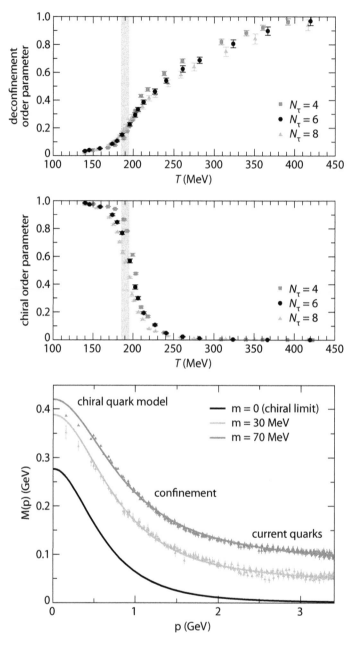

**Figure 8.47.** Order parameters for deconfinement (top) and chiral symmetry restoration (middle), as a function of temperature. The physical temperature $T = a/N_\tau$, where $a$ is the lattice spacing. Agreement for several values of $N_\tau$ thus indicates that discretization effects from the lattice are under control (from Ref. [84]). The dressed-quark mass function $M(p)$ (bottom panel), computed by Dyson–Schwinger techniques, compared with numerical simulations of lattice-QCD for several values of the current-quark mass $m$ (adapted from Ref. [85]).

Let us consider, as a low-energy effective field theory, a modified standard model with a single generation of massless fermions and no scalar sector: an $SU(3)_c \otimes SU(2)_L \otimes U(1)_Y$ theory of massless up and down quarks, plus massless electron and neutrino. The chiral quark fields are (cf. section 7.1)

$$L_q^a = \begin{pmatrix} u^a \\ d^a \end{pmatrix}_L, \qquad u_R^a, \quad d_R^a, \qquad (8.9.1)$$

with $(SU(3)_c, SU(2)_L)_Y$ quantum numbers $(\mathbf{3}, \mathbf{2})_{1/3}$, $(\mathbf{3}, \mathbf{1})_{4/3}$, and $(\mathbf{3}, \mathbf{1})_{-2/3}$, respectively, and color index $a = 1, 2, 3$. The chiral lepton fields are

$$L_e = \begin{pmatrix} \nu \\ e \end{pmatrix}_L, \qquad e_R, \qquad (8.9.2)$$

with quantum numbers $(\mathbf{1}, \mathbf{2})_{-1}$ and $(\mathbf{1}, \mathbf{1})_{-2}$, respectively. Electroweak-singlet "sterile" neutrinos $N_R$, with quantum numbers $(\mathbf{1}, \mathbf{1})_0$, may be added at will.

Provided that the $SU(3)_c$ interaction (QCD) is dominant, so that we may treat the $SU(2)_L \otimes U(1)_Y$ interactions as a perturbation, these models have the striking property that the QCD quark condensates dynamically break electroweak symmetry.

For vanishing quark masses (and with electroweak interactions turned off), QCD displays an exact $SU(2)_L \otimes SU(2)_R$ chiral symmetry. At an energy scale $\sim \Lambda_{QCD}$, the strong interactions become strong and $\langle \bar{q}q \rangle \equiv \langle \sum_{a=1}^N \bar{q}^a q^a \rangle$ quark condensates appear. Vacuum-alignment arguments [88] imply that the condensates are not only color neutral, but also electrically neutral, of the form $\langle \bar{u}u \rangle$ and $\langle \bar{d}d \rangle$. In the limit of vanishing $SU(2)_L \otimes U(1)_Y$ interactions, $\langle \bar{u}u \rangle = \langle \bar{d}d \rangle$. Formation of the quark condensates spontaneously breaks the chiral symmetry to the familiar isospin flavor symmetry,

$$SU(2)_L \otimes SU(2)_R \to SU(2)_V. \qquad (8.9.3)$$

The chiral symmetry breaking is associated with the dynamical generation of equal "constituent" masses for the up and down quarks. Three Nambu–Goldstone bosons appear, one for each broken generator of the original chiral invariance. These were identified by Nambu [89] as three massless pions (cf. section 5.2).

The $\langle \bar{q}q \rangle = \langle \bar{q}_L q_R + \bar{q}_R q_L \rangle$ condensate links left-handed and right-handed quarks, which transform differently under $SU(2)_L \otimes U(1)_Y$. It follows from the weak-isospin and weak-hypercharge quantum numbers of the left- and right-handed $u$ and $d$ quarks that the condensate transforms as a weak isodoublet with weak hypercharge $|Y| = 1$ and hence breaks $SU(2)_L \otimes U(1)_Y \to U(1)_{EM}$.

The broken generators correspond to three axial currents whose couplings to pions are measured by the pion decay constant $f_\pi$, which measures the strength of the pion-to-vacuum transition mediated by the charged weak current.[‡] The measured value of the charged-pion lifetime determines the real-world value of the decay constant, $f_\pi \approx 92.2$ MeV [24]. Within chiral perturbation theory, Gasser and Leutwyler [90] have estimated that $f_\pi$ would decrease by about 6% relative to its

---

[‡] The standard definition of $f_\pi$ in the literature on this topic differs by a factor of $1/\sqrt{2}$ from the usage in section 5.4.

experimental value if the $u$ and $d$ quark masses were reduced to zero from the real-world value (8.8.23). We adopt $\tilde{f}_\pi \approx 87$ MeV as an estimate of the corresponding parameter in the world of massless light quarks.

When we turn on the $SU(2)_L \otimes U(1)_Y$ electroweak interaction with couplings $g$ and $g'/2$, the electroweak gauge bosons couple to the axial currents and acquire masses of order $\sim g \tilde{f}_\pi$. In the interplay between the electroweak gauge interactions and QCD, the would-be massless pions disappear from the hadron spectrum, having become the longitudinal components of the weak gauge bosons. The mass-squared matrix,

$$\mathcal{M}^2 = \begin{pmatrix} g^2 & 0 & 0 & 0 \\ 0 & g^2 & 0 & 0 \\ 0 & 0 & g^2 & gg' \\ 0 & 0 & gg' & g'^2 \end{pmatrix} \frac{\tilde{f}_\pi^2}{4} \tag{8.9.4}$$

(where the rows and columns correspond to the three weak-isospin gauge bosons $b_1$, $b_2$, $b_3$, and the weak-hypercharge gauge boson $\mathcal{A}$, as introduced in section 6.3) has the same structure as the mass-squared matrix for gauge bosons in the standard electroweak theory.

Diagonalizing the matrix (8.9.4), we find that the photon, corresponding as in the standard model to the combination $A = (g\mathcal{A} + g'b_3)/\sqrt{g^2 + g'^2}$, emerges massless. Two charged gauge bosons, $W^\pm = (b_1 \mp ib_2)/\sqrt{2}$, acquire mass-squared $\overline{M}_W^2 = g^2 \tilde{f}_\pi^2/4$, and the neutral gauge boson, $Z = (-g'\mathcal{A} + gb_3)/\sqrt{g^2 + g'^2}$, obtains $\overline{M}_Z^2 = (g^2 + g'^2) \tilde{f}_\pi^2/4$. The ratio

$$\frac{\overline{M}_Z^2}{\overline{M}_W^2} = \frac{g^2 + g'^2}{g^2} = \frac{1}{\cos^2 \theta_W} , \tag{8.9.5}$$

where $\theta_W$ is the weak mixing angle, echoes the standard-model result, because the $SU(2)_L \otimes U(1)_Y$ transformation properties of the quark condensate ensure a custodial $SU(2)$ symmetry.

Here the symmetry breaking is dynamical and automatic; it can be traced, through spontaneous chiral symmetry breaking and confinement, to the asymptotic freedom (and infrared slavery) of QCD. Electroweak symmetry breaking determined by preexisting dynamics stands in contrast to the standard electroweak theory, in which spontaneous symmetry breaking results from the ad hoc choice of $\mu^2 < 0$ for the coefficient of the quadratic term in the Higgs potential (6.3.19).

Despite the structural similarity to the standard model, the chiral symmetry breaking of QCD does not yield a satisfactory theory of the weak interactions. To illustrate this statement quantitatively, we must specify the values of the gauge couplings. In the case of QCD, we have assumed that the scale $\Lambda_{QCD}$ retains its real-world value. In that spirit, it is reasonable to take the values of the $SU(2)_L \otimes U(1)_Y$ couplings at $\overline{M}_Z$ to be the same as those measured in the real world at $M_Z$: $g \approx 0.65$ and $g' \approx 0.34$ (Ref. [24]).

The masses acquired by the intermediate bosons are some 2 800 times smaller than required for a successful low-energy phenomenology, because their scale is set by $\bar{f}_\pi \approx 87$ MeV instead of $v \approx 246$ GeV [91]:

$$\overline{M}_W \approx 28 \text{ MeV}; \quad \overline{M}_Z \approx 32 \text{ MeV}. \tag{8.9.6}$$

The change in scale from the real-world values [24] of $M_W = 80.399 \pm 0.023$ GeV and $M_Z = 91.1876 \pm 0.0021$ GeV would have many far-reaching implications.

The example of QCD-induced electroweak symmetry breaking serves as a reminder that the agent that endows the fermions with mass need not be the same one responsible for hiding the $SU(2)_L \otimes U(1)_Y$ symmetry and giving mass to the weak bosons. The possible division of labor between electroweak symmetry breaking and the generation of gauge-boson masses on the one hand and the generation of fermion masses on the other hand is to be kept in mind as we carry out experiments to explore the 1-TeV scale.

Within QCD, hypothetical exotic (color $6, 8, 10, \ldots$) quarks would interact more strongly through than the normal color triplets, so the chiral-symmetry breaking in exotic quark sectors would occur at much larger mass scales than the standard chiral-symmetry breaking we have just reviewed. If those mass scales were sufficiently high, exotic-quark condensates could break $SU(2)_L \otimes U(1)_Y \to U(1)_{EM}$ dynamically and yield phenomenologically viable $W^\pm$ and $Z^0$ masses [92]. No exotic quarks have yet been detected, either by direct observation or in the evolution of the strong coupling constant, $\alpha_s$ (Ref. [25]).

The observation that QCD dynamically breaks electroweak symmetry (but at too low a scale) inspired the invention of analogous no-Higgs theories in which dynamical symmetry breaking is accomplished by the formation of a condensate of new fermions subject to a new, asymptotically free, vectorial gauge interaction (often called technicolor) that becomes strongly coupled at the TeV scale [93]. The technifermion condensates that dynamically break the electroweak symmetry produce masses for the $W^\pm$ and $Z^0$ bosons but do not directly give masses to the standard-model fermions. To endow the quarks and leptons with mass, it is necessary to embed technicolor in a larger *extended technicolor* framework containing degrees of freedom that communicate the broken electroweak symmetry to the (technicolor-singlet) standard-model fermions [94].

Other suggestive work in the area of dynamical symmetry breaking also builds on the metaphor of the BCS theory of superconductivity but attributes a special role to quarks of the third generation or beyond. A rich line, based on the notion that a top-quark condensate drives electroweak symmetry breaking, was initiated in Ref. [95]. The idea that condensation of a strongly coupled fourth generation of quarks could trigger electroweak symmetry breaking has been a lively area of contemporary research [96].

## 8.10 THE $1/N$ EXPANSION

The search for small parameters that can play the role of expansion parameters is a central element of the process of systematic approximation and model building that characterizes theoretical physics. In many physical situations, extremes of energy or distance suggest highly accurate and readily improved approximation schemes. In classical electrodynamics, for example, the indispensable far-field approximation is applicable when the size of a radiator is negligible compared to the distance between radiator and receiver. The Born approximation for the scattering of charged-particle beams from atomic electrons is trustworthy for beam energies greatly in excess of the atomic binding energy. In quantum chromodynamics, a perturbative treatment—an expansion in terms of the strong coupling $\alpha_s(Q^2)$—is expected to be reliable when the invariant momentum transfer $\sqrt{Q^2}$ is large compared to a characteristic mass scale set by $\Lambda_{QCD}$ (cf. figure 8.22 and the example applications discussed in section 8.4-8.7).

At least at first sight, no similar expansion in powers of the coupling constant would seem to apply to the problem of hadron structure. The light quarks are essentially massless, and all the relevant energies of the problem are on the same order as the naturally occurring scale, $\Lambda_{QCD}$. In a typical light hadron, the separation of the quarks is simply the hadronic scale of approximately 1 fm—hardly a regime in which perturbative QCD makes sense. It is not productive to imagine artificially varying the strong coupling constant $g$ because it is absorbed, through dimensional transmutation (cf. section 8.8), in the definition of the mass scale of light hadrons. Even for the heaviest quarkonium family, $\Upsilon$, perturbation theory based on a power expansion in $\alpha_s$ cannot describe the spectrum. A fully nonperturbative treatment by means of the lattice formulation of QCD addresses an expanding range of questions. It is, nevertheless, important to seek other approximations to the complete content of QCD that give insight into the results we see in nature or in lattice calculations. In the absence of an obvious free parameter, we must seek a hidden parameter in order to make progress.

A trick that simplifies the analysis of hadron structure is to generalize the color gauge group from $SU(3)_c$ to $SU(N)_c$ and to consider the limit in which $N$ becomes very large [97, 98]. Although $SU(N)_c$ is in general richer than $SU(3)_c$, the hadron structure problem is simplified by two observations: (1) At any order in the coupling constant, some classes of diagrams are combinatorially negligible, at large $N$. (2) The remaining diagrams have common consequences.

The strategy of the $1/N$ expansion is a familiar one. When confronted with a problem we cannot solve, we invent a related problem that we can solve. If this is done adroitly, the new problem will not only be simpler, but it will capture the physical essence of the original one. The $1/N$ expansion represents an attempt to introduce a parameter that permits a simplification of the calculation at hand. Problem 8.28 introduces an elementary example in nonrelativistic quantum mechanics.

This technique does not entirely free us from the constraints of perturbative analysis in $\alpha_s$. Because we shall find, by inspection, that entire classes of combinatorially favored diagrams have common features at all orders in the coupling constant, we shall have to assume that the relevant content of the theory is accurately represented by the set of all diagrams. We also assume that the privileged diagrams, both for $SU(3)_c$ and for $SU(N)_c$, sum up to give a confining theory. We infer the reliability of the $1/N$ expansion as a token for QCD from the fact that $SU(N)_c$-QCD has features in common with the world we observe. Clear introductions to the method, with allusions to other physical situations, are given by Coleman [99], Witten [100], and Buras [101].

The combinatorial analysis of $SU(N)_c$-QCD is most transparent in terms of the double-line notation introduced for this purpose by 't Hooft (Ref. [97]),

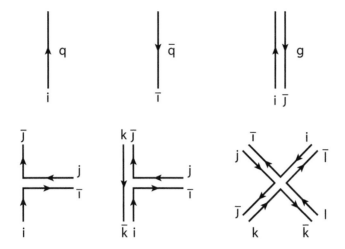

Figure 8.48. Double-line notation appropriate to $1/N$ analysis for propagators and vertices in QCD. The theory contains $N$ colors of quarks and $N^2 - 1 \approx N^2$ gluons. Color labels are to be read flowing into the vertex. Compare the diagrammatic methods for computing color factors in appendix A.5.

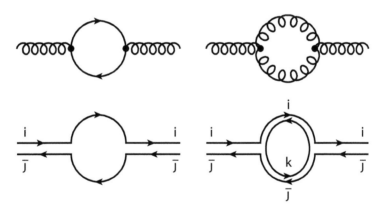

Figure 8.49. One-loop corrections to the gluon propagator in conventional (upper panel) and double-line (lower panel) notation.

which is illustrated in figure 8.48. Several examples will suffice to make the main points.

Consider first the lowest-order vacuum polarization contributions to the gluon propagator, which arise from the quark and gluon loops shown (in conventional notation) in the upper panel of figure 8.49. These are redrawn in the double-line notation in the lower panel of figure 8.49. For an initial gluon that carries color indices $i\bar{\jmath}$, only a single color combination is possible for the quark-loop intermediate state: a quark of color $i$ and an antiquark of color $\bar{\jmath}$. For the gluon loop, however, the index $k$ is free to take on any value $1, 2, \ldots, N$. Thus a combinatorial factor of $N$ is associated to the gluon-loop diagram. This illustrates a general rule that gluon loops dominate over quark loops by a factor of $N$, as $N \to \infty$.

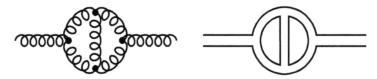

**Figure 8.50.** A two-loop diagram in conventional (left panel) and double-line (right panel) notation.

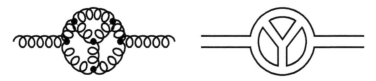

**Figure 8.51.** A three-loop diagram in conventional (left panel) and double-line (right panel) notation.

The presence of the factor $N$ would seem to imply that the gluon-loop diagram's contribution diverges as $N \to \infty$. This outcome can be avoided by choosing the coupling constant to be $g/\sqrt{N}$, with $g$ held fixed as $N \to \infty$. Then for any value of $N$, the contribution of the gluon loop goes as

$$\left(\frac{g}{\sqrt{N}}\right)^2 \cdot N \to g^2, \tag{8.10.1}$$

a smooth limit. With this definition, the $q\bar{q}g$ and three-gluon vertices carry weight $1/\sqrt{N}$, whereas the four-gluon vertex carries weight $1/N$. After rescaling, the contribution of the quark loop in figure 8.49 goes as

$$\left(\frac{g}{\sqrt{N}}\right)^2 \cdot 1 \to \frac{g^2}{N} \tag{8.10.2}$$

and so becomes negligible in the limit $N \to \infty$. It is easy to see that there is a suppression by $1/N$ for each internal quark loop.

An analysis of diagrams containing more than one loop indicates that this device resolves the combinatorial divergence issue in general. The two-loop diagram depicted in figure 8.50 is immediately seen to be proportional to

$$\left(\frac{g}{\sqrt{N}}\right)^4 \cdot N^2 \to g^4. \tag{8.10.3}$$

Similarly, the three-loop diagram of figure 8.51 evidently goes as

$$\left(\frac{g}{\sqrt{N}}\right)^6 \cdot N^3 \to g^6. \tag{8.10.4}$$

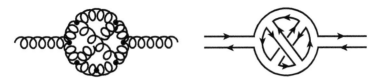

**Figure 8.52.** A nonplanar diagram in conventional (left panel) and double-line (right panel) notation.

The situation is different for nonplanar graphs, however. The simplest such graph is shown in figure 8.52. The double-line notation makes it apparent that this graph contains but a single, tangled color loop and, therefore, scales as

$$\left(\frac{g}{\sqrt{N}}\right)^6 \cdot N \to \frac{g^6}{N^2}, \tag{8.10.5}$$

it is, therefore, suppressed by $1/N^2$ compared to its planar counterpart at the same order in $g^2$. It is generally the case that nonplanar graphs are suppressed by $1/N^2$ as $N \to \infty$.

These combinatorial arguments select planar graphs without internal quark loops as an important subclass. To evaluate and sum all the graphs thus selected is no trivial task. Instead, we may identify their common features and speculate that these survive confinement. It is possible in this way to establish the following qualitative insights in the large-$N$ limit:

1. Mesons are free, stable, and noninteracting. For each allowed combination of $J^{PC}$ and flavor quantum numbers, there is an infinite number of resonances.
2. Meson decay amplitudes are proportional to $1/\sqrt{N}$, so mesons are narrow structures.
3. Multibody decays of unstable mesons are dominated by resonant, quasi-two-body channels whenever they are open. The partial width of an intrinsically $k$-body final state goes as $1/N^{k-1}$.
4. The meson–meson elastic scattering amplitude is proportional to $1/N$.
5. The Okubo [102]–Zweig [103]–Iizuka [104] (OZI) rule that decays corresponding to connected quark-line diagrams are allowed, but decays that correspond to disconnected quark-line diagrams are forbidden, is exact. [The original motivation arose from the observation that $\Gamma(\phi \to K\bar{K})/\Gamma(\phi \to \pi^+\pi^-\pi^0) \approx 5.4$, notwithstanding the tiny phase space available for the $K\bar{K}$ decay.] Singlet-octet mixing (through virtual annihilations) and meson-glue mixing are suppressed. Mesons are pure $(q\bar{q})$ states, with no quark-antiquark sea.
6. Meson–meson bound states, which would include particles with exotic quantum numbers, are absent.
7. For each allowed $J^{PC}$, there are infinitely many glueball states, with widths of order $1/N^2$. They are, thus, more stable than $(q\bar{q})$ mesons, interact feebly with $(q\bar{q})$ mesons, and mix only weakly with $(q\bar{q})$ states.

Until QCD is actually solved, we shall not know how closely the $N \to \infty$ limit of SU$(N)_c$-QCD resembles the case of interest, which is SU$(3)_c$. The preceding

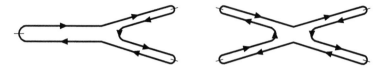

**Figure 8.53.** Mechanism for light-meson decay (left panel) and meson-meson scattering (right panel) in double-line notation. Strokes on arcs denote projection onto color-singlet states.

list of large-($N$) results does bear a striking (if idealized) resemblance to the real world. To the extent that the $1/N$ expansion faithfully represents the consequences of QCD, much of the traditional phenomenology is explained, and many model approximations are justified.

It is worth noting that the success of lattice-QCD calculations that omit additional quark-antiquark pairs in accounting for the light-hadron spectrum justifies, in some measure, the quark-model idealization that treats mesons as pure quark-antiquark states. The most mature quenched calculation, by the CP-PACS Collaboration [105], yields 10% agreement for the light-hadron masses—an impressive achievement in a strongly coupled quantum field theory. As we saw in figure 8.46, a full calculation that incorporates dynamical fermions improves the agreement to the 5% level and also permits a systematic evaluation of uncertainties. A useful physical picture [106] holds that some effects of the sea quarks should be absorbable into shifts of the bare coupling and masses, as in a dielectric. In effect, the quark model, with its "constituent quarks," tries to absorb even more of QCD's complicated dynamics into simpler objects.

To see how conclusions 1–7 may be reached, let us examine a few simple cases. A schematic representation of the decay $M \to M_1 M_2$ of a light meson into two light mesons is shown in the left panel of figure 8.53. In the double-line diagram the ends of the quark and antiquark lines are tied together to emphasize that the mesons are color singlets. The strokes on the arcs joining $(q\bar{q})$ pairs into mesons represent the projection onto a color-singlet state, which induces a factor $1/\sqrt{N}$ for each external meson. Consequently, the decay amplitude is proportional to

$$\left(\frac{1}{\sqrt{N}}\right)^3 \cdot N = \frac{1}{\sqrt{N}}. \tag{8.10.6}$$

A straightforward generalization leads to the conclusion that the decay rate for a multibody decay is further suppressed:

$$\Gamma(M \to M_1 M_2 \cdots M_k) \propto \frac{1}{N^{k-1}}. \tag{8.10.7}$$

After verifying that the same conclusions derive from the dominant planar diagrams at every order, we establish properties 2 and 3.

It is no more difficult to examine the meson-meson scattering diagram in the right panel of figure 8.53 and to see that the scattering amplitude scales as

$$\left(\frac{1}{\sqrt{N}}\right)^4 \cdot N = \frac{1}{N}, \tag{8.10.8}$$

which puts us on the path to establishing property 4.

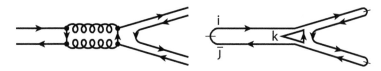

**Figure 8.54.** Mechanism for OZI-forbidden meson decay in conventional (left panel) and double-line (right panel) notation.

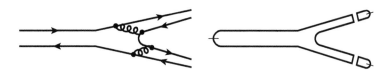

**Figure 8.55.** OZI-allowed decay of a meson, at order $g^4$, in conventional and double-line notation.

The $1/N$ derivation of the OZI rule is only a little more involved. A possible mechanism for the OZI-forbidden decay of a $(q\bar{q})$ state is shown in figure 8.54, the process

$$(q\bar{q}) \to gg \to q'\bar{q}' \to \text{mesons.} \qquad (8.10.9)$$

This is shown in standard and in double-line notation in figure 8.54. The Zweig-forbidden decay amplitude contains two color loops. It therefore scales as

$$\left(\frac{1}{\sqrt{N}}\right)^3 \cdot \left(\frac{g}{\sqrt{N}}\right)^4 \cdot N^2 \to \frac{g^4}{N^{3/2}}. \qquad (8.10.10)$$

At the same order in the strong coupling constant, the allowed decay is illustrated in figure 8.55. In the double-line representation, it is seen to contain three color loops. The allowed amplitude is, therefore, proportional to

$$\left(\frac{1}{\sqrt{N}}\right)^3 \cdot \left(\frac{g}{\sqrt{N}}\right)^4 \cdot N^3 \to \frac{g^4}{N^{1/2}}, \qquad (8.10.11)$$

in harmony with what we found in (8.10.6). Thus at each order in perturbation theory, the Zweig-forbidden decay is down by one power of $N$ in magnitude compared with the allowed decay. Because this reasoning does not rely on the smallness of the strong coupling constant, which does control some aspects of $\psi$ and $\Upsilon$ decays, it is an appealing argument for the inhibition of $\phi \to \rho\pi$ relative to $K\bar{K}$. The $1/N$ expansion has also been applied to the problem of baryon structure—baryons are $N$-quark composites in $SU(N)_c$-QCD—by Witten (Ref. [100]). Extensions of the $1/N$ formalism have enabled quantitative applications to CP violation and other phenomena (Ref. [101]). Dualities between conformal large-$N$ gauge theories in four dimensions and string (gravity) theories in anti-de Sitter space have yielded new insights into the structure of both classes of theories and suggest characteristics of QCD in the strong-coupling regime [107].

# 8.11 STRONG-INTERACTION SYMMETRIES

Quantum chromodynamics is based on an exact $SU(3)_c$ symmetry of the strong inter-actions. However, as is well known, many other symmetries and conservation laws are respected by the strong interactions. Among these are the exact conservation of additive quantum numbers such as electric charge, baryon number, strangeness, and the like, the flavor symmetries isospin and $SU(3)_{\text{flavor}}$, and discrete symmetries under parity, charge conjugation, and time-reversal invariance. A considerable body of work exploits approximate chiral symmetry, soft-pion theorems, and related aspects of current algebra.

It is of clear interest to investigate the origins of these symmetries and approximate symmetries: Why do they hold and how do they arise? Before the formulation of QCD, it was common practice to construct effective strong-interaction Lagrangians embodying the important symmetry principles. In such a framework, there was no possibility to ask whether symmetries might be preserved beyond the lowest order in a perturbative calculation. Moreover, before the development of the electroweak theory, it was not possible to contemplate a systematic inquiry into the influence of the (less symmetrical) weak interactions upon the strong. Now that we possess calculable theories of the strong and electroweak interactions, the power of effective field theories that capture the apt degrees of freedom and dynamics over a limited range of energies is greatly enhanced. We shall mention a few of the important areas under active study. The discussion is abbreviated, in part because it would be necessary to develop an extensive phenomenology as prerequisite to a thorough treatment.

The spectroscopic systematics that were so important for the development of the quark model emerge from the rule that hadrons must be color singlets (now justified by QCD) and the absence of additional quark–antiquark pairs from the static wave functions of hadrons. The relative unimportance of the sea is indeed suggested by the $1/N$ expansion and by studies on the lattice, as we have just seen in section 8.10.

Although the strong interactions among quarks are governed by color charges and are thus independent of quark flavor, the masses of the quarks do influence the behavior of the hadrons assembled by QCD. For example, the (transient) importance of kinematics to the shape of Regge trajectories is given [108] by the string model of the hadron spectrum (cf. section 8.1.2).

Hadrons that contain heavy quarks ($b$, $c$, and, in some settings, $s$), exhibit simplified behavior. In a bound state that contains a single heavy quark, the identity of the heavy quark has only a secondary influence on the hadron dynamics because the heavy quark sits essentially at rest inside the hadron [109]. The center of mass of the hadron and heavy quark essentially coincide, whereas the light degrees of freedom depend little on the identity of the heavy quark. Approximate symmetries associated with the heavy-quark flavor and spin result [110]. In a meson with a heavy quark and corresponding antiquark, the two orbit each other. The velocity depends on the heavy-quark mass, but the spin decouples (to leading order), in a manner reminiscent of QED applied to atomic physics [111].

Because the notion of flavor seems incidental to QCD, it is natural to ask whether the goodness of isospin symmetry is accidental [112]. We saw in chapter 7 that (in the standard electroweak theory, if not in nature) quark masses originate in the spontaneous breakdown of the $SU(2)_L \otimes U(1)_Y$ symmetry of the weak and electromagnetic interactions and, specifically, in the arbitrary Yukawa interaction term that couples quarks to the Higgs boson. From this perspective, the goodness of isospin symmetry as a survivor of spontaneously broken chiral symmetry breaking is a consequence of the unexplained fact that

$$m_u \approx m_d \approx 0. \tag{8.11.1}$$

The small (and also unexplained) quark mass difference,

$$m_d > m_u > 0, \tag{8.11.2}$$

seems to lie at the origin of the nonelectromagnetic (or "explicit") isospin breaking required to explain the sign of the neutron–proton mass difference.

In contrast, QCD does provide some insight into the successes of current algebra and the long-standing problem of the masses of the isoscalar pseudoscalar mesons known as the U(1) problem. Generalizing from the discussion in section 8.9 and the end of section 8.8.2, in the limit of vanishing quark masses, the QCD Lagrangian displays an exact global $SU(n_f)_L \otimes SU(n_f)_R$ chiral symmetry operating independently on the left-handed and right-handed parts of the quark fields for the $n_f$ massless flavors of quarks. In the less extreme limit of zero masses for the up, down, and strange quarks, the $SU(3)_L \otimes SU(3)_R$ symmetry of the QCD Lagrangian implies a flavor octet of exactly conserved axial currents. To describe the real world, the corresponding chiral symmetry must be spontaneously broken along the lines described by Nambu and Jona-Lasinio [113] in two extremely prescient papers. Accordingly, in the world of three massless quark flavors, there should be eight massless Nambu–Goldstone bosons, which we identify with the pseudoscalar octet. Because the up, down, and strange quarks acquire small masses, thanks to spontaneous electroweak symmetry breaking, it follows that the $\pi$, $K$, and $\eta$ are only approximately massless, although they are presumed to retain some memory of their chiral symmetry–breaking origin. The chiral quark model [114] is a very useful effective field theory inspired by these insights.

The QCD Lagrangian also possesses a vectorial U(1) symmetry associated with baryon-number conservation and an axial U(1) that leads to the puzzle of the mass of the flavor-singlet pseudoscalar meson. If the corresponding ninth axial current, which corresponds to a flavor singlet, is conserved, the pion should have a light partner $\eta_1$ with mass [115]

$$m_{\eta_1} \leq m_\pi \sqrt{3}. \tag{8.11.3}$$

The "U(1) problem" refers to the fact that the predicted light state does not exist: the obvious candidate, $\eta(958)$, is by no means light. A phenomenological explanation invokes the influence on the spectrum of states composed of glue alone. A formal solution to the U(1) problem was given by 't Hooft [116], who argued that because of the effects of so-called instanton solutions, the U(1) current has an anomaly that leads to the physical nonconservation of the ninth axial charge. This removes the raison d'être for a ninth light pseudoscalar. Other aspects of chiral symmetry breaking have been similarly illuminated by quantum chromodynamics.

**Figure 8.56.** Electroweak corrections to strong $n$-point functions.

Let us now turn to the question of the strong-interaction symmetries that are not respected by the weak and electromagnetic interactions. Parity invariance and strangeness conservation, for example, hold to quite excellent approximation in strong-interaction processes [117]. However, it is apparent that strong-interaction $n$-point functions will, in general, receive corrections from processes involving electroweak gauge bosons, as indicated schematically in figure 8.56. Such electroweak corrections entail potentially dangerous loop integrals; beyond these appearances, we may be put on guard by the phenomenon of nonleptonic enhancement in weak decays, which raises the possibility that symmetry violations might somehow be amplified. A lengthy analysis by Weinberg [118] shows this not to be the case. Electroweak effects at order $\alpha^1$ occur only as corrections to the quark mass matrix. As such, they necessarily conserve parity, strangeness, and so on, and produce only isovector departures from exact isospin invariance. One facet of the discrete symmetry problem, which may be posed as Why are the strong interactions not severely polluted by the discrete- and internal-symmetry violations of the electroweak interactions? is thus dealt with.

We have not explained, however, why the strong interactions should respect discrete symmetries in the first place, just as we offered no explanation for the parity-violating left-handed form of the charged-current weak interactions. In this case, a challenge arises from an unexpected quarter. QCD, as formulated, seems unable to protect itself against self-inflicted *strong* CP violations. Consider adding to the QCD Lagrangian (8.1.2) an additional term of the form

$$\mathcal{L}' = \text{constant} \times \text{tr} G_{\mu\nu}\,{}^*G^{\mu\nu}, \tag{8.11.4}$$

where the dual of the gluon field-strength tensor [cf. (3.2.14)] is

$${}^*G^{\mu\nu} = -\tfrac{1}{2}\varepsilon^{\mu\nu\alpha\beta}G_{\alpha\beta}. \tag{8.11.5}$$

In electromagnetism, an analogous term would be of the form

$$\text{tr}(F_{\mu\nu}\,{}^*F^{\mu\nu}) = 4\mathbf{E}\cdot\mathbf{B}, \tag{8.11.6}$$

which manifestly breaks parity, under which

$$\mathsf{P}\mathbf{E} = -\mathbf{E} \qquad \mathsf{P}\mathbf{B} = \mathbf{B}, \tag{8.11.7}$$

but is charge-conjugation invariant because

$$\mathbf{CE} = -\mathbf{E} \qquad \mathbf{CB} = -\mathbf{B}. \tag{8.11.8}$$

Consequently the presence of (8.11.4) in the QCD Lagrangian would give rise to violations of CP and P invariance.

Because such a term is gauge invariant and renormalizable, our only basis for excluding it is that it is CP violating. There the matter would rest for a weak-coupling theory. However, this procedure is not tenable for QCD—at least not without further argumentation. The reason is that, because QCD is a theory of the strong interactions, we are obliged to consider field configurations in which the fields are intense.

The existence of "large" gauge transformations [119] and nontrivial topological structures in QCD means that the vacuum cannot be represented simply as a state of vanishing gauge fields but must be given by a more complex structure reminiscent of a Bloch wave in condensed-matter physics:

$$|\theta\rangle \equiv \sum_{n=-\infty}^{\infty} e^{in\theta} |n\rangle, \tag{8.11.9}$$

where $|n\rangle$ is the minimum-energy state with topological winding number

$$n = \frac{1}{8\pi^2} \int d^4x \, \mathrm{tr} G_{\mu\nu}{}^* G^{\mu\nu}. \tag{8.11.10}$$

The parameter $\theta$ is a new and arbitrary parameter of QCD. The effect of the $\theta$-vacuum may be represented by a new term in the Lagrangian, of the form

$$\mathcal{L}_\theta = \frac{\theta}{16\pi^2} \mathrm{tr} G_{\mu\nu}{}^* G^{\mu\nu}. \tag{8.11.11}$$

Although this has the look of a total derivative, or surface term, it cannot be ignored, because the gauge fields do not necessarily vanish at infinity. The phase of the quark mass matrix and the coefficient $\theta$—two quantities with distinctly separate origins—combine to cause effects that violate CP invariance. To respect the upper limit on the neutron's electric dipole moment [120],

$$|d_n| \lesssim 0.29 \times 10^{-25} \, e \text{ cm at } 90\% \text{ CL}, \tag{8.11.12}$$

the parameter $\theta$ must be extraordinarily small [121]:

$$|\theta| \lesssim 10^{-10}. \tag{8.11.13}$$

The mystery of the exquisite smallness of $\theta$ is the *strong CP problem*. The most promising strategy consists in adding a second Higgs doublet and an additional U(1) symmetry to the standard model Lagrangian [122]. The new U(1) symmetry is spontaneously broken in the course of electroweak symmetry breaking, and minimizing the Higgs potentials enforces $\theta = 0$. The breaking of the U(1) symmetry implies the existence of a new pseudo-Nambu–Goldstone boson called the axion. No signal for an axion has yet been found; many imaginative searches are ongoing.

## 8.12 ASSESSMENT

Four decades after the synthesis of quarks, partons, and color into the QCD Lagrangian, quantum chromodynamics has been tested and validated up to energies of 1 TeV. QCD elegantly incorporates many of the observed systematics of the strong interactions. Thanks to the property of asymptotic freedom, the theory allows us to make reliable perturbative calculations for many processes of experimental interest. The total cross section in electron–positron annihilation into hadrons, the evolution of structure functions in deeply inelastic lepton–nucleon scattering, the production of jets in high-energy collisions, and the pointlike structure of the photon are prime examples. Great advances have also come in the nonperturbative regime. The lattice formulation of QCD has helped to understand why free quarks and gluons are not observed and has enabled calculations of hadron properties, while explaining the absence of unobserved species and pointing toward new varieties of hadrons, such as glueballs and $q\bar{q}g$ hybrids. There is important progress toward deriving the interactions among hadrons as a collective effect of the interactions among constituents.

Like the electroweak theory, QCD can reasonably be regarded as a new law of nature. In contrast to the electroweak theory, QCD is internally consistent up to very high energies and so could be a complete theory of the strong interactions [123]. Whether QCD is the final answer for the strong interactions is a subject for continuing experimental tests. Those tests are already being extended in early experimentation at the Large Hadron Collider (cf. problem 8.20). Beyond the comparison of perturbative calculations with experiment, it remains critically important to test the confinement hypothesis by searching for free quarks, or for signatures of unconfined color. Sensitive negative searches for quarks continue to be interesting, and the definitive observation of free quarks would be revolutionary. Breakdowns of factorization would compromise the utility of perturbative QCD. Other discoveries that would require small or large revisions to QCD include the observation of new kinds of colored matter beyond quarks and gluons (and perhaps their superpartners), the discovery that quarks are composite, or evidence that $SU(3)_c$ gauge symmetry is the vestige of a larger, spontaneously broken, color symmetry.

Although QCD does not exhibit any structural problems, the challenge of the CP-violating phase $\theta$ persists. Should the solution lie in the Peccei–Quinn symmetry and its implication of axions, the ramifications for the dark-matter question and for cosmology in general would be profound.

Most physicists do not expect the LHC to reveal big surprises in the structure of QCD. Instead, quantum chromodynamics will serve as basic knowledge, following the precedent of quantum electrodynamics, enabling discoveries beyond the standard model of particle physics. In this arena, future research will focus on techniques for evaluating parton amplitudes with increasingly many real and virtual particles, for both signals and backgrounds. The higher energies of the scattering processes will continue to implicate multiple scales (several TeV compared to the top-quark mass, for example) and will thus require new tools such as the soft-collinear effective theory [124]. Future experimental studies of *B* decays will also continue to rely on QCD to pin down aspects of the weak interactions and any new interactions of quarks. The aspiration here is to compute many simple amplitudes with total uncertainties at the level of 1% or smaller [125], to enable the precise extraction of electroweak parameters. Calculations of similar difficulty are related to moments of the parton distributions. Reliable lattice-QCD calculations would pin down predictions of signals and backgrounds at the LHC. The most crucial in this regard—and the most challenging computationally—are moments of the gluon density inside the proton [126].

The strong interactions comprise a richer field than the set of phenomena that we have learned to describe in terms of perturbative QCD or the (near-) static nonperturbative domain of lattice QCD. The technology by which we apply QCD is incomplete and still evolving. Many aspects of hadron phenomenology and spectroscopy, including the proton–proton total cross section and the spectrum of excited hadrons, are not yet calculable beginning from the QCD Lagrangian. Much analysis of experimental information relies on highly stylized, truncated pictures of the implications of the theory. While expanding the horizons, it is important to distinguish tests of QCD from tests of auxiliary assumptions.

QCD is unquestionably a triumph of reductionist science, distilling the immense variety of the hadronic zoo into a simple Lagrangian field theory. But physics does not live by reductionism alone: The idea of emergent behavior is that there are in nature phenomena, or regularities, or even very precise laws, that cannot readily be recognized by starting with the Lagrangian of the universe. These include situations that arise in the many-body problems of condensed-matter physics, but also situations in which a simple perturbation-theory analysis is not sufficient to see what will happen. This notion of emergence is ubiquitous in particle physics. What is quark confinement in QCD, the theory of the strong interactions, if not emergent behavior? You could do perturbation theory for a very long time and not demonstrate the phenomenon of confinement. As QCD becomes strongly coupled, new phenomena emerge—not only confinement, but also chiral symmetry breaking and the appearance of Goldstone bosons—that we wouldn't have anticipated by staring at the Lagrangian. This is part of the motivation for investigating heavy-ion collisions at high energies. The very lack of simplicity may push us into realms of QCD where we can't guess the answers by simple analysis.

The phenomenon of AdS/CFT correspondence is one of the most startling and potentially revealing insights obtained from string theory. It describes uncanny dualities between gauge theories and theories containing gravity. The archetype is an exact equivalence conjectured by Maldacena [127] between type IIB string theory compactified on $\text{AdS}^5 \otimes \text{S}_5$ (five-dimensional anti-de Sitter space times the five-sphere) with four-dimensional super-Yang–Mills theory. [An anti-de Sitter space refers to a maximally symmetric solution to Einstein's equations with a negative (hence the "anti") cosmological constant.] Holographic QCD [128] is an attempt to find such a gravity dual for quantum chromodynamics, with the initial aim of giving insight into the low-energy properties of hadrons. It reproduces familiar aspects of hadron physics, including the emergence of massless pions as Nambu–Goldstone bosons. Intriguing attempts are being made to extract further insights, and even semiquantitative predictions, from holographic QCD and the AdS/CFT correspondence. Within the framework of Bjorken's hydrodynamics of heavy-ion collisions [129], the AdS/CFT correspondence has been employed to try to predict the viscosity of the quark–gluon fluid [130].

The common mathematical structure of QCD and the electroweak theory, combined with the asymptotic freedom of the $SU(3)_c$ and $SU(2)_L$ interactions, encourages the hope that a unified theory of the strong, weak, and electromagnetic interactions may be within reach. The resulting program is the subject of chapter 9.

## PROBLEMS

8.1. If the quark colors are designated as red $(R)$, green $(G)$, and blue $(B)$, the gluons may be represented conveniently as $\bar{R}G$, $\bar{R}B$, $\bar{G}R$, $\bar{G}B$, $\bar{B}R$, $\bar{B}G$, $(\bar{R}R - \bar{G}G)/\sqrt{2}$, and $(\bar{R}R + \bar{G}G - 2\bar{B}B)/\sqrt{6}$. The last two are color-preserving forms that are orthogonal to the color-singlet combination $(\bar{R}R + \bar{G}G + \bar{B}B)/\sqrt{3}$. The elementary quark–gluon interactions will be of the form

$$\text{red quark} + \bar{R}G \text{ gluon} \rightarrow \text{green quark},$$

and so on. Repeat in this language the maximally attractive channel analysis of section 8.1, and show that the color-singlet $q\bar{q}$ and $qqq$ configurations are energetically favored. [References: R. P. Feynman, in *Weak and Electromagnetic Interactions at High Energy*, 1976 Les Houches Lectures, ed. R. Balian and C. H. Llewellyn Smith, North-Holland, Amsterdam, 1977, p. 120; C. Quigg, in *Techniques and Concepts of High-Energy Physics*, ed. T. Ferbel, Plenum, New York, 1981, p. 143, §6.A, [http://j.mp/eV81cR]] .

8.2. Consider the representations of the color gauge group $SU(N > 2)$ that correspond to diquark and quark–antiquark configurations. For the symmetric and antisymmetric diquark representations and for the singlet and adjoint $q\bar{q}$ representations, calculate the dimension of the representation and evaluate the quadratic Casimir operator $\langle \mathbf{T}^2 \rangle$, where the $T_i$ are the normalized generators of $SU(N)$.

8.3. For quark–antiquark states governed by an $SU(N)$ color gauge group, calculate the quantity $\langle \mathbf{T}^{(1)} \cdot \mathbf{T}^{(2)} \rangle$ that characterizes the interaction energy to lowest order in perturbation theory. What is the state of lowest energy?

8.4. To investigate the main features of the hadron spectrum, recast the "interaction energies" for $(qq)$ and $(q\bar{q})$ systems introduced in section 8.1.1 as Coulomb potentials of the form

$$V(r) = \frac{\alpha_s}{r} \langle \mathbf{T}^{(1)} \cdot \mathbf{T}^{(2)} \rangle,$$

with expectation values given in table 8.3, so that

$$V_{(q\bar{q})_1}(r) = -\tfrac{4}{3}\alpha_s r, \quad V_{(q\bar{q})_8}(r) = +\tfrac{1}{6}\alpha_s r,$$
$$V_{(qq)_{3*}}(r) = -\tfrac{2}{3}\alpha_s r, \quad V_{(qq)_6}(r) = +\tfrac{1}{3}\alpha_s r.$$

These interactions underlie a rough description of $L = 0$ meson and baryon masses as the sum of "constituent-quark" masses plus the sum of two-body interactions. This picture distinguishes hadrons with different flavor content but does not account for mass differences between states that contain the same quarks, but in different spin configurations.

To improve the description of hadron masses, introduce a color hyperfine interaction in analogy to the form familiar from the Coulomb problem of atomic physics:

$$\mathcal{H}^{\text{hf}} = \nabla^2 V \frac{\boldsymbol{\sigma}^{(1)} \cdot \boldsymbol{\sigma}^{(2)}}{6m_1 m_2},$$

where $\boldsymbol{\sigma}$ is a Pauli spin matrix.

(a) Show that in the one-gluon-exchange picture of hadron structure, the corresponding energy shift between two quarks in a relative $s$-wave becomes

$$\Delta E^{\text{hf}} = -\frac{2\pi\alpha_s |\psi(0)|^2}{3m_1 m_2} \langle \mathbf{T}^{(1)} \cdot \mathbf{T}^{(2)} \boldsymbol{\sigma}^{(1)} \cdot \boldsymbol{\sigma}^{(2)} \rangle,$$

where $|\psi(0)|^2$ represents the square of the two-body wave function at zero separation. Verify that

$$\Delta E^{\text{hf}}_{(q\bar{q})_1} = \frac{8\pi\alpha_s}{9m_1 m_2} |\psi(0)|^2 \langle \boldsymbol{\sigma}^{(1)} \cdot \boldsymbol{\sigma}^{(2)} \rangle \quad \text{(mesons)},$$

$$\Delta E^{\text{hf}}_{(qq)_{3^*}} = \frac{4\pi\alpha_s}{9m_1 m_2} |\psi(0)|^2 \langle \boldsymbol{\sigma}^{(1)} \cdot \boldsymbol{\sigma}^{(2)} \rangle \quad \text{(baryons)}.$$

(b) Using the techniques employed in section 8.1.1 to evaluate expectation values of scalar products of color matrices, compute the strength of the hyperfine interactions for spin-singlet and spin-triplet $L = 0$ mesons. Show that the hyperfine interaction is attractive for $J^P = 0^-$ mesons and repulsive for $J^P = 1^-$ mesons and that the $K^*$-$K$ splitting is smaller than the $\rho$-$\pi$ splitting.

(c) Now apply similar reasoning to the $L = 0$ baryon states, for which the relevant expectation value will be

$$\sum_{i<j} \frac{\langle \boldsymbol{\sigma}^{(i)} \cdot \boldsymbol{\sigma}^{(j)} \rangle}{m_i m_j}.$$

Compute the hyperfine contribution to the masses of the $J^P = \frac{1}{2}^+$ and $\frac{3}{2}^+$ baryons, and show that the $\frac{3}{2}^+$ decimet states lie above their $\frac{1}{2}^+$ octet counterparts.

(d) Consider the ground-state baryons $\Lambda$ and $\Sigma$. From the requirement that fermion wavefunctions be antisymmetric under the exchange of space $\times$ spin $\times$ isospin $\times$ color, deduce that the nonstrange quarks must couple to spin 0 in the $\Lambda$ and to spin 1 in the $\Sigma$. Then compute the hyperfine contributions to the $\Lambda$ and $\Sigma^0$ masses, taking into account all the pairwise hyperfine interactions and idealizing $|\psi(0)|^2$ to be universal. Show that the $\Lambda$-$\Sigma^0$ splitting is understood as a consequence of the color hyperfine interaction. [A. De Rújula, H. Georgi, and S. L. Glashow, *Phys. Rev.* D**12**, 147 (1975).]

8.5. The one-gluon-exchange arguments of section 8.1.1 give no indication that tetraquark ($qq\bar{q}\bar{q}$) mesons, pentaquark ($4q\bar{q}$) baryons, or ($6q$) dibaryons should be appreciably bound by the color force. Let us investigate whether a strong hyperfine interaction might cause some specific configurations to be bound. It is convenient to compute the parameter of interest in a multiquark state, $\mathcal{C} \equiv \langle \sum_{i<j} \mathbf{T}^{(i)} \cdot \mathbf{T}^{(j)} \, \boldsymbol{\sigma}^{(i)} \cdot \boldsymbol{\sigma}^{(j)} \rangle$, using a colorspin technique. An SU(6) color $\otimes$ spin algebra can be constructed out of the 35 operators

$$\mathbf{G}^{(6)} \equiv \begin{cases} T_a \otimes \dfrac{\sigma_b}{\sqrt{2}}, \\[2mm] T_a \otimes \dfrac{1_{2\times 2}}{\sqrt{2}}, \\[2mm] 1_{3\times 3} \otimes \dfrac{\sigma_b}{2\sqrt{3}}, \end{cases},$$

which are normalized so that $\mathrm{tr}(G_i^{(6)} G_j^{(6)}) = \frac{1}{2}\delta_{ij}$. In similar notation, the normalized generators of SU(2) and SU(3) are $\mathbf{G}^{(2)} = \boldsymbol{\sigma}/2$ and $\mathbf{G}^{(3)} = \mathbf{T}$ [cf. (8.1.16)]. Values of $\mathbf{G}^{(6)2}$ for some representations of interest are as follows [131].

| SU(6)$_{\text{colorspin}}$ | | (SU(3)$_c$, SU(2)$_{\text{spin}}$) | $\mathbf{G}^{(6)2}$ |
|---|---|---|---|
| □ | 1 | $(\mathbf{3}, \mathbf{2})$ | $\frac{35}{12}$ |
| | 21 | $(\mathbf{6}, \mathbf{3}) \oplus (\mathbf{3}^*, \mathbf{1})$ | $\frac{20}{3}$ |
| | 15 | $(\mathbf{6}, \mathbf{1}) \oplus (\mathbf{3}^*, \mathbf{3})$ | $\frac{14}{3}$ |
| | 56 | $(\mathbf{8}, \mathbf{2}) \oplus (\mathbf{10}, \mathbf{4})$ | $\frac{45}{4}$ |
| | 70 | $(\mathbf{1}, \mathbf{2}) \oplus (\mathbf{8}, \mathbf{4}) \oplus (\mathbf{8}, \mathbf{2}) \oplus (\mathbf{10}, \mathbf{2})$ | $\frac{33}{4}$ |
| | 20 | $(\mathbf{1}, \mathbf{4}) \oplus (\mathbf{8}, \mathbf{2})$ | $\frac{21}{4}$ |
| | 490 | $(\mathbf{1} \oplus \mathbf{27} \oplus \mathbf{28}, \mathbf{1}) \oplus \ldots$ | 18 |

(a) Following the example of (8.1.18) to evaluate scalar products of generators, show that

$$\mathcal{C} = \mathbf{G}^{(6)2}_{\text{tot}} - \tfrac{1}{2}\mathbf{G}^{(3)2}_{\text{tot}} - \tfrac{1}{3}s(s+1) - \sum_i \mathbf{G}^{(6)2}_i + \tfrac{1}{2}\sum_i \mathbf{G}^{(3)2}_i + \tfrac{1}{3}\sum_i s_i(s_i+1).$$

(b) Now specialize to color-singlet states, for which $\langle \mathbf{G}^{(3)2}_{\text{tot}} \rangle = 0$. Show that for $n$ quarks in a color singlet,

$$\mathcal{C}(n:1) = \mathbf{G}^{(6)2}_{\text{tot}} - \tfrac{1}{3}s(s+1) - 2n.$$

For $s$-wave baryons, the SU(3)$_{\text{flavor}} \otimes$ SU(6)$_{\text{colorspin}}$ wave functions must be antisymmetric. Determine the colorspin representations that must be paired with the the flavor representations of the octet and decimet baryons. Then compute the colorspin parameter $\mathcal{C}$ and show that the color hyperfine interaction raises the spin-$\frac{3}{2}$ flavor-decimet states above the spin-$\frac{1}{2}$ flavor-octet states, as derived in problem 8.4(c).

(c) The one-gluon exchange picture implies no residual color force between two spin-singlet baryons. Evaluate the colorspin parameter $\mathcal{C}$ for two baryons, which is simply twice the value for one baryon. It is interesting to ask whether the color hyperfine interaction might be more attractive for a different six-quark configuration, so that it might be bound. Compute the colorspin parameter $\mathcal{C}$ for a spin-zero color-singlet flavor-singlet $(uds)^2$ state contained in the colorspin **490** representation, and discuss the implications. [R. L. Jaffe, *Phys. Rev.* D**15**, 267, 281 (1977); *Phys. Rev. Lett.* **38**, 197 (1977).]

8.6. Modify the calculation (problem 6.10) of the process $e^+e^- \rightarrow \gamma\gamma$ to describe the reaction $q\bar{q} \rightarrow gg$ in quantum chromodynamics, In this case, the analogues of the two diagrams shown in problem 6.10 are not by themselves gauge invariant. (a) Show that in QCD the quantities $k_{1\nu}(A^{\mu\nu} + \tilde{A}^{\mu\nu})$ and $k_{2\mu}(A^{\mu\nu} + \tilde{A}^{\mu\nu})$ are proportional to $[\lambda^a, \lambda^b]$, where $a$ and $b$ are the $SU(3)_c$ indices of the two gluons. (b) What is the resolution of this noninvariance? (c) For the full gauge-invariant amplitude described by $\epsilon_{1\nu}^* \epsilon_{2\mu}^* T^{\mu\nu}$, under what conditions are the requirements $k_{1\nu} T^{\mu\nu} = 0 = k_{2\mu} T^{\mu\nu}$ fulfilled?

8.7. Repeat the one-loop calculation of the charge renormalization in scalar electrodynamics using the Pauli–Villars regularization. Relate the relative magnitudes of the charge renormalization in scalar and spinor electrodynamics to the ratio $\sigma(e^+e^- \rightarrow \sigma^+\sigma^-)/\sigma(e^+e^- \rightarrow \mu^+\mu^-)$ developed in problems 1.4 and 1.5.

8.8. Compute the one-loop charge renormalization in (spinor) QED using the method of dimensional regularization.

8.9. Consider the one-loop modifications to Coulomb scattering in the limit of low momentum transfer $-q^2 \ll m^2$. Beginning from (8.2.31), show that the scattering amplitude is modified by a factor

$$\left[ 1 - \frac{\alpha_R}{15\pi} \frac{q^2}{m^2} + O(\alpha_R^2) \right].$$

Show that this corresponds in the position-space potential to an additional interaction of the form

$$\frac{4}{15} \frac{\alpha_R}{m^2} \delta^3(\mathbf{x}),$$

and estimate the first-order shift in the energy levels of the hydrogen atom (Ref. [17]).

8.10. Using the Feynman rules for the Faddeev–Popov ghost given in figure 8.10 and section 8.3, verify that the modification to the gluon propagator due to one ghost loop, as shown in figure 8.18, is given by (8.3.15).

8.11. (a) Show that an arbitrary $N \times N$ matrix $\mathbf{M}$ can be expressed as a linear combination of the identity matrix and the generators $T^a$ of the fundamental representation (defined for SU(3)$_c$ in (8.1.16)) as

$$M = \frac{1}{N}\text{tr}(M)\mathbb{I} + 2\,\text{tr}(MT^a)\,T^a.$$

(b) Now write the preceding equation in component form and obtain the Fierz identity for the color algebra (A.5.14),

$$T^a_{\alpha\beta}\,T^a_{\gamma\delta} = \tfrac{1}{2}\delta_{\alpha\delta}\delta_{\beta\gamma} - \frac{1}{2N}\delta_{\alpha\beta}\delta_{\gamma\delta}.$$

(c) Express the result graphically and show that it encapsulates the double-line representation of the gluon in the large-$N$ limit (cf. section 8.10).

8.12. Using the Fierz identity for the color algebra derived in problem 8.11, characterize the binding energy resulting from one-gluon-exchange between a quark and an antiquark. Show that for a color-singlet pair, the interaction is attractive, with strength set by $C_F$, whereas for a color-octet pair the interaction is repulsive, with strength set by $1/2N$. This calculation reproduces the results obtained by more pedestrian means for table 8.3. Evaluate the corresponding interaction energies for isoscalar and isovector pairs in SU(2) Yang–Mills theory, and recover the results of (4.3.5).

8.13. Use the double-line notation of (A.5.16) for the coupling of a gluon to a pair of quarks to make a purely graphical evaluation of the color factor for the one-loop quark self-energy diagram, and recover the result of (8.3.20).

8.14. Consider a gauge theory of the strong interactions based on the color symmetry group SO(3), in which both quarks and gluons are assigned to the adjoint representation. By appropriately modifying the color factors entering the expressions leading to (8.3.29), evaluate the running coupling constant in one-loop order. What is the condition for asymptotic freedom in this theory?

8.15. Consider the dissociation of an electron of momentum $p$ into a photon with momentum $\mathbf{k} = (\vec{k}_\perp, zp)$ and an electron in QED. Calculate the square of the matrix element for dissociation, and thus the probability to find such a photon associated with the electron beam. (Weizsäcker and Williams, Ref. [41].)

8.16. Following the method of Altarelli and Parisi, compute the splitting function $P_{q\leftarrow q}(z)$ for a theory of colored quarks interacting by means of *scalar* gluons for the color group SU($N$). Assume that the theory has a fixed coupling constant $\alpha_s^*$. Calculate the $Q^2$-evolution of the nonsinglet moments, and predict the slope of the logarithmic ratio

$$\frac{\ln(\Delta_n(\tau)/\Delta_n(0)]}{\ln(\Delta_k(\tau)/\Delta_k(0)]}$$

for $(n, k) = (5, 3)$ and $(6, 4)$. [M. Glück and E. Reya, *Phys. Rev. D* **16**, 3242 (1977); D. Bailin and A. Love, *Nucl. Phys. B* **75**, 159 (1974).]

8.17. (a) Express the integral

$$I_G(Q^2) = \int_0^1 dx \frac{F_2^p(x, Q^2) - F_2^n(x, Q^2)}{x}$$

in terms of parton densities. Use isospin invariance to relate the neutron and proton structure functions through the interchange $u \leftrightarrow d$.

(b) Separating the quark parton densities into valence and sea components as $q_i = q_{iv} + q_{is}$, show that

$$I_G(Q^2) = \tfrac{1}{3},$$

under the assumption that the light-quark sea is flavor symmetric: the Gottfried sum rule [K. Gottfried, *Phys. Rev. Lett.* **18**, 1174 (1967)].

(c) By isospin invariance and the symmetry of the splitting function $P_{q \leftarrow g}(z) = P_{q \leftarrow g}(1 - z)$ [cf. (8.5.29)], perturbative evolution cannot induce a flavor asymmetry in the sea or a distinction between sea quarks and antiquarks. A toy model shows how a Gottfried sum rule defect could arise as a quasistatic property of the nucleon. A simplified picture of a nucleon appropriate at low values of $Q^2$ would contain only valence up- and down-quark distributions. In the chiral quark model (Ref. [114]), these valence quarks may emit a pion, as in these examples:

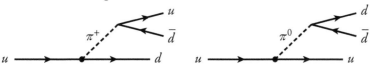

Let $a$ designate the probability for a valence up quark to fluctuate into a valence down quark plus a $\pi^+$, which contains at this approximation an up quark and an antidown quark. The $u$ quark can also emit a $\pi^0$ (with $u\bar{u}$ and $d\bar{d}$ components) and remain a $u$ quark. Suppose that the probability of emission is small enough to be treated as a perturbation. Compute the final state resulting from single pion emission by an up quark, and then use isospin symmetry to derive the corresponding result for a down quark.

(d) Taking account of the probability for zero-pion emission, calculate the proton composition after one interaction, and obtain the neutron composition by applying isospin invariance.

(e) Evaluate the Gottfried integral (expressed in terms of parton densities) and show that the Gottfried sum rule defect is

$$\Delta I_G = -\frac{2a}{3} \neq 0.$$

[E. J. Eichten, I. Hinchliffe, and C. Quigg, *Phys. Rev. D* **45**, 2269 (1992). A Gottfried sum rule defect was first observed by P. Amaudruz et al. [New Muon Collaboration], *Phys. Rev. Lett.* **66**, 2712 (1991).]

8.18. Expand the definition (8.6.1) of rapidity for an object with mass $m$, under the assumption that $p \gg m$, to show that as $m/p \to 0$, $y \to \eta \equiv -\ln \tan(\theta/2)$. (See §39.5 of the 2010 *Review of Particle Physics*, Ref. [24].)

8.19. Compute the partonic cross section $\hat{\sigma}(ud \to ud)$ for the scattering of unlike quarks in QCD. Work in the extreme relativistic limit in which masses are negligible.
(a) Show that

$$\hat{\sigma}(ud \to ud) = \frac{4\pi\alpha_s^2}{9\hat{s}^2} \cdot \frac{\hat{s}^2 + \hat{u}^2}{\hat{t}^2},$$

where $\hat{s}, \hat{t}, \hat{u}$ are the usual Mandelstam invariants [132] for the parton–parton collision.
(b) Compare with the cross section for $\mu^- e^-$ scattering in QED.
(c) How would the result change if the quarks were color sextets instead of color triplets?

8.20. (a) Express the $ud \to ud$ cross section computed in problem 8.19 in terms of c.m. angular variables, and note that the angular distribution is reminiscent of that for Rutherford scattering, $d\sigma/d\Omega^* \propto 1/\sin^4(\theta^*/2)$ (cf. Problem 3.3).
(b) In the search for new interactions, the angular distribution for quark–quark scattering, inferred from dijet production in $p^\pm p$ collisions, is a sensitive diagnostic. Show that when reexpressed in terms of the variable $\chi = (1 + \cos\theta^*)/(1 - \cos\theta^*)$, the angular distribution for $ud$ scattering is $d\sigma/d\chi \propto$ constant.
(c) The rapidity variable, $y = \frac{1}{2}\ln[(E + p_z)/(E - p_z)]$, is useful in the study of high-energy collisions because it shifts simply under Lorentz boosts. Show that in the extreme relativistic limit, measuring the jet rapidities in the reaction $p^\pm p \to \text{jet}_1 + \text{jet}_2$ leads directly to a determination of the variable $\chi$ for parton–parton scattering as $\chi = \exp(y_1 - y_2)$. [For early applications to data from the Large Hadron Collider, see G. Aad et al. [ATLAS Collaboration], *Phys. Lett. B* **694**, 327 (2011); V. Khachatryan et al. [CMS Collaboration], *Phys. Rev. Lett.* **106**, 201804 (2011); S. Chatrchyan et al. [CMS Collaboration], *JHEP* **1205**, 055 (2012).]

8.21. Consider the reaction $p^\pm p \to \text{jet}_1 + \text{jet}_2 + \text{anything at c.m. energy } \sqrt{s}$. Denote the rapidity of the dijet system as $y_{\text{boost}} \equiv \frac{1}{2}(y_1 + y_2)$ and the individual jet rapidity in the dijet rest frame as $y^* \equiv \frac{1}{2}(y_1 - y_2)$. (a) Neglecting the invariant masses of the individual jets with respect to $p_\perp$, show that the invariant mass of the dijet system, and thus of the colliding partons, is $\sqrt{\hat{s}} = 2p_\perp \cosh y^*$.
(b) Deduce the momentum fractions carried by the two colliding partons [cf. (8.6.3)]. Show that $x_{a,b} = \sqrt{\tau}e^{\pm y_{\text{boost}}}$, where $\tau \equiv \hat{s}/s$.

8.22. Express the Mandelstam invariants for two-body parton scattering in terms of $\cos\theta = (1 - 4p_\perp^2/\hat{s})^{1/2}$, the cosine of the scattering angle in the parton–parton c.m., as

$$\hat{t} = -\frac{\hat{s}}{2}(1 - \cos\theta), \qquad \hat{u} = -\frac{\hat{s}}{2}(1 + \cos\theta).$$

Evaluate the elementary cross sections given in the following table for 90° scattering in the c.m., which is relevant for the production of jets with maximum transverse momentum, and compute the ratio to the reference cross section $\hat{\sigma}(ud \to ud)$ calculated in problem 8.19. Verify the entries in the right column of the table.

| Process | $\hat{\sigma}$ | 90° Ratio to $\hat{\sigma}(ud \to ud)$ |
|---------|----------------|----------------------------------------|
| $ud \to ud$ | $\dfrac{4\alpha_s^2}{9\hat{s}}\dfrac{\hat{s}^2 + \hat{u}^2}{\hat{t}^2}$ | $1$ |
| $q\bar{q} \to q'\bar{q}'$ | $\dfrac{4\alpha_s^2}{9\hat{s}}\dfrac{\hat{t}^2 + \hat{u}^2}{\hat{s}^2}$ | $\dfrac{1}{10} = 0.1$ |
| $uu \to uu$ | $\dfrac{4\alpha_s^2}{9\hat{s}}\left(\dfrac{\hat{s}^2 + \hat{u}^2}{\hat{t}^2} + \dfrac{\hat{s}^2 + \hat{t}^2}{\hat{u}^2} - \dfrac{2}{3}\dfrac{\hat{s}^2}{\hat{u}\hat{t}}\right)$ | $\dfrac{22}{15} \approx 1.47$ |
| $q\bar{q} \to q\bar{q}$ | $\dfrac{4\alpha_s^2}{9\hat{s}}\left(\dfrac{\hat{s}^2 + \hat{u}^2}{\hat{t}^2} + \dfrac{\hat{t}^2 + \hat{u}^2}{\hat{s}^2} - \dfrac{2}{3}\dfrac{\hat{u}^2}{\hat{s}\hat{t}}\right)$ | $\dfrac{7}{6} \approx 1.17$ |
| $q\bar{q} \to gg$ | $\dfrac{8\alpha_s^2(\hat{t}^2 + \hat{u}^2)}{3\hat{s}}\left(\dfrac{4}{9\hat{t}\hat{u}} - \dfrac{1}{\hat{s}^2}\right)$ | $\dfrac{7}{15} \approx 0.47$ |
| $gg \to q\bar{q}$ | $\dfrac{3\alpha_s^2(\hat{t}^2 + \hat{u}^2)}{8\hat{s}}\left(\dfrac{4}{9\hat{t}\hat{u}} - \dfrac{1}{\hat{s}^2}\right)$ | $\dfrac{21}{320} \approx 0.66$ |
| $gq \to gq$ | $\dfrac{\alpha_s^2(\hat{s}^2 + \hat{u}^2)}{\hat{s}}\left(\dfrac{1}{\hat{t}^2} - \dfrac{4}{9\hat{s}\hat{u}}\right)$ | $\dfrac{11}{4} = 2.75$ |
| $gg \to gg$ | $\dfrac{9\alpha_s^2}{2\hat{s}}\left(3 - \dfrac{\hat{t}\hat{u}}{\hat{s}^2} - \dfrac{\hat{s}\hat{u}}{\hat{t}^2} - \dfrac{\hat{s}\hat{t}}{\hat{u}^2}\right)$ | $\dfrac{2187}{160} \approx 13.67$ |

[Reference for cross sections: J. F. Owens, E. Reya, and M. Glück, *Phys. Rev. D* **18**, 1501 (1978). See also E. Eichten, I. Hinchliffe, K. D. Lane, and C. Quigg, *Rev. Mod. Phys.* **56**, 579 (1984), §III.B.]

8.23. In the valence approximation for the reactions $\bar{p}p \to W^\pm + \text{anything}$, the rapidity asymmetry,

$$A_W(y) \equiv \frac{d\sigma(W^+)/dy - d\sigma(W^-)/dy}{d\sigma(W^+)/dy + d\sigma(W^-)/dy},$$

is sensitive to the relative shapes of the $u$ and $d$ quark distributions within the proton. Express the asymmetry in terms of the valence–quark distributions $u_p(x) = \bar{u}_{\bar{p}}(x)$ and $d_p(x) = \bar{d}_{\bar{p}}(x)$ and then in terms of $R(x) \equiv d_p(x)/u_p(x)$. Taking the direction of the incident proton to define positive rapidity, verify that the fractional momenta carried by the quark and antiquark, respectively, are $x_{q,\bar{q}} = \sqrt{\tau}\exp(\pm y)$, where $\sqrt{\tau} \equiv M_W/\sqrt{s}$. Now make a Taylor expansion to approximate $x_{q,\bar{q}}$ and $R(x_q, x_{\bar{q}})$ for small values of the rapidity, $y$. Thereby show that

$$A_W(y) \approx -y\sqrt{\tau}\frac{R'(\sqrt{\tau})}{R(\sqrt{\tau})}.$$

[E. L. Berger et al., *Phys. Rev. D* **40**, 83 (1989). See also §9.4 of R. K. Ellis, W. J. Stirling, and B. R. Webber, *QCD and Collider Physics*, Cambridge University Press, Cambridge, 1996. The link between the produced $W^{\pm}$ and observed $\ell^{\pm}$ rapidity distributions is explained in C. Quigg, *Rev. Mod. Phys.* **49**, 297 (1977). For early evidence from LHC experiments, see G. Aad et al. [ATLAS Collaboration], *Phys. Lett. B* **701**, 31 (2011); S. Chatrchyan et al. [CMS Collaboration], *JHEP* **1104**, 050 (2011).]

8.24. Examine the production of a standard-model Higgs boson in two-photon collisions at a lepton collider, $\ell^{+}\ell^{-} \to \ell^{+}\ell^{-} H^{0}$. For a Higgs boson of mass $M_H = 130$ GeV, use (8.7.8) to compute the cross section as a function of (symmetric) lepton beam energy, from $E = 250$ GeV to 3 TeV, for both electrons and muons in the initial state. Take $\Gamma(H^0 \to \gamma\gamma) = 11$ keV [cf. table B.3 of S. Dittmaier et al. [LHC Higgs Cross Section Working Group], "Handbook of LHC Higgs Cross Sections: 1. Inclusive Observables," arXiv:1101.0593.] Also express the cross section in units of $\sigma(e^+e^- \to \mu^+\mu^-)$.

8.25. The principal mechanism for the production of the standard-model Higgs boson at the Large Hadron Collider is the gluon-fusion process $gg \to H$, where $q_i$ represents a quark of flavor $i$, with mass $m_i$:

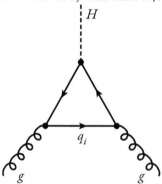

(a) Compute the cross section for the reaction $pp \to H + $ anything and show that it can be written in the form

$$\frac{d\sigma}{dy}(pp \to H + X) = \frac{G_F \alpha_s^2}{32\pi\sqrt{2}} \frac{M_H^2}{s} |\eta|^2\, g(x_a, M_H^2) g(x_b, M_H^2),$$

where $g(x, Q^2)$ is the gluon parton distribution function, $x_{a,b} = (M_H/\sqrt{s})e^{\pm y}$, and

$$\eta = \sum_i \int_0^1 dx \int_0^{1-x} \frac{dy(1 - 4xy)}{1 - (xy M_H^2/m_i^2)}.$$

(b) With the definition $\epsilon \equiv 4m_i^2/M_H^2$, show that for a single quark flavor

$$\eta = \frac{\epsilon}{2}[1 + (\epsilon - 1)\varphi(\epsilon)],$$

where

$$\varphi(\epsilon) = \begin{cases} - \left[\sin^{-1}(1/\sqrt{\epsilon})\right]^2, & \epsilon > 1 \\ \frac{1}{4} \left[\ln(\zeta_+/\zeta_-) + i\pi\right]^2, & \epsilon < 1, \end{cases}$$

and $\zeta_{\pm} = 1 \pm \sqrt{1 - \epsilon}$.

(c) Plot the quantity $|\eta(\epsilon)|^2$ as a function of $\epsilon$. For a Higgs boson with mass $M_H = 120$ GeV, which of the known quarks $u, d, s, c, t, b$ contribute significantly to the loop diagram?

(d) How would the production rate be modified by the presence of a fourth generation of standard-model quarks heavier than top?

[H. M. Georgi et al., *Phys. Rev. Lett.* **40**, 692 (1978). The loop calculation is done in two ways in §9.2 of R. K. Ellis et al., *Phys. Rep.* **518**, 141 (2012)].

8.26. Consider the nonsinglet photon structure function in the parton-model approximation.

(a) Referring to (8.7.24) for the splitting function $P_{q \leftarrow \gamma}$, verify that

$$A_n(q \leftarrow \gamma) = \int_0^1 dz z^{n-1} P_{q \leftarrow \gamma}(z) = 3 \left[ \frac{n^2 + n + 2}{n(n+1)(n+2)} \right]. \qquad (8.7.35)$$

(b) Referring to the definitions (8.7.30) and (8.7.29), show that the parton-model prediction for the $n$th moment of the nonsinglet distribution is given by

$$\Delta_n^{(i)}(\tau) = \frac{\alpha(e_i^2 - \langle e^2 \rangle) A_n(q \leftarrow \gamma)\tau}{2\pi},$$

with $\langle e^2 \rangle$ given by (8.7.32).

8.27. The model of M.-Y. Han and Y. Nambu [*Phys. Rev.* **149B**, 1006 (1965)] was proposed as an integer-charge alternative to the fractional-charge quark model, with charges assigned as

| Color | Flavor | | |
|-------|---|---|---|
|  | $u$ | $d$ | $s$ |
| $R$ | 0 | $-1$ | $-1$ |
| $G$ | 1 | 0 | 0 |
| $B$ | 1 | 0 | 0 |

The photon thus couples to the combination $\gamma_{\mathrm{HN}} \sim u_G \bar{u}_G + u_B \bar{u}_B - d_R \bar{d}_R - s_R \bar{s}_R$.

(a) Resolve $\gamma_{\mathrm{HN}}$ into color-singlet and color-octet components.

(b) Show that below a hypothetical threshold for the production of colored states, the ratio

$$R \equiv \frac{\sigma(e^+ e^- \to \text{hadrons})}{\sigma(e^+ e^- \to \mu^+ \mu^-)}$$

is $R_{\mathrm{HN}} = 2$, as in the fractional-charge picture, and that $R_{\mathrm{HN}}^* = 4$ if color can be liberated.

(c) Now consider the reaction $\gamma\gamma \to$ hadrons, viewed as $\gamma\gamma \to q\bar{q}$. Show that with fractionally charged quarks

$$\sigma(\gamma\gamma \to \text{ hadrons}) \propto \sum_{i=u,d,s} e_i^4 = \frac{2}{3},$$

whereas in the Han–Nambu model

$$\sigma(\gamma\gamma \to \text{ hadrons}) \propto \begin{cases} \dfrac{4}{3} & \text{below color threshold,} \\ 4 & \text{above color threshold.} \end{cases}$$

[F. E. Close, *An Introduction to Quarks and Partons*, Academic, New York, 1979, chap. 8; M. Chanowitz, "Color and Experimental Physics," in *Particles and Fields—1975*, ed. H. J. Lubatti and P. M. Mockett, University of Washington, Seattle, 1976, p. 448 [http://j.mp/h0QRX5].]

8.28. Consider the Schrödinger equation for $s$-wave bound states of a $1/r$ potential in $N$ space dimensions,

$$\left[\nabla^2 + 2\mu\left(E + \frac{\alpha}{r}\right)\right]\psi(r) = 0.$$

(a) Show that the radial equation is

$$\left[\frac{d^2}{dr^2} + \frac{(N-1)}{r}\frac{d}{dr} + 2\mu\left(E + \frac{\alpha}{r}\right)\right]\psi(r) = 0.$$

(b) Now take the limit of large $N$, so that $(N-1) \to N$. Introduce a reduced radial wave function $u = r^{N/2}\psi$ and a scaled radial coordinate $R = r/N^2$. Show that the radial equation becomes

$$\frac{1}{N^2}\frac{d^2u}{dR^2} - \frac{u}{4R^2} + 2\mu\left(N^2 E + \frac{\alpha}{R}\right)u = 0.$$

(c) Apart from the factor $N^2$ that sets the scale of $E$, this equation describes a particle with effective mass $\mu N^2$ moving in an effective potential

$$V_{\text{eff}} = \frac{1}{8\mu R^2} - \frac{\alpha}{R}.$$

Find the energy of the ground state in the limit as $N \to \infty$, for which the kinetic energy vanishes. Show that it is given by the absolute minimum of $V_{\text{eff}}$,

so that

$$E_{N \to \infty} = -\frac{2\mu\alpha^2}{N^2}.$$

Corrections to this result may be obtained by expanding $V_{\text{eff}}$ about the minimum and treating the additional terms as perturbations.

(d) The solution to the exact eigenvalue equation of part (a) is easily verified to be

$$E_{\text{exact}} = -\frac{2\mu\alpha^2}{(N-1)^2}.$$

Show that it can be recast in the form of an expansion in inverse powers of $N$ as

$$E_{\text{exact}} = -\frac{2\mu\alpha^2}{N^2} \sum_{j=1}^{\infty} j N^{1-j} = E_{N \to \infty} \left( 1 + \sum_{j=2}^{\infty} j N^{1-j} \right),$$

so that the $N \to \infty$ result may form the basis for a systematic approximation scheme. How many terms must be retained to obtain a 1% approximation for $N = 3$? [L. D. Mlodinow and N. Papanicolaou, *Ann. Phys. (NY)* **128**, 314 (1980).]

## ⅠⅠⅠⅠⅠⅠⅠⅠⅠⅠⅠ FOR FURTHER READING ⅠⅠⅠⅠⅠⅠⅠⅠⅠⅠⅠ

**QCD Resource Letter.** For an extensive annotated biography, with explanations of the central points, see

A. S. Kronfeld and C. Quigg, *Am. J. Phys.* **78**, 1081 (2010).

**Gentle introductions.** Expositions of QCD for undergraduate students and general readers include

Y. Nambu, *Quarks: Frontiers in Elementary Particle Physics*, World Scientific, Singapore, 1985,

G. 't Hooft, *In Search of the Ultimate Building Blocks*, Cambridge University Press, Cambridge and New York, 1997,

A. Watson, *The Quantum Quark*, Cambridge University Press, Cambridge, 2004,

F. Wilczek, *The Lightness of Being: Mass, Ether, and the Unification of Forces*, Basic Books, New York, 2008.

**Textbooks.** Many books treat quantum chromodynamics, in whole or in part, from a modern point of view:

R. K. Ellis, W. J. Stirling, and B. R. Webber, *QCD and Collider Physics*, Cambridge University Press, Cambridge, 1996;

G. Dissertori, I. G. Knowles, and M. Schmelling, *Quantum Chromodynamics: High Energy Experiments and Theory*, Oxford University Press, Oxford and New York, 2003;

F. J. Ynduráin, *The Theory of Quark and Gluon Interactions*, 4th ed., Springer, Berlin and New York, 2006;

W. Greiner, S. Schramm, and E. Stein, *Quantum Chromodynamics*, 3rd ed., Springer, Berlin, 2007;

T. Muta, *Foundations of Quantum Chromodynamics: An Introduction to Perturbative Methods in Gauge Theories*, 3rd ed., World Scientific, Singapore and River Edge, NJ, 2009;

B. L. Ioffe, V. S. Fadin, and L. N. Lipatov, *Quantum Chromodynamics: Perturbative and Nonperturbative Aspects*, Cambridge University Press, Cambridge and New York, 2010;

J. Collins, *Foundations of Perturbative QCD*, Cambridge University Press, Cambridge, 2011.

Many nontrivial applications, especially to the physics of jets, are treated in

Yu. L. Dokshitzer, V. A. Khoze, A. H. Mueller, and S. I. Troyan, *Basics of Perturbative QCD*, Éditions Frontières, Singapore, 1991, available at http://j.mp/hpmPDC.

Among many fine field-theory textbooks,

G. Sterman, *An Introduction to Quantum Field Theory*, Cambridge University Press, Cambridge and New York, 1993,

is particularly inclined toward QCD and the issue of factorization.

**Origins of QCD.** The idea of a vector gluon theory may be found in

Y. Nambu, in *Preludes in Theoretical Physics in Honor of V. F. Weisskopf*, ed. A. De Shalit, H. Feshbach, and L. Van Hove, North-Holland, Amsterdam, 1966, p. 133,

and the path from currents to a gauge theory of the strong interactions is laid out in

H. Fritzsch and M. Gell-Mann, "Current Algebra: Quarks and What Else?" in *Proceedings of the XVI International Conference on High Energy Physics*, vol. 2, ed. J. D. Jackson and A. Roberts, National Accelerator Laboratory, Batavia, IL, 1972, p. 135, hep-ph/0208010.

Physical arguments in favor of the $SU(3)_c$ gauge theory are collected in

H. Fritzsch, M. Gell-Mann, and H. Leutwyler, *Phys. Lett.* **B47**, 365 (1973).

A clear formulation of the theory, after the recognition of asymptotic freedom, appears in

S. Weinberg, *Phys. Rev. Lett.* **31**, 494 (1973).

**Regge phenomenology.** The systematics of hadron spectra and two-body reactions as expressed in terms of singularities in the complex angular momentum plane provided the backdrop for the invention of QCD. Some useful reviews of that approach include

P.D.B. Collins, *Phys. Rept.* **1**, 103 (1971),

A. C. Irving and R. P. Worden, *Phys. Rept.* **34**, 117 (1977),

A. B. Kaidalov, "Regge poles in QCD," In *At the Frontier of Particle Physics: Handbook of QCD*, ed. M. Shifman, World Scientific, Singapore, 2001, vol. 1, p. 603, arXiv:hep-ph/0103011,

V. N. Gribov, *The Theory of Complex Angular Momenta*, Cambridge University Press, Cambridge, 2003.

**Quantization of gauge theories.** Technical aspects and questions of consistency are treated by

E. S. Abers and B. W. Lee, *Phys. Rept.* **9C**, 1 (1973),

R. P. Feynman, in *Weak and Electromagnetic Interactions at High Energy*, 1976 Les Houches Lectures, ed. R. Balian and C. H. Llewellyn Smith, North-Holland, Amsterdam, 1977, p. 120,

L. D. Faddev and A. A. Slavnov, *Gauge Fields, Introduction to Quantum Theory*, Benjamin, Reading, MA, 1980,

C. Itzykson and J.-B. Zuber, *Quantum Field Theory*, McGraw-Hill, New York, 1980,

M. E. Peskin and D. V. Schroeder, *An Introduction To Quantum Field Theory*, Westview Press, Boulder, CO, 1995,

B. de Wit and J. Smith, *Field Theory in Particle Physics*, North-Holland, Amsterdam, 1986,

S. Weinberg, *The Quantum Theory of Fields*, (two volumes) Cambridge University Press, Cambridge, 1995, 1996,

J. Zinn-Justin, *Quantum Field Theory and Critical Phenomena*, 4th ed., Oxford University Press, Oxford, 2002,

M. Srednicki, *Quantum Field Theory*, Cambridge University Press, Cambridge, 2007,

T. Banks, *Modern Quantum Field Theory: A Concise Introduction*, Cambridge University Press, Cambridge, 2008,

A. Zee, *Quantum Field Theory in a Nutshell*, 2nd ed., Princeton University Press, Princeton, NJ, 2010.

**Charge screening in QED.** An early reference is the paper

V. F. Weisskopf, *Phys. Rev.* **56**, 72 (1939).

A classic commentary on the evolution of the coupling constant is given in

L. D. Landau, in *Niels Bohr and the Development of Physics*, ed. W. Pauli, Pergamon, London, 1955, p. 52.

The sign of the dielectric constant is the subject of the review article

O. V. Dolgov, D. A. Kirzhnits, and E. G. Maksimov, *Rev. Mod. Phys.* **53**, 81 (1981).

Experimental consequences of the Uehling term in the atomic potential may be seen in the review of muonic atoms,

E. T. Borie and G. A. Rinker, *Rev. Mod. Phys.* **54**, 67 (1982).

**Asymptotic freedom.** For derivations and discussions of the asymptotic freedom of non-Abelian gauge theories, see

V. B. Berestetski, *Usp. Fiz. Nauk* **120**, 439 (1976) [English translation: *Sov. Phys.-Uspekhi* **19**, 934 (1976)],

T.-P. Cheng and L.-F. Li, *Gauge Theory of Elementary Particle Physics*, Oxford University Press, Oxford, 1984,

C. Itzykson and J.-B. Zuber, *Quantum Field Theory*, McGraw-Hill, New York, 1980.

H. D. Politzer, *Phys. Rep.* **14C**, 129 (1974),

E. Reya. *Phys. Rep.* **69**, 195 (1981),

Nobel Foundation, "Asymptotic freedom and quantum chromodynamics: the key to the understanding of the strong nuclear forces," http://j.mp/iuOn9x,

D. J. Gross, *Rev. Mod. Phys.* **77**, 837 (2005),

H. D. Politzer, *Rev. Mod. Phys.* **77**, p. 851,

F. Wilczek, *Rev. Mod. Phys.* **77**, p. 857.

For another view of the historical setting, see
> G. 't Hooft, *Nucl. Phys. Proc. Suppl.* **74**, 413 (1999).

Many theories that proved not to be asymptotically free were analyzed by
> A. Zee, *Phys. Rev. D* **7**, 3630 (1973).

The proof that the only renormalizable theories that are asymptotically free in four dimensions are non-Abelian gauge theories is due to
> S. Coleman and D. J. Gross, *Phys. Rev. Lett.* **31**, 851 (1973).

Some insight into the physical origin of antiscreening is provided by calculations in (noncovariant) Coulomb and axial gauge, including
> J. Frenkel and J. C. Taylor, *Nucl. Phys.* **B109**, 439 (1976); **B117**, 546E (1976),
> J. D. Bjorken, "Elements Of Quantum Chromodynamics," SLAC-PUB-2372, http://j.mp/j5YdTN.

**Renormalization group methods.** The renormalization group as a technique for summing to all orders in perturbation theory in electrodynamics was invented by
> E.C.G. Stueckelberg and A. Peterman, *Helv. Phys. Acta* **26**, 499 (1953),
> M. Gell-Mann and F. E. Low, *Phys. Rev.* **95**, 1300 (1954).

A thorough review of early applications appears in the textbook by
> N. N. Bogoliubov and D. V. Shirkov, *Introduction to the Theory of Quantized Fields,* Wiley-Interscience, New York, 1959, chap. 8.

The modern formulation of the renormalization group equations is due to
> C. G. Callan, Jr., *Phys. Rev. D* **2**, 1541 (1970),
> K. Symanzik, *Commun. Math. Phys.* **18**, 227 (1970).

The power of renormalization group methods for a wide range of physical problems was recognized by
> K. G. Wilson, *Phys. Rev. D* **3**, 1818 (1971); *Sci. Am.* **241**, (8) 140 (August, 1979).

A fascinating survey with many references is
> K. G. Wilson, *Rev. Mod. Phys.* **55**, 583 (1983).

Newcomers to the subject would be well advised to begin their studies with
> S. R. Coleman, "Dilatations," in *Aspects of Symmetry,* Cambridge University Press, Cambridge, 1985, p. 67.

Applications of the renormalization group to QCD are stressed in
> D. J. Gross, in *Methods in Field Theory [Méthodes en Théorie des Champs],* edited by R. Balian and J. Zinn-Justin, North-Holland, Amsterdam and New York, 1981, p. 140,
> A. Peterman, *Phys. Rept.* **53**, 157 (1979),
> J. C. Collins, *Renormalization,* Cambridge University Press, Cambridge and New York, 1984.

**Cancellation of infrared divergences.** The classic treatment of the "infrared catastrophe" is due to
> F. Bloch and A. Nordsieck, *Phys. Rev.* **52**, 54 (1937).

For modern treatments with extensive references to the literature, see
> D. R. Yennie, S. C. Frautschi, and H. Suura, *Annals Phys.* **13**, 379 (1961),
> G. Grammer, Jr., and D. R. Yennie, *Phys. Rev.* **D8**, 4332 (1973).

(The generalization to massless fermions was made by Kinoshita and Lee and Nauenberg, Ref. [34].) The infrared-finiteness of Yang–Mills theories was shown by
> T. Appelquist, J. Carazzone, H. Kluberg-Stern, and M. Roth, *Phys. Rev. Lett.* **36**, 768 (1976).

**Jet physics in $e^+e^-$ annihilation.** The Sterman–Weinberg predictions for the energy dependence of the size of a quark jet were extended to the evolution of gluon jets by
> K. Shizuya and S. H. Tye, *Phys. Rev. Lett.* **41**, 787 (1978) [Erratum: *Phys. Rev. Lett.* **41**, 1195 (1978)],
> M. B. Einhorn, B. G. Weeks, *Nucl. Phys.* **B146**, 445 (1978).

For retrospective accounts of the gluon discovery, see
> P. Söding, B. Wiik, G. Wolf, and S. L. Wu, "The First Evidence for Three-Jet Events in $e^+e^-$ Collisions at PETRA: First Direct Observation of the Gluon," in *International Europhysics Conference on High Energy Physics (HEP 95)*, ed. J. Lemonne, C. Vander Velde, and F. Verbeure, World Scientific, Singapore, 1996, p. 3, http://j.mp/UaxRG1,
> P. Söding, *Eur. Phys. J.* **H35**, 3 (2010).

A valuable perspective appears in
> A. Ali and G. Kramer, *Eur. Phys. J. H* **36**, 245 (2011).

**Deeply inelastic scattering.** Early applications in the context of the operator product expansion were made by
> H. Georgi and H. D. Politzer, *Phys. Rev. D* **9**, 416 (1974), D. J. Gross and F. Wilczek, *Phys. Rev. D* **9**, 980 (1974).

A systematic review of QCD effects beyond leading order was given by
> A. J. Buras, *Rev. Mod. Phys.* **52**, 199 (1980).

A "physical gauge" approach is emphasized by
> Y. L. Dokshitzer, D. Diakonov, and S. I. Troian, *Phys. Rept.* **58**, 269 (1980).

For comprehensive general surveys of the comparison with experiment, see
> R. G. Roberts, *The Structure of the Proton: Deep Inelastic Scattering*, Cambridge University Press, Cambridge and New York, 1990,
> R. Devenish and A. Cooper-Sarkar, *Deep Inelastic Scattering*, Oxford University Press, Oxford and New York, 2004,
> C. Diaconu, T. Haas, M. Medinnis, K. Rith, and A. Wagner, *Ann. Rev. Nucl. Part. Sci.* **60**, 101 (2010).

**Parton distribution functions.** The state of the art is reviewed in
> A. De Roeck and R. S. Thorne, *Prog. Part. Nucl. Phys.* **66**, 727 (2011).

Recommendations for the use of parton distribution functions for the Large Hadron Collider and assessments of the uncertainties are given in
> M. Botje et al, "The PDF4LHC Working Group Interim Recommendations," arXiv:1101.0538.

Convenient access to many sets of parton distributions. with facilities for online plotting and calculation, is available through the
> Durham HEPData Project Online, http://j.mp/lgXz6U.

A library providing a common interface to many modern sets of parton distributions is
> M. R. Whalley and A. Buckley, "LHAPDF: the Les Houches Accord Parton Distribution Function Interface," http://j.mp/jqYdvM.

A set of parton distribution functions incorporating QED corrections (and thus photons as partons) is

> A. D. Martin, R. G. Roberts, W. J. Stirling, and R. S. Thorne, *Eur. Phys. J.* **C39**, 155 (2005).

**Jet physics at hadron colliders.** Modern definitions of jets—taking infrared safety, calculability, and ease of measurement into account, as well as the extension to hadronic collisions—are surveyed in the lecture course by

> G. P. Salam, *Eur. Phys. J. C* **67**, 637 (2010).

For a summary of recent QCD studies at the Tevatron, see

> A. Bhatti and D. Lincoln, *Ann. Rev. Nucl. Part. Sci.* **60**, 267 (2010).

An essay on comparisons between theory and experiment is presented in

> J. M. Campbell, J. W. Huston, and W. J. Stirling, *Rept. Prog. Phys.* **70**, 89 (2007).

**Modern computational methods.** QCD amplitudes are simpler than the individual diagrams might suggest. A simple form for the $n$-gluon amplitude was exhibited in

> S. J. Parke and T. R. Taylor, *Phys. Rev. Lett.* **56**, 2459 (1986),
> M. L. Mangano and S. J. Parke, *Phys. Rept.* **200**, 301 (1991).

The power of recursive methods (in the number of gluons) is on display in

> F. A. Berends and W. T. Giele, *Nucl. Phys.* **B306**, 759 (1988).

The striking simplifications can be related to deep connections between non-Abelian gauge theories and string theory, as discussed in

> Z. Bern and D. A. Kosower, *Nucl. Phys.* **B379**, 451 (1992),
> E. Witten, *Commun. Math. Phys.* **252**, 189 (2004) [hep-th/0312171].

The power of unitarity constraints is elaborated in

> Z. Bern, L. J. Dixon, D. C. Dunbar, and D. A. Kosower, *Nucl. Phys.* **B425**, 217 (1994).

Reviews of on-shell methods for evaluating multiparton amplitudes are given by

> Z. Bern, L. J. Dixon, and D. A. Kosower, *Annals Phys.* **322**, 1587 (2007);
> C. F. Berger and D. Forde, *Ann. Rev. Nucl. Part. Sci.* **60**, 181 (2010).

For a comprehensive guide to one-loop calculations, see

> R. K. Ellis, Z. Kunszt, K. Melnikov, and G. Zanderighi, *Phys. Rep.* **518**, 141 (2012).

For a general overview, see

> Z. Bern, L. Dixon, and D. Kosower, "Loops, Trees and the Search for New Physics," *Sci. Am.* **306** (5), 24 (May 2012).

Pedagogical introductions appear in

> L. J. Dixon, "Calculating Scattering Amplitudes Efficiently," in *QCD & Beyond, Proceedings of TASI '95*, ed. D. E. Soper, World Scientific, Singapore and River Edge, N.J., 1996, p. 539, arXiv:hep-ph/9601359,
> M. E. Peskin, "Simplifying Multi-Jet QCD Computation," arXiv:1101.2414.

**Hadronic component of the photon.** Many manifestations are treated in the comprehensive article by

T. H. Bauer, R. D. Spital, D. R. Yennie, and F. M. Pipkin, *Rev. Mod. Phys.* 50, 260 (1978) [Erratum: *Rev. Mod. Phys.* **51**, 407 (1979)].

The vector-meson-dominance philosophy is expounded in

J. J. Sakurai, *Currents and Mesons*, University of Chicago Press, Chicago, 1969.

**Two-photon processes.** Detailed accounts of theoretical prospects were given by

H. Terazawa, *Rev. Mod. Phys.* **45**, 615 (1973),
V. M. Budnev, I. F. Ginzburg, G. V. Meledin, V. G. Serbo, *Phys. Rept.* **15**, 181 (1975).

For a compact summary of experimental progress, see

M. Przybycien, *Nucl. Phys. Proc. Suppl.* **179**, 54 (2008).

**Photon structure function.** The proposal to measure the photon structure function is due to

S. J. Brodsky, T. Kinoshita, and H. Terazawa, *Phys. Rev. Lett.* **27**, 280 (1971),
T. F. Walsh, *Phys. Lett.* **36B**, 121 (1971).

The QCD prediction for the pointlike component was introduced by

E. Witten, *Nucl. Phys.* **B120**, 189 (1977).

Applications of the leading-order formalism to a variety of reactions are given in

C. H. Llewellyn Smith, *Phys. Lett. B* **79**, 83 (1978).

Higher-order corrections were first evaluated by

W. A. Bardeen and A. J. Buras, *Phys. Rev. D* **20**, 166 (1979).

The NNLO corrections were computed by

S. Moch, J.A.M. Vermaseren, and A. Vogt, *Nucl. Phys. B* **621**, 413 (2002).

The current state of the art is summarized in

A. J. Buras, *Acta Phys. Polon. B* **37**, 609 (2006).
A. Vogt, S. Moch, and J. Vermaseren, *Acta Phys. Polar. B*, **37**, 683 (2006).

The experimental situation is reviewed in

R. Nisius, *Phys. Rept.* **332**, 165 (2000);
M. Krawczyk, A. Zembrzuski, M. Staszel, *Phys. Rept.* **345**, 265 (2001).

**Confinement.** The increase of the QCD coupling constant, $\alpha_s(Q^2)$, at long distances or low energies points to the confinement of quarks and gluons into color-singlet hadrons, as explained in

Y. Nambu, "The Confinement of Quarks," *Sci. Am.* **235** (11), 48 (November, 1976).

A useful review, with many references to the early literature, is

M. Bander, *Phys. Rept.* **75**, 205 (1981).

**Lattice QCD.** The lattice formulation of QCD was created to gain insight into color confinement. How it began is the subject of
> K. G. Wilson, *Nucl. Phys. Proc. Suppl.* **140**, 3 (2005).

The essential ideas are explained in
> C. Rebbi, "The Lattice Theory of Quark Confinement," *Sci. Am.* **248** (2), 54 (February, 1983),
> D. H. Weingarten, "Quarks by Computer," *Sci. Am.* **274** (2), 116 (February, 1996),
> K. G. Wilson, *Phys. Rev. D* **10**, 2445 (1974).

Reviews of recent analytical and numerical work are
> R. Alkofer and J. Greensite, *J. Phys. G* **34**, S3 (2007),
> A. S. Kronfeld, "Twenty-first Century Lattice Gauge Theory: Results from the QCD Lagrangian," arXiv:1203.1204,
> Z. Fodor and C. Hoelbling, *Rev. Mod. Phys.* **84**, 449 (2012).

The latest textbooks include
> T. DeGrand and C. DeTar, *Lattice Methods for Quantum Chromodynamics,*World Scientific, Singapore, 2006,
> C. Gattringer and C. B. Lang, *Quantum Chromodynamics on the Lattice*, Springer, Berlin, 2010,
> H. J. Rothe, *Lattice Gauge Theories: An Introduction*, 4th ed., World Scientific, Singapore, 2012.

The path to an understanding of hadron masses is traced in
> A. S. Kronfeld, "Latice Gauge Theory and the Origin of Mass," arXiv:1209.3468.

**The dielectric analogy to confinement.** The picture was formulated by
> J. B. Kogut and L. Susskind, *Phys. Rev. D* **9**, 3501 (1974).

A variation appears in
> T. D. Lee, *Particle Physics and Introduction to Field Theory*, Harwood Academic, Chur, Switzerland, 1981, chap. 17.

For attempts to deduce an effective dia-electric theory from QCD, see
> S. L. Adler, *Phys. Rev. D* **23**, 2905 (1981); **24**, 1063E (1981),
> H. B. Nielsen, A. Patkós, *Nucl. Phys.* **B195**, 137 (1982).

**The MIT bag model.** In this useful caricature of confinement, massless quarks were confined within a finite radius by fiat, as explained in
> K. A. Johnson, "The Bag Model of Quark Confinement," *Sci. Am.* **241** (7), 112 (July, 1979),
> A. Chodos, R. L. Jaffe, K. Johnson, C. B. Thorn, and V. F. Weisskopf, *Phys. Rev. D* **9**, 3471 (1974),
> C. E. DeTar and J. F. Donoghue, *Ann. Rev. Nucl. Part. Sci.* **33**, 235 (1983).

**Confinement and deconfinement.** If QCD is indeed a confining theory, it may be of interest to ask whether spontaneous breaking of the color symmetry could lead to a liberation of color. This question was investigated in
> A. De Rújula, R. C. Giles, and R. L. Jaffe, *Phys. Rev. D* **17**, 285 (1978); *Phys. Rev. D* **22**, 227 (1980)
> H. Georgi, *Phys. Rev. D* **22**, 225 (1980),

L. B. Okun and M. A. Shifman, *Z. Phys.* **C8**, 17-20 (1981),

R. Slansky, J. T. Goldman, and G. L. Shaw, *Phys. Rev. Lett.* **47**, 887 (1981),

and recently revisited by

S. Nussinov and R. Shrock, *Phys. Rev. D* **82**, 034031 (2010).

**Current algebra and chiral symmetry.** The proposal that the charges associated with electroweak currents could be identified with SU(3)$_{flavor}$ symmetry operators is enunciated in

M. Gell-Mann, *Physics* **1**, 63 (1964).

Current algebra proved immensely fruitful for interactions involving pseudoscalar mesons. In QCD, an SU(3)$_{flavor}$ symmetry appears in the limit that the $u, d, s$ quark masses can be neglected. Excellent books from the golden age of the subject include

S. L. Adler and R. F. Dashen, *Current Algebras and Applications to Particle Physics*, W. A. Benjamin, New York, 1968,

S. B. Treiman, R. W. Jackiw, and D. Gross, *Lectures on Current Algebra and Its Applications*, Princeton University Press, Princeton, NJ, 1972,

B. W. Lee, *Chiral Dynamics*, Gordon and Breach, New York, 1972.

The spontaneous breaking is driven by the formation of a condensate of the light quarks, measured by the vacuum expectation value $\langle 0|\bar{q}q|0\rangle$. For an evaluation of the chiral condensate from $(2 + 1)$-flavor lattice QCD, see

H. Fukaya et al. [JLQCD Collaboration], *Phys. Rev. Lett.* **104**, 122002 (2010).

Other important aspects of chiral symmetry breaking in QCD are treated in

H. Pagels, *Phys. Rev. D* **19**, 3080 (1979),

G. 't Hooft, in *Recent Developments in Gauge Theories,* ed. G. 't Hooft et al, Plenum, New York, 1980, p. 135,

S. R. Coleman and E. Witten, *Phys. Rev. Lett.* **45**, 100 (1980).

An excellent place to start learning the modern perspective on chiral perturbation theory is

H. Leutwyler, *Ann. Phys.* **235**, 165 (1994).

**The U(1) problem.** Clear statements of the problem that the ninth pseudoscalar meson is too massive to be identified as a Goldstone boson in the massless-quark limit were formulated by

M. Gell-Mann, R. J. Oakes, and B. Renner, *Phys. Rev.* **175**, 2195 (1968),

S. Weinberg, *Phys. Rev. D* **11**, 3583 (1975).

The mixing with gluonic intermediate states is discussed at a phenomenological level in

A. De Rújula, H. Georgi, and S. L. Glashow, *Phys. Rev. D* **12**, 147 (1975),

N. Isgur, *Phys. Rev. D* **13**, 122 (1976).

Instantons yield a deeper understanding of the way in which the axial-U(1) symmetry of the massless-quark limit is broken by quantum corrections. A reexamination of the U(1) problem in the language of AdS/QCD is given by

E. Katz and M. D. Schwartz, *JHEP* **0708**, 077 (2007).

**Instantons.** The topological structure of the SU(3)$_c$ potentials and the possibility of tunneling between distinct vacua was identified by

A. M. Polyakov, *Phys. Lett.* **59B**, 82 (1075),

A. A. Belavin, A. M. Polyakov, A. S. Schwartz, and Y. S. Tyupkin, *Phys. Lett.* **59B**, 85 (1975).

A classic discussion can be found in

S. R. Coleman, "The Uses of Instantons," in *Aspects of Symmetry*, Cambridge University Press, Cambridge, 1985, p. 265.

How instantons solve the U(1) problem is developed by

G. 't Hooft, *Phys. Rept.* **142**, 357 (1986),
P. Di Vecchia and G. Veneziano, *Nucl. Phys.* **B171**, 253 (1980),
M. A. Shifman, *Phys. Rept.* **209**, 341 (1991).

**Emergent behavior.** Physics is rich in phenomena that arise as cooperative properties of large numbers of particles—perhaps in interaction with their environment—that are not readily anticipated from the behavior of a single particle in isolation. The rich structures evolved by cellular automata, many condensed-matter systems, and living organisms display emergence. A sketch of the role of emergence is given by

P. Coleman, *Nature* **446**, 379 (2007) [http://j.mp/j1Fv0v].

See also

E. Witten, "Emergent Behavior in Condensed-Matter and Particle Physics," Albanova/NORDITA Colloquium (2008), http://j.mp/kOdErc,
S. Weinberg, *Int. J. Mod. Phys.* **A23**, 1627 (2008).

For combative takes on emergence vis-à-vis reductionism, see

P. W. Anderson, *Science* **177**, 393 (1972);
D. Pines and R. B. Laughlin, *Proc. Natl. Acad. Sci. USA* **97**, 28 (2000); http://emergentuniverse.org.

For an excellent introductory chapter on the ubiquity of emergent phenomena, see

D. Pines, "Emergent Behavior in Quantum Matter," http://j.mp/mEHpdn, in the online course, *Physics for the 21st Century*, http://j.mp/9HuEDb.

Behavior that may emerge from cellular automata is the subject of

S. Wolfram, *A New Kind of Science*, Wolfram Media, Champaign, IL, 2002.

The possibility that the statistical character of quantum mechanics is emergent is explored by

S. L. Adler, *Quantum Theory as an Emergent Phenomenon*, Cambridge University Press, Cambridge, 2004.
G. 't Hooft, "Emergent Quantum Mechanics and Emergent Symmetries," *AIP Conf. Proc.* **957**, 154 (2007), arXiv:0707.4568.

Some emergent properties of nucleon structure are treated in

A. W. Thomas and W. Weise, *The Structure of the Nucleon*, Wiley-VCH, Berlin, 2001,
C. D. Roberts, *Few Body Syst.* **52**, 345 (2012), arXiv:1109.6325.

How the nuclear force and nuclear chiral dynamics emerge from low-energy QCD is the subject of

W. Weise, *Prog. Theor. Phys. Suppl.* **170**, 161 (2007).

Progress toward deriving the nuclear force from lattice QCD is reported in

T. Inoue et al. [HAL QCD Collaboration], *Prog. Theor. Phys.* **124**, 591 (2010).

**Electroweak symmetry breaking through strong dynamics.** For a compact survey, see

S. Chivukula, M. Narain, and J. Womersley, "Dynamical Electroweak Symmetry Breaking," in Ref. [24], p. 1340.

For reviews of the "technicolor" approach to electroweak symmetry breaking, see

K. Lane, "Two lectures on technicolor," arXiv:hep-ph/0202255,

C. T. Hill and E. H. Simmons, *Phys. Rept.* **381**, 235 (2003) [Erratum: *Phys. Rept.* **390**, 553 (2004), arXiv:hep-ph/0203079],

F. Sannino, *Acta Phys. Polon.* **B40**, 3533 (2009),

J. R. Andersen et al., *Eur. Phys. J. Plus* **126**, 81 (2011).

*Gedanken* worlds in which QCD is the source of electroweak symmetry breaking were investigated in

C. Quigg and R. Shrock, *Phys. Rev. D* **79**, 096002 (2009).

**The $1/N$ Expansion.** The use of the $N \to \infty$ limit in SU$(N)_c$ was pioneered by

G. 't Hooft, *Nucl. Phys.* **B72**, 461 (1974); *Nucl. Phys.* **B75**, 461.

Clear introductions to the method, with allusions to different physical situations, are given by

E. Witten, *Nucl. Phys.* **B160**, 57 (1979); "Quarks, atoms, and the $1/N$ expansion," *Phys. Today* **33** (7), 38 (July 1980),

S. R. Coleman, "$1/N$," in *Aspects of Symmetry*, Cambridge University Press, Cambridge, 1985, p. 351,

S. R. Das, *Rev. Mod. Phys.* **59**, 235 (1987).

Applications to the baryon spectrum are further developed in

T. D. Cohen, *Rev. Mod. Phys.* **68**, 599 (1996),

E. Jenkins, *Ann. Rev. Nucl. Part. Sci.* **48**, 81 (1998).

For additional applications in atomic physics, see

L. D. Mlodinow and N. Papanicolaou, *Annals Phys.* **128**, 314 (1980).

The equivalence between the large-$N$ limits of quantum theories and classical limits is developed in

L. G. Yaffe, *Rev. Mod. Phys.* **54**, 407 (1982).

**AdS/CFT connection.** The evolution of the link among large-$N$ field theories, string theory, and gravity can be traced through

N. Seiberg and E. Witten, *Nucl. Phys.* **B426**, 19 (1994),

E. Witten, *Adv. Theor. Math. Phys.* **2**, 505 (1998),

O. Aharony et al., *Phys. Rept.* **323**, 183 (2000),

I. R. Klebanov, "TASI lectures: Introduction to the AdS/CFT correspondence," in *Strings, Branes and Gravity: TASI 99*, ed. J. Harvey, S. Kachru, and E. Silverstein, p. 615, World Scientific, Singapore, 2001, arXiv:hep-th/0009139.

E. D'Hoker and D. Z. Freedman, "Supersymmetric gauge theories and the AdS / CFT correspondence," arXiv:hep-th/0201253.

J. M. Maldacena, "TASI 2003 lectures on AdS/CFT," arXiv:hep-th/0309246.

For applications to heavy-ion collisions, see

J. Casalderrey-Solana et al., "Gauge/String Duality, Hot QCD and Heavy Ion Collisions," arXiv:1101.0618.

**The strong CP problem.** The threat to CP-invariance posed by the $\theta$-vacuum was developed in

G. 't Hooft, *Phys. Rev. Lett.* **37**, 8 (1976); *Phys. Rev. D* **14**, 3432 (1976),

R. Jackiw and C. Rebbi, *Phys. Rev. Lett.* **37**, 172 (1976),

C. Callan, R. Dashen, and D. Gross, *Phys. Lett.* **63B**, 334 (1976).

The possibility of resolving the problem by imposing a new quasisymmetry was raised in Ref. [122] and

R. D. Peccei and H. R. Quinn, *Phys. Rev. D* **16**, 1791 (1977).

The Peccei-Quinn solution requires a new particle, the axion, with several implications for particle physics and cosmology, as noted by

S. Weinberg, *Phys. Rev. Lett.* **40**, 223 (1978),

F. Wilczek, *Phys. Rev. Lett.* **40**, 279 (1978).

For reviews of the strong CP problem and possible solutions, see

H. R. Quinn, "CP Symmetry Breaking, or the Lack of It, in the Strong Interactions," SLAC-PUB-10698 (2004), http://j.mp/hKIVLe.

M. Dine, "The Strong CP Problem," in *Flavor Physics for the Millennium*, ed. J. L. Rosner, World Scientific, Singapore, 2000, hep-ph/0011376.

A pedagogical description of the axion is given by

P. Sikivie, *Phys. Today* **49N12**, 22 (December 1996), hep-ph/9506229.

The status of axion searches is reviewed in

S. J. Asztalos, L. J. Rosenberg, K. van Bibber, P. Sikivie, and K. Zioutas, *Ann. Rev. Nucl. Part. Sci.* **56**, 293 (2006).

M. Kuster, G. Raffelt, and B. Beltrán, *Axions: Theory, Cosmology, and Experimental Searches*, Lecture notes in physics, **741**, Springer-Verlag, Berlin and Heidelberg, 2008.

J. E. Kim and G. Carosi, "Axions and the Strong CP Problem," *Rev. Mod. Phys.* **82**, 557 (2010).

C. Hagmann et al., "Axions and Other Similar Particles", in Ref. [24], p. 496.

**Heavy-quark effective theory.** The simpler dynamics of heavy-quark systems lend themselves to effective field theories. An excellent general reference on the strategy of effective field theories is

H. Georgi, *Ann. Rev. Nucl. Part. Sci.* **43**, 209 (1993).

For heavy-light hadrons (those with one heavy quark), this insight led to the development of the heavy-quark effective theory (HQET) in

E. Eichten, *Nucl. Phys. Proc. Suppl.* **4**, 170 (1988),

E. Eichten and B. Hill, *Phys. Lett.* **B234**, 511 (1990); *Phys. Rev.* **B243**, 427 (1990),

H. Georgi, *Phys. Lett.* **B240**, 447 (1990),

B. Grinstein, *Nucl. Phys.* **B339**, 253 (1990).

Extensive development was launched by the work of Ref. [110]. Some pedagogical reviews are

B. Grinstein, *Ann. Rev. Nucl. Part. Sci.* **42**, 101 (1992),

M. Neubert, *Phys. Rept.* **245**, 259 (1994),

I. Bigi, M. Shifman, and N. Uraltsev, *Ann. Rev. Nucl. Part. Sci.* **47**, 591 (1997),

and a wide-ranging textbook is

A. V. Manohar and M. B. Wise, *Heavy Quark Physics*, Cambridge University Press, Cambridge and New York, 2000.

IIIIIIIIIIIIIIIIIII REFERENCES IIIIIIIIIIIIIIIIIIIIIIIII

1. C. N. Yang and R. L. Mills, *Phys. Rev.* **96**, 191 (1954).
2. J. J. Sakurai, *Ann. Phys. (NY)* **11**, 1 (1960).
3. Y. Ne'eman, *Nucl. Phys.* **26**, 222 (1961).
4. F. Englert and R. Brout, *Phys. Rev. Lett.* **13**, 321 (1964).
5. G. 't Hooft, *Nucl. Phys.* **B35**, 167 (1971).
6. G. Altarelli and G. Parisi, *Nucl. Phys. B* **126**, 298 (1977). See also V. N. Gribov and L. N. Lipatov, *Sov. J. Nucl. Phys.* **15**, 438 (1972); Yu. L. Dokshitzer, *Sov. Phys. JETP* **46**, 641 (1977).
7. Y. Nambu, in *Preludes in Theoretical Physics in Honor of V. F. Weisskopf,* ed. A. DeShalit, H. Feshbach, and L. Van Hove, North-Holland, Amsterdam, 1966, pp. 133.
8. O. W. Greenberg and D. Zwanziger, *Phys. Rev.* **150**, 1177 (1966); H. J. Lipkin, *Phys. Lett.* **45B**, 267 (1973); H. Fritzsch, M. Gell-Mann, and H. Leutwyler, *Phys. Lett.* **47B**, 365 (1973).
9. M. Gell-Mann, "The Eightfold Way," Caltech Synchrotron Report No. CTSL-20 (1961), http://j.mp/xC6Gqw.
10. Useful references are R. L. Jaffe, *Phys. Rev. D* **15**, 281 (1977), in which the normalizations differ from those adopted here, and J. L. Rosner, in *Techniques and Concepts of High Energy Physics,* St. Croix, 1980, ed. T. Ferbel, Plenum, New York, 1981, p. 1.
11. Simulations in the lattice formulation of QCD show that, in the absence of light quarks, the potential energy between static sources of color rises linearly at large separations, in agreement with the QCD string picture of hadrons [cf. Y. Nambu, *Phys. Rev. D* **10**, 4262 (1974)]. In the presence of light quarks, the string breaks at separations around 1 fm into two color singlets. See G. S. Bali, H. Neff, T. Duessel, T. Lippert, and K. Schilling [SESAM Collaboration], *Phys. Rev. D* **71**, 114513 (2005).
12. Demonstrating confinement through the mathematical existence of non-Abelian gauge theory and a mass gap has been put forward as a Millennium Prize Problem by the Clay Institute of Mathematics. For elaboration, see Arthur Jaffe and Edward Witten, "Quantum Yang–Mills Theory," available at http://j.mp/egYI8w.
13. For reviews of free-quark searches, see L. Lyons, *Phys. Rep.* **129**, 225 (1985); P. F. Smith, *Ann. Rev. Nucl. Part. Sci.* **39**, 73 (1989); M. L. Perl, E. R. Lee, and D. Loomba, *Ann. Rev. Nucl. Part. Sci.* **59**, 47 (2009).
14. Y. Nambu, *Phys. Rev. D* **10**, 4262 (1974).
15. The diquark literature is extensive. A recent example is A. Selem and F. Wilczek, "Hadron systematics and emergent diquarks," in *Ringberg 2005, New Trends in HERA Physics,* ed. G. Grindhammer, et al., World Scientific, Singapore, 2006, p. 337, arXiv:hep-ph/0602128. For a review of earlier work, see M. Anselmino et al., *Rev. Mod. Phys.* **65**, 1199 (1993).
16. W. Pauli and F. Villars, *Rev. Mod. Phys.* **21**, 434 (1949).
17. E. A. Uehling, *Phys. Rev.* **48**, 55 (1935).
18. S. Mele, "Measurement of the running of the electromagnetic coupling at LEP," *PoS* **HEP2005**, 286 (2006), arXiv:hep-ex/0601045.
19. H. Burkhardt and B. Pietrzyk, *Phys. Lett. B* **513**, 46 (2001).
20. G. 't Hooft and M. Veltman, *Nucl. Phys.* **B44**, 189 (1972).
21. The factors of $i$ by which these rules appear to differ from those given in figure 6.14 for the electroweak theory are due to the use of a Cartesian basis for the gluon color indices and a spherical basis for the intermediate-boson charges.
22. That the result is indeed gauge invariant may be verified from the expressions given previously.

23. D. J. Gross and F. Wilczek, *Phys. Rev. Lett.* **30**, 1343 (1973); H. D. Politzer, *Phys. Rev. Lett.* **30**, p. 1346. See also the work of G. 't Hooft, *Phys. Lett.* **61B**, 455 (1973), **62B**, 444 (1973) and the interesting calculation by I. B. Khriplovich, *Yad. Fiz.* **10**, 409 (1969) [English translation: *Sov. J. Nucl. Phys.* **10**, 235 (1970)], in which the possibility of an antiscreening effect in non-Abelian gauge theories was noted.

24. K. Nakamura et al. (Particle Data Group), *J. Phys. G* **37**, 075021 (2010). Consult J. Beringer et al. [Particle Data Group], *Phys. Rev. D* **86**, 010001 (2012) for updates.

25. The progress is reviewed and critically evaluated in S. Bethke, *Eur. Phys. J. C* **64**, 689 (2009).

26. T. van Ritbergen, J.A.M. Vermaseren, and S. A. Larin, *Phys. Lett. B* **400**, 379 (1997).

27. A. S. Kronfeld and C. Quigg, *Am. J. Phys.* **78**, 1081 (2010).

28. W. A. Bardeen, A. J. Buras, D. W. Duke, and T. Muta, *Phys. Rev. D* **18**, 3998 (1978).

29. K. G. Chetyrkin, J. H. Kühn, and A. Kwiatkowski, *Phys. Rept.* **277**, 189 (1996).

30. See §3.1 of R. K. Ellis, W. J. Stirling, and B. R. Webber, *QCD and Collider Physics*, Cambridge University Press, Cambridge, 1996, for a thorough discussion of how the infrared divergences are controlled. The $O(\alpha_s)$ computation is carried out explicitly in §5.1 (and again in §6.1) of T. Muta, *Foundations of Quantum Chromodynamics*, 2nd ed., World Scientific, Singapore, 1998.

31. R. Jost and J. M. Luttinger, *Helv. Phys. Acta* **23**, 201 (1950).

32. T. Appelquist and H. Georgi, *Phys. Rev. D* **8**, 4000 (1973); A. Zee, *Phys. Rev. D* **8**, 4038. Continuation of the running coupling into the timelike regime is discussed in R. P. Feynman, *Phys. Rev.* **76**, 769 (1949).

33. At $O(\alpha_s^2)$: K. G. Chetyrkin, A. L. Kataev, and F. V. Tkachov, *Phys. Lett. B* **85**, 277 (1979); M. Dine and J. R. Sapirstein, *Phys. Rev. Lett.* **43**, 668 (1979); W. Celmaster and R. J. Gonsalves, *Phys. Rev. Lett.* **44**, 560 (1980). At $O(\alpha_s^3)$: S. G. Gorishny, A. L. Kataev, and S. A. Larin, *Phys. Lett.* **259**, 144 (1991); L. R. Surguladze and M. A. Samuel, *Phys. Rev. Lett.* **66**, 560, 2416(E) (1991).

34. T. Kinoshita, *J. Math. Phys.* **3**, 650 (1962); T. D. Lee and M. Nauenberg, *Phys. Rev.* **133**, B1549 (1964).

35. J. R. Ellis, M. K. Gaillard and G. G. Ross, *Nucl. Phys. B* **111**, 253 (1976) [Erratum: *Nucl. Phys. B* **130**, 516 (1977)]. See also T. A. DeGrand, Y. J. Ng, and S.H.H. Tye, *Phys. Rev. D* **16**, 3251 (1977).

36. R. Brandelik et al. [TASSO Collaboration], *Phys. Lett.* **B86**, 243 (1979); C. Berger et al. [PLUTO Collaboration], *Phys. Lett.* **B86**, 418; W. Bartel et al. [JADE Collaboration], *Phys. Lett.* **B91**, 142 (1980); D. P. Barber et al. [Mark J Collaboration], *Phys. Rev. Lett.* **43**, 830 (1979). An excellent overview is given in §7.3 of K. H. Mess and B. H. Wiik, "Recent Results in Electron-Positron and Lepton-Hadron Interactions," in *Théories de jauge en physique des hautes énergies / Gauge Theories in High Energy Physics*, ed. M. K. Gaillard and R. Stora, North-Holland, Amsterdam, 1983, p. 865 [DESY 82-011 (March 1982) http://j.mp/i80Y1D].

37. G. F. Sterman and S. Weinberg, *Phys. Rev. Lett.* **39**, 1436 (1977). For an explicit computation, see §5.4 of Muta, Ref. [30].

38. G. P. Salam, *Eur. Phys. J. C* **67**, 637 (2010).

39. B. Rossi, *High-Energy Particles*, Prentice Hall, Englewood Cliffs, NJ, 1952, chap. 5

40. M. Acciarri et al. [L3 Collaboration], *Phys. Lett.* **B438**, 363 (1998); G. Abbiendi, et al. [OPAL Collaboration], *Eur. Phys. J.* **C11**, 409 (1999); P. Abreu et al. [DELPHI Collaboration], *Eur. Phys. J.* **C19**, 15 (2001).

41. The spirit of this method may be traced to E. Fermi, *Z. Phys.* **29**, 315 (1924), who employed it to calculate the ionization of atoms by $\alpha$-particles. The development for radiation theory is due to C. F. von Weizsäcker, *Z. Phys.* **88**, 612 (1934) and E. J. Williams, *Phys. Rev.* **45**, 729 (1934).

42. See, for example, E. Reya, *Phys. Rep.* **69**, 195 (1981).

43. For a general reference on kinematics, consult §39 of the 2010 *Review of Particle Physics*, Ref. [24].

44. An elegant method to reconstruct structure functions from their moments is presented in F. J. Ynduráin, *Phys. Lett. B* **74**, 68 (1978).

45. H. Abramowicz et al. (CDHS Collaboration), *Z. Phys. C* **13**, 199 (1982).

46. Target-mass corrections are made by using the definition of moments due to O. Nachtmann, *Nucl. Phys. B* **117**, 50 (1976).

47. F. D. Aaron et al. (H1 Collaboration and ZEUS Collaboration), *JHEP* **1001**, 109 (2010).

48. H. L. Lai et al., *Phys. Rev. D* **82**, 074024 (2010).

49. M. Tzanov et al. [NuTeV Collaboration], *Phys. Rev. D* **74**, 012008 (2006).

50. E. D. Bloom et al., *Phys. Rev. Lett.* **23**, 930 (1969); M. Breidenbach et al., *Phys. Rev. Lett.* **23**, 935.

51. A. D. Martin, W. J. Stirling, R. S. Thorne, and G. Watt, *Eur. Phys. J. C* **63**, 189 (2009), http://j.mp/iWjDI9.

52. P. L. McGaughey, J. M. Moss, and J. C. Peng, *Ann. Rev. Nucl. Part. Sci.* **49**, 217 (1999); S. Kumano, *Phys. Rept.* 303, 183 (1998).

53. W. Vogelsang, *J. Phys. G* **34**, S149 (2007); S. D. Bass, *The Spin Structure of the Proton*, World Scientific, Singapore, 2007; S. E. Kuhn, J. P. Chen, and E. Leader, *Prog. Part. Nucl. Phys.* **63**, 1 (2009).

54. C. G. Callan and D. J. Gross, *Phys. Rev. Lett.* **22**, 156 (1969).

55. G. Altarelli and G. Martinelli, *Phys. Lett.* **76B**, 89 (1978); see also A. Zee, F. Wilczek, and S. B. Treiman, *Phys. Rev. D* **10**, 2881 (1974). G. Curci, W. Furmanski, R. Petronzio, *Nucl. Phys.* **B175**, 27 (1980), treated the nonsinglet case at next-to-leading order. The calculation has been extended to third order by S. Moch, J.A.M. Vermaseren, and A. Vogt, *Phys. Lett.* **B606**, 123 (2005).

56. F. D. Aaron et al. [H1 Collaboration], "Measurement of the Inclusive $e^{\pm}p$ Scattering Cross Section at High Inelasticity $y$ and of the Structure Function $F_L$," arXiv:1012.4355; S. Chekanov et al. [ZEUS Collaboration], *Phys. Lett.* **B682**, 8 (2009).

57. P. M. Nadolsky et al., *Phys. Rev. D* **78**, 013004 (2008).

58. M. Diehl, *Phys. Rept.* **388**, 41 (2003); X. Ji, *Ann. Rev. Nucl. Part. Sci.* **54**, 413 (2004); A. V. Belitsky and A. V. Radyushkin, *Phys. Rept.* **418**, 1 (2005).

59. L. V. Gribov, E. M. Levin, and M. G. Ryskin, *Phys. Rept.* **100**, 1 (1983); A. H. Mueller and J.-W. Qiu, *Nucl. Phys.* **B268**, 427 (1986).

60. B. Badełek, M. Krawczyk, K. Charchula, and J. Kwiecinski, *Rev. Mod. Phys.* **64**, 927 (1992); B. Badełek and J. Kwiecinski, *Rev. Mod. Phys.* **68**, 445 (1996); L. N. Lipatov, *Phys. Rept.* **286**, 131 (1997); E. Levin and K. Tuchin, *Nucl. Phys.* **B573**, 833 (2000). Implications of the HERA observations for future experiments are explored in L. Frankfurt, M. Strikman, and C. Weiss, *Ann. Rev. Nucl. Part. Sci.* **55**, 403 (2005).

61. J. C. Collins, D. E. Soper, and G. F. Sterman, "Factorization of Hard Processes in QCD," in *Perturbative QCD*, ed. A. H. Mueller, World Scientific, Singapore, 1989, p. 1.

62. V. M. Abazov et al. [D0 Collaboration], *Phys. Rev. Lett.* **101**, 062001 (2008). See also A. Abulencia et al. [CDF Collaboration], *Phys. Rev. D75*, 092006 (2007).

63. For some recent examples see W. T. Giele and G. Zanderighi, *JHEP* **0806**, 038 (2008); C. F. Berger et al., *Phys. Rev. Lett.* **106**, 092001 (2011).

64. F. E. Low, *Phys. Rev.* **120**, 582 (1960); a misprint occurs in the expression for $f(x)$.

65. L. D. Landau and E. M. Lifshitz, *Phys. Z. Sowjetunion* **6**, 244 (1934); E. J. Williams, *Det. Kgl. Danske Videnskab. Selskab Mat-Fyz.-Med.* **XIII**, No. 4 (1935).

66. For observations of the exclusive reactions $\bar{p}p \rightarrow \bar{p}p\ell^+\ell^-$ at $\sqrt{s} = 1.96$ TeV, see A. Abulencia *et al.* [CDF Collaboration], *Phys. Rev. Lett.* **98**, 112001 (2007);

T. Aaltonen *et al.* [CDF Collaboration], *Phys. Rev. Lett.* **102**, 242001 (2009). For a perspective, including strong-interaction analogues of the two-photon process, see M. G. Albrow, "Central Exclusive Production at the Tevatron," *AIP Conf. Proc.* **1105**, 3 (2009).

67. Illustrative tables for resonance production, along with useful discussion, can be found in F. J. Gilman, "Production of Resonances in Photon – Photon Collisions," SLAC-PUB-2461 (1980).

68. The most recent measurements are reported by K. M. Ecklund et al. [CLEO Collaboration], *Phys. Rev. D* **78**, 091501 (2008); K. Abe et al. [Belle Collaboration], *Phys. Lett. B* **540**, 33 (2002).

69. S. Uehara et al. [Belle Collaboration], *Phys. Rev. Lett.* **96**, 082003 (2006); B. Aubert et al. [BABAR Collaboration], *Phys. Rev. D* **81**, 092003 (2010).

70. See, for example, W. Bartel et al. [JADE Collaboration], *Phys. Lett.* **107B**, 163 (1981); R. Bradelik et al. [TASSO Collaboration], *Phys. Lett.* **107B**, 290. For a digest of LEP results, see M. Przybycien, *Nucl. Phys. Proc. Suppl.* **179**, 54 (2008).

71. E. Witten, *Nucl. Phys.* **B120**, 189 (1977).

72. R. L. Kingsley, *Nucl. Phys.* **B60**, 45 (1973); T. F. Walsh, and P. M. Zerwas, *Phys. Lett. B* **44**, 195 (1973).

73. R. J. DeWitt, L. M. Jones, J. D. Sullivan, D. E. Willen, and H. W. Wyld, *Phys. Rev. D* **19**, 2046 (1979) [Erratum: *Phys. Rev. D* **20**, 1751 (1979)]; C. Peterson, T. F. Walsh, and P. M. Zerwas, *Nucl. Phys. B* **174**, 424 (1980); A. Nicolaidis, *Nucl. Phys. B* **163**, 156 (1980).

74. W. A. Bardeen, "Two Photon Physics," in *Proceedings of The 1981 International Symposium on Lepton and Photon Interactions at High Energies*, ed. W. Pfeil p. 432, http://j.mp/yt7eOH.

75. F. Cornet, P. Jankowski, and M. Krawczyk, *Phys. Rev. D* **70**, 093004 (2004).

76. G. Abbiendi et al. [OPAL Collaboration], *Phys. Lett.* **B533**, 207 (2002).

77. See figure 16.14 of the 2010 *Review of Particle Physics*, Ref. [24]. A comprehensive compilation and critical review appears in M. Krawczyk, A. Zembrzuski, M. Staszel, *Phys. Rept.* **345**, 265 (2001). Consult J. Beringer et al. [Particle Data Group], *Phys. Rev. D* **86**, 010001 (2012) for updates.

78. For an elementary argument, see L. D. Landau and E. M. Lifshitz, *Electrodynamics of Continuous Media*, 2nd ed., Elsevier Butterworth-Heinemann, Burlington, Mass., 1984.

79. E. Eichten, K. Gottfried, T. Kinoshita, K. D. Lane, and T. M. Yan, *Phys. Rev. D* **17**, 3090 (1978) [Erratum *Phys. Rev. D* **21**, 313 (1980)]; E. Eichten, K. Gottfried, T. Kinoshita, K. D. Lane and T. M. Yan, *Phys. Rev. D* **21**, 203 (1980).

80. G. S. Bali, *Phys. Rept.* **343**, 1 (2001).

81. S. Dürr et al. [BMW Collaboration], *Science* **322**, 1224 (2008). See also S. Aoki et al. [PACS-CS Collaboration], *Phys. Rev. D* **79**, 034503 (2009); A. Bazavov et al., *Rev. Mod. Phys.* **82**, 1349 (2010).

82. T. Ishikawa et al. [CP-PACS and JLQCD Collaborations], *Phys. Rev. D* **78**, 011502 (2008).

83. For elaboration of this point, see F. Wilczek, *Phys. Today* **52**, 11 (November 1999); *Phys. Today* **53**, 13 (January 2000); *Mod. Phys. Lett.* **A21**, 701 (2006); C. Quigg, *Rept. Prog. Phys.* **70**, 1019 (2007).

84. M. Cheng et al., *Phys. Rev. D* **77**, 014511 (2008). A. Bazavov et al., *Phys. Rev. D* **80**, 014504 (2009).

85. M. S. Bhagwat, I. C. Cloet, and C. D. Roberts, "Covariance, Dynamics and Symmetries, and Hadron Form Factors," arXiv:0710.2059. See also L. Chang and C. D. Roberts, "Hadron Physics: The Essence of Matter," arXiv:1003.5006.

86. V. L. Ginzburg and L. D. Landau, *Zh. Eksp. Teor. Fiz.* **20**, 1064 (1950) (English translation: *Men of Physics: L. D. Landau,* ed. D. ter Haar, Pergamon, New York, 1965, vol. I, p. 138).

87. J. Bardeen, L. N. Cooper, and J. R. Schrieffer, *Phys. Rev.* **108**, 1175 (1957).

88. R. F. Dashen, *Phys. Rev. D* **3**, 1879 (1971).

89. Y. Nambu, *Phys. Rev. Lett.* **4**, 380 (1960).

90. J. Gasser and H. Leutwyler, *Nucl. Phys. B* **250**, 465 (1985).

91. M. Weinstein, *Phys. Rev. D* **8**, 2511 (1973).

92. W. J. Marciano, *Phys. Rev. D* **21**, 2425 (1980).

93. S. Weinberg, *Phys. Rev. D* **13**, 974 (1976); *Phys. Rev. D* **19**, 1277 (1979); L. Susskind, *Phys. Rev. D* **20**, 2619 (1979).

94. S. Dimopoulos and L. Susskind, *Nucl. Phys. B* **155**, 237 (1979); E. Eichten and K. D. Lane, *Phys. Lett. B* **90**, 125 (1980).

95. Y. Nambu, "Model Building Based on Bootstrap Symmetry Breaking," in *New Trends in Strong Coupling Gauge Theories,* ed. M. Bando, T. Muta, and K. Yamawaki, World Scientific, Singapore, 1989, p. 1; V. A. Miransky, M. Tanabashi, and K. Yamawaki, *Mod. Phys. Lett.* **A4**, 1043 (1989); W. A. Bardeen, C. T. Hill, and M. Lindner, *Phys. Rev. D* **41**, 1647 (1990).

96. G. Burdman, L. Da Rold, O. Eboli, and R. D'E. Matheus, *Phys. Rev. D* **79**, 075026 (2009).

97. G. 't Hooft, *Nucl. Phys. B* **72**, 461 (1974). Note the diagrammatic representation of the Fierz identity for gluon exchange following (A.5.14).

98. G. 't Hooft, *Nucl. Phys. B* **75**, 461 (1974).

99. S. Coleman, "1/N," in *Aspects of Symmetry,* Cambridge University Press, Cambridge, 1985, chap. 8, SLAC-PUB-2484.

100. E. Witten, *Nucl. Phys. B* **160**, 57 (1979).

101. A. J. Buras, *Nucl. Phys. Proc. Suppl.* **10A**, 199 (1989).

102. S. Okubo, *Phys. Lett.* **5**, 165 (1963).

103. G. Zweig, "An SU(3) Model for Strong Interaction Symmetry and Its Breaking," CERN-TH-401 (1964), http://j.mp/AFfsaG; "An SU(3) Model for Strong Interaction Symmetry and Its Breaking 2," CERN-TH-412 (1964), http://j.mp/AvBGHM.

104. J. Iizuka, *Suppl. Prog. Theor. Phys.* **37–38**, 21 (1966).

105. S. Aoki et al. [CP-PACS Collaboration], *Phys. Rev. D* **67**, 034503 (2003).

106. D. Weingarten, "Evidence for the Observation of a Glueball," in *Continuous Advances in QCD 1996,* ed. M. I. Polikarpov, World Scientific, Singapore, 1996, p. 227, hep-ph/9607212.

107. I. R. Klebanov, "Introduction to the AdS/CFT Correspondence," in *Strings, Branes, and Gravity: TASI99,* ed. J. Harvey, S. Kachru, and E. Silverstein, World Scientific, Singapore, 2001, p. 615, hep-th/0009139.

108. Y. Nambu, *Phys. Rev. D* **10**, 4262 (1974).

109. E. V. Shuryak, *Nucl. Phys.* **B198**, 83 (1982); M. A. Shifman and M. B. Voloshin, *Sov. J. Nucl. Phys.* **45**, 292 (1987).

110. N. Isgur and M. B. Wise, *Phys. Lett.* **B232**, 113 (1989); *Phys. Lett.* **B237**, 527 (1990).

111. W. E. Caswell and G. P. Lepage, *Phys. Lett.* **B167**, 437 (1986).

112. For a provocative essay, see S. Weinberg, "Symmetry: A 'Key to Nature's Secrets,'" *New York Review of Books* **58** (16), 69 (October 27, 2011).

113. Y. Nambu and G. Jona-Lasinio, *Phys. Rev.* **122**, 345 (1961); **124**, 246 (1961). See also Y. Nambu, *Phys. Rev. Lett.* **4**, 380 (1960).

114. H. Georgi and A. Manohar, *Nucl. Phys.* **B234**, 189 (1984).

115. S. Weinberg, *Phys. Rev. D* **11**, 3583 (1975).

116. G. 't Hooft, *Phys. Rev. Lett.* **37**, 8 (1976).

117. The new phases of nuclear matter that may be created in heavy-ion collisions may exhibit violations of parity and time-reversal invariance. See B. Müller, *Physics* **2**, 104 (2009), http://j.mp/wKQ6XH, for a brief introduction.

118. S. Weinberg, *Phys. Rev. D* **8**, 605, 4482 (1973).

119. V. N. Gribov, "Instability of Nonabelian Gauge Theories and Impossibility of Choice of Coulomb Gauge," in *The Gribov Theory of Quark Confinement,* ed. J. Nyiri, World Scientific, Singapore, 2001, p. 24, SLAC-Trans-176 (1977), http://j.mp/gK1l3U.

120. C. A. Baker et al., *Phys. Rev. Lett.* **97**, 131801 (2006).

121. See C. Hagmann et al., "Axions and Other Similar Particles," in Ref. [24], p. 496.

122. R. D. Peccei and H. R. Quinn, *Phys. Rev. Lett.* **38**, 1440 (1977).

123. F. Wilczek, *Nucl. Phys. A* **663**, 3 (2000).

124. For an early example, see C. W. Bauer et al., *Phys. Rev. D* **66**, 014017 (2002).

125. M. Antonelli et al., *Phys. Rept.* **494**, 197 (2010).

126. P. Hagler, *Phys. Rept.* **490**, 49 (2010); D. B. Renner, *PoS* **LAT2009**, 018 (2009).

127. J. M. Maldacena, *Adv. Theor. Math. Phys.* **2**, 231 (1998) [*Int. J. Theor. Phys.* **38**, 1113 (1999)], arXiv:hep-th/9711200.

128. T. Sakai and S. Sugimoto, *Prog. Theor. Phys.* **113**, 843 (2005), arXiv:hep-th/0412141.

129. J. D. Bjorken, *Phys. Rev. D* **27** (1983) 140.

130. G. Policastro, D. T. Son and A. O. Starinets, *Phys. Rev. Lett.* **87**, 081601 (2001). For a review, see D. T. Son and A. O. Starinets, *Ann. Rev. Nucl. Part. Sci.* **57**, 95 (2007).

131. J. Patera and D. Sankoff, *Tables of Branching Rules for Representations of Simple Lie Algebras,* Presses de l'Université de Montréal, Montreal, 1973. For a $d$-dimensional representation, the quantity $\ell$ listed in their Table II is related to the SU($N$) Casimir operators required in Problem 8.5 by $G^{(N)2} = (N^2 - 1)\ell/2d$.

132. For a two-body reaction with incoming and outgoing four-momenta labeled by $p_1$, $p_2$ and $q_1$, $q_2$, so that momentum conservation is expressed by $p_1 + p_2 = q_1 + q_2$, the Mandelstam invariants are $s = (p_1 + p_2)^2$, $t = (p_1 - q_1)^2$, $u = (p_1 - q_2)^2$.

# Nine ||||||||||||||||||||||||||||||||||||||||||||||||||||||||||||||||||||||||||||||||||||||||||||

## Unified Theories

In the early chapters of this book we stressed the economy and elegance of the gauge principle as a guide to constructing theories of the fundamental interactions among the elementary constituents of matter. Subsequently, we put those ideas into practice, and a satisfying picture emerged of the weak and electromagnetic interactions as well as the strong interactions among quarks and gluons. Together, the electroweak theory and quantum chromodynamics account for all the prominent experimental observations in subnuclear physics and provide a large measure of understanding of the relationships among different phenomena. Three decades of lively interplay between theory and experiments have validated both QCD and the electroweak theory as laws of nature, with the usual proviso that future observations or insights may require revisions or inspire extensions. Although both of these gauge theories face many more tests, the $SU(3)_c \otimes SU(2)_L \otimes U(1)_Y$ structure serves as a theoretical paradigm and a framework in which we consider experimental prospects and analyze experimental results. Indeed, the unfinished business of the standard model largely sets the agenda for contemporary experimentation and theoretical invention.

In preceding discussions we have emphasized the logic, consequences, and successes of these theories, while remarking from time to time on the incompleteness or arbitrariness of the descriptions they provide. An important lesson to be drawn from these considerations—both theoretical and phenomenological—is that no obstacles have yet arisen to the general program of constructing interactions from local gauge symmetries and that nothing seems to prevent extending the program. In this chapter, we shall not continue to celebrate the successes of the $SU(3)_c \otimes SU(2)_L \otimes U(1)_Y$ picture but shall instead begin by exploring some of its shortcomings. Some of these will be seen to invite a further unification of the strong, weak, and electromagnetic interactions. Accordingly, most of this chapter will be devoted to a brief exposition of the simplest theory that unifies these three forces: the theory of Georgi and Glashow [1] based on the gauge group SU(5) [2].

The standard model based on $SU(3)_c \otimes SU(2)_L \otimes U(1)_Y$ gauge symmetry encapsulates much of what we know and describes many observations, but it leaves other significant facts unexplained. Both the success and the incompleteness of the standard model encourage us to look beyond it to a more comprehensive understanding. One attractive way to proceed is by *enlarging the gauge group,* which we may attempt either by accreting new symmetries or by unifying the symmetries we have already recognized.

Left-right symmetric models [3], such as those based on the gauge symmetry $SU(3)_c \otimes SU(2)_L \otimes SU(2)_R \otimes U(1)_{B-L}$, follow the first path. (Here $B$ and $L$ stand for baryon number and lepton number.) The defining characteristic of such models is that the observed maximal parity violation in the weak interactions does not reflect a fundamental asymmetry of the laws of nature, but a circumstance based on the pattern of electroweak symmetry breaking. To respect real-world limits on right-handed charged currents and a $B - L$ gauge force, the $SU(2)_R \otimes U(1)_{B-L}$ symmetry must be broken to $U(1)_Y$ at a scale well above 1 TeV. This means that the right-handed $W^{\pm}$-bosons must acquire a much higher mass than the known left-handed $W^{\pm}$, so the low-energy charged-current interaction would be left-handed to the high accuracy established experimentally [4, 5]. Furthermore, the mixing among the neutral-gauge–boson eigenstates must be such as to reproduce the neutral-current properties of the standard model to high precision.

An appealing feature of the left-right–symmetric theory is that the electric charge operator takes an elegant form in terms of generators of both weak-isospin gauge groups and the fundamental quantities baryon number $B$ and lepton number $L$, namely, $Q = I_{3L} + I_{3R} + \frac{1}{2}(B - L)$. What the standard model identifies as weak hypercharge is given, in the $SU(3)_c \otimes SU(2)_L \otimes SU(2)_R \otimes U(1)_{B-L}$ world, by $Y = 2I_{3R} + (B - L)$, an expression that makes sense of the curious weak hypercharge assignments (6.3.7), (7.1.2), and (7.1.4) in the $SU(2)_L \otimes U(1)_Y$ electroweak theory. Left-right symmetric theories also open new possibilities, including transitions that induce neutron–antineutron oscillations [6], a mechanism for spontaneous CP violation [7], and a natural place for Majorana neutrinos.

More generally, enlarging the gauge group by accretion seeks to add a missing element or to explain additional observations.

Unified theories, on the other hand, seek to find a symmetry group $\mathcal{G} \supset SU(3)_c \otimes SU(2)_L \otimes U(1)_Y$ (usually a simple group, to maximize the predictive power) that contains the known interactions. This approach is motivated by the desire to unify quarks and leptons and to reduce the number of independent coupling constants, the better to understand the relative strengths of the strong, weak, and electromagnetic interactions at laboratory energies. Supersymmetric unified theories, which we shall mention only in passing, bring the added ambitions of incorporating gravity and relating constituents and forces.

Two very potent ideas are at play here. The first is the program of unification itself: what Feynman has called *amalgamation* [8], which is the central notion of *generalization and synthesis* that scientific explanation represents. Examples from the history of physics include Maxwell's joining of electricity and magnetism and light; the atomic hypothesis, which situates thermodynamics and statistical mechanics within the realm of Newtonian mechanics; and the links among atomic structure, chemistry, and quantum mechanics. This is a powerful impulse, and we shall argue from our knowledge of the gauge theories of the strong, weak, and electromagnetic interactions for the inevitability, subject to experimental check, of a unified theory of the fundamental interactions. The expanding reach of scientific explanation is the traditional focus of unified theories, but another aspect deserves our attention.

That second idea is that the human scale of space and time is not privileged for understanding nature and may even be disadvantaged. Not only in physics, but across the sciences, this has been a growing recognition since the quantum-mechanical revolution of the 1920s. To understand why a rock is solid or why a metal gleams, we must discern its structure on a scale $10^9$ times smaller than the human scale, and we must understand the rules that prevail there. What has matured recently is our comfort at cruising between different scales of momentum or distance and our understanding—through the renormalization group—of how one scale relates to another. Certain scales may well be privileged for understanding certain globally important aspects of the universe. For example, a unification scale might be privileged for understanding how it came to be that the fine structure constant $\alpha \approx \frac{1}{137}$ in the long-wavelength limit and the strong coupling constant $\alpha_s \approx \frac{1}{5}$ at energies characteristic of the $\Upsilon$ resonances.

> I believe that the discovery that *the human scale is not preferred* is as important as the discoveries that the human location is not privileged (Copernicus) and that there is no preferred inertial frame (Einstein) and will prove to be as influential.

The $\mathrm{SU(3)}_c \otimes \mathrm{SU(2)}_L \otimes \mathrm{U(1)}_Y$ standard model has been extensively validated by experiment; it contains a great measure of truth. In contrast, a unified theory is still largely an exercise in conjecture, in musing about things we would like to be true. The SU(5) model is by no means the only imaginable "grand unified" theory, nor will it be the answer to all our prayers, but it will nicely illustrate the general strategy and analytical techniques of unification without introducing encumbering complications. Furthermore, it entails a number of predictions that are interesting both as prototypes and as specific targets for experiment. These include fixing the weak mixing parameter $\sin^2 \theta_W$ and the expectation that the nucleon be unstable, with possible implications for understanding the baryon number of the universe. On the other hand, unification of the strong, weak, and electromagnetic interactions does nothing by itself to resolve the hierarchy problem we encountered in section 7.11. An extended gauge symmetry spontaneously broken at a high scale greatly exacerbates the vacuum energy problem of the electroweak theory (cf. §7.12): the SU(5) counterpart of (7.12.2) implies a vacuum energy density more than 100 orders of magnitude greater than observed!

## 9.1  WHY UNIFY?

With quantum chromodynamics and the standard model of weak and electromag-netic interactions in hand, what remains to be understood? If both theories are correct, can they also be complete? Actually, many observations are explained only in part, or not at all, by the separate gauge theories of the strong and the electroweak interactions. It is instructive to list the most prominent among these.

- The weak mixing parameter $x_W = \sin^2 \theta_W$ is arbitrary, and there are three distinct coupling constants, which may be characterized by $\alpha_s$, $\alpha_{EM}$, and

$\sin^2 \theta_W$. This reflects the fact that the gauge group is not simple but is given by the direct product $SU(3)_c \otimes SU(2)_L \otimes U(1)_Y$. Could the number of independent parameters be reduced to two or one?

- Both quarks and leptons are spin-$\frac{1}{2}$ particles that are structureless at present resolution. Are they related in any way?
- What is the meaning of electroweak universality [9], embodied in the matching left-handed doublets,

$$\begin{pmatrix} \nu_e \\ e \end{pmatrix}_L \text{ and } \begin{pmatrix} u \\ d_\theta \end{pmatrix}_L \cdots, \tag{9.1.1}$$

of quarks and leptons? This both requires and makes possible the anomaly cancellation that is a prerequisite for the renormalizability of the theory and suggests the grouping of quark and lepton doublets into fermion "generations." Why should this pattern hold? How many fermion generations are there?

- Why is electric charge quantized? Why is $Q(e) + Q(p) \equiv 0$ [10]? Why is $Q(\nu) - Q(e) \equiv Q(u) - Q(d)$? Why is $Q(d) = (1/3)Q(e)$? Why is $Q(\nu) + Q(e) + 3[Q(u) + Q(d)] \equiv 0$?
- Fermion masses and mixings arise outside the standard model. Higgs-boson self-interactions do not arise from a gauge principle. Leaving aside the QCD vacuum phase and possible Majorana-neutrino phases, the theories entail the 3 coupling parameters noted before, 6 mass parameters for the 6 quarks plus 3 quark-mixing angles, 1 CP violating quark phase; 10 analogous mass, mixing, and phase parameters for the leptons; and 2 parameters for the Higgs potential, for a total of 25 independent—and seemingly arbitrary—parameters.
- Gravitation is absent.

We may also note that in the standard electroweak model, the strengths of weak and electromagnetic interactions become comparable for $s$ (or $Q^2$) $\gg M_W^2$. For example, the cross section for the reaction

$$\bar{\nu}_e e \rightarrow \bar{\nu}_\mu \mu \tag{9.1.2}$$

is asymptotically

$$\sigma(\bar{\nu}_e e \rightarrow \bar{\nu}_\mu \mu) \xrightarrow{s \rightarrow \infty} \frac{G_F^2 M_W^4}{3\pi s} = \frac{\pi \alpha^2}{6 x_W^2 s}, \tag{9.1.3}$$

which is of the same size as the $s$-channel photon contribution to electron–positron annihilation into muons,

$$\sigma_\gamma(e^+ e^- \rightarrow \mu^+ \mu^-) \xrightarrow{s \rightarrow \infty} \frac{4\pi \alpha^2}{3s}. \tag{9.1.4}$$

Recall also (cf. the discussion surrounding figure 7.14) that the neutral-current and charged-current cross sections for deeply inelastic lepton–nucleon scattering become comparable at large values of $Q^2$. Moreover, in view of the evolution of the strong coupling constant given by (8.3.28), it is conceivable that $\alpha_s(Q^2)$ itself approaches the electroweak couplings for very large values of $Q^2$.

Some of these questions and observations argue for a qualitative quark–lepton connection. Others inspire a more complete unification of weak and electromagnetic interactions, perhaps in the form of a larger gauge symmetry group

$$G \supset SU(2)_L \otimes U(1)_Y \tag{9.1.5}$$

that would fix $x_W$. Still others suggest a grand unification of the strong, weak, and electromagnetic interactions, which would automatically complete the electroweak unification. Finally, it is possible to envisage a "superunification" that would include gravitation as well, perhaps through the medium of a supersymmetric field theory.

It is natural to base a unified description of the strong, weak, and electromagnetic interactions upon a simple group

$$G \supset SU(3)_c \otimes SU(2)_L \otimes U(1)_Y, \tag{9.1.6}$$

in order that interactions be determined by one single coupling constant. The scale at which the larger symmetry is attained and the unification is realized is set by the energy at which the running coupling constants coincide, which is approximately $10^{15}$ GeV in the SU(5) example we shall study. The enlarged gauge group $G$ will imply extra gauge bosons beyond the photon, the intermediate bosons, and the gluons. These new bosons, which will carry both flavor and color charges, are, presumably, extremely massive, because their effects are unfamiliar to us. We expect that all colored gauge bosons will be confined, as quarks and gluons are, into color singlets.

Once unification is undertaken, there is no reason not to assign quarks and leptons to the same representation of the symmetry group. Indeed, the equality of the proton and positron charges argues for extended quark–lepton families. In the absence of symmetry principles that require the exact conservation of lepton number and baryon number, quark–lepton transitions may be expected to occur. What consequences these will have will depend on the details of the unification, but proton decay is a natural, and extremely interesting, outcome.

Why unify? Why not?

## 9.2  THE SU(5) MODEL

The choice of a unifying gauge group $G$ is to be guided by requirements implicit in our discussion of the motivation for unification. It is worthwhile to formulate these requirements somewhat more precisely. It will be necessary to assume, as indicated in (9.1.6), that $G$ contains $SU(3)_c \otimes SU(2)_L \otimes U(1)_Y$. Next, because of the fermion content we wish to build into the theory, $G$ must admit complex representations. To see why this is so, let us examine the $SU(3)_c \otimes SU(2)_L \otimes U(1)_Y$ content of the "first generation" of fermions $u$, $d$, $\nu_e$, and $e$. (There is no need to introduce the complication of flavor mixing at this stage.)

It will be convenient to express all the fermions in terms of left-handed fields only. We have frequently used the chiral decomposition of a Dirac field,

$$\psi = \tfrac{1}{2}(1 - \gamma_5)\psi + \tfrac{1}{2}(1 + \gamma_5)\psi = \psi_L + \psi_R, \qquad (9.2.1)$$

which is particularly useful because the gauge-field couplings preserve chirality. If the field $\psi$ is understood to annihilate a particle, the charge-conjugate field

$$\psi^c \equiv C\bar{\psi}^T, \qquad (9.2.2)$$

where T designates transpose, annihilates an antiparticle. In the (standard) representation of Dirac matrices that we have adopted, the charge-conjugation matrix is given by

$$C = i\gamma^2\gamma^0 = -C^{-1} = -C^\dagger = -C^T, \qquad (9.2.3)$$

for which

$$C\gamma_\mu C^{-1} = -\gamma_\mu^T. \qquad (9.2.4)$$

It is then easy to see that the charge conjugate of a right-handed field is left-handed, for

$$\begin{aligned}
\psi_L^c &= \tfrac{1}{2}(1 - \gamma_5)\psi^c = \tfrac{1}{2}(1 - \gamma_5)C\bar{\psi}^T \\
&= C\tfrac{1}{2}(1 - \gamma_5)\bar{\psi}^T = C[\bar{\psi}\tfrac{1}{2}(1 - \gamma_5)]^T = C(\bar{\psi}_R)^T;
\end{aligned} \qquad (9.2.5)$$

similarly,

$$\bar{\psi}_L^c = \psi_R^T C = -\psi_R^T C^{-1}. \qquad (9.2.6)$$

Thus we enumerate the fermion fields of the first generation in terms of their $SU(3)_c \otimes SU(2)_L \otimes U(1)_Y$ quantum numbers as

$$\left.
\begin{aligned}
u_L, d_L &: (\mathbf{3}, \mathbf{2})_{Y=1/3} \\
d_L^c &: \quad (\mathbf{3}^*, \mathbf{1})_{2/3} \\
u_L^c &: \quad (\mathbf{3}^*, \mathbf{1})_{-4/3} \\
\nu_L, e_L &: \quad (\mathbf{1}, \mathbf{2})_{-1} \\
e_L^c &: \quad (\mathbf{1}, \mathbf{1})_2
\end{aligned}
\right\}. \qquad (9.2.7)$$

To incorporate a massive neutrino (cf. §6.7), we could add

$$N_L^c : (\mathbf{1}, \mathbf{1})_0. \qquad (9.2.8)$$

**TABLE 9.1**
Branching Rules for SU(5) → SU(3)$_c$ ⊗ SU(2)$_L$ ⊗ U(1)$_Y$

| Young Tableau | Dimension | (SU(3)$_c$, SU(2)$_L$)$_Y$ Decomposition |
|:---:|:---:|:---|
| □ | 5 | $(\mathbf{3}, \mathbf{1})_{-2/3} \oplus (\mathbf{1}, \mathbf{2})_1$ |
| (two-box column) | 10 | $(\mathbf{3}, \mathbf{2})_{1/3} \oplus (\mathbf{3^*}, \mathbf{1})_{-4/3} \oplus (\mathbf{1}, \mathbf{1})_2$ |
| (two-box row) | 15 | $(\mathbf{6}, \mathbf{1})_{-4/3} \oplus (\mathbf{3}, \mathbf{2})_{1/3} \oplus (\mathbf{1}, \mathbf{3})_2$ |
| (adjoint tableau) | $24 = 24^*$ | $(\mathbf{8}, \mathbf{1})_0 \oplus (\mathbf{3}, \mathbf{2})_{-5/3} \oplus (\mathbf{3^*}, \mathbf{2})_{5/3}$ $\oplus (\mathbf{1}, \mathbf{3})_0 \oplus (\mathbf{1}, \mathbf{1})_0$ |
| (tableau) | 45 | $(\mathbf{8}, \mathbf{2})_{-1} \oplus (\mathbf{6}, \mathbf{1})_{2/3} \oplus (\mathbf{3^*}, \mathbf{3})_{2/3}$ $\oplus (\mathbf{3}, \mathbf{2})_{7/3} \oplus (\mathbf{3^*}, \mathbf{1})_{2/3}$ $\oplus (\mathbf{3}, \mathbf{1})_{-8/3} \oplus (\mathbf{1}, \mathbf{2})_{-1}$ |
| (2×2 tableau) | 50 | $(\mathbf{8}, \mathbf{2})_{-1} \oplus (\mathbf{6^*}, \mathbf{1})_{-8/3} \oplus (\mathbf{6}, \mathbf{3})_{2/3}$ $\oplus (\mathbf{3}, \mathbf{2})_{7/3} \oplus (\mathbf{3^*}, \mathbf{1})_{2/3} \oplus (\mathbf{1}, \mathbf{1})_4$ |

Evidently this set of representations is not equal to its complex conjugate, so we must require that $G$ possess complex representations to accommodate these fermions. As a final requirement, we demand that $G$ have only a single coupling constant. For the purposes of this discussion, that will be what is meant by unification.

We may search systematically for candidate-unifying groups that meet these (or indeed, other) criteria. The smallest such group is SU(5). Let us now see how it meets the requirements we have set out. To analyze the structure of the model, it is helpful to refer to the SU(3)$_c$ ⊗ SU(2)$_L$ ⊗ U(1)$_Y$ decomposition of some of the low-dimensioned representations of SU(5) given in table 9.1. The **15**-dimensional representation, which might be the first hope to accommodate the 15 fermions of the first generation (putting aside a right-handed neutrino), contains color-sextet quarks and, thus, is not acceptable. Similarly, the **45**-dimensional representation, which could be hoped to contain three such fermion generations, contains color-octet and color-antisextet elements. However, the first-generation fermions may be assigned to the **5**$^*$ and **10** representations as

$$\mathbf{5^*}: \quad \psi_{jL} = \begin{pmatrix} d_1^c \\ d_2^c \\ d_3^c \\ \hline e \\ -\nu_e \end{pmatrix}_L \qquad (9.2.9)$$

and

$$
\mathbf{10}: \quad \psi_{\mathrm{L}}^{jk} = \frac{1}{\sqrt{2}}
\left(
\begin{array}{ccc|cc}
0 & u_3^c & -u_2^c & -u_1 & -d_1 \\
-u_3^c & 0 & u_1^c & -u_2 & -d_2 \\
u_2^c & -u_1^c & 0 & -u_3 & -d_3 \\
\hline
u_1 & u_2 & u_3 & 0 & -e^c \\
d_1 & d_2 & d_3 & e^c & 0
\end{array}
\right)_{\mathrm{L}},
\qquad (9.2.10)
$$

where the quark colors (red, green, blue) have been denoted $(1, 2, 3)$. The factor $1/\sqrt{2}$ in (9.2.10) is a convenient normalization, and many of the signs are matters of convention. The identification of $(e, -\nu_e)^{\mathrm{T}}$ as a $\mathbf{2}^*$ of weak isospin follows from the assignment of $(\nu_e, e)$ as a $\mathbf{2}$. An additional (right-handed) neutrino state (9.2.8) could be assigned as an SU(5) singlet.* Although it would have been pleasing to assign all the particles of the first generation to a single irreducible representation, as can be done in the closely analogous group SO(10) (see problem 9.1), there is nothing intrinsically objectionable about this assignment.

We see that the SU(5) model can accommodate, if not predict, the known elementary fermions in terms of 15-member generations assigned to the (reducible) $\mathbf{5}^* \oplus \mathbf{10}$ representations. These assignments bring with them some agreeable features. First, because the electric charge operator $Q$ is to be a generator of SU(5), the sum of electric charges over any representation must be zero. (Indeed, this requirement influenced the assignments presented earlier.) This means, in particular, that

$$
Q(d^c) = (-1/3) Q(e), \qquad (9.2.11)
$$

which explains the quantization of electric charge. In fact, although it appears here almost as an arithmetic triviality, charge quantization is deeply related to the existence of magnetic monopoles [11] in unified theories.

We know that the $\mathrm{SU}(3)_{\mathrm{c}} \otimes \mathrm{SU}(2)_{\mathrm{L}} \otimes \mathrm{U}(1)_{\mathrm{Y}}$ subgroup upon which the nonunified gauge theories of strong, weak, and electromagnetic interactions are founded is anomaly free for the representation we have chosen. This was in fact part of the motivation for identifying quark and lepton doublets as belonging to the same fermion generation. It remains to verify that the unified theory is also anomaly-free. This may be done either by direct computation or by using the fact [12] that the anomaly for a representation $R$ may be characterized by

$$
\mathrm{tr}\big(\{T^a, T^b\} T^c\big) = A(R) d^{abc}, \qquad (9.2.12)
$$

where the $T^a$ are the normalized generators of the representation $R$ and the $d^{abc}$ are the symmetric structure constants of the Lie algebra. For a representation that is given by the completely antisymmetric product of $p$ fundamental representations of

---

* For most of what follows, we will suppress the right-handed neutrino because it is sterile with respect to the established gauge interactions.

SU($N$), the coefficient $A(R)$ is given by

$$A(R) = \frac{(N-3)!(N-2p)}{(N-p-1)!(p-1)!}. \tag{9.2.13}$$

Thus for the **5**\* ($p = 4$) we find

$$A(5^*) = -1, \tag{9.2.14}$$

and for the **10** ($p = 2$) we have

$$A(\mathbf{10}) = 1. \tag{9.2.15}$$

Thus, remarkably, the **5**\* and **10** have equal and opposite anomalies and the unified theory is entirely anomaly free.

Let us now verify that the gauge-boson content is also what is required by the known phenomenology. Constructing a gauge theory by standard methods, we shall encounter 24 gauge bosons corresponding to the 24 elements of the adjoint representation of SU(5). Using the decomposition of table 9.1, we may identify the gauge bosons as follows:

$$\left. \begin{array}{l} (\mathbf{8}, \mathbf{1})_0 \leftrightarrow \text{gluons} \\ (\mathbf{1}, \mathbf{3})_0 \leftrightarrow W^+, \; W^-, \; W_3 \\ (\mathbf{1}, \mathbf{1})_0 \leftrightarrow \mathcal{A} \\ (\mathbf{3}, \mathbf{2})_{-5/3} \leftrightarrow X^{-4/3}, \; Y^{-1/3} \\ (\mathbf{3}^*, \mathbf{2})_{5/3} \leftrightarrow X^{4/3}, \; Y^{1/3} \end{array} \right\}. \tag{9.2.16}$$

The first 12 of these are precisely the known gauge bosons of the SU(3)$_c$ ⊗ SU(2)$_L$ ⊗ U(1)$_Y$ theory: the 8 gluons and 4 electroweak bosons that become, upon spontaneous symmetry breaking, the physical $W^+$, $W^-$, $Z^0$, and photon. The last 12 objects are new in the SU(5) theory. They are often called *leptoquark bosons* because they mediate transitions between quarks and leptons. The SU(5) gauge bosons may be displayed in matrix form as

$$\mathsf{H}\sqrt{2} = \left( \begin{array}{ccc|cc} & & & X_1 & Y_1 \\ & \text{gluons} & & X_2 & Y_2 \\ & & & X_3 & Y_3 \\ \hline X_{\bar 1} & X_{\bar 2} & X_{\bar 3} & W_3/\sqrt{2} & W^+ \\ Y_{\bar 1} & Y_{\bar 2} & Y_{\bar 3} & W^- & -W_3/\sqrt{2} \end{array} \right)$$

$$+ \frac{\mathcal{A}}{\sqrt{30}} \left( \begin{array}{ccc|cc} -2 & & & & \\ & -2 & & & 0 \\ & & -2 & & \\ \hline & & & 3 & \\ & 0 & & & 3 \end{array} \right), \tag{9.2.17}$$

where the matrix H is written in terms of the gauge bosons and the normalized generators of the adjoint representation $t^a$ as

$$\mathsf{H} \equiv \sum_{a=1}^{24} t^a H^a. \tag{9.2.18}$$

We have separated the $U(1)_Y$ gauge field to emphasize a crucial difference between the $SU(5)$ theory and the $SU(2)_L \otimes U(1)_Y$ model, which is that the hypercharge coupling is no longer a free parameter. Its strength, as we shall see explicitly, is fixed by its position in the $SU(5)$ gauge group.

Until now, we have constructed only theories in which the fermions lie in the fundamental representation of the gauge group. (A modest exception to this restriction occurred in problem 8.14.) The fermion assignments we have chosen for this $SU(5)$ theory necessitate an extension of our methods to include fermions in the $5^*$ and $10$ representations. The transformation properties of these representations and the construction of minimal interaction terms are, however, easily worked out. For each fermion generation, the interaction term in the $SU(5)$ Lagrangian may be written

$$\begin{aligned}
\mathcal{L}_{\text{int}} = &-\frac{g_5}{2} G_\mu^a (\bar{u}\gamma^\mu \lambda^a u + \bar{d}\gamma^\mu \lambda^a d) \\
&- \frac{g_5}{2} W_\mu^i (\bar{L}_u \gamma^\mu \tau^i L_u + \bar{L}_e \gamma^\mu \tau^i L_e) \\
&- \frac{g_5}{2} \frac{3}{5} \mathcal{A}_\mu \sum_{\substack{\text{fermion} \\ \text{species}}} \bar{f}\gamma^\mu Y f \\
&- \frac{g_5}{\sqrt{2}} [X_{\mu;\alpha}^- (\bar{d}_R^\alpha \gamma^\mu e_R^c + \bar{d}_L^\alpha \gamma^\mu e_L^c + \varepsilon_{\alpha\beta\gamma} \bar{u}_L^{c\gamma} \gamma^\mu u_L^\beta) + \text{h.c.}] \\
&+ \frac{g_5}{\sqrt{2}} [Y_{\mu;\alpha}^- (\bar{d}_R^\alpha \gamma^\mu \nu_R^c + \bar{u}_L^\alpha \gamma^\mu e_L^c + \varepsilon_{\alpha\beta\gamma} \bar{u}_L^{c\beta} \gamma^\mu d_L^\gamma) + \text{h.c.}],
\end{aligned} \tag{9.2.19}$$

where the notation $\lambda^a$ has been used for $SU(3)$ matrices and gluons are noted by $G_\mu^a$. The color index $a$ runs from 1 to 8, and the color indices $\alpha$, $\beta$, and $\gamma$ run over $1, 2, 3$ or red, green, blue. The $SU(2)_L$ doublets are

$$L_u = \begin{pmatrix} u \\ d \end{pmatrix}_L , \quad L_e = \begin{pmatrix} \nu_e \\ e \end{pmatrix}_L . \tag{9.2.20}$$

(Substitute $d \to d_\theta$ in (9.2.19) and (9.2.20) to incorporate quark mixing.)

The first three terms of (9.2.19) are the standard interactions of the unbroken $SU(3)_c \otimes SU(2)_L \otimes U(1)_Y$ model, provided that we identify the coupling constants $g_{\text{strong}} = g_3$ and $g_{SU(2)_L} = g_2$ with $g_5$. In the weak-hypercharge term we must identify

$$g' = g_5 \sqrt{\frac{3}{5}}, \tag{9.2.21}$$

where $g'$ corresponds to the canonical definition in the electroweak theory, as expressed in (6.3.13). The numerical factor arose, as earlier remarked, from the requirement that the field $\mathcal{A}_\mu$ couple to the current of a normalized generator of SU(5). This result may be retrieved in more pedestrian fashion as follows. Because the electric charge $Q$ is a generator of the group, it may be written in the form

$$Q = T_3 + \kappa T_0, \tag{9.2.22}$$

where $T_3$ is a generator of $SU(2)_L$ and $T_0$ is a weak-isosinglet generator of SU(5). For $SU(N)$, we then have that

$$\sum_{\text{representation}} Q^2 = (1 + \kappa^2) \sum_{\text{representation}} T_3^2. \tag{9.2.23}$$

An explicit calculation for the **5\*** then yields

$$\tfrac{4}{3} = (1 + \kappa^2)\tfrac{1}{2}, \tag{9.2.24}$$

so that

$$\kappa^2 = \tfrac{5}{3}. \tag{9.2.25}$$

This shows that the normalization of $T_0$ differs by a factor of $\sqrt{\tfrac{3}{5}}$ from the conventional U(1) hypercharge operator $Y$.

The important consequence of this information is that the weak-hypercharge coupling is no longer an independent parameter but is given by

$$g'^2 = \tfrac{3}{5}g_2^2. \tag{9.2.26}$$

Recalling the definition (6.3.50) of the weak mixing angle, we find that

$$x_W = \sin^2 \theta_W = \frac{g'^2}{g_2^2 + g'^2} = \frac{3}{8} \tag{9.2.27}$$

in the unbroken SU(5) theory.

The fourth and fifth terms in the interaction Lagrangian correspond to new transitions that occur in the unified theory. The new vertices, summarized in figure 9.1, mediate processes such as proton decay, which proceeds by the three elementary transitions shown in figure 9.2. Each of these changes both baryon number and lepton number by $-1$ unit and changes the number of elementary fermions by $-4$.

The intermediate bosons $W^+$, $W^-$, and $Z^0$ are expected to acquire masses according to something resembling the usual mechanism of spontaneous symmetry breaking. The leptoquark bosons $X$ and $Y$ must a fortiori be endowed with enormous masses ($\sim 10^{15}$ GeV) by means of a similar scheme for the theory to

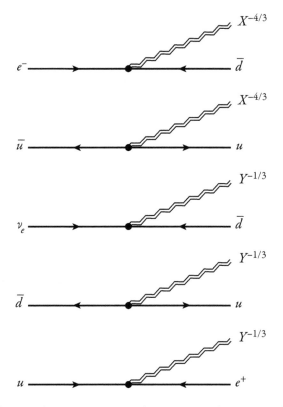

**Figure 9.1.** New fermion–fermion transitions that appear in the SU(5) unified theory.

survive the existing bounds on the proton lifetime, including [13]

$$\begin{aligned}
\tau(p \to e^+ \pi^0) &> 8.2 \times 10^{33} \text{ years}, \\
\tau(p \to \mu^+ \pi^0) &> 6.6 \times 10^{33} \text{ years},
\end{aligned} \tag{9.2.28}$$

and [14]

$$\tau(p) > 2.1 \times 10^{29} \text{ years (``invisible'' modes)}, \tag{9.2.29}$$

from a search for a $\gamma$ ray signaling the transition $^{16}\text{O} \to {}^{15}\text{N}^* + \text{unseen}$.

The necessary symmetry breaking can be achieved in two steps. First, a **24** of real scalar fields is introduced to break

$$\text{SU}(5) \xrightarrow{\mathbf{24}} \text{SU}(3)_\text{c} \otimes \text{SU}(2)_\text{L} \otimes \text{U}(1)_\text{Y}. \tag{9.2.30}$$

At this point, the $X$ and $Y$ leptoquark bosons acquire mass. Next, a 5-dimensional representation of complex scalar fields is employed to accomplish the breaking

$$\text{SU}(3)_\text{c} \otimes \text{SU}(2)_\text{L} \otimes \text{U}(1)_\text{Y} \xrightarrow{\mathbf{5}} \text{SU}(3)_\text{c} \otimes \text{U}(1)_\text{EM}. \tag{9.2.31}$$

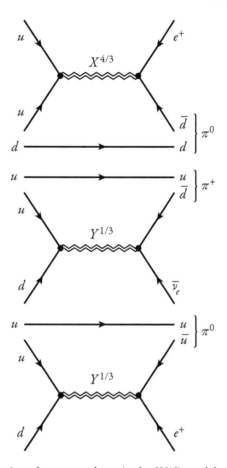

**Figure 9.2.** Some mechanisms for proton decay in the SU(5) model of unification.

This is the straightforward extension of the spontaneous symmetry breaking in the electroweak theory described in section 6.3, in which a complex scalar $SU(2)_L$ doublet breaks $SU(2)_L \otimes U(1)_Y$ down to $U(1)_{EM}$. The spontaneous symmetry breaking gives rise to many physical Higgs bosons. It is instructive to consider the logic of the breakdown in some detail.

To accomplish the symmetry breaking at the leptoquark scale, we assume that the adjoint **24** representation of real scalar bosons acquires a vacuum expectation value

$$\langle \mathbf{\Phi}_{24} \rangle_0 = \begin{pmatrix} \begin{array}{ccc|cc} 1 & & & & \\ & 1 & & & 0 \\ & & 1 & & \\ \hline & 0 & & -\frac{3}{2} & \\ & & & & -\frac{3}{2} \end{array} \end{pmatrix} \cdot v_{24}, \qquad (9.2.32)$$

where $v_{24}$ is the parameter that will set the scale of masses of the $X$ and $Y$ bosons. It is obvious that this preserves separately the SU(3) symmetry of the first three indices and the SU(2) symmetry of the last two indices. However, the symmetry under the SU(5) generators that couple the color indices to the $SU(2)_L$ indices will be broken, and the corresponding gauge bosons will acquire mass. Referring to the $SU(3)_c \otimes SU(2)_L \otimes U(1)_Y$ decomposition of the **24** given in table 9.1, we see that the $X$ and $Y$ bosons corresponding to the $(\mathbf{3}, \mathbf{2})_{-5/3}$ and $(\mathbf{3^*}, \mathbf{2})_{5/3}$ representations of the subgroup acquire masses, whereas the gluons $(\mathbf{8}, \mathbf{1})_0$, incipient $W$ bosons $(\mathbf{1}, \mathbf{3})_0$, and hypercharge boson $(\mathbf{1}, \mathbf{1})_0$ remain massless. These are accompanied by 12 physical scalar particles, or Higgs bosons, corresponding to the $(\mathbf{8}, \mathbf{1})_0$, $(\mathbf{1}, \mathbf{3})_0$, and $(\mathbf{1}, \mathbf{1})_0$ representations. The masses of these Higgs bosons will be comparable to those acquired by the leptoquark bosons. The **24** will not endow the fermions with similar masses because it does not occur in the $\bar{L}R$ products [cf. (6.3.20) and (7.1.32)],

$$
\begin{aligned}
\mathbf{5^*} \otimes \mathbf{10} &= \mathbf{5} \oplus \mathbf{45}, \\
\mathbf{10} \otimes \mathbf{10} &= \mathbf{5^*} \oplus \mathbf{45^*} \oplus \mathbf{50^*},
\end{aligned}
\tag{9.2.33}
$$

that generate renormalizable mass terms for the quarks and charged leptons.

The complex **5** contains the complex $(\mathbf{1}, \mathbf{2})_1$ representation used for spontaneous symmetry breaking in the standard model of the weak and electromagnetic interactions. It is, therefore, appropriate to choose the vacuum expectation value

$$
\langle \phi_5 \rangle_0 = \begin{pmatrix} 0 \\ 0 \\ 0 \\ 0 \\ 1 \end{pmatrix} v_5/\sqrt{2},
\tag{9.2.34}
$$

where $v_5 = (G_F\sqrt{2})^{-1/2} \approx 246$ GeV [cf. (6.3.45)]. Of the 10 real fields of the complex **5**, only 3 are absorbed into the longitudinal components of the now-massive $W^\pm$ and $Z^0$. In addition to the normal Higgs scalar $H^0$, still with unknown mass, there are 6 additional physical scalar particles, the color triplets $h^{\pm 1/3}$, corresponding to the

$$
(\mathbf{3}, \mathbf{1})_{-2/3} \quad \text{and} \quad (\mathbf{3^*}, \mathbf{1})_{2/3}
\tag{9.2.35}
$$

representations. These are potentially troublesome objects because they can mediate nucleon decay through processes such as

$$
u + d \to h^{1/3} \to \begin{cases} e^+ + \bar{u}, \\ \bar{\nu}_e + \bar{d}. \end{cases}
\tag{9.2.36}
$$

The Yukawa couplings are, as usual, closely tied to the fermion masses. Unless these vanish, the only way to suppress the rate for $h^{\pm 1/3}$-mediated proton decay is to arrange that the color-triplet scalar particles receive masses comparable to those of the $X$ and $Y$ bosons. This may be achieved [15] by introducing locally gauge-invariant interactions among $\Phi_{24}$ and $\phi_5$, which would, in any event, arise from quantum corrections to the lowest-order interaction terms. However, maintaining a ratio of 13 orders of magnitude between the electroweak scale characterized by $v_5$ and the leptoquark scale characterized by $v_{24}$ requires an exceedingly delicate tuning of parameters in the Higgs potentials. This balancing act is neither natural nor likely to survive radiative corrections. Its precarious nature may well be a symptom of the incompleteness of this minimal example of a unified theory.

We noted earlier, in our discussion of the symmetry-breaking effects of the 24, that scalars belonging to the fundamental 5 representation could generate fermion masses in the manner familiar from the electroweak theory. The necessary Yukawa term in the Lagrangian is

$$\mathcal{L}_{\text{Yukawa}} = \left[ \zeta_d \bar{\psi}_{Rj}^c \psi_L^{jk} \phi_k^\dagger + \zeta_u \varepsilon_{jklmn} \bar{\psi}_L^{cjk} \psi_L^{lm} \phi^n \right] + \text{h.c.} \tag{9.2.37}$$

for a single generation. If we write the vacuum expectation value of the scalar field as

$$\langle \phi_k \rangle_0 = \frac{\delta_k^5 v_5}{\sqrt{2}}, \tag{9.2.38}$$

we see that the first of the Yukawa terms becomes, simply,

$$\begin{aligned}
\mathcal{L}_d &= \frac{\zeta_d v_5}{\sqrt{2}} (\bar{\psi}_{Rj}^c \psi_L^{j5} + \text{h.c.}) \\
&= -\frac{\zeta_d v_5}{\sqrt{2}} (\bar{d}_R^\alpha d_L^\alpha + \bar{e}_R^c e_L^c + \text{h.c.}) \\
&= -\frac{\zeta_d v_5}{\sqrt{2}} (\bar{d}d + \bar{e}e), \tag{9.2.39}
\end{aligned}$$

where $\alpha$ is a color index for the quarks. The masses of the electron and the down quark are, therefore, given by

$$m_d = m_e = \frac{\zeta_d v_5}{\sqrt{2}}. \tag{9.2.40}$$

That is, the SU(5) symmetry requires the down quark and the electron mass matrices to be the same. In similar fashion, for several generations we obtain, in addition the predictions,

$$m_s = m_\mu, \tag{9.2.41}$$
$$m_b = m_\tau. \tag{9.2.42}$$

The mass of the up quark, generated by the second term in (9.2.37), continues to be a free parameter of the theory. Although the relations (9.2.40)–(9.2.42) support in a qualitative fashion our identification of the quarks and leptons that make up each fermion generation, they are not quantitatively successful. However, as we have seen earlier for coupling constants, masses in field theory are to be interpreted as parameters that depend on the momenta at which they are measured. The equalities

(9.2.40)–(9.2.42) should then be interpreted as predictions that apply for $Q^2 \gtrsim M_X^2$. At lower values of $Q^2$, the equalities are then broken by the differing rates of evolution of the fermion mass operators, with results that are at least in schematic agreement with experiment (Ref. [15]). We shall return to this question in section 9.6 but turn first to the subject of coupling constants and their evolution.

## 9.3 COUPLING-CONSTANT UNIFICATION

In motivating the unification of the strong, weak, and electromagnetic interactions, we remarked on the possibility that the running coupling constants of the $SU(3)_c$, $SU(2)_L$, and $U(1)_Y$ gauge groups might evolve to a common value at some large value of $Q^2$. We have now seen that in the unbroken $SU(5)$ theory, the equality of the coupling constants indeed emerges when the $U(1)_Y$ coupling is suitably normalized. This was exhibited in (9.2.19) and reflected in the prediction (9.2.27) for the weak mixing parameter $\sin^2 \theta_W$. It is now appropriate to ask at what value of $Q^2$ does this equality obtain, or, in other terms, what have the predictions of the unified theory to do with the low-energy world in which we live? The general analysis of these questions was given by Georgi, Quinn, and Weinberg [16].

The evolution of running coupling constants has been investigated in sections 8.2 and 8.3 for Abelian and non-Abelian gauge theories, respectively. In both cases, the coupling-constant evolution is influenced by the spectrum of particles that can appear in quantum corrections to the gauge-boson propagator. For QCD, for example, the number of quark flavors influences the rate of change of the coupling constant. Before transcribing the appropriate results from chapter 8, we need only cite an important technical result from field theory [17]: at each momentum or mass scale $\mu$, particles that have masses $M \gg \mu$ effectively decouple from matrix elements involving ordinary external particles, with masses $\lesssim \mu$. This means, for present purposes, that we may ignore the contribution of a particle to the coupling-constant evolution when the particle's mass exceeds the momentum scale of interest. The practical effect of this decoupling theorem is to excuse us from worrying about the effects of superheavy particles [18].

Let us now consider the evolution of the three coupling constants in turn. Writing, as usual, $\alpha \equiv g^2/4\pi$, we have from (8.3.29) that the strong $SU(3)_c$ coupling constant behaves as

$$\frac{1}{\alpha_3(Q^2)} = \frac{1}{\alpha_3(\mu^2)} + b_3 \ln\left(\frac{Q^2}{\mu^2}\right), \tag{9.3.1}$$

with

$$4\pi b_3 = 11 - \frac{2n_f}{3}, \tag{9.3.2}$$

where $n_f$ is the number of flavors with masses less than $\sqrt{Q^2}$. It will be more apt for present purposes to count the number of fermion generations $n_g$ rather than the number of individual flavors, so that

$$4\pi b_3 = 11 - \frac{4n_{\mathrm{g}}}{3}. \tag{9.3.3}$$

Now referring to (8.3.28), we may write the evolution of the $SU(2)_{\mathrm{L}}$ weak-isospin coupling constant as

$$\frac{1}{\alpha_2(Q^2)} = \frac{1}{\alpha_2(\mu^2)} + b_2 \ln\left(\frac{Q^2}{\mu^2}\right), \tag{9.3.4}$$

with

$$4\pi b_2 = \frac{22 - 4n_{\mathrm{g}}}{3}, \tag{9.3.5}$$

neglecting a small contribution from Higgs bosons. The first term in $b_2$ is the standard gauge-boson contribution for an $SU(2)$ theory. The second is the contribution of fermion loops, which is composed as follows. In each generation there are four left-handed fermion doublets: one lepton doublet and the quark doublet in three colors. Each of these doublets contributes the standard QED value $(-\frac{4}{3})$ times $\frac{1}{2}$ from the group factor

$$\mathrm{tr}\left(\frac{\tau^a}{2} \cdot \frac{\tau^b}{2}\right) = \frac{\delta^{ab}}{2}, \tag{9.3.6}$$

analogous to (8.3.10), times a spin factor $\frac{1}{2}$ because only left-handed fermions appear in the loop (cf. problem 9.8). Thus we have a total fermion contribution of

$$4 \cdot n_{\mathrm{g}} \cdot \left(-\frac{4}{3}\right) \cdot \left(\frac{1}{2}\right) \cdot \left(\frac{1}{2}\right) = -\frac{4n_{\mathrm{g}}}{3}. \tag{9.3.7}$$

Finally, for the $U(1)_{\mathrm{Y}}$ weak-hypercharge coupling, which evolves as

$$\frac{5}{3} \cdot \frac{1}{\alpha_1(Q^2)} = \frac{1}{\alpha_{\mathrm{Y}}(Q^2)} = \frac{1}{\alpha_{\mathrm{Y}}(\mu^2)} + b_{\mathrm{Y}} \ln\left(\frac{Q^2}{\mu^2}\right), \tag{9.3.8}$$

we simply have the QED contribution of the fermion loops weighted by the square of the hypercharge rather than the square of the electric charge for each species:

$$\begin{aligned}
4\pi b_{\mathrm{Y}} &= \left(-\frac{4}{3}\right)\left[\frac{1}{2}\mathrm{tr}\left(Y_{\mathrm{L}}^2\right) + \frac{1}{2}\mathrm{tr}\left(Y_{\mathrm{R}}^2\right)\right] \\
&= \left(-\frac{4}{3}\right) \cdot \frac{5}{3} \cdot n_{\mathrm{g}} = -\frac{20n_{\mathrm{g}}}{9}.
\end{aligned} \tag{9.3.9}$$

The factor $\frac{5}{3}$ recovered here simply reflects the now-familiar ratio (9.2.21) between the normalizations of $\alpha_{\mathrm{Y}}$ and $\alpha_1$.

**TABLE 9.2**

Coefficients $(b_N, \bar{b}_N)$ for running couplings in the (standard, supersymmetric) SU(5) unified theory [19]: $1/\alpha_N(Q^2) = 1/\alpha_N(\mu^2) + b_N \ln(Q^2/\mu^2)$. The number of fermion generations is $n_g$, and the number of Higgs doublets is $N_H$.

| Standard SU(5) ($n_g = 3$, $N_H = 1$ for minimal model): |
|---|

$$4\pi \begin{pmatrix} b_3 \\ b_2 \\ b_1 \end{pmatrix} = \begin{pmatrix} 11 \\ \frac{22}{3} \\ 0 \end{pmatrix} + \begin{pmatrix} -\frac{4}{3} \\ -\frac{4}{3} \\ -\frac{4}{3} \end{pmatrix} n_g + \begin{pmatrix} 0 \\ -\frac{1}{6} \\ -\frac{1}{10} \end{pmatrix} N_H$$

| Supersymmetric SU(5) ($n_g = 3$, $N_H = 2$ for minimal model): |
|---|

$$4\pi \begin{pmatrix} \bar{b}_3 \\ \bar{b}_2 \\ \bar{b}_1 \end{pmatrix} = \begin{pmatrix} 9 \\ 6 \\ 0 \end{pmatrix} + \begin{pmatrix} -2 \\ -2 \\ -2 \end{pmatrix} n_g + \begin{pmatrix} 0 \\ -\frac{1}{2} \\ -\frac{3}{10} \end{pmatrix} N_H$$

The leading-order evolution of $\alpha_3$, $\alpha_2$, and $\alpha_1$ is summarized in table 9.2 for normal and supersymmetric versions of the SU(5) unified theory.

If we now impose the equality of the SU(3)$_c$, SU(2)$_L$, and U(1)$_Y$ coupling constants at the unification scale $u$, we have, in the conventional SU(5) unified theory,

$$\frac{1}{\alpha_3(Q^2)} - \frac{1}{\alpha_2(Q^2)} = \ln\left(\frac{Q^2}{u^2}\right) \cdot \begin{cases} (11/12\pi), \\ (23/24\pi), \end{cases} \qquad (9.3.10)$$

where the two coefficients omit (to make contact with much of the literature) and include the standard-model Higgs-boson contribution. Then, with the definition of the electromagnetic coupling constant introduced in (6.3.49),

$$\frac{1}{\alpha} \equiv \frac{1}{\alpha_Y} + \frac{1}{\alpha_2}, \qquad (9.3.11)$$

which has only a formal meaning for momenta above the scale of electroweak symmetry breaking, we may write

$$\begin{aligned}
\frac{1}{\alpha(Q^2)} &= \frac{1}{\alpha_Y(u^2)} + \frac{1}{\alpha_2(u^2)} + (b_Y + b_2) \ln\left(\frac{Q^2}{u^2}\right) \\
&= \frac{8}{3} \cdot \frac{1}{\alpha_3(u^2)} + (b_Y + b_2) \ln\left(\frac{Q^2}{u^2}\right).
\end{aligned} \qquad (9.3.12)$$

We may then form the combination

$$\begin{aligned}
\frac{8}{3} \cdot \frac{1}{\alpha_3(Q^2)} - \frac{1}{\alpha(Q^2)} &= \left(\frac{8b_3}{3} - b_Y - b_2\right) \ln\left(\frac{Q^2}{u^2}\right) \\
&= \ln\left(\frac{Q^2}{u^2}\right) \cdot \begin{cases} (11/2\pi), \\ (67/12\pi), \end{cases}
\end{aligned} \qquad (9.3.13)$$

(excluding and including Higgs-boson contributions), which does not depend on the number of (light) fermion generations. Let us now use the evolution equations just derived, together with experimentally determined values of the strong and electromagnetic coupling constants, to estimate the scale at which unification—in the sense of coincident values of the coupling constants—is achieved.[**]

The unification scale $u$ is exponentially sensitive to the values of the combination of couplings on the left-hand side of (9.3.13), so it is important to take into account the $Q^2$-evolution of $\alpha$ itself, which we considered in the discussion surrounding figure 8.13. At the electroweak scale characterized by $M_Z = (91.1875 \pm 0.0021)$ GeV [20], the inverse of the fine structure constant is [21]

$$\alpha^{-1}(M_Z^2) = 128.957 \pm 0.020. \tag{9.3.14}$$

Then, with the value of the strong coupling noted in section 8.3,

$$\alpha_s(M_Z^2) = 0.1184 \pm 0.0007 \tag{8.3.31}$$

(so that $\alpha_s^{-1}(M_Z^2) \approx 8.446$), we deduce that the unification scale is

$$u \approx 1.44 \times 10^{15} \text{ GeV}. \tag{9.3.15}$$

Although this is enormous on the scale of laboratory energies (a fact that underlines the riskiness of insouciant extrapolations), it lies well below the Planck mass

$$M_{Pl} \equiv \left( \frac{\hbar c}{G_{Newton}} \right)^{1/2} = 1.22 \times 10^{19} \text{ GeV}, \tag{9.3.16}$$

at which the gravitational interactions become relevant for individual particles. This may be seen to justify a posteriori our decision to defer the inclusion of gravity in the construction of the unified theory. How two mass scales as different as $u$ and $M_Z$ may arise and be preserved remains mysterious (cf. §7.11). It is worth noting (cf. figure 7.26) that, depending on the mass of a standard-model Higgs boson, consistency of the electroweak theory might require new physics to intervene below the putative SU(5) unification scale.

Having roughly estimated the unification mass, we are now in a position to calculate the values of the $SU(3)_c \otimes SU(2)_L \otimes U(1)_Y$ coupling constants at low energy, in the *idealized* SU(5) unified theory. These are shown, in one-loop approximation, for three fermion generations in figure 9.3. There we show the evolution of the $SU(3)_c$, $SU(2)_L$, and $U(1)$ couplings $\alpha_3$, $\alpha_2$, and $\alpha_1$, which would attain a common value at the unification scale $u$, as well as the weak hypercharge coupling $\alpha_Y$ and, as a derived quantity, the electromagnetic coupling constant $\alpha$ defined through (9.3.11). Problem 9.11 tests this version of coupling-constant unification and explores how the evolution of couplings would be changed by the introduction of supersymmetric partners of the known particles. On current

---

[**] A refined treatment would make use of evolution equations beyond leading-logarithmic approximation as well as attention to the renormalization scheme in which the couplings are defined, but a leading-order approach allows us to extract the essential results with a minimum of fuss.

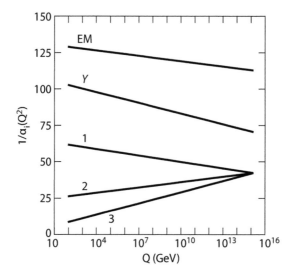

**Figure 9.3.** Idealized evolution of running coupling constants in leading logarithmic approximation is the SU(5) model. Three fermion generations are assumed.

evidence, gauge-coupling unification offers suggestive evidence in favor of TeV-scale supersymmetry, but no superpartners have yet been observed.

Of special interest is the weak mixing parameter

$$x_W = \sin^2 \theta_W = \frac{\alpha}{\alpha_2}, \tag{9.3.17}$$

which is plotted in figure 9.4. A convenient expression for this quantity may be obtained as follows. From the definition (9.3.11) of the electromagnetic coupling constant, subtract an amount

$$\frac{8}{3} \cdot \frac{1}{\alpha_2} = \frac{8}{3} \cdot \frac{x_W}{\alpha} \tag{9.3.18}$$

to obtain

$$\left[ 1 - \frac{8 x_W(Q^2)}{3} \right] \cdot \frac{1}{\alpha(Q^2)} = \frac{1}{\alpha_Y(Q^2)} - \left( \frac{5}{3} \right) \cdot \frac{1}{\alpha_2(Q^2)}$$

$$= \left( b_Y - \frac{5 b_2}{3} \right) \ln \left( \frac{Q^2}{u^2} \right). \tag{9.3.19}$$

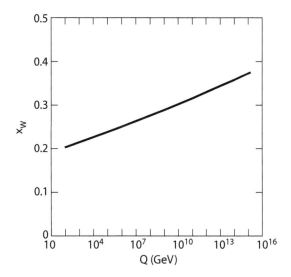

**Figure 9.4.** Evolution of the weak mixing parameter $x_W = \sin^2 \theta_W$ in the idealized SU(5) model (same assumptions as figure 9.3).

This may be solved at once to yield

$$x_W(Q^2) = \frac{3}{8} - \alpha(Q^2)\frac{(3b_Y - 5b_2)}{8} \ln\left(\frac{Q^2}{u^2}\right)$$

$$= \frac{3}{8} + \frac{55\alpha(Q^2)}{48\pi} \ln\left(\frac{Q^2}{u^2}\right). \tag{9.3.20}$$

The implied value of the weak mixing parameter at the electroweak scale,

$$x_W(M_Z^2) \approx 0.203, \tag{9.3.21}$$

approaches, but does not quite match, the value extracted from experiments. In the "on-shell" scheme, we determine from measured masses of the electroweak gauge bosons that

$$x_W^{(os)}(M_Z^2) \equiv 1 - M_W^2/M_Z^2 = 0.2226. \tag{9.3.22}$$

On the assumption that we have already discovered the relevant particle spectrum, the running of the coupling constants reduces the SU(5) prediction from the value $x_W = \frac{3}{8}$ at the unification scale to the low-energy value given in (9.3.21). This is a remarkable qualitative triumph—but quantitative failure—of the minimal unification hypothesis. Problem 9.13 explores how the prediction for $x_W$ would be changed by the introduction of superpartners at a low scale.

## 9.4 NUCLEON DECAY

We have seen, in a minimal unified theory, the possibility of sharpening the low-energy predictions of the $SU(2)_L \otimes U(1)_Y$ electroweak theory by predicting the value of the weak mixing parameter. This has definite implications for the properties of the intermediate vector bosons of the weak interactions. Beyond the aesthetic arguments for unification, this represents an important advance in predictive power, although—as we shall discuss at greater length in section 9.7—the theory is not yet free of arbitrariness. However, the most dramatic consequence of unification, in a qualitative sense, is the prediction that the nucleon be unstable.

We already remarked in connection with problem 3.7 that the exact conservation of baryon number is not implied by any of the known symmetry principles.[†] The unified theory goes beyond merely reminding us of this fact by providing specific mechanisms by which nucleons may decay. Without serious calculation, we may note from the resemblance of the muon-decay process to the diagrams relevant to proton decay shown in figures 9.1 and 9.2 that the proton lifetime must be approximately given by

$$\tau_p \approx \tau_\mu \left( \frac{m_\mu}{m_p} \right)^5 \left( \frac{u}{M_W} \right)^4 , \tag{9.4.1}$$

where all coupling constants have been assumed equal and the unification scale has been taken to represent the mass of the leptoquark bosons. Using the measured muon and proton masses in the somewhat schematic phase-space factor, together with the observed muon lifetime

$$\tau_\mu = 7 \times 10^{-14} \text{ year} \tag{9.4.2}$$

and the extremely crude conclusion (9.3.15) that

$$u/M_W \approx 10^{13}, \tag{9.4.3}$$

we estimate that

$$\tau_p \approx 10^{34} \text{ years.} \tag{9.4.4}$$

Clearly, this estimate is both very uncertain and exquisitely (but not exponentially) sensitive to the unification scale. This very rough prediction comfortably exceeds the experimental lower bound [23], $\tau_p > 3 \times 10^{23}$ years, that was known at the time SU(5) unification was proposed, so the theory was not immediately ruled out. As we shall see momentarily, more detailed estimates lie in the range of a few $\times 10^{32}$ years, which set an inviting target for experiment.

Recalling that $1 \text{ cm}^3$ of water contains Avogadro's number $(6 \times 10^{23})$ of nucleons, we readily find that a year's careful scrutiny of a cube of water 10 m

---

[†] In the electroweak theory, baryon number is conserved at every finite order in perturbation theory but is broken nonperturbatively [22].

on a side ($6 \times 10^{32}$ nucleons) is the magnitude of effort required to reach the decay rate suggested by the SU(5) theory. The largest proton-decay instrument exploited so far, the Super-Kamiokande detector, has a fiducial volume of 32,000 m$^3$ [24].

The detailed computation of nucleon decay rates is a rather involved undertaking, which requires the development of more technology than is appropriate to this introduction. Apart from the desirablity of estimating the unification scale with greater care, there is the need to compute transition matrix elements within physical hadrons, which requires understanding the strong-interaction or bound-state corrections to the elementary vertices of figure 9.2. This necessitates the intervention of models for hadronic structure as well as a more systematic application of the renormalization group so that elementary vertices computed at the unification scale may be related to transition amplitudes relevant on the scale of hadronic dimensions [25]. Some specialized reviews are included in the bibliography to this chapter. Steps toward a quark-based description of nucleon structure are taken in problems 4.6 and 4.7.

In contrast, it is quite straightforward to enumerate the allowed channels for nucleon decay. The leptoquark terms in the interaction Lagrangian (9.2.19) generate an effective four-fermion Lagrangian in the local (large unification scale) limit, which may be written as

$$\mathcal{L}_{\text{eff}} = \frac{4G_u}{\sqrt{2}} \varepsilon_{\alpha\beta\gamma} [\bar{u}_L^{c\gamma} \gamma^\mu u_L^\beta (\bar{e}_L^c \gamma_\mu d_L^\alpha + \bar{e}_R^c \gamma_\mu d_R^\alpha)$$
$$+ \bar{u}_L^{c\beta} \gamma^\mu d_L^\gamma (\bar{\nu}_R^c \gamma_\mu d_R^\alpha + \bar{e}_L^c \gamma_\mu u_L^\alpha)] + \text{h.c.} \tag{9.4.5}$$

where, in analogy with the Fermi constant defined in (6.3.44) by

$$\frac{G_F}{\sqrt{2}} = \frac{g_2^2}{8 M_W^2}, \tag{9.4.6}$$

we have introduced

$$\frac{G_u}{\sqrt{2}} = \frac{g_5^2}{8u^2}. \tag{9.4.7}$$

As usual, the indices $\alpha$, $\beta$, and $\gamma$ run over the three quark colors. After a Fierz rearrangement of the final term in (9.4.5), the effective Lagrangian takes the form

$$\mathcal{L}_{\text{eff}} = \frac{4G_u}{\sqrt{2}} \varepsilon_{\alpha\beta\gamma} [\bar{u}_L^{c\gamma} \gamma^\mu u_L^\beta (2\bar{e}_L^c \gamma_\mu d_L^\alpha + \bar{e}_R^c \gamma_\mu d_R^\alpha) + \bar{u}_L^{c\beta} \gamma^\mu d_L^\gamma \bar{\nu}_R^c \gamma_\mu d_R^\alpha] + \text{h.c.} \tag{9.4.8}$$

for a single generation, ignoring the effects of mixing between generations.

The possible semifinal states for proton decay are, therefore,

$$p \to d\bar{d}e^+, \ u\bar{u}e^+, \ u\bar{d}\bar{\nu}_e, \tag{9.4.9}$$

and, generalizing to a second generation,

$$p \to u\bar{s}\bar{\nu}_\mu, \ d\bar{s}\mu^+. \tag{9.4.10}$$

Similarly, for neutron decay we have

$$n \to d\bar{u}e^+, \ d\bar{d}\bar{\nu}_e, \ d\bar{s}\bar{\nu}_\mu. \tag{9.4.11}$$

Nucleon decay always leads, in the SU(5) model, to antileptons and nonnegative strangeness in the final state. The conservation of $B - L$, the difference of baryon number and lepton number, is more generally valid in unified theories [26]. In specific SU(5) calculations, the most important decays are

$$p \to e^+\pi^0, \ e^+\rho^0, \quad \text{and} \quad e^+\omega^0 \tag{9.4.12}$$

and

$$n \to e^+\pi^-, \ e^+\rho^-, \ \text{and} \quad \bar{\nu} + \text{anything}. \tag{9.4.13}$$

Those that lead to photons and charged particles in the final state can be detected with good efficiency in water-Cherenkov detectors.

A recent explicit calculation within the SU(5) model yields a partial lifetime [27]

$$\tau(p \to e^+\pi^0)|_{\text{th}} = 1.9 \times 10^{32} \text{ years}, \tag{9.4.14}$$

assuming the existence of three generations of quarks and leptons and a single electroweak Higgs multiplet. This prediction is on the high end of estimates in the minimal SU(5) theory but still is more than 40 times shorter than the observed lower bound, $8.2 \times 10^{33}$ years, quoted in (9.2.28). This conflict constitutes part of the evidence that minimal SU(5) unification is excluded.

The current status of the search for proton decay is summarized in figure 9.5, which compares lower bounds on the proton partial lifetimes with the expectations of a variety of unification schemes. The most straightforward version of SU(5) unification—which fails to unify the couplings $\alpha_1$, $\alpha_2$, $\alpha_3$—is excluded. A minimal supersymmetric SU(5) model [28], which does unify the gauge couplings, is strongly disfavored by the bounds on strange decay modes. Plans are being developed for megaton-class detectors that would extend the searches to $O(10^{35})$ year, which would challenge more unification models [29]. Observation of proton decay would constitute powerful evidence in favor of quark–lepton unification. The experimental challenges, which include the need to master backgrounds from cosmic-ray muons and radioactive contaminants along with the fact that the reach in unification scale grows only as the fourth root of the reach in lifetime, are extreme.

## 9.5 THE BARYON NUMBER OF THE UNIVERSE

Although the prediction of baryon decay is a dramatic consequence of unification, the difficulty of studying leptoquark transitions in the laboratory is clear. We live in a world in which energies are low compared with the unification scale. However, the discovery [30] of the cosmic microwave background radiation together with many supporting pieces of evidence makes it extremely likely that the early universe was characterized by extraordinarily high energy density. Many aspects of the

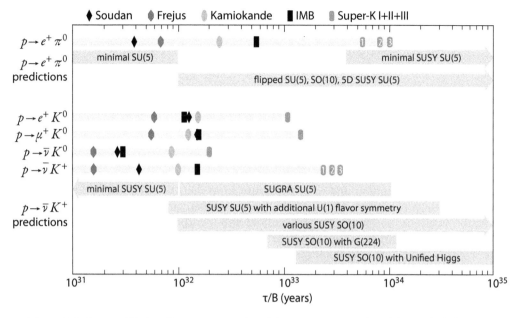

**Figure 9.5.** Proton lifetime limits compared to lifetime ranges predicted by unified theories. The upper section is for $p \to e^+ \pi^0$, a prominent mode in minimal SU(5). The lower section pertains to supersymmetric unified theories, which commonly predict decay modes with kaons in the final state. The symbols indicate published limits by experiments, as indicated by the sequence on top of the figure (from Ref. [29].)

observed universe find natural explanations in terms of this standard cosmological model [31]. What is particularly striking is how much can be understood on the basis of laboratory experience and the hypothesis that the universe has remained in thermal equilibrium since its temperature was about 1 MeV.

Other features of the universe are not so easily understood in terms of common experience and innocuous hypotheses. Prominent among these is the net baryon number of the universe. These more difficult questions are of special interest for the window they may provide on the early thermal history of the universe and the interactions among fundamental particles at exceedingly high energies. This realization, catalyzed by the prediction of baryon-number violation in unified theories, has stimulated a new symbiosis between elementary particle physics and cosmology. Our current ignorance in both domains precludes a complete and credible explanation of the baryon asymmetry of the universe. Nevertheless, the issue is of such interest—it has to do with our very existence—that it is appropriate to include in our brief study of unified theories an outline of the principal ingredients of a calculation.

We wish to understand why matter dominates over antimatter in the universe, or at least in our region of the universe. To be slightly more specific, observations indicate that the density of antibaryons is negligible, whereas the average density of matter, characterized by the baryon-to-photon ratio in the present universe,

$$5.1 \times 10^{-10} \leq \frac{n_B}{n_\gamma} \leq 6.5 \times 10^{-10} \ (95\% \ \text{CL}), \tag{9.5.1}$$

is small but nonzero [32]. It is natural to assume that the present universe evolved from a symmetric state of zero baryon number, either at "the beginning" or at the period of reheating following inflation. But, then, how did the present asymmetric universe evolve?

Three necessary elements for the evolution to a state with net baryon number are (1) the existence of fundamental processes that violate baryon number; (2) microscopic CP violation; and (3) a departure from thermal equilibrium during the epoch in which baryon-number-violating processes were important. Unified theories entail transitions that violate baryon number and incorporate—at a minimum—the CP violation that arises from the phase in the quark-mixing matrix (cf. §7.1 and problem 7.2). The processes that do not conserve baryon number may be forced out of thermal equilibrium by the expansion of the universe. Let us see why these three "Sakharov conditions" are required and then examine their relevance to unified theories.

The need for $B$-violating interactions requires no comment. That CP violation must be present in the $B$-nonconserving processes may be argued as follows. Suppose that a heavy boson $X$ decays into two channels characterized by baryon numbers $B_1$ and $B_2$, with branching ratios

$$
\begin{aligned}
\frac{\Gamma(X \to B_1)}{\Gamma(X \to \text{all})} &\equiv f, \\
\frac{\Gamma(X \to B_2)}{\Gamma(X \to \text{all})} &= 1 - f.
\end{aligned}
\tag{9.5.2}
$$

The antiparticle $\overline{X}$ decays into channels with baryon numbers $-B_1$ and $-B_2$, with branching ratios

$$
\begin{aligned}
\frac{\Gamma(\overline{X} \to -B_1)}{\Gamma(\overline{X} \to \text{all})} &\equiv \bar{f}, \\
\frac{\Gamma(\overline{X} \to -B_2)}{\Gamma(\overline{X} \to \text{all})} &= 1 - \bar{f}.
\end{aligned}
\tag{9.5.3}
$$

The equality of the total decay rates,

$$\Gamma(X \to \text{all}) = \Gamma(\overline{X} \to \text{all}), \tag{9.5.4}$$

is a consequence of CPT invariance. Decays from an initial state with equal numbers of bosons and antibosons will then lead to a net baryon number

$$
\begin{aligned}
\Delta B &= (f - \bar{f})B_1 + [(1 - f) - (1 - \bar{f})]B_2 \\
&= (f - \bar{f})(B_1 - B_2).
\end{aligned}
\tag{9.5.5}
$$

For this to be nonzero requires $B_1 \neq B_2$, which is to say baryon-number nonconservation in the decays, as was already apparent, and also

$$f \neq \bar{f}. \tag{9.5.6}$$

However, CP invariance requires the equality of the $S$-matrix elements

$$S(X \to B_1) = S(\overline{X} \to \bar{B}_1), \tag{9.5.7}$$

which would imply the equality of $f$ and $\bar{f}$. Hence CP violation is seen to be a prerequisite to the generation of net baryon number in the decay process.

To see that equilibrium of the environment prevents the development of a baryon–antibaryon asymmetry, let us consider the implications of unitarity and CPT invariance. The unitarity of the $S$-matrix may be expressed as

$$SS^\dagger = S^\dagger S = I. \tag{9.5.8}$$

Writing

$$S_{ba} = S(a \to b), \tag{9.5.9}$$

we then have

$$
\begin{aligned}
(SS^\dagger)_{aa} &= \sum_i S(a \to i) S^*(a \to i) \\
&= \sum_i |S(a \to i)|^2 = \sum_i |S(i \to a)|^2,
\end{aligned}
\tag{9.5.10}
$$

where the summation over intermediate states includes both particle and antiparticle states. Invariance under CPT implies that

$$S(a \to b) = S(\bar{b} \to \bar{a}), \tag{9.5.11}$$

whereupon

$$\sum_i |S(a \to i)|^2 = \sum_i |S(i \to a)|^2 = \sum_{\bar{i}} |S(\bar{i} \to \bar{a})|^2. \tag{9.5.12}$$

Because the sum runs over both particle and antiparticle states, we may relabel $i \to \bar{i}$ in the last term, whereupon the identity

$$\sum_i |S(i \to a)|^2 = \sum_i |S(i \to \bar{a})|^2 \tag{9.5.13}$$

is obtained as a consequence of CPT and unitarity alone, although it would also be implied by exact CP invariance. In thermal equilibrium, all the states $i$ corresponding to a given energy are equally populated, so the total rates leading to particle and antiparticle final states are equal. Accordingly, no baryon number asymmetry can develop unless the system is out of thermal equilibrium. This can occur if the expansion rate of the universe, and thus its cooling rate, is large compared with the reaction rates involved in baryon-number-violating processes.

We have now identified the basic elements required to generate a net baryon number in the universe. Specific calculations are sensitive to the early thermal history of the universe as well as the details of baryon number nonconservation and CP violation in the unified theory. For example, we have already seen from very general considerations that departures from CP invariance must persist up to the unification scale if this sort of mechanism is to succeed. When the first examples of unified theories were constructed, it was tempting to think that an explanation of the matter excess in the universe might be at hand. Those hopes have not yet been realized, in part because postinflationary reheating tends to erase any baryon excess established at very early times. At this time, two interesting paths toward an explanation of the baryon-to-photon ratio are electroweak (sphaleron-induced) baryogenesis (Ref. [22]) and a scenario that seeds electroweak baryogenesis with an early lepton–antilepton asymmetry arising from decays of heavy Majorana neutrinos [33]. Neither relies on quark–lepton unification, but each is missing a crucial ingredient. Electroweak baryogenesis requires a source of CP violation beyond the phase in the quark-mixing matrix, whereas leptogenesis rests on neutrino properties that have not been established. Evidently, we have much more to learn before we can explain the origin of matter in our universe!

## 9.6 THE PROBLEM OF FERMION MASSES

Unraveling the origins of electroweak symmetry breaking will not necessarily give insight into the origin and pattern of fermion masses, because they are set by the Yukawa couplings $\zeta_i$, of unknown provenance, that we first met in (6.3.20). The puzzling pattern of quark masses is depicted in figure 9.6. The fact that masses— like coupling constants—are scale dependent might encourage us to hope that what looks like an irrational pattern at low scales might reveal an underlying order at some other scale.

To illustrate the possibilities, let us adopt the specific framework of SU(5) unification, with the two-step spontaneously symmetry breaking we introduced in section 9.2. At a high scale, a **24** of scalars breaks SU(5) → SU(3)$_c$ ⊗ SU(2)$_L$ ⊗ U(1)$_Y$, giving extremely large masses to the leptoquark gauge bosons $X^{\pm 4/3}$ and $Y^{\pm 1/3}$. As we have already observed, the **24** does not occur in the $\bar{L}R$ products that generate fermion masses, so quarks and leptons escape large tree-level masses. At the electroweak scale, a **5** of scalars (containing the standard-model Higgs fields) breaks SU(3)$_c$ ⊗ SU(2)$_L$ ⊗ U(1)$_Y$ → SU(3)$_c$ ⊗ U(1)$_{EM}$ and endows fermions with mass. This approach relates quark and lepton masses at the unification scale,

$$
\left.\begin{array}{r}
m_e = m_d, \\
m_\mu = m_s, \\
m_\tau = m_b,
\end{array}\right\} \text{ at } u \text{ ; plus separate parameters } \left\{\begin{array}{l}
m_u, \\
m_c, \\
m_t,
\end{array}\right. \tag{9.6.1}
$$

with implications for the observed masses that we will now elaborate.

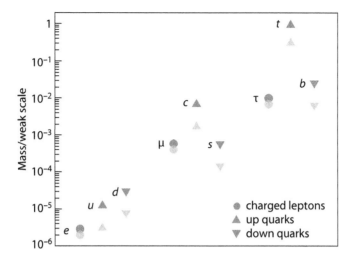

**Figure 9.6.** Running masses of the quarks and charged leptons (from Ref. [34]), evaluated in the $\overline{\text{MS}}$ renormalization scheme, in units of the electroweak scale $v/\sqrt{2} \approx$ 174 GeV. The upper (labeled) points are low-energy values: heavy-quark $(c, b, t)$ masses and the $\tau$-lepton mass are evaluated at the particle masses, $\overline{m}_f(m_f^2)$, whereas the light-quark $(u, d, s)$ and electron and muon masses are evaluated at 1 GeV. The lower points show masses evaluated at a high-energy scale, $10^{15}$ GeV.

The fermion masses evolve from the unification scale $u$ to the experimental scale $\mu$:

$$\ln\left[m_{u,c,t}(\mu^2)\right] \approx \ln\left[m_{u,c,t}(u^2)\right] + \frac{12}{33 - 2n_f} \ln\left(\frac{\alpha_3(\mu^2)}{\alpha_u}\right) \tag{9.6.2}$$

$$+ \frac{27}{88 - 8n_f} \ln\left(\frac{\alpha_2(\mu^2)}{\alpha_u}\right) - \frac{3}{10n_f} \ln\left(\frac{\alpha_1(\mu^2)}{\alpha_u}\right),$$

$$\ln\left[m_{d,s,b}(\mu^2)\right] \approx \ln\left[m_{d,s,b}(u^2)\right] + \frac{12}{33 - 2n_f} \ln\left(\frac{\alpha_3(\mu^2)}{\alpha_u}\right) \tag{9.6.3}$$

$$+ \frac{27}{88 - 8n_f} \ln\left(\frac{\alpha_2(\mu^2)}{\alpha_u}\right) + \frac{3}{20n_f} \ln\left(\frac{\alpha_1(\mu^2)}{\alpha_u}\right),$$

$$\ln\left[m_{e,\mu,\tau}(\mu^2)\right] \approx \ln\left[m_{e,\mu,\tau}(u^2)\right] \tag{9.6.4}$$

$$+ \frac{27}{88 - 8n_f} \ln\left(\frac{\alpha_2(\mu^2)}{\alpha_u}\right) - \frac{27}{20n_f} \ln\left(\frac{\alpha_1(\mu^2)}{\alpha_u}\right),$$

where we have omitted a small Higgs-boson contribution to keep the formulas short. The classic success of SU(5) unification is the predicted relation between $m_b$ and $m_\tau$

(Ref. [15]). Combining (9.6.3) and (9.6.4), we have

$$\ln\left[\frac{m_b(\mu^2)}{m_\tau(\mu^2)}\right] \approx \ln\left[\frac{m_b(u^2)}{m_\tau(\mu^2)}\right] + \frac{12}{33 - 2n_f}\ln\left(\frac{\alpha_3(\mu^2)}{\alpha_u}\right) - \frac{3}{2n_f}\ln\left(\frac{\alpha_1(\mu^2)}{\alpha_u}\right), \quad (9.6.5)$$

where the first term on the right-hand side vanishes. Choosing, for illustration, $n_f = 6$, $1/\alpha_u = 40$, $1/\alpha_3(\mu^2) = 5$, and $1/\alpha_1(\mu^2) = 65$, we compute at a low scale

$$m_b = 2.91 m_\tau \approx 5.16 \text{ GeV}, \qquad (9.6.6)$$

in suggestive agreement with experiment. The factor-of-three ratio arises because the quark masses, influenced by QCD, evolve more rapidly than the lepton masses.

The example of $b$-$\tau$ unification raises the hope that all fermion masses arise on high scales and show simple patterns there. The other cases are not so encouraging (cf. problem 9.18).

The prospect of finding order among the fermion masses has spawned a lively theoretical industry [35]. The essential strategy comprises four steps: (1) Begin with supersymmetric SU(5), which has advantages for $\sin^2\theta_W$, coupling-constant unification, and the proton lifetime, or with supersymmetric SO(10), which accommodates a massive neutrino gracefully. (2) Find "textures"—simple patterns of Yukawa matrices—that lead to successful predictions for masses and mixing angles. (3) Interpret the textures in terms of symmetry breaking patterns or inter-family ("horizontal") symmetries. (4) Seek a derivation—or at least a motivation—for the winning entry.

## 9.7 ASSESSMENT

The appeal that a unified theory of the strong, weak, and electromagnetic interactions seems to hold was reviewed in section 9.1. Having now looked at the basic elements of the simplest example of a unified theory, we are in a position to take stock of what has been accomplished and what remains to be done. The minimal SU(5) theory has numerous desirable attributes and has successfully dealt with a number of the issues that motivate the search for a unified theory.

- It contains the standard $SU(3)_c \otimes SU(2)_L \otimes U(1)_Y$ gauge group and, thus, brings with it the attractive attributes of quantum chromodynamics and the electroweak theory. The number of independent gauge couplings is reduced from three to one.
- It puts quarks and leptons on a common footing, although more than one representation is required to contain a generation of fermions. As a consequence, the charged-current weak interactions are correctly described.
- Electric charge is quantized.
- The ratio of the $b$-quark and $\tau$-lepton masses is qualitatively understood.
- The masses of the predicted leptoquark bosons, which mediate new sorts of interactions, can be made very large, so the exotic processes are feeble effects at low energies. The leptoquark masses are small enough, compared with the Planck mass, that the neglect of gravitation seems justified.

- The implied quark–lepton transitions make proton decay possible and could give insight into the baryon excess in the universe.

The SU(5) model thus provides an existence proof for unified theories. It appears to show that a satisfying unification of the strong, weak, and electromagnetic interactions could meaningfully be achieved without gravitation. There are, however, several areas in which accomplishments fall short of the announced aspirations, and there are a number of specific problems to be faced.

- No particular insight has been gained into the nature of fermion masses or mixing angles. The SU(5) theory does not predict—or even set the scale for—neutrino masses. The meaning of CP violation in the weak interactions remains obscure.
- The number of parameters of the theory has not been materially reduced. The loss of two gauge coupling parameters and three mass parameters (corresponding to the charged leptons) is approximately compensated by the increased number of parameters in the Higgs sector of the theory.
- The three gauge couplings $\alpha_1$, $\alpha_2$, and $\alpha_3$ do not appear to unify at a single scale in the minimal SU(5) theory. The prediction for $\sin^2 \theta_W(M_Z^2)$ is close to the observed value but not in precise agreement.
- Although the idea that fermion generations are meaningful gains support, no understanding of the pattern has emerged. We still do not know why generations repeat or how many generations there are.
- It seems inelegant that each generation is assigned to a reducible representation of the gauge group.
- Gravitation is omitted, although the unification scale is only four orders of magnitude from the Planck mass.
- The most serious structural problem is the requirement that there be a dozen orders of magnitude between the electroweak and leptoquark mass scales. Is it possible to maintain the result $M_Z/M_X \ll 1$ beyond low orders of perturbation theory?
- What are the steps to unification? Is there just one, or are there several?
- Is perturbation theory a reliable guide to coupling-constant unification? How could a unified theory be constructed, if new strong dynamics should be the mechanism for electroweak symmetry breaking?
- Is the proton unstable? How does it decay?
- What sets the mass scale for the additional gauge bosons in a unified theory? For the additional Higgs bosons?
- How can we incorporate gravity?
- How are the quark doublets matched with the lepton doublets?

So far as experimental implications are concerned, unified theories provide the important reminder that we do not understand the basis of baryon-number and lepton-number conservation. Searches for nucleon decay and lepton-flavor violation, together with detailed investigations of neutrino masses, mixing, and perhaps CP violation, make up an exciting and significant complement to the exploration of the TeV scale in collider experiments.

# PROBLEMS

9.1. Consider a unified theory based on the gauge group SO(10).

(a) By referring to the $SU(5) \otimes U(1)$ decomposition of the representations of SO(10), show that each fermion generation can be accommodated in an irreducible **16**-dimensional representation, which also has a place for a left-handed antineutrino.

(b) Show that the adjoint **45** representation contains the gauge bosons of the SU(5) theory. To do so, give the transformation properties of the SO(10) gauge bosons under $SU(3) \otimes SU(2) \otimes SU(2)$, and identify the SU(5) gauge bosons among them.

(c) Now examine the branching of SO(10) into $SU(4) \otimes SU(2) \otimes SU(2)$ and the subsequent branching of SU(4) into $SU(3) \otimes U(1)$. Use the $SU(3) \otimes SU(2) \otimes SU(2)$ decomposition of the fermion representation to show that SO(10) contains the left–right symmetric electroweak group as a subgroup.

[H. Georgi, "The State of the Art—Gauge Theories," *AIP Conf. Proc.* **23**, 575 (1975); H. Fritzsch and P. Minkowski, *Ann. Phys. (NY)* **93**, 193 (1975).]

9.2. Elements of the **16** spinor representation of SO(10) can be characterized by their transformation properties under the diagonal $SO(2) \otimes SO(2) \otimes SO(2) \otimes SO(2) \otimes SO(2)$ embedding.

(a) Show that each fermion can be associated with a set of on or off (1 or 0) bits for five color charges (red, green, blue, orange, and purple), subject to the requirement that the number of on bits is even. In the notation of (9.2.7) and (9.2.8),

| Fermion | $R$ | $G$ | $B$ | $O$ | $P$ |
|---------|-----|-----|-----|-----|-----|
| $u_1$ | 1 | 0 | 0 | 1 | 0 |
| $u_2$ | 0 | 1 | 0 | 1 | 0 |
| $u_3$ | 0 | 0 | 1 | 1 | 0 |
| $d_1$ | 1 | 0 | 0 | 0 | 1 |
| $d_2$ | 0 | 1 | 0 | 0 | 1 |
| $d_3$ | 0 | 0 | 1 | 0 | 1 |
| $u_1^c$ | 0 | 1 | 1 | 0 | 0 |
| $u_2^c$ | 1 | 0 | 1 | 0 | 0 |
| $u_3^c$ | 1 | 1 | 0 | 0 | 0 |
| $d_1^c$ | 0 | 1 | 1 | 1 | 1 |
| $d_2^c$ | 1 | 0 | 1 | 1 | 1 |
| $d_3^c$ | 1 | 1 | 0 | 1 | 1 |
| $\nu$ | 1 | 1 | 1 | 1 | 0 |
| $e$ | 1 | 1 | 1 | 0 | 1 |
| $e^c$ | 0 | 0 | 0 | 1 | 1 |
| $N^c$ | 0 | 0 | 0 | 0 | 0 |

(b) Now verify that the conventional weak-hypercharge assignments are reproduced with the identification

$$Y = -\tfrac{2}{3}(R + G + B) + (O + P).$$

[F. Wilczek and A. Zee, *Phys. Rev. D* **25**, 553 (1982).]

9.3. In the usual extension of SU(5) to SO(10) discussed in problem 9.1, the standard-model gauge group $SU(3)_c \otimes SU(2)_L \otimes U(1)_Y$ is embedded directly in the Georgi–Glashow SU(5). If, instead, we consider SO(10) as a unifying group on its own, two distinct paths $SO(10) \rightarrow SU(5) \otimes U(1) \rightarrow SU(3)_c \otimes SU(2)_L \otimes U(1)_Y$ lead to the standard-model gauge group above the electroweak scale. Consider the symmetry breaking pattern $SO(10) \rightarrow SU(5)_Z \otimes U(1)_X$, with $SU(5)_Z \rightarrow SU(3)_c \otimes SU(2)_L \otimes U(1)_Z$. The weak hypercharge must be a linear combination of the additive quantum numbers associated with $U(1)_X$ and $U(1)_Z$, $Y = \alpha Z + \beta X$. $X$ commutes with the generators of SU(5) and so is proportional to the identity, $X = \text{diag}(1, 1, 1, 1, 1)$. For the $5^*$ representation, $Z = \text{diag}(\frac{2}{3}, \frac{2}{3}, \frac{2}{3}, -1, -1)$ commutes with the generators of $SU(3)_c \otimes SU(2)_L$. Evidently the Georgi–Glashow solution corresponds to $(\alpha, \beta) = (1, 0)$ and $Z \equiv Y$, with assignments of the fermions to the SU(5) representations given in (9.2.9) and (9.2.10). Find a second $SU(5)' \otimes U(1)_X$ solution for which $U(1)_Y$ is embedded in $U(1)_Z \otimes U(1)_X$, with $(\alpha, \beta) = (-\frac{1}{5}, \frac{1}{5})$. Show that for this alternative case, the fermion assignments to $SU(5)'$ multiplets are flipped from the Georgi–Glashow assignments by the interchanges $u \leftrightarrow d$, $e_L \leftrightarrow \nu_L$, and $e_L^c \leftrightarrow N_L^c$. [S. M. Barr, *Phys. Lett. B* **112**, 219 (1982).]

9.4. The groups SO(N) are generated by antisymmetric tensors $T_{ij} = -T_{ji}$ for $(i, j = 1, \ldots, N)$, which satisfy the commutation relations $[T_{ij}, T_{kl}] = i(\delta_{ik}T_{jl} - \delta_{il}T_{jk} - \delta_{jk}T_{il} + \delta_{jl}T_{ik})$. Show that for $N > 6$, the quantity $\text{tr}(\{T_{ij}, T_{kl}\}T_{mn})$ must vanish, so that all representations of $SO(N \geq 7)$ are anomaly free. This helps to explain the apparently miraculous anomaly cancellation in the SU(5) model. [H. Georgi and S. L. Glashow, *Phys. Rev. D* **6**, 429 (1972).]

9.5. As we saw in section 9.2, the 24 gauge bosons of the SU(5) unified theory include 12 new force carriers beyond the 12 familiar gluons and electroweak gauge bosons of the standard model. The double simplex [C. Quigg, "The Double Simplex," in *CP Violation and the Flavor Puzzle,* ed. D. Emmanuel-Costa et al., Poligrafia Inspektoratu, Krakow, 2006, p. 193, arXiv:hep-ph/0509037], shown here for one generation of quarks and leptons (including a right-handed neutrino),

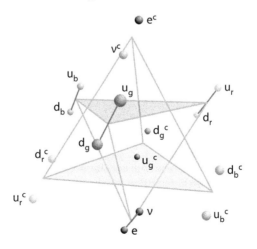

is a graphical representation that employs SO(10) as a classification symmetry. The primitive element is an SU(4) quartet, represented as a tetrahedron, embodying the suggestive notion of lepton number as the "fourth color." [J. C. Pati and A. Salam, *Phys. Rev. D* **10**, 275 (1974) [Erratum: *Phys. Rev. D* **11**, 703 (1975)].

(a) Draw links between fermions to represent the new transitions mediated by superheavy gauge bosons among members of the **5*** representation.

(b) Draw links between fermions to represent the new transitions mediated by superheavy gauge bosons among members of the **10** representation.

9.6. In the SO(10) unified theory, the quarks and leptons of each generation lie in a single representation, the **16**. Again referring to the double simplex illustrated in Problem 9.5, draw links between fermions to illustrate the transitions mediated by superheavy SO(10) gauge bosons that do not occur in either the $SU(3)_c \otimes SU(2)_L \otimes U(1)_Y$ standard model or in the SU(5) unified theory.

9.7. Consider the first stage of spontaneous symmetry breaking in the SU(5) model, namely, $SU(5) \xrightarrow{24} SU(3)_c \otimes SU(2)_L \otimes U(1)_Y$. Suppose that the effective potential for the scalars is given by

$$V(\mathbf{\Phi}^2) = \frac{\mu^2}{2} \mathrm{tr}(\mathbf{\Phi}^2) + \frac{a}{4}[\mathrm{tr}(\mathbf{\Phi}^2)]^2 + \frac{b}{2} \mathrm{tr}(\mathbf{\Phi}^4),$$

where $\mathbf{\Phi} \in \mathbf{24}$.

(a) Show that for $\mu^2 < 0$, the asymmetric vacuum state characterized by (9.2.32) corresponds to the absolute minimum of the classical potential, provided that $15a + 7b > 0$ and $b > 0$.

(b) Express the parameter $v_{24}$ of the vacuum expectation value in terms of $\mu$, $a$, and $b$.

(c) Compute the masses of the superheavy gauge bosons and Higgs scalars that appear at this stage in the symmetry breaking. [A. J. Buras, J. Ellis, M. K. Gaillard, and D. V. Nanopoulos, *Nucl. Phys.* **B135**, 66 (1978).]

9.8. Calculate separately the contributions of loops containing left-handed and right-handed electrons to the renormalization of the photon propagator.

9.9. (a) Referring to (8.3.28), which omits a small Higgs-boson contribution, characterize the variation of the $SU(2)_L$ coupling, $1/\alpha_2(Q^2)$, over the range $1\,\text{GeV} \leq Q \leq 300\,\text{GeV}$. Make use of the relation

$$1/\alpha_2 = \sin^2 \theta_W / \alpha$$

to fix the value of $1/\alpha_2(M_Z^2)$, and plot the evolution of $1/\alpha_2(Q^2)$. Explain the prominent features.

(b) Now use what is known about the evolution of $1/\alpha(Q^2)$ (e.g., S. Mele, "Measurement of the running of the electromagnetic coupling at LEP," *PoS* **HEP2005**, 286 (2006), arXiv:hep-ex/0601045; cf. figure 8.13) to predict the variation of $\sin^2 \theta_W(Q^2)$ over the same range of $Q$, and plot the result. Comment on the comparison with data presented in figure 10.1 of

Ref. [5]. [J. Erler and M. J. Ramsey-Musolf, *Phys. Rev.* D **72**, 073003 (2005).]

9.10. Suppose that experiments specify the values of the $SU(3)_c$ and $SU(2)_L$ couplings at the Z-boson mass. Neglecting the Higgs-sector contribution to coupling-constant evolution for simplicity, denote by $u$ the energy at which $1/\alpha_3(Q^2)$ and $1/\alpha_2(Q^2)$ coincide in the standard $SU(5)$ unified theory. In supersymmetric $SU(5)$, show that the energy $\tilde{u}$ at which the two couplings coincide satisfies $\tilde{u}/M_Z = (u/M_Z)^{11/9} > u/M_Z$. [S. Dimopoulos, S. Raby, and F. Wilczek, *Phys. Rev.* D **24**, 1681 (1981).]

9.11. Take as inputs the current determinations of $\alpha_1(M_Z^2)$, $\alpha_2(M_Z^2)$, and $\alpha_3(M_Z^2)$.
(a) Using the leading-logarithmic evolution equations for the minimal $SU(5)$ unified theory captured in table 9.2, plot $\alpha_1^{-1}$, $\alpha_2^{-1}$, and $\alpha_3^{-1}$ from the Z-boson scale up to $10^{16}$ GeV and comment on the degree to which the coupling constants unify.
(b) Repeat the exercise using the supersymmetric evolution equations summarized in table 9.2, first neglecting the Higgs-sector contributions by setting $N_H = 0$.
(c) Now repeat the exercise for the minimal supersymmetric $SU(5)$ case, with $N_H = 2$. Comment on the implications of the unification scale, should you find one. [U. Amaldi, W. de Boer, and H. Fürstenau, *Phys. Lett.* B **260**, 447 (1991).]

9.12. Take as inputs the current determinations of $\alpha_1(M_Z^2)$, $\alpha_2(M_Z^2)$, and $\alpha_3(M_Z^2)$. Now consider augmenting minimal $SU(5)$ unification by incorporating several Higgs doublets.
(a) Using the leading-logarithmic evolution equations captured in table 9.2, plot $\alpha_1^{-1}$, $\alpha_2^{-1}$, and $\alpha_3^{-1}$ from the Z-boson scale up to $10^{16}$ GeV and comment on the degree to which the coupling constants unify, for $N_H = 3, 6, 9$.
(b) Comment on the implications of the unification scale, should you find one, and on challenges raised by the presence of many physical Higgs bosons. [S. Willenbrock, *Phys. Lett.* B **561**, 130 (2003).]

9.13. (a) Neglecting small Higgs-scalar contributions to the evolution of $\alpha_1$ and $\alpha_2$ and working to leading logarithmic approximation, show that the $SU(5)$ unified theory predicts

$$SU(5): \quad \sin^2\theta_W(M_Z^2) = \frac{1}{6} + \frac{5}{9} \cdot \frac{\alpha(M_Z^2)}{\alpha_s(M_Z^2)}.$$

A straightforward strategy is to seek solutions $(A, B)$ to the general relation,

$$\frac{A}{\alpha_2(M_Z^2)} - \frac{1}{\alpha(M_Z^2)} = \frac{B}{\alpha_s(M_Z^2)}.$$

(b) Using current determinations of $\alpha(M_Z^2)$ and $\alpha_s(M_Z^2)$, compute the $SU(5)$ prediction for $\sin^2\theta_W(M_Z^2)$ and compare with experiment.

(c) Now repeat the exercise for supersymmetric SU(5), including the contributions of the minimal two Higgs doublets. Show that

$$\text{SUSY SU(5):} \quad \sin^2\theta_W(M_Z^2) = \frac{1}{5} + \frac{7}{15}\cdot\frac{\alpha(M_Z^2)}{\alpha_s(M_Z^2)}.$$

(d) Again, use current determinations of $\alpha(M_Z^2)$ and $\alpha_s(M_Z^2)$ to compute the prediction for $\sin^2\theta_W(M_Z^2)$ and compare with experiment.

9.14. Imagine that the standard-model couplings $\alpha_1$, $\alpha_2$, and $\alpha_3$ unify at an energy scale $u$ in the SU(5) unified theory. Compute the evolution of the unified coupling $\alpha_5^{-1}(Q^2)$ above the unification scale, that is, for $Q \geq u$.

9.15. Suppose that a unified theory, SU(5) for definiteness, fixes the value of the unification scale, $u$, and the common strength of the couplings, $\alpha_u$, at that scale. The values of the coupling constants that we measure on a low scale have encrypted in them information about the spectrum of particles between our energy scale and $u$. Assume that there are no particles in that range beyond those we know from the standard model. How is the value of the strong coupling constant $\alpha_s = \alpha_3$ at low energies influenced by the mass of the top quark? What is the effect on the proton mass? [C. Quigg, *Phys. Today* 50N5, 20–26 (1997), hep-ph/9704332.]

9.16. (a) Working to leading logarithmic approximation, characterize the evolution of the strong coupling $\alpha_3^{-1}(Q^2)$ below and above a sharp threshold $\tilde{Q}$ at which a full complement of superpartners becomes active.
(b) Suppose now that $\tilde{Q} = 1$ TeV. Describe an experimental strategy to measure the change in slope of the evolution equation for $\alpha_3^{-1}(Q^2)$ at the Large Hadron Collider.

9.17. (a) Working to leading logarithmic approximation, characterize the evolution of the weak mixing parameter $\sin^2\theta_W(Q^2)$ below and above a sharp threshold $\tilde{Q}$ at which a full complement of superpartners becomes active.
(b) Suppose now that $\tilde{Q} = 1$ TeV. Describe an experimental strategy to measure the change in slope of the evolution equation for $\sin^2\theta_W(Q^2)$ at the Large Hadron Collider.

9.18. (a) Choosing an observation scale $\mu \approx 1$ GeV, compute the low-energy values of $m_s/m_\mu$ and $m_d/m_e$ in the SU(5) unified theory described in section 9.2, and compare with experiment. Make use of the (renormalization group) evolution equations quoted in (9.6.2)–(9.6.4).
(b) A more elaborate symmetry breaking scheme that adds a 45 of scalars can change the relation for $m_e/m_d$ at the unification scale and lead to a more agreeable result at low energies. Show that the relations $m_s = \frac{1}{3}m_\mu$, $m_d = 3m_e$ at the unification scale lead to the low-energy predictions, $m_s \approx \frac{4}{3}m_\mu$ and $m_d \approx 12m_e$.

9.19. One generation of the known quarks and leptons (including a right-handed neutrino) fills out the spinor 16 representation of SO(10). A larger unifying group could inspire the search for "exotic" fermions, beyond those already known. Consider the exceptional group $E_6$, an anomaly-free group that contains SO(10). Beginning from the SO(10) decomposition of the fundamental representation of $E_6$,

$$E_6: \quad \mathbf{27} = \mathbf{16} \oplus \mathbf{10} \oplus \mathbf{1} \quad : SO(10),$$

assign the exotic fermions to representations of SU(5), and characterize their quantum numbers $(SU(3)_c, SU(2)_L)_Y$. Specify the color, electric charge, and weak isospin of each new fermion. [J. L. Rosner, *Phys. Rev. D* **61**, 097303 (2000).]

## ⅢⅢⅢⅢⅢ FOR FURTHER READING ⅢⅢⅢⅢⅢⅢⅢ

**Unified theories.** Many influential early papers are reprinted in
R. N. Mohapatra and C. H. Lai (Editors), *Selected Papers on Gauge Theories of the Fundamental Interactions,* World Scientific, Singapore, 1981.

Among recent reviews of specific models and their predictions, see
G. G. Ross, *Grand Unified Theories,* Westview Press, Boulder, CO, 2003,
R. N. Mohapatra, *Unification and Supersymmetry,* 3rd ed., Springer, Heidelberg, 2003,
S. Raby, "Grand Unified Theories," in Ref. [5], p. 193,
J. C. Pati, *Int. J. Mod. Phys.* **D15**, 1677 (2006),
P. Langacker, *The Standard Model and Beyond,* CRC Press, Boca Raton, FL, 2010.

**SO(10) unification.** For a look at minimal SO(10) grand unification, after the first wave of searches for proton decay, see
L. Lavoura and L. Wolfenstein, *Phys. Rev. D* **48**, 264 (1993).

Very explicit analyses of SO(10) gauge bosons and interactions appear in
A. D. Özer, "SO(10)-Grand Unification and Fermion Masses," Dissertation, LMU München: Faculty of Physics (2005), http://j.mp/ywxjUb.

**$E_6$ unification.** Unification based on the exceptional group $E_6$ was proposed in
F. Gürsey and P. Sikivie, *Phys. Rev. Lett.* **36**, 775 (1976).

For an extensive discussion of the extra interactions that occur in $E_6$-based models, see
D. London and J. L. Rosner, *Phys. Rev. D* **34**, 1530 (1986).

**Supersymmetric SU(5) unification.** An influential investigation of the consequences of supersymmetry for the unification scale is
S. Dimopoulos, S. Raby, and F. Wilczek, *Phys. Rev. D* **24**, 1681 (1981).

For an insightful discussion of coupling constant unification including supersymmetric SU(5), see
S. Dimopoulos, S. A. Raby, and F. Wilczek, "Unification of Couplings," *Phys. Today* **44N10**, 25 (1991).

**Mass scales in unified theories.** A general analysis for SU(5), SO(10), and $E_6$ unification theories is given in
R. Robinett and J. L. Rosner, *Phys. Rev. D* **26**, 2396 (1982).

**New gauge bosons.** When broken down to the residual $SU(3)_c \otimes U(1)_{EM}$ symmetry, unifying groups larger than $SU(5)$ typically lead to additional $U(1)$ factors that imply new neutral gauge bosons, perhaps accessible to accelerator experiments. A general formalism was set out by

H. Georgi and S. Weinberg, *Phys. Rev. D* **17**, 275 (1978).

Some possibilities may be traced through

P. Langacker, *Rev. Mod. Phys.* **81**, 1199 (2009),

P. Langacker, R. W. Robinett, and J. L. Rosner, *Phys. Rev. D* **30**, 1470 (1984),

J. L. Hewett and T. G. Rizzo, *Phys. Rept.* **183**, 193 (1989),

M. S. Carena, A. Daleo, B. A. Dobrescu, and T.M.P. Tait, *Phys. Rev. D* **70**, 093009 (2004).

**Partial unification.** Various scenarios for reducing the three standard-model coupling constants to two independent parameters by embedding the $SU(3)_c \otimes SU(2)_L \otimes U(1)_Y$ gauge group in a product $G_s \otimes G_w$ of strong and weak gauge groups have been studied in

P. Q. Hung, J. D. Bjorken, and A. J. Buras, *Phys. Rev. D* **25**, 805 (1982).

**Supersymmetry.** An accessible introduction to supersymmetry, which relates fermionic and bosonic degrees of freedom, is given by

J. Wess and J. Bagger, *Supersymmetry and Supergravity*, 2nd ed., Princeton University Press, Princeton, NJ, 1992.

For a modern take on supersymmetry, see

S. P. Martin, "A Supersymmetry Primer," version 6, arXiv:hep-ph/9709356 (September 2011).

For compact surveys, see

H. E. Haber, "Supersymmetry (Theory)," p. 1292, and J.-F. Grivaz, "Supersymmetry (Experiment)," p. 1309, in Ref. [5].

Among recent monographs, see

I. J. R. Aitchison, *Supersymmetry in Particle Physics*, Cambridge University Press, Cambridge, 2007,

J. Terning, *Modern Supersymmetry*, Oxford University Press, Oxford, 2005,

H. Baer and X. Tata, *Weak Scale Supersymmetry*, Cambridge University Press, Cambridge, 2006,

M. Drees, P. Roy, and R. M. Godbole, *Theory and Phenomenology of Sparticles*, World Scientific, Singapore, 2004,

S. Weinberg, *The Quantum Theory of Fields* vol. 3, Cambridge University Press, Cambridge, 2005.

**Group theory for unified model building.** Group theory texts that reflect a contemporary physicist's orientation include

W.-K. Tung, *Group Theory in Physics*, World Scientific, Singapore, 1985,

R. N. Cahn, *Semi-Simple Lie Algebras and Their Representations*, Dover Publications, Mineola, NY, 2006,

P. Ramond, *Group Theory: A Physicist's Survey*, Cambridge University Press, Cambridge, 2010.

Convenient sources for the properties of groups and their representations are

J. Patera and D. Sankoff, *Tables of Branching Rules for Representations of Simple Algebras*, Université de Montréal, Montréal, 1973,

W. McKay and J. Patera, *Tables of Dimensions, Indices, and Branching Rules for Representations of Simple Algebras,* Dekker, New York, 1981.

Specific discussions of unification are given by
R. Slansky, *Phys. Rep.* **79**, 1 (1981),
H. Georgi, *Lie Groups in Particle Physics,* Benjamin, Reading, MA, 1982.

A very useful group-theoretical analysis of spontaneous symmetry breaking appears in
L.-F. Li, *Phys. Rev. D* **9**, 1723 (1974).

**Gauge hierarchies.** The problem of sustaining widely different mass scales in spontaneously broken theories is studied in
E. Gildener, *Phys. Rev. D* **14**, 1667 (1976),
S. Weinberg, *Phys. Lett.* **82B**, 387 (1979),
L. Susskind, *Phys. Rev. D* **20**, 2619 (1979).

**Nucleon instability.** Informative discussions of baryon-number and lepton-number conservation include
M. Goldhaber, P. Langacker, and R. Slansky, *Science* **210**, 851 (1980),
"Note on Nucleon Decay" in L. Montanet et al., [Particle Data Group], *Phys. Rev. D* **50**, 1173 (1994),
P. Nath and P. Fileviez Perez, *Phys. Rept.* **441**, 191 (2007).

How to deal with backgrounds to sensitive experiments underground is the subject of
J. A. Formaggio and C. J. Martoff, *Ann. Rev. Nucl. Part. Sci.* **54**, 361 (2004).

The evolution of detectors may be traced through the series of
International Workshops on Next generation Nucleon Decay and Neutrino Detectors, http://j.mp/WsODBD.

A proposal for a next-generation detector is set out in
K. Abe et al., "Letter of Intent: The Hyper-Kamiokande Experiment—Detector Design and Physics Potential—," arXiv:1109.3262.

**Cosmology.** For detailed accounts of inflationary cosmology, see
E. W. Kolb and M. S. Turner, *The Early Universe,* Addison–Wesley, Redwood City, CA, 1990,
A. R. Liddle and D. H. Lyth, *Cosmological Inflation and Large-Scale Structure,* Cambridge University Press, Cambridge, 2000,
J. A. Peacock, *Cosmological Physics,* Cambridge University Press, Cambridge, 1998.

An authoritative popular account is given by
A. Guth, *The Inflationary Universe,* Basic Books, New York, 1998.

Valuable introductions to the *concordance cosmology* are given by
S. Dodelson, *Modern Cosmology,* Academic Press, New York, 2003,
S. Weinberg, *Cosmology,* Oxford University Press, Oxford, 2008.

For a crisp critique of the incompleteness of the standard cosmology, see
P. J. Steinhardt, "The Inflation Debate," *Sci. Am.* **304** (4), 36 (April 2011).

**Baryon number of the universe.** The link with baryon- and lepton-number nonconservation has been understood in general terms for some time. See, for example, the remark on p. 482 of

S. Weinberg, in *Lectures in Particles and Field Theory*, 1964 Brandeis Lectures, ed. S. Deser and K. Ford, Prentice Hall, Englewood Cliffs, NJ, 1965, p. 405.

A specific scenario for baryogenesis was given by

A. D. Sakharov, *ZhETF Pis'ma* **5**, 32 (1967) [English translation: *JETP Lett.* **5**, 24 (1967)].

The issue was reopened in the context of gauge theories by

M. Yoshimura, *Phys. Rev. Lett.* **41**, 281 (1978); *Phys. Rev. Lett.* **42**, 746E (1979).

The importance of the nonequilibrium condition was emphasized in

D. Toussaint, S. B. Treiman, F. Wilczek, and A. Zee, *Phys. Rev. D* **19**, 1036 (1979).

For an interesting detailed investigation, see

E. W. Kolb and S. Wolfram, *Nucl. Phys.* **B172**, 224 (1980).

For recent reviews, see

A. Riotto and M. Trodden, *Ann. Rev. Nucl. Part. Sci.* **49**, 35 (1999),
J. M. Cline, "Baryogenesis," in *Particle Physics and Cosmology: The Fabric of Spacetime,* ed. F. Bernardeau, C. Grojean, and J. Dalibard, Elsevier, Amsterdam, 2007, p. 53, arXiv:hep-ph/0609145.

A popular account is given by

H. R. Quinn and Y. Nir, *The Mystery of the Missing Antimatter,* Princeton University Press, Princeton, 2007.

**Electroweak baryogenesis.** For an early review, see

M. Trodden, *Rev. Mod. Phys.* **71**, 1463 (1999).

Recent developments may be traced in

A. Menon, D. E. Morrissey, and C.E.M. Wagner, *Phys. Rev. D* **70**, 035005 (2004),
J. Kang, P. Langacker, T.-j. Li, and T. Liu, *Phys. Rev. Lett.* **94**, 061801 (2005),
T. Konstandin, T. Prokopec, M. G. Schmidt, and M. Seco, *Nucl. Phys.* **B738**, 1 (2006).

**Leptogenesis.** The hypothesis that a primordial lepton–antilepton generated early in the history of the universe is the ultimate cause for the observed baryon excess is under very active development. For surveys, see

W. Buchmüller, R. D. Peccei, and T. Yanagida, *Ann. Rev. Nucl. Part. Sci.* **55**, 311 (2005),
M.-C. Chen, "TASI 2006 Lectures on Leptogenesis," in *Colliders and Neutrinos,* ed. S. Dawson and R. N. Mohapatra, World Scientific, Singapore, 2008, p. 123, arXiv:hep-ph/0703087,
S. Davidson, E. Nardi and Y. Nir, *Phys. Rept.* **466**, 105 (2008).

**Magnetic monopoles.** An excellent account of the Dirac monopole is given by

E. Amaldi and N. Cabibbo, *Aspects of Quantum Theory*, ed. A. Salam and E. P. Wigner, Cambridge University Press, Cambridge, 1972, p. 183.

The monopoles that appear in unified theories are reviewed by

P. Goddard and D. Olive, *Rep. Prog. Phys.* **41**, 1357 (1978).

General theoretical background and the state of monopole searches can be found in

K. A. Milton, *Rept. Prog. Phys.* **69**, 1637 (2006),
D. Milstead and E. J. Weinberg, "Magnetic Monopoles," in Ref. [5], p. 1285.

## ⅢⅢⅢⅢⅢⅢⅢⅢ REFERENCES ⅢⅢⅢⅢⅢⅢⅢⅢⅢ

1. H. Georgi and S. L. Glashow, *Phys. Rev. Lett.* **32**, 438 (1974).
2. Issues of unification and proton stability were raised in the context of $SU(2)_L \otimes SU(2)_R \otimes SU(4)_c$ gauge theories by J. C. Pati and A. Salam, *Phys. Rev. D* **8**, 1240 (1973); *Phys. Rev. Lett.* **31**, 661 (1973); *Phys. Rev. D* **10**, 275 (1974).
3. R. N. Mohapatra and J. C. Pati, *Phys. Rev. D* **11**, 566 (1975), *Phys. Rev. D* **11**, 2558 (1975); G. Senjanovic and R. N. Mohapatra, *Phys. Rev. D* **12**, 1502 (1975).
4. C. A. Gagliardi, R. E. Tribble, and N. J. Williams, *Phys. Rev. D* **72**, 073002 (2005); R. Bayes et al., [TWIST Collaboration], *Phys. Rev. Lett.* **106**, 041804 (2011); J. F. Bueno et al., [TWIST Collaboration], *Phys. Rev. D* **84**, 032005 (2011). See also the minireview on $W'$ searches by M.-C. Chen and B. A. Dobrescu in Ref. [5].
5. K. Nakamura et al., [Particle Data Group], *J. Phys. G* **37**, 075021 (2010). Consult J. Beringer et al., [Particle Data Group], *Phys. Rev. D* **86**, 010001 (2012) for updates.
6. For a recent theoretical review, see R. N. Mohapatra, *J. Phys. G* **36**, 104006 (2009), arXiv:0902.0834. For a review of free-neutron oscillation experiments, see D. Dubbers, *Progr. Part. Nucl. Phys.* **26**, 173 (1991). See also M. Baldo-Ceolin et al., *Z. Phys. C* **63**, 409 (1994). Future experiments are described by W. M. Snow, *Nucl. Instr. Meth.* **A611**, 144 (2009). A recent search is described in K. Abe et al., [Super-Kamiokande Collaboration], "The search for $n$-$\bar{n}$ oscillation in Super-Kamiokande I," arXiv:1109.4227.
7. For compact reviews, see §9 of Y. Nir, "CP violation in and beyond the standard model," in *Proceedings of the 27th SLAC Summer Institute on CP Violation: In and Beyond the Standard Model*, SLAC Report SLAC-R-719, chap. 1, arXiv:hep-ph/9911321; and §17.1 of I. I. Bigi and A. I. Sanda, *CP Violation*, 2nd ed., Cambridge University Press, Cambridge, 2009.
8. R. P. Feynman, *Lectures on Physics* (Basic Books, New York, 2010), vol. 1, chap. 2.
9. N. Cabibbo, E. C. Swallow, and R. Winston, *Annu. Rev. Nucl. Part. Sci.* **53**, 39 (2003), review the evidence for universality derived from the study of semileptonic hyperon decays. A. Pich, *Nucl. Phys. B, Proc. Suppl.* **181–182**, 300 (2008), takes the measure of lepton universality tests. Relevant measurements are collected in Ref. [5].
10. On the assumption that the neutron charge is equal to the sum of the proton and electron charges, H. F. Dylla and G. King, *Phys. Rev. A* **7**, 1224 (1973), have demonstrated the neutrality of matter at the level of $|Q(p) + Q(e)| < 10^{-21}e$. Their article contains a review of earlier experimental work.
11. This is explained in S. Coleman, "Classical Lumps and Their Quantum Descendants," in *Aspects of Symmetry*, Cambridge University Press, Cambridge, 1985, p. 185. See, in particular, §3.
12. J. Banks and H. Georgi, *Phys. Rev. D* **14**, 1159 (1976). See also S. Okubo, *Phys. Rev. D* **16**, 3528 (1977). Very general and convenient methods for evaluating Casimir

operators are due to A. M. Perelomov and V. S. Popov, *Yad. Fiz.* **5**, 693 (1967); **7**, 460 (1968) [English translation: *Sov. J. Nucl. Phys.* **5**, 489 (1967); **7**, 290 (1968)].

13. H. Nishino et al., [Super-Kamiokande Collaboration], *Phys. Rev. Lett.* **102**, 141801 (2009).

14. S. N. Ahmed et al., [SNO Collaboration], *Phys. Rev. Lett.* **92**, 102004 (2004).

15. A. J. Buras, J. Ellis, M. K. Gaillard, and D. V. Nanopoulos, *Nucl. Phys.* **B135**, 66 (1978).

16. H. Georgi, H. R. Quinn, and S. Weinberg, *Phys. Rev. Lett.* **33**, 351 (1974).

17. T. Appelquist and J. Carazzone, *Phys. Rev. D* **11**, 2856 (1975).

18. A less brutal treatment of the thresholds is clearly desirable. For early precise formulations of the evolution problem, see D. A. Ross, *Nucl. Phys.* **B140**, 1 (1978); I. Antoniadis, C. Bouchiat, and J. Iliopoulos, *Phys. Lett.* **97B**, 367 (1980). A recent careful treatment, in the context of supersymmetric unification, is given by S. Raby, M. Ratz, K. Schmidt-Hoberg, *Phys. Lett.* **B687**, 342 (2010).

19. A convenient source for leading-order evolution (and beyond) is M. B. Einhorn and D. R. T. Jones, *Nucl. Phys. B* **196**, 475 (1982).

20. The ALEPH, DELPHI, L3, OPAL, SLD Collaborations, the LEP Electroweak Working Group, the SLD Electroweak and Heavy Flavour Groups, *Phys. Rept.* **427**, 257 (2006). For updates to the analysis of precision electroweak measurements, see http://j.mp/lu9g5s.

21. F. Jegerlehner, *Nucl. Phys. Proc. Suppl.* **181–182**, 135 (2008), arXiv:0807.4206.

22. G. 't Hooft, *Phys. Rev. Lett.* **37**, 8 (1976). See also V. A. Kuzmin, V. A. Rubakov, M. E. Shaposhnikov, *Phys. Lett.* **B155**, 36 (1985). A strong first-order thermal phase transition in the early universe is generally required.

23. G. N. Flerov, D. S. Klochkov, V. S. Skobkin, and V. V. Terent'ev, *Sov. Phys. Dokl.* **3**, 79 (1958); F. W. Dix, "Search for proton decay as a test of baryon conservation," Case Western Reserve University thesis (1970).

24. Y. Fukuda et al., *Nucl. Instrum. Meth.* **A501**, 418 (2003).

25. For recent lattice-QCD evaluations of the proton-decay matrix elements, with summaries of earlier work, see Y. Aoki, C. Dawson, J. Noaki and A. Soni, *Phys. Rev. D* **75**, 014507 (2007); Y. Aoki et al., [RBC–UKQCD Collaboration], *Phys. Rev. D* **78**, 054505 (2008). See also V. M. Braun et al., *Phys. Rev. D* **79**, 034504 (2009).

26. See, for example, the operator analyses by S. Weinberg, *Phys. Rev. Lett.* **43**, 1566 (1979), and by F. Wilczek and A. Zee, *Phys. Rev. Lett.* **43**, 1571 (1979).

27. P. Cooney, "Proton decay matrix elements from lattice QCD," Edinburgh Thesis (2010), http://j.mp/imLSKR.

28. For a discussion of the new (dimension-5) operators that mediate proton decay in supersymmetric models, see S. Dimopoulos, S. Raby, and F. Wilczek, *Phys. Lett.* **B112**, 133 (1982). General analyses of proton decay in the context of supersymmetric unified theories are given by S. Weinberg, *Phys. Rev. D* **26**, 287 (1982); N. Sakai and T. Yanagida, *Nucl. Phys.* **B197**, 533 (1982).

29. M. Bass et al., [Long-Baseline Neutrino Experiment Science Collaboration Physics Working Group and others], "A Study of the Physics Potential of the Long-Baseline Neutrino Experiment Project with an Extensive Set of Beam, Near Detector and Far Detector Configurations," LBNE-PWG-002, INT-PUB-11-002, http://j.mp/ifaCL2. For surveys of the theoretical environment, see S. Raby et al., "DUSEL Theory White Paper," arXiv:0810.4551; G. Senjanovic, "Proton Decay and Grand Unification," *AIP Conf. Proc.* **1200**, 131 (2010), arXiv:0912.5375.

30. A. A. Penzias and R. W. Wilson, *Astrophys. J.* **142**, 419 (1965); the interpretation is due to R. H. Dicke, P.J.E. Peebles, P. G. Roll, and D. T. Wilkinson, *Astrophys. J.* **142**, 414 (1965). See also the Nobel lectures by A. A. Penzias, *Rev. Mod. Phys.* **51**, 425 (1979) and R. W. Wilson, *Rev. Mod. Phys.* **51**, 433 (1979).

31. For accessible introductions see S. Singh, *Big Bang*, HarperCollins, New York, 2004; J. Silk, *The Big Bang*, 3rd ed., Holt, New York, 2001; S. Weinberg, *The First Three Minutes*, 2nd ed., Basic Books, New York, 1993. The original "hot big bang" picture must now be supplemented with an inflationary epoch, dark matter, and dark energy.

32. The observational case for the absence of antimatter is reviewed by P. Coppi, "How Do We Know Antimatter Is Absent?," in *Proceedings of the 32nd SLAC Summer Institute on Particle Physics (SSI 2004)*, p. L017 (2004), http://j.mp/jz4ZjB. The bounds on the baryon-to-photon ratio are due to B. D. Harris and S. Sarkar, "Big-Bang Nucleosynthesis," in Ref. [5], p. 241. See G. Steigman, *Ann. Rev. Nucl. Part. Sci.* 57, 463–491 (2007) for elaboration of the nucleosynthesis constraints. The Alpha Magnetic Spectrometer on the International Space Station promises to search for antimatter near Earth with greatly increased sensitivity: R. Battiston et al., [AMS-02 Collaboration], *Nucl. Instr. Meth.* A588, 227 (2008).

33. M. Fukugita and T. Yanagida, *Phys. Lett.* B174, 45 (1986).

34. Z.-z. Xing, H. Zhang, and S. Zhou, *Phys. Rev. D* 77, 113016 (2008). See also A. V. Manohar and C. T. Sachrajda, "Quark Masses," in Ref. [5], p. 583.

35. For recent reviews of unified models for fermion masses and mixings, with an emphasis on supersymmetric examples, see M.-C. Chen and K. T. Mahanthappa, *Int. J. Mod. Phys. A* 18, 5819 (2003), arXiv:hep-ph/0305088; K. S. Babu, "TASI Lectures on Flavor Physics," arXiv:0910.2948. Examples from the original literature include K. S. Babu, J. C. Pati, and F. Wilczek, *Nucl. Phys.* B566, 33 (2000), and M. Albrecht, W. Altmannshofer, A. J. Buras, D. Guadagnoli, and D. M. Straub, *JHEP* 0710, 055 (2007).

# Epilogue ||||||||||||||||||||||||||||||||||||||||||||||||||||||||||||||||||||||||||||||||||||||||||||||||||||

The preceding chapters have exhibited the promise and power of gauge theories and shown our description of the fundamental particles and the interactions among them to be in a very provocative state. Gauge theories unquestionably provide us with an extraordinarily unified and unifying language for the description of natural phenomena. Two new laws of nature—the electroweak theory and quantum chromodynamics—summarize a simple and coherent conception of an unprecedented range of natural phenomena, but that new picture raises captivating questions. We stand on the threshold of a higher level of understanding, with the nature of electroweak symmetry breaking virtually certain to be revealed by experiments on the 1-TeV scale.

The widespread accord between observations and expectations may cause us to wonder at the unreasonable effectiveness of the standard model of particle physics. At the same time, the incompleteness of the electroweak theory argues that we have much more to learn. We have only begun to explore the richness of QCD under diverse conditions. Moreover, the universe around us testifies to phenomena that—for now—lie outside the explanatory reach of our theories. We have encountered many outstanding issues in the concluding sections of the chapters on the electroweak theory, quantum chromodynamics, and unified theories. It is worth citing a few big questions—far from an exhaustive list—that particle physics can address in the near future.

1. Are quarks and leptons elementary?
2. What makes a top quark a top quark, an electron an electron, and a neutrino a neutrino?
3. What do fermion generations signify? Will flavor or family symmetries give insights into fermion masses and mixings? Are there additional generations of quarks and leptons?
4. Does the different behavior of left-handed and right-handed fermions with respect to charged-current weak interactions reflect a fundamental asymmetry in the laws of nature? Can we detect right-handed weak interactions? Is the electroweak theory part of some larger edifice?
5. What hides the electroweak symmetry? Specifically, is there a Higgs boson? Might there be several? Is the Higgs boson elementary or composite? How does the Higgs boson interact with itself? What triggers electroweak symmetry breaking?
6. Does the Higgs field give mass to fermions, or only to the weak bosons? What sets the masses and mixings of the quarks and leptons? (How) is fermion mass related to the electroweak scale?

7.  At what scale are neutrino masses set? What is the nature of the right-handed neutrino? Is the neutrino its own antiparticle?
8.  Why is empty space so nearly weightless?
9.  What separates the electroweak scale from distant scales?
10. Where are the flavor-changing neutral currents?
11. What new phenomena are to be found in strong interactions?
12. What resolves the strong CP problem?
13. What is the relationship of quarks to leptons?
14. What is the (grand) unifying symmetry? What are the steps to unification? Is there just one, or are there several? Do new gauge interactions link quarks and leptons?
15. What lessons does electroweak symmetry breaking hold for unified theories of the strong, weak, and electromagnetic interactions?
16. Is the proton unstable? How does it decay? What explains the baryon asymmetry of the universe?
17. What symmetry will next be recognized in nature? Is nature supersymmetric? If so, why is the world built of fermions, not bosons—that is, quarks, not squarks, and leptons, not sleptons?
18. Are there novel kinds of matter? Are there new gauge interactions? New forces of a novel kind?
19. Can we observe additional spacetime dimensions?
20. What constitutes the dark matter?

To these I would add an essential metaquestion: How are we prisoners of conventional thinking?

I hope in these chapters not only to have communicated a few facts and conveyed a point of view, but also to have evoked an awareness that there is much to be done that is significant and exciting. In theory and experiment alike, there are many opportunities to contribute to the numinous intellectual adventure in which we are privileged to share.

# Appendix A ||||||||||||||||||||||||||||||||||||||||||||||||||||||||||||||||||||||||||||||||||||||||||||||||||||||

## Notations and Conventions

This book conforms closely to the conventions of Bjorken and Drell [1], except for the normalization of Dirac spinors [2]. It will generally be advantageous to adopt natural units in which the quantities $\hbar$ and $c$ are set equal to unity. Appendix B sets out phase-space formulas and the path from Feynman rules to observables. Numerical values and conversion factors are given in appendix C.

## A.1 FOUR-VECTORS AND SCALAR PRODUCT

Four-vectors are printed in italic type: $a$; (spatial) three-vectors are in boldface: $\mathbf{a}$. Spacetime coördinates $(t; x, y, z) = (t; \mathbf{x})$ are denoted by the contravariant four-vector

$$x^\mu \equiv (t; x, y, z) \equiv (x^0; x^1, x^2, x^3). \tag{A.1.1}$$

The metric tensor

$$g_{\mu\nu} = \begin{pmatrix} 1 & 0 & 0 & 0 \\ 0 & -1 & 0 & 0 \\ 0 & 0 & -1 & 0 \\ 0 & 0 & 0 & -1 \end{pmatrix} \tag{A.1.2}$$

generates the covariant four-vector

$$x_\mu \equiv (x_0; x_1, x_2, x_3) \equiv g_{\mu\nu}x^\nu = (t; -x, -y, -z). \tag{A.1.3}$$

Unless specifically indicated to the contrary, repeated indices are summed.
The scalar product is

$$x^2 \equiv x^\mu \cdot x_\mu = t^2 - \mathbf{x}^2. \tag{A.1.4}$$

Thus a general scalar product is

$$a \cdot b = a^\mu b_\mu = a^0 b_0 - \mathbf{a} \cdot \mathbf{b}. \tag{A.1.5}$$

Momentum vectors are written as

$$p^\mu = (E; p_x, p_y, p_z). \tag{A.1.6}$$

A convenient notation for the four-gradient is

$$\partial^\mu = \frac{\partial}{\partial x_\mu} \quad \text{or} \quad \partial_\mu = \frac{\partial}{\partial x^\mu}, \tag{A.1.7}$$

in terms of which the position–space momentum operator is

$$\mathsf{p}^\mu = i\partial^\mu = (i\partial/\partial t; -i\nabla). \tag{A.1.8}$$

Thus

$$\mathsf{p}^\mu \mathsf{p}_\mu = -\partial^\mu \partial_\mu = -[(\partial^2/\partial t^2) - \nabla^2] = -\Box, \tag{A.1.9}$$

where $\nabla^2$ is the Laplacian and $\Box$ is the d'Alembertian operator.

## A.2 DIRAC MATRICES

The Dirac $\gamma$-matrices satisfy the anticommutation relations

$$\{\gamma^\mu, \gamma^\nu\} \equiv \gamma^\mu \gamma^\nu + \gamma^\nu \gamma^\mu = 2g^{\mu\nu}, \tag{A.2.1}$$

so that

$$\gamma^\mu \gamma_\mu = 4 \cdot \mathbb{I}, \tag{A.2.2}$$

where $\mathbb{I}$ is the $4 \times 4$ identity matrix. In contexts in which the matrix character of $\mathbb{I}$ is obvious, it will be convenient simply to write 1. Other useful identities follow at once from the commutation relations:

$$[\gamma^\mu \gamma^\nu, \gamma^\rho] \equiv \gamma^\mu \gamma^\nu \gamma^\rho - \gamma^\rho \gamma^\mu \gamma^\nu = 2(\gamma^\mu g^{\nu\rho} - \gamma^\nu g^{\mu\rho}); \tag{A.2.3}$$

$$\gamma^\mu \gamma_\nu \gamma_\mu = -2\gamma_\nu; \tag{A.2.4}$$

$$\gamma^\mu \gamma_\nu \gamma_\rho \gamma_\mu = 4g_{\nu\rho}; \tag{A.2.5}$$

$$\gamma^\mu \gamma_\nu \gamma_\rho \gamma_\sigma \gamma_\mu = -2\gamma_\sigma \gamma_\rho \gamma_\nu; \tag{A.2.6}$$

$$\gamma^\mu \gamma_\nu \gamma_\rho \gamma_\sigma \gamma_\tau \gamma_\mu = 2(\gamma_\tau \gamma_\nu \gamma_\rho \gamma_\sigma + \gamma_\sigma \gamma_\rho \gamma_\nu \gamma_\tau). \tag{A.2.7}$$

When an explicit representation is required, we shall adopt

$$\gamma^0 = \begin{pmatrix} I & 0 \\ 0 & -I \end{pmatrix}, \quad \boldsymbol{\gamma} = \begin{pmatrix} 0 & \boldsymbol{\sigma} \\ -\boldsymbol{\sigma} & 0 \end{pmatrix}, \tag{A.2.8}$$

where $I$ is the $2 \times 2$ identity matrix and the $2 \times 2$ Pauli matrices are given by

$$\sigma_x = \begin{pmatrix} 0 & 1 \\ 1 & 0 \end{pmatrix}; \quad \sigma_y = \begin{pmatrix} 0 & -i \\ i & 0 \end{pmatrix}; \quad \sigma_z = \begin{pmatrix} 1 & 0 \\ 0 & -1 \end{pmatrix}. \tag{A.2.9}$$

The Pauli matrices satisfy commutation and anticommutation relations,

$$[\sigma_i, \sigma_j] = 2i\varepsilon^{ijk}\sigma_k, \tag{A.2.10}$$

$$\{\sigma_i, \sigma_j\} = 2\delta_{ij}\, I, \tag{A.2.11}$$

where the antisymmetric three-index symbol takes the values

$$\varepsilon^{ijk} = \begin{cases} +1, & \text{for even permutations of 123,} \\ -1, & \text{for odd permutations,} \\ 0, & \text{otherwise.} \end{cases} \tag{A.2.12}$$

Summing (A.2.10) and (A.2.11) leads to the useful identity,

$$\boldsymbol{\sigma} \cdot \mathbf{a}\, \boldsymbol{\sigma} \cdot \mathbf{b} = \mathbf{a} \cdot \mathbf{b} + i\boldsymbol{\sigma} \cdot \mathbf{a} \times \mathbf{b}, \tag{A.2.13}$$

The spin tensor is

$$\sigma^{\mu\nu} = (i/2)[\gamma^\mu, \gamma^\nu] = i(\gamma^\mu\gamma^\nu - g^{\mu\nu}), \tag{A.2.14}$$

for which

$$[\sigma^{\mu\nu}, \gamma^\rho] = 2i(\gamma^\mu g^{\nu\rho} - \gamma^\nu g^{\mu\rho}). \tag{A.2.15}$$

In the standard representation, the spin tensor has components (Latin indices run over 1, 2, 3)

$$\sigma^{ij} = \varepsilon^{ijk}\begin{pmatrix} \sigma^k & 0 \\ 0 & \sigma^k \end{pmatrix}, \tag{A.2.16}$$

with

$$\sigma^{0j} = \begin{pmatrix} 0 & i\sigma^j \\ i\sigma^j & 0 \end{pmatrix}. \tag{A.2.17}$$

The remaining important combination is

$$\begin{aligned} \gamma^5 &\equiv i\gamma^0\gamma^1\gamma^2\gamma^3 = -i\gamma_0\gamma_1\gamma_2\gamma_3 \equiv \gamma_5 \\ &= (i/4!)\varepsilon_{\mu\nu\rho\sigma}\gamma^\mu\gamma^\nu\gamma^\rho\gamma^\sigma \\ &= (i/4!)\varepsilon^{\mu\nu\rho\sigma}\gamma_\mu\gamma_\nu\gamma_\rho\gamma_\sigma, \end{aligned} \tag{A.2.18}$$

where the Levi-Cività tensor is defined as

$$\varepsilon_{\mu\nu\rho\sigma} = \begin{cases} +1, & \text{for even permutations of 0123,} \\ -1, & \text{for odd permutations,} \\ 0, & \text{otherwise,} \end{cases} \tag{A.2.19}$$

and $\varepsilon^{\mu\nu\rho\sigma} = -\varepsilon_{\mu\nu\rho\sigma}$. Evidently

$$(\gamma_5)^2 = \mathbb{I} \tag{A.2.20}$$

and

$$\{\gamma^5, \gamma^\mu\} = 0. \tag{A.2.21}$$

It follows from the definition (A.2.18) that

$$\gamma^5\gamma^\sigma = (i/3!)\varepsilon^{\mu\nu\rho\sigma}\gamma_\mu\gamma_\nu\gamma_\rho. \tag{A.2.22}$$

In the standard representation,

$$\gamma_5 = \begin{pmatrix} 0 & I \\ I & 0 \end{pmatrix}. \tag{A.2.23}$$

The frequently encountered scalar product of a four-vector and a $\gamma$-matrix is denoted by

$$\gamma \cdot a \equiv \gamma_\mu a^\mu \equiv \mathbf{\not{a}} = \gamma^0 a^0 - \boldsymbol{\gamma} \cdot \mathbf{a}, \tag{A.2.24}$$

so that, in particular,

$$\gamma \cdot p \equiv i\gamma \cdot \partial = i\gamma_\mu \partial^\mu = i\mathbf{\not{\partial}}. \tag{A.2.25}$$

## A.3  TRACE THEOREMS AND TENSOR CONTRACTIONS

In evaluating the traces of products of $\gamma$-matrices that occur in the computation of transition matrix elements, the following theorems are useful:

$$\text{tr}(\mathbb{I}) = 4, \tag{A.3.1}$$

$$\text{tr}(AB) = \text{tr}(BA), \tag{A.3.2}$$

$$\text{tr}(\gamma_\mu) = 0, \tag{A.3.3}$$

$$\text{tr}(\text{odd number of } \gamma\text{-matrices}) = 0, \tag{A.3.4}$$

$$\text{tr}(\gamma_\mu\gamma_\nu) = 4g_{\mu\nu}, \tag{A.3.5a}$$

$$\text{tr}(\mathbf{\not{a}}\mathbf{\not{b}}) = 4a \cdot b, \tag{A.3.5b}$$

$$\text{tr}(\gamma_\mu\gamma_\nu\gamma_\rho\gamma_\sigma) = 4\left[g_{\mu\nu}g_{\rho\sigma} - g_{\mu\rho}g_{\nu\sigma} + g_{\mu\sigma}g_{\nu\rho}\right], \tag{A.3.6a}$$

$$\text{tr}(\mathbf{\not{a}}\mathbf{\not{b}}\mathbf{\not{c}}\mathbf{\not{d}}) = 4\left[(a \cdot b)(c \cdot d) - (a \cdot c)(b \cdot d) + (a \cdot d)(b \cdot c)\right], \tag{A.3.6b}$$

$$\mathrm{tr}(\gamma_5) = 0, \tag{A.3.7}$$

$$\mathrm{tr}(\gamma_5\gamma_\mu) = 0, \tag{A.3.8}$$

$$\mathrm{tr}(\gamma_5\gamma_\mu\gamma_\nu) = 0, \tag{A.3.9}$$

$$\mathrm{tr}(\gamma_5\gamma_\mu\gamma_\nu\gamma_\rho) = 0, \tag{A.3.10}$$

$$\mathrm{tr}(\gamma_5\gamma_\mu\gamma_\nu\gamma_\rho\gamma_\sigma) = 4i\varepsilon_{\mu\nu\rho\sigma}, \tag{A.3.11a}$$

$$\mathrm{tr}(\gamma_5 \not{a}\not{b}\not{c}\not{d}) = 4i\varepsilon_{\mu\nu\rho\sigma}a^\mu b^\nu c^\rho d^\sigma. \tag{A.3.11b}$$

Derivations and extensions of these results may be found in many places, including section 7.2 of Bjorken and Drell (Ref. [1]) and section 22 of the "Landau and Lifshitz" QED volume [3].

Some useful results from tensor calculus are:

$$g^{\lambda\mu}g_{\mu\nu} = \delta^\lambda_\nu = \begin{cases} 1, & \lambda = \nu, \\ 0, & \text{otherwise,} \end{cases} \tag{A.3.12}$$

$$-\varepsilon^{\alpha\lambda\mu\nu}\varepsilon_{\alpha\rho\sigma\tau} = \delta^\lambda_\rho(\delta^\mu_\sigma\delta^\nu_\tau - \delta^\mu_\tau\delta^\nu_\sigma) - \delta^\lambda_\sigma(\delta^\mu_\rho\delta^\nu_\tau - \delta^\mu_\tau\delta^\nu_\rho) + \delta^\lambda_\tau(\delta^\mu_\rho\delta^\nu_\sigma - \delta^\mu_\sigma\delta^\nu_\rho), \tag{A.3.13}$$

$$-\varepsilon^{\alpha\beta\mu\nu}\varepsilon_{\alpha\beta\sigma\tau} = 2(\delta^\mu_\sigma\delta^\nu_\tau - \delta^\mu_\tau\delta^\nu_\sigma), \tag{A.3.14}$$

$$-\varepsilon^{\alpha\beta\gamma\nu}\varepsilon_{\alpha\beta\gamma\tau} = 6\delta^\nu_\tau, \tag{A.3.15}$$

$$-\varepsilon^{\alpha\beta\gamma\delta}\varepsilon_{\alpha\beta\gamma\delta} = 24. \tag{A.3.16}$$

## A.4  DIRAC EQUATION AND DIRAC SPINORS

A free spin-$\frac{1}{2}$ particle of mass $m$ with four-momentum $p = (\sqrt{\mathbf{p}^2 + m^2}; \mathbf{p})$ and spin $s$ is described by the positive-energy spinor $u(p, s)$. The four-vector $s$ satisfies $s \cdot p = 0$, $s^2 = -1$. In the rest frame of the particle, it is the polarization vector

$$s^\mu = (0; \hat{\mathbf{s}}), \quad \hat{\mathbf{s}} \cdot \hat{\mathbf{s}} = 1. \tag{A.4.1}$$

The positive-energy spinor satisfies the Dirac equation

$$(\not{p} - m)u(p, s) = 0, \tag{A.4.2}$$

whereas the adjoint spinor

$$\bar{u}(p, s) \equiv u^\dagger(p, s)\gamma^0 \tag{A.4.3}$$

satisfies

$$\bar{u}(p, s)(\not{p} - m) = 0. \tag{A.4.4}$$

The negative-energy solutions $v(p, s)$ and $\bar{v}(p, s) \equiv v^\dagger(p, s)\gamma^0$ correspond to antiparticles. They satisfy the Dirac equations

$$(\not{p} + m)v(p, s) = 0, \tag{A.4.5}$$

$$\bar{v}(p, s)(\not{p} + m) = 0. \tag{A.4.6}$$

It is frequently a convenience to work in a helicity basis in which the spinors $u_\lambda(p)$ are eigenstates of the operator $\gamma_5\not{s}$ (defined for convenience as twice the helicity), with eigenvalues $\lambda = \pm 1$ for spin aligned parallel or antiparallel to the direction of motion. The spinors are normalized such that

$$\bar{u}_\lambda(p)u_\mu(p) = 2m\delta_{\lambda\mu}, \tag{A.4.7}$$

$$\bar{v}_\lambda(p)v_\mu(p) = -2m\delta_{\lambda\mu}, \tag{A.4.8}$$

$$\bar{u}_\lambda(p)v_\mu(p) = 0 = \bar{v}_\lambda(p)u_\mu(p). \tag{A.4.9}$$

The projection operators $\Lambda$ that occur in the evaluation of matrix elements are then

$$2m\Lambda(p) \equiv \sum_\lambda u_\lambda(p)\bar{u}_\lambda(p) = m + \not{p}, \tag{A.4.10}$$

$$2m\Lambda(-p) \equiv -\sum_\lambda v_\lambda(p)\bar{v}_\lambda(p) = m - \not{p}, \tag{A.4.11}$$

$$2m\Lambda_\lambda(p) \equiv u_\lambda(p)\bar{u}_\lambda(p) \text{ not summed} = \tfrac{1}{2}(m + \not{p})(1 + \lambda\gamma_5\not{s}), \tag{A.4.12}$$

$$2m\Lambda_\lambda(-p) \equiv -v_\lambda(p)\bar{v}_\lambda(p) \text{ not summed} = \tfrac{1}{2}(m - \not{p})(1 - \lambda\gamma_5\not{s}). \tag{A.4.13}$$

These imply the completeness relation

$$\Lambda(p) + \Lambda(-p) = \mathbb{I}. \tag{A.4.14}$$

It is sometimes essential to have an explicit form for the spinor. The positive-energy spinor with momentum $|\mathbf{p}|$ along the positive $z$-axis and helicity $\lambda/2$ is

$$u_\lambda(p) = \sqrt{E + m} \begin{pmatrix} \chi_\lambda \\ \dfrac{\lambda |\mathbf{p}|}{E + m}\chi_\lambda \end{pmatrix}, \tag{A.4.15}$$

where

$$\chi_+ = \begin{pmatrix} 1 \\ 0 \end{pmatrix}, \quad \chi_- = \begin{pmatrix} 0 \\ 1 \end{pmatrix}. \tag{A.4.16}$$

For the corresponding antiparticle spinor, it is convenient to substitute

$$v_\lambda(p) = -\lambda \gamma_5 u_{-\lambda}(p). \tag{A.4.17}$$

The spinor appropriate to a particle with momentum $\mathbf{p}$ such that $\hat{\mathbf{p}} \cdot \hat{\mathbf{z}} = \cos\theta$ is obtained by a rotation about the $y$-axis:

$$u_+(p) = \sqrt{E+m} \begin{pmatrix} \cos(\theta/2) \\ \sin(\theta/2) \\ \cos(\theta/2)\, |\mathbf{p}|\,/(E+m) \\ \sin(\theta/2)\, |\mathbf{p}|\,/(E+m) \end{pmatrix}; \tag{A.4.18}$$

$$u_-(p) = \sqrt{E+m} \begin{pmatrix} -\sin(\theta/2) \\ \cos(\theta/2) \\ \sin(\theta/2)\, |\mathbf{p}|\,/(E+m) \\ -\cos(\theta/2)\, |\mathbf{p}|\,/(E+m) \end{pmatrix}. \tag{A.4.19}$$

The operators $\frac{1}{2}(1 \pm \gamma_5)$ are spin projection operators in the limit $m \to 0$. Thus, the equations

$$\tfrac{1}{2}(1 \pm \gamma_5)u_\lambda(p) = \tfrac{1}{2}\left(1 \pm \frac{\lambda\,|\mathbf{p}|}{E+m}\right)\sqrt{E+m}\begin{pmatrix} \chi_\lambda \\ \pm\chi_\lambda \end{pmatrix} \tag{A.4.20}$$

become, for massless particles,

$$\tfrac{1}{2}(1 + \gamma_5)u_\pm(p) = \begin{cases} u_+(p), \\ 0, \end{cases} \tag{A.4.21a}$$

$$\tfrac{1}{2}(1 - \gamma_5)u_\pm(p) = \begin{cases} 0, \\ u_-(p). \end{cases} \tag{A.4.21b}$$

Hermitian conjugates of matrix elements are encountered routinely in calculations. They may be reexpressed as

$$[\bar{u}(p')\mathcal{O}u(p)]^\dagger = \bar{u}(p)\overline{\mathcal{O}}u(p'), \tag{A.4.22}$$

where $\overline{\mathcal{O}} \equiv \gamma^0\mathcal{O}^\dagger\gamma^0$ is the Dirac conjugate of the operator $\mathcal{O}$. Simple and commonly occurring examples are

$$\bar{1} = \gamma^0 1^\dagger \gamma^0 = 1, \tag{A.4.23}$$

$$\bar{\gamma}^\mu = \gamma^0 \gamma^{\mu\dagger} \gamma^0 = \gamma^\mu, \tag{A.4.24}$$

from which

$$\overline{\not q_1 \not q_2 \dots \not q_n} = \not q_n \not q_{n-1} \dots \not q_1, \tag{A.4.25}$$

$$\bar{\sigma}^{\mu\nu} = \gamma^0 \sigma^{\mu\nu\dagger} \gamma^0 = \sigma^{\mu\nu}, \tag{A.4.26}$$

$$\bar{\gamma}^5 = \gamma^0 \gamma^{5\dagger} \gamma^0 = -\gamma^5, \tag{A.4.27}$$

and

$$\overline{\gamma^\mu \gamma^5} = \gamma^0 \gamma^{5\dagger} \gamma^{\mu\dagger} \gamma^0 = \gamma^\mu \gamma^5. \tag{A.4.28}$$

The Fierz reordering transformation [4] is of value in computing amplitudes represented by a sum of Feynman graphs. In the operator basis defined by $\mathcal{O}_i = (1, \gamma_\mu, \sigma_{\mu\nu}, i\gamma_\mu \gamma_5, \gamma_5)$, a transition among four arbitrary spinors $u_i$ may be expressed as

$$\bar{u}_3 \mathcal{O}_i u_2 \bar{u}_1 \mathcal{O}_i u_4 = \sum_{j=1}^{5} \lambda_{ij} \bar{u}_1 \mathcal{O}_j u_2 \bar{u}_3 \mathcal{O}_j u_4, \tag{A.4.29}$$

where

$$\lambda_{ij} = \frac{1}{4} \begin{pmatrix} 1 & 1 & 1 & 1 & 1 \\ 4 & -2 & 0 & 2 & -4 \\ 6 & 0 & -2 & 0 & 6 \\ 4 & 2 & 0 & -2 & -4 \\ 1 & -1 & 1 & -1 & 1 \end{pmatrix}. \tag{A.4.30}$$

Because the spins are arbitrary, we may replace $u_4 \to \gamma_5 u_4$ and recover the same result for matrix elements of the form $\bar{u}_1 \mathcal{O}_i u_2 \bar{u}_3 \mathcal{O}_i \gamma_5 u_4$, and so on. Of particular utility is the result

$$\bar{u}_3 \gamma_\mu (1 - \gamma_5) u_2 \bar{u}_1 \gamma^\mu (1 - \gamma_5) u_4 = -\bar{u}_1 \gamma_\mu (1 - \gamma_5) u_2 \bar{u}_3 \gamma^\mu (1 - \gamma_5) u_4. \tag{A.4.31}$$

## A.5 COLOR ALGEBRA

Any Feynman diagram in QCD may be expressed as the product of color factors and Lorentz structure. Although the color factor can be evaluated by algebraic means, a graphical representation of the color structure may make it easier to recognize relationships and to see simplifications [5].

In color space, the quark (8.3.9) and gluon (8.3.7) propagators are simply identity matrices for the fundamental and adjoint representations, respectively:

$$\tag{A.5.1}$$

Loops with no external legs amount to traces over the color indices. For a quark loop,

$$\text{tr}(\mathbb{I}_F) = \delta_{\alpha\alpha} = N : \qquad \bigcirc \qquad = N \,, \tag{A.5.2}$$

whereas for a gluon loop,

$$\text{tr}(\mathbb{I}_A) = \delta_{aa} = N^2 - 1 : \qquad \bigcirc \qquad = \delta_{aa} = N^2 - 1 \,. \tag{A.5.3}$$

Three-point functions correspond to generators of the color algebra [recall from (8.1.16) that $T^a = \frac{1}{2}\lambda^a$].

$$T^a_{\alpha\beta} \tag{A.5.4}$$

$$if^{abc} \tag{A.5.5}$$

Accordingly, the color factor for a quark loop with one external gluon (a quark tadpole) is

$$\text{tr}(T^a) = 0 : \qquad a \qquad = 0 \,. \tag{A.5.6}$$

For a gluon tadpole, we have

$$f^{abb} = 0 : \qquad a \qquad = 0 \,. \tag{A.5.7}$$

For a quark loop with two external gluons, the color factor is [cf. (8.1.8)]

$$\text{tr}\left(T^a T^b\right) = T_R \delta_{ab} = \tfrac{1}{2}\delta_{ab} :$$

$$a \;\text{———}\; b = T_R \; a \;\text{———}\; b. \tag{A.5.8}$$

A gluon loop with two external gluons contributes with weight

$$f^{acd} f^{bcd} = C_A \delta_{ab} = N\delta_{ab} :$$

$$a \;\text{———}\; b = C_A \; a \;\text{———}\; b. \tag{A.5.9}$$

A quark self-energy loop carries a color factor [cf. (8.3.20)]

$$(T^a T^a)_{\alpha\beta} = C_F \delta_{\alpha\beta} = \frac{N^2 - 1}{2N}\delta_{\alpha\beta} :$$

$$\alpha \;\longrightarrow\; \beta = C_F \; \alpha \;\longrightarrow\; \beta. \tag{A.5.10}$$

The commutation relations [cf. (8.1.9)] for the generators in the fundamental representation have a simple graphical interpretation:

$$[T^a, T^b] = i f^{abc} T^c,$$

$$\tag{A.5.11}$$

In the graphical depiction of the commutation relations in the adjoint representation,

$$\tag{A.5.12}$$

we immediately recognize the Jacobi identity:

$$= 0. \tag{A.5.13}$$

The Fierz relation derived in problem 8.11,

$$T^a_{\alpha\beta} T^a_{\gamma\delta} = \tfrac{1}{2}\delta_{\alpha\delta}\delta_{\beta\gamma} - \frac{1}{2N}\delta_{\alpha\beta}\delta_{\gamma\delta}, \tag{A.5.14}$$

may be expressed graphically as

which makes manifest the replacement of a gluon as a quark-antiquark pair, as in the $1/N$ construction of section 8.10:

$$\tag{A.5.15}$$

Evidently the quark–quark–gluon vertex may now be expressed in double-line notation as

$$: T^a. \tag{A.5.16}$$

The diagram weighted by $1/N$, for which the quark and antiquark in the hairpin must carry the same color, ensures the tracelessness of the generator $T^a$. The overall factor $1/\sqrt{2}$ reflects the normalization of the generators characterized by $T_R$.

In similar fashion, the three-gluon vertex is represented by

$$: f^{abc}. \tag{A.5.17}$$

As in the $1/N$ expansion, a closed loop of color, which stands for the trace of the unit matrix in the fundamental representation, yields a factor of $N$ [6].

The color factors for vertex corrections at one loop are readily evaluated using graphical methods. For the first diagram in figure 8.19(b), we may either apply (A.5.14) and (A.5.6) or the double-line calculus of (A.5.16) to find

$$(T^b T^a T^b) = -\frac{1}{2N} T^a = (C_F - \tfrac{1}{2} C_A) T^a$$

$$(A.5.18)$$

To evaluate the color factor for the second diagram in figure 8.19(b), we employ the commutation relations expressed in (A.5.11) to replace the mixed quark–gluon–gluon loop with the gluon bubble (A.5.9). In this way we find

$$i f^{abc}(T^b T^c) = \frac{C_A}{2} T^a = \frac{N}{2} T^a$$

$$(A.5.19)$$

## A.6 WEYL–VAN DER WAERDEN SPINORS

Four-component Dirac spinors combine distinct irreducible representations of the Lorentz group. This traditional approach is graceful for quantum electrodynamics and quantum chromodynamics, parity-conserving theories for which the gauge symmetries $U(1)_{EM}$ and $SU(3)_c$ do not distinguish left-handed and right-handed fermions. The electroweak theory, based on $SU(2)_L \otimes U(1)_Y$ gauge symmetry, does distinguish left-handed and right-handed quarks and leptons. In unified theories of the strong, weak, and electromagnetic interactions such as the $SU(5)$ unified theory discussed in chapter 9, left-handed and right-handed components of a fermion species may lie in different representations of the unifying gauge group. The chiral nature of quark and lepton quantum numbers and interactions indicates that the fundamental fermionic degrees of freedom are two-component spinors [7]. Two-component spinors also enter naturally in the treatment of Majorana [8] fermions and in supersymmetric field theories [9].

A comprehensive textbook treatment of two-component fermions and their quantization appears in Ref. [10]. For an extensive presentation of modern applications of the two-component formalism, including Feynman rules, many example calculations, and the translation to the four-component formalism, see Ref. [11]. Majorana neutrinos are formulated in terms of two-component spinors in Ref. [12].

## ⅢⅢⅢⅢⅢ REFERENCES ⅢⅢⅢⅢⅢⅢⅢ

1. J. D. Bjorken and S. D. Drell, *Relativistic Quantum Mechanics,* McGraw-Hill, New York, 1964; *Relativistic Quantum Fields,* McGraw-Hill, New York, 1965.
2. A helpful translation dictionary appears in appendix F of M. Veltman, *Diagrammatica: The Path to Feynman Diagrams,* Cambridge University Press, Cambridge, 1994.
3. E. M. Lifshitz, V. B. Berestetskii, and L. P. Pitaevski, *Quantum Electrodynamics,* 2nd ed., Butterworth-Heinemann, Oxford, 1982.
4. M. Fierz, *Z. Phys.* **88**, 161 (1934); see also Ref. [3], §28.
5. For an introduction to graphical methods for dealing with the color algebra, see Yu. L. Dokshitzer, "Perturbative QCD (and Beyond)," in *Lectures on QCD: Applications,* ed. F. Lenz, H. Grießhammer, and D. Stoll, *Lecture Notes in Physics* **460**, 87 (1997), Springer, Berlin and Heidelberg, http://j.mp/lOR8E5. See P. Cvitanovic, *Phys. Rev.* **D14**, 1536 (1976) and T. van Ritbergen, A. N. Schellekens, and J. A. M. Vermaseren, *Int. J. Mod. Phys. A* **14**, 41 (1999), arXiv:hep-ph/9802376, for general treatments of group theory factors for Feynman diagrams.
6. The color Feynman rules in double-line notation are presented, along with a few examples, in P. Nason, "Introduction to Perturbative QCD," http://j.mp/lOmkq6 (2001); and in M. L. Mangano, "Introduction to QCD in Hadronic Collisions," http://j.mp/if3Bqc (2005).
7. H. Weyl, *The Theory of Groups and Quantum Mechanics*, Dover, New York, 1931, 1950 / Kessinger Publishing, Whitefish, MT, 2007; B. L. van der Waerden, *Group Theory and Quantum Mechanics,* Springer-Verlag, Berlin, 1974, 1986.
8. E. Majorana, *Nuovo Cim.* **14**, 171 (1937).
9. For an excellent pedagogical introduction, see S. P. Martin, "A Supersymmetry Primer," version 6, arXiv:hep-ph/9709356 (September 2011).
10. M. Srednicki, *Quantum Field Theory,* Cambridge University Press, Cambridge, 2007; see also Ref. [3], especially part III.
11. H. K. Dreiner, H. E. Haber, and S. P. Martin, *Phys. Rept.* **494,** 1 (2010).
12. K. M. Case, *Phys. Rev.* **107**, 307 (1957); R. N. Mohapatra and P. B. Pal, *Massive Neutrinos in Physics and Astrophysics,* 3rd ed., World Scientific, River Edge, NJ, 2004, chap. 4.

# Appendix B ||||||||||||||||||||||||||||||||||||||||||||||||||||||||||||||||||||||||||||||||||||||||||||

## Observables and Feynman Rules

In this volume, Feynman rules will be given for the evaluation of the invariant amplitude $\mathcal{M}$, the specific form of which depends on the dynamics of a process. The connection between invariant amplitude and observables is then entirely kinematic. This appendix summarizes the rules for writing down the invariant amplitude, some methods for evaluating integrals over undetermined loop momenta, and the computation of the simplest measurable quantities.

## B.1 PHASE-SPACE FORMULAS: DECAY RATES AND CROSS SECTIONS

It is straightforward to derive the following useful relations between amplitudes and observables.

For a general decay process $\alpha \to (1, 2, \ldots, n) \equiv \beta$, the differential decay rate is

$$
d\Gamma_{\beta\alpha} = \frac{(2\pi)^4 \delta^{(4)}(p_\beta - p_\alpha) \left| \mathcal{M}_{\beta\alpha} \right|^2}{2 p_\alpha^0} \prod_{i=1}^{n} \frac{\tilde{d} p_i}{(2\pi)^3} \cdot S , \tag{B.1.1}
$$

where

$$
\tilde{d} p_i = \frac{d^3 \mathbf{p}_i}{2\sqrt{\mathbf{p}_i^2 + m_i^2}} \tag{B.1.2}
$$

is the Lorentz-invariant phase-space volume element and the statistical weight $S$ contains a factor $1/n!$ if there are $n$ identical particles in the final state:

$$
S = \prod_{k=\text{species}} \frac{1}{n_k!}. \tag{B.1.3}
$$

For the important special case $\alpha \to (1, 2) \equiv \beta$ of the two-body decay of a particle at rest, the differential decay rate is

$$
\frac{d\Gamma_{\beta\alpha}}{d\Omega_{\text{cm}}} = \frac{\left| \mathcal{M}_{\beta\alpha} \right|^2}{64\pi^2} \frac{S_{12}}{m_\alpha^2} \cdot S. \tag{B.1.4}
$$

Here the quantity

$$S_{12} = \left[ m_\alpha^2 - (m_1 + m_2)^2 \right]^{1/2} \left[ m_\alpha^2 - (m_1 - m_2)^2 \right]^{1/2} \tag{B.1.5}$$

is equal to $2m_\alpha$ times the three-momentum of the products.

The cross section for a general two-body collision $\alpha \equiv (1, 2) \to (3, 4, \ldots, n) \equiv \beta$ is written as

$$d\sigma_{\beta\alpha} = \frac{(2\pi)^4 \delta^{(4)}(p_\beta - p_\alpha) |\mathcal{M}_{\beta\alpha}|^2}{2S_{12}} \prod_{i=3}^{n} \frac{\tilde{d}p_i}{(2\pi)^3} \cdot S. \tag{B.1.6}$$

Again it is the case of a two-body final state $\alpha \equiv (1, 2) \to (3, 4) \equiv \beta$ that is of particular interest. The differential cross section is

$$\frac{d\sigma_{\beta\alpha}}{d\Omega_{\rm cm}} = \frac{|\mathcal{M}_{\beta\alpha}|^2}{64\pi^2 s} \frac{S_{34}}{S_{12}} \cdot S, \tag{B.1.7}$$

or, in terms of the invariant momentum transfer $t = (p_1 - p_3)^2 = (p_2 - p_4)^2$,

$$\frac{d\sigma_{\beta\alpha}}{dt} = \frac{|\mathcal{M}_{\beta\alpha}|^2}{16\pi S_{12}^2} \cdot S. \tag{B.1.8}$$

In all the preceding expressions, an optional summation over final spins and average over initial spins is implicit.

## B.2 FEYNMAN RULES: GENERALITIES

The invariant amplitude $-i\mathcal{M}$ that corresponds to a Feynman diagram is given as a product of factors associated with external lines (particles in the initial or final state), internal lines, and interactions. For lowest-order perturbation calculations, the essential elements are these:

### EXTERNAL LINES:

1. For each external spinless boson, a factor 1.
2. For absorption or emission of a spin-1 vector boson, a factor $\epsilon_\mu(k, \lambda)$ or $\epsilon_\mu(k, \lambda)^*$, where $\epsilon$ is the polarization four-vector for a boson with momentum $k$ and helicity $\lambda$. Only the transverse polarization states of massless vector bosons are populated. In the case of propagation along the $z$-direction, the polarization vectors are, thus, $\epsilon_\mu(k, \pm 1) = (0; 1, \pm i, 0)/\sqrt{2}$. A massive vector at rest is also permitted the longitudinal state of polarization specified by $\epsilon_\mu(k, 0) = (0; 0, 0, -1)$. All these polarization vectors satisfy $\epsilon_\mu(k, \lambda)\epsilon^\mu(k, \lambda')^* = -\delta_{\lambda\lambda'}$. For the massless case, $k \cdot \epsilon = 0$ as well.
3. For a spin-$\frac{1}{2}$ fermion with momentum $p$ and helicity $\lambda/2$ in the initial state, a factor $u_\lambda(p)$ on the right; in the final state, a factor $\bar{u}_\lambda(p)$ on the left.

4. For a spin-$\frac{1}{2}$ antifermion with momentum $p$ and helicity $\lambda/2$ in the initial state, a factor $\bar{v}_\lambda(p)$ on the left; in the final state, a factor $v_\lambda(p)$ on the right.

## INTERNAL LINES:

A propagator is associated with each internal line representing a particle with four-momentum $q$ and mass $m$. It is given by $i \times$ the inverse of the operator that multiplies the field in the free-particle equation of motion, expressed in momentum space.

1. For each internal spin-zero boson, a factor

$$\frac{i}{q^2 - m^2 + i\varepsilon}.$$

2. For each internal photon, a factor (in Feynman gauge)

$$\frac{-ig_{\mu\nu}}{q^2 + i\varepsilon}.$$

3. For each internal massive vector boson with mass $M_V$, a factor (in unitary gauge)

$$\frac{-i(g_{\mu\nu} - q_\mu q_\nu / M_V^2)}{q^2 - M_V^2 + i\varepsilon}.$$

4. For each internal fermion line, a factor

$$\frac{i(\not{q} + m)}{q^2 - m^2 + i\varepsilon}.$$

An internal antifermion is regarded as a fermion of negative four-momentum $(-q)$.

## LOOPS AND COMBINATORICS:

1. For each momentum $k$ not fully determined by the requirement of momentum conservation at the vertices, a factor

$$\int \frac{d^4k}{(2\pi)^4},$$

where the integral runs over all values of the loop momentum. Techniques for dealing with the resulting integrals are reviewed in section B.3.

2. For each closed loop of fermions, a factor $-1$.

3. A factor $1/n!$ for each closed loop containing $n$ identical boson lines.

VERTEX FACTORS:

Each vertex represents a factor given by $i\times$ the corresponding interaction term in the momentum-space Lagrangian, with all fields amputated. Four-momentum conservation is implied.

## B.3 FEYNMAN INTEGRALS

In general, the integral over an undetermined loop momentum $k$ in a Feynman diagram takes the form

$$\mathcal{I} = (2\pi)^{-4} \int \frac{d^4k\, F(k; p_i; m_j)}{a_1 a_2 \cdots a_n}, \tag{B.3.1}$$

where

$$a_k = (k - s_k)^2 - m_k^2 + i\varepsilon, \tag{B.3.2}$$

$s_k$ is a linear combination of external momenta $p_i$, $m_j$ are (internal and external) masses in the problem, and the function $F$ is a polynomial in the components of $k$. Convenient general methods for evaluating such integrals were devised by R. P. Feynman [1]. Careful elaborations appear in many textbooks (see in particular Jauch and Rohrlich [2]). The form of the $k$-space integral may be simplified by the introduction of auxiliary parameters. This may be done in two ways, either of which may be advantageous in particular applications.

The first method relies upon the identity

$$\xi_n \equiv \frac{1}{a_1 a_2 \cdots a_n} = (n-1)! \int_0^1 dz_1 \int_0^1 dz_2 \cdots \int_0^1 dz_n \frac{\delta(\sum_{i=1}^n z_i - 1)}{[\sum_{i=1}^n a_i z_i]^n} \tag{B.3.3}$$

or the related forms

$$\xi_n = (n-1)! \int_0^1 dz_1 \int_0^{z_1} dz_2 \cdots \int_0^{z_{n-2}} dz_{n-1} \tag{B.3.4}$$

$$\times [a_1 + z_1(a_2 - a_1) + \cdots + z_{n-1}(a_n - a_{n-1})]^{-n},$$

and

$$\xi_n = (n-1)! \int_0^1 dz_1^{n-2} \int_0^1 dz_2^{n-3} \cdots \int_0^1 dz_{n-1} \tag{B.3.5}$$

$$\times [z_1 z_2 \cdots z_{n-1}(a_1 - a_2) + z_1 z_2 \cdots z_{n-2}(a_2 - a_3) + \cdots + a_n]^{-n}.$$

Some other useful relationships of the same kind are

$$\frac{1}{n A^n B} = \int_0^1 \frac{dx\, x^{n-1}}{[Ax + B(1 - x)]^{n+1}}, \tag{B.3.6}$$

$$\frac{1}{A^n} - \frac{1}{B^n} = -\int_0^1 \frac{dx\, n(A-B)}{[(A-B)x+B]^{n+1}}. \tag{B.3.7}$$

An integral characterized by (B.3.1) can always be brought to the schematic form

$$\mathcal{I} = \int \underset{\substack{\text{auxiliary} \\ \text{parameters}}}{\cdots \cdots} \int (2\pi)^{-4} \int \frac{d^4k\, F(k; p_i; m_j)}{[(k-R)^2 - a^2]^n}, \tag{B.3.8}$$

by such devices, the precise form depending on the identities employed. If the $k$-space integral is at worst logarithmically divergent, a change of variable

$$k' = k - R \tag{B.3.9}$$

will not affect the value of the integral (nor add any finite number to the logarithmic divergence). Hence the $k$-space integral can be rewritten as

$$(2\pi)^{-4} \int \frac{d^4k\, F(k+R; p_i; m_j)}{[k^2 - a^2]^n}. \tag{B.3.10}$$

Because of the symmetry of the range of integration, the odd powers of $k_\mu$ in $F$ do not contribute. Symmetric integration requires an average over the directions of $k_\mu$, which amounts to the substitutions

$$k_\mu k_\nu \rightarrow \tfrac{1}{4} g_{\mu\nu} k^2 \,; \tag{B.3.11a}$$

$$k_\mu k_\nu k_\rho k_\sigma \rightarrow \frac{1}{4!}(g_{\mu\nu}g_{\rho\sigma} + g_{\mu\rho}g_{\nu\sigma} + g_{\mu\sigma}g_{\nu\rho})(k^2)^2 \,, \tag{B.3.11b}$$

and so on. Therefore, the general form to be evaluated is

$$\mathcal{I}_{mn} = (2\pi)^{-4} \int \frac{d^4k\, (k^2)^{m-2}}{(k^2 - a^2)^n} = \frac{B(m, n-m)}{16\pi^2 i (a^2)^{n-m}}, \tag{B.3.12}$$

where the Euler beta function, $B(x, y) = \Gamma(x)\Gamma(y)/\Gamma(x+y)$, is finite provided $n > m > 0$. It remains to perform the integrals over auxiliary parameters. In many calculations this last step accounts for the bulk of the labor, and an extensive calculus has been developed.

An alternative procedure for combining denominators is based upon the representation

$$\frac{i}{q^2 - m^2 + i\varepsilon} = \int_0^\infty d\alpha \, \exp[i\alpha(q^2 - m^2 + i\varepsilon)], \tag{B.3.13}$$

which leads to

$$\xi_n = \frac{1}{a_1 a_2 \cdots a_n} = \prod_{j=1}^{n} \left[ -i \int_0^\infty d\alpha_j \, \exp(i\alpha_j a_j) \right] \tag{B.3.14}$$

$$= (-i)^n \int_0^\infty d\alpha_1 \cdots \int_0^\infty d\alpha_n \, \exp\left( i \sum_{j=1}^{n} \alpha_j a_j \right).$$

By completing the square in the exponential, we may always bring the integral over loop momentum to the form

$$(2\pi)^{-4} \int d^4k \, F(k + R; p_i; m_j) \exp(i\lambda k^2), \tag{B.3.15}$$

which may be evaluated using standard Gaussian integral techniques. (Simple illustrations of these manipulations are given in §8.2.1 and problem 8.7.) Note in particular that

$$(2\pi)^{-4} \int d^4k \, \exp(i\lambda k^2) = \frac{1}{16\pi^2 i\lambda^2} \, . \tag{B.3.16}$$

Differentiation with respect to $\lambda$ then yields the general result

$$(2\pi)^{-4} \int d^4k \, (k^2)^m \exp(i\lambda k^2) = \frac{(m+1)!}{16\pi^2 i\lambda^2 (-i\lambda)^m} \, . \tag{B.3.17}$$

## B.4 REGULARIZATION PROCEDURES

It is frequently the case that the Feynman graphs to be calculated suffer from severe ultraviolet divergences. Several procedures have been developed to regularize the integrals without losing the gauge invariance of the theory. The simplest such method is the Pauli–Villars procedure [3] described in section 8.2 in the calculation of the vacuum polarization in electrodynamics. We add unphysical fields of adjustable mass to the Lagrangian in a gauge-invariant manner. After gauge-invariant renormalization, the unphysical masses are taken to be infinite, and the renormalized quantities are shown to be finite.

This device is ill suited to deal with non-Abelian theories. An alternative scheme, known as dimensional regularization, was applied to this problem by 't Hooft and Veltman [4]. Stated simply, dimensional regularization consists in defining the loop integral in exponential form in $(d - 1) > 3$ space dimensions and 1 time dimension, while restricting the external momenta and polarization vectors to the physical $(3 + 1)$-dimensional space. After performing the $(d - 4)$-dimensional integration in the space orthogonal to the physical space, we analytically continue the result in $d$. For sufficiently small values of $d$ or for complex values of $d$, the subsequent 4-dimensional integrals are convergent. Divergences may then be isolated systematically as singularities of the form $(d-4)^{-1}$, and so on. The following results are sufficient to deal with integrals encountered in this book:

$$(2\pi)^{-d} \int d^d k \exp(i\lambda k^2) = \frac{-\exp(-i\pi d/4)}{(4\pi\lambda)^{d/2}i}, \tag{B.4.1}$$

$$(2\pi)^{-d} \int d^d k (k^2)^m \exp(i\lambda k^2) = \frac{-\exp(-i\pi d/4)\Gamma(m+d/2)}{(4\pi\lambda)^{d/2}i(-i\lambda)^m\Gamma(d/2)}, \tag{B.4.2}$$

where symmetric integration yields

$$k_\mu k_\nu \to \frac{k^2}{d}g_{\mu\nu}, \dots. \tag{B.4.3}$$

## B.5 FEYNMAN RULES: ELECTRODYNAMICS

The Feynman rules for spinor and scalar electrodynamics are needed for several illustrative examples and problems. Complete rules and their derivations are given in many places; only the elementary vertices deduced in section 3.6 are required here.

For spinor electrodynamics of a fermion with charge $q$, the only interaction is the emission or absorption of a photon,

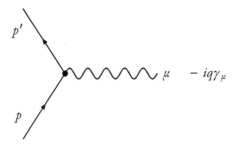

In the case of a spinless boson with charge $q$, the one- and two-photon vertices are

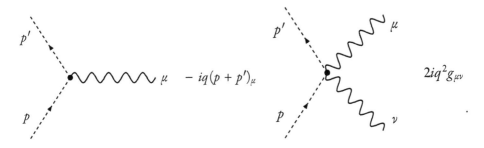

The needed Feynman rules for other theories are given in the text where the theories are introduced.

## |||||||||||||||||| FOR FURTHER READING ||||||||||||||||||

**Regularization and renormalization.** Most of the material in this appendix is standard fare for textbooks in relativistic quantum mechanics and quantum field theory. One exception is the method of dimensional regularization, here introduced only briefly. A further discussion is to be found in

G. 't Hooft and M. Veltman, "Diagrammar," CERN Yellow Report 73-9; http://j.mp/xt3NiV,

M. Veltman, *Diagrammatica: The Path to Feynman Diagrams*, Cambridge University Press, Cambridge, 1994, §6.2 and especially appendix C.

The problems of regularization and renormalization of non-Abelian gauge theories are treated at some length in the books by

T. P. Cheng and L.-F. Li, *Gauge Theory of Elementary Particle Physics*, Oxford University Press, Oxford, 1984; *Gauge Theory of Elementary Particle Physics: Problems and Solutions*, Oxford University Press, Oxford, 2000,

J. C. Collins, *Renormalization: An Introduction to Renormalization, the Renormalization Group and the Operator-Product Expansion*, Cambridge University Press, Cambridge, 1985,

C. Itzykson and J.-B. Zuber, *Quantum Field Theory*, Dover, New York, 2006,

M. E. Peskin and D. V. Schroeder, *An Introduction to Quantum Field Theory*, Westview Press, Boulder, CO, 1995,

S. Pokorski, *Gauge Field Theories*, 2nd ed., Cambridge University Press, Cambridge, 2000.

P. Ramond, *Field Theory: A Modern Primer*, 2nd ed., Westview Press, Boulder, CO, 1990.

A. Zee, *Quantum Field Theory in a Nutshell*, 2nd ed., Princeton University Press, Princeton, NJ, 2010.

**Evaluation of Feynman integrals.** For a thorough grounding in modern methods, consult

S. Weinzierl, "Introduction to Feynman Integrals," arXiv:1005.1855,

R. K. Ellis, Z. Kunszt, K. Melnikov and G. Zanderighi, *Phys. Rep.* **518**, 141 (2012).

A compendium of one-loop integrals may be found in

R. K. Ellis and G. Zanderighi, *JHEP* **0802**, 002 (2008) and http://qcdloop.fnal.gov.

**Computer-enabled techniques.** Prodigious effort has gone into the development of automated techniques for generating Feynman rules from Lagrangians, generating Feynman diagrams, and evaluating amplitudes. A highly versatile symbolic manipulation system, well adapted to Dirac algebra, is FORM:

J.A.M. Vermaseren, "New features of FORM,"arXiv:math-ph/0010025. The program and documentation are available at http://www.nikhef.nl/~form/.

The program LanHEP and the *Mathematica*® package FeynRules automatically generate Feynman rules from the Lagrangian.

A. V. Semenov, *Comput. Phys. Commun.* **180**, 431 (2009) [arXiv:0805.0555]. For the LanHEP package, see http://j.mp/z99yvV,

N.D. Christensen and C. Duhr, "FeynRules—Feynman Rules Made Easy," arXiv:0806.4194. The package is available at http://feynrules.phys.ucl.ac.be.

MadGraph automatically generates tree-level diagrams (and code to evaluate helicity amplitudes) from a Lagrangian. With its companion MadEvent, it makes it possible to generate amplitudes and events for processes arising in any model.

J. Alwall, M. Herquet, F. Maltoni, O. Mattelaer, and T. Stelzer, "MadGraph 5 : Going Beyond," *JHEP* **1106**, 128 (2011) [arXiv:1106.0522]. Programs and documentation may be found at http://madgraph.hep.uiuc.edu.

CompHEP and CalcHEP are packages for evaluation of Feynman diagrams, integration over multi-particle phase space and event generation.

E. Boos et al. [CompHEP Collaboration],"CompHEP 4.4: Automatic Computations from Lagrangians to Events,"*Nucl. Instrum. Meth. A* **534**, 250 (2004), arXiv:hep-ph/0403113; A. Pukhov et al., "CompHEP: A package for evaluation of Feynman Diagrams and Integration over Multi-Particle Phase Space. User's Manual for Version 3.3," arXiv:hep-ph/9908288. CompHEP programs and documentation are available at http://comphep.sinp.msu.ru. For CalcHEP and its documentation, see http://j.mp/xSPXAb.

The package GRACE is a system for generating Feynman diagrams through one loop and creating code for their evaluation.

G. Bélanger et al., *Phys. Rept.* **430**, 117 (2006), Documentation and programs are available at http://j.mp/wrxnw2.

FeynArts is a *Mathematica*® package for the generation and visualization of Feynman diagrams and amplitudes, which can be evaluated up to one loop by the companion package, FormCalc.

T. Hahn, *Comput. Phys. Commun.* **140**, 418 (2001), arXiv:hep-ph/0012260. Packages and documentation are available at http://www.feynarts.de.

FeynCalc is a *Mathematica*® package for algebraic calculations in elementary particle physics:

R. Mertig, M. Bohm, and A. Denner, *Comput. Phys. Commun.* **64**, 345 (1991). For information and programs, see http://www.feyncalc.org.

**Helicity amplitude formalism.** The calculation of transition amplitudes and cross sections or decay rates at lowest order in perturbation theory is straightforward for processes involving four or fewer external particles, *e.g.*, $1 \to 2$ or $1 \to 3$ decays, or $2 \to 2$ scattering. The computational complexity escalates as the number of external legs grows and the number of Feynman diagrams increases. The traditional textbook approach of squaring amplitudes and carrying out the sum over final-state spins rapidly becomes inefficient, because of the large number of terms at intermediate stages of the calculation. Amplitudes involving non-Abelian interactions bring their own combinatorial complications. Evaluating amplitudes for specific polarization (helicity) configurations leads to significant economies, in part because discrete symmetries constrain or relate different helicity amplitudes. For an introduction to the helicity amplitude method and other advances, see

L. J. Dixon, "Calculating scattering amplitudes efficiently," in *QCD & Beyond, Proceedings of TASI '95*, ed. D. E. Soper, World Scientific, Singapore and River Edge, N.J., 1996, p. 539, arXiv:hep-ph/9601359.

The classic papers on the definition and properties of helicity amplitudes are

M. Jacob and G. C. Wick, *Annals Phys.* **7**, 404 (1959),
T. L. Trueman and G. C. Wick, *Annals Phys.* **26**, 322 (1964).

**Two-component spinor formalism.** It is often advantageous to carry out calculations in terms of two-component (Weyl–van der Waerden) fermions, rather than four-component (Dirac) fermions. A clear presentation of two-component methods for the computation of QED helicity amplitudes is given in

S. Dittmaier, *Phys. Rev. D* **59**, 016007 (1999).

For an authoritative treatment of two-component spinor techniques and Feynman rules for the $SU(3)_c \otimes SU(2)_L \otimes U(1)_Y$ standard model and its supersymmetric extensions, see

H. K. Dreiner, H. E. Haber, and S. P. Martin, *Phys. Rept.* **494,** 1 (2010).

## |||||||||||||| REFERENCES ||||||||||||||||||

1. R. P. Feynman, *Phys. Rev.* **76,** 769 (1949), appendix.
2. J. M. Jauch and F. Rohrlich, *The Theory of Photons and Electrons,* 2nd ed., Springer-Verlag, New York, 1976. For applications to gauge theories, see G. 't Hooft and M.J.G. Veltman, *Nucl. Phys. B* **153,** 365 (1979).
3. W. Pauli and F. Villars, *Rev. Mod. Phys.* **21,** 434 (1949); see also S. N. Gupta, *Proc. Phys. Soc. (London)* **66A,** 129 (1953).
4. G. 't Hooft and M. Veltman, *Nucl. Phys.* B44, 189 (1972).

# Appendix C

## Physical Constants

| | |
|---|---|
| Speed of light *in vacuo* | $c = 2.997\,924\,58 \times 10^{10}$ cm/s |
| Planck's constant, reduced | $\hbar = 6.582\,119\,28(15) \times 10^{-22}$ MeV·s |
| | $\hbar c = 1.973\,269\,718(44) \times 10^{-11}$ MeV·cm |
| | $\quad\quad = 197.326\,9718(44)$ MeV·fm |
| | $(\hbar c)^2 = 0.389\,379\,338(17)$ GeV$^2$·mb |
| Fine structure constant | $\alpha = e^2/4\pi\hbar c = 7.297\,352\,5698(24) \times 10^{-3}$ |
| | $\quad\quad\quad\quad = 1/137.035\,999\,074(44)$ |
| Fermi constant | $G_F = 1.166\,3787(6) \times 10^{-5}$ GeV$^{-2}$ |
| | $G_F^2 \approx 5.297 \times 10^{-38}$ cm$^2$/GeV$^2$ |
| Gravitational constant | $G_N = 6.673\,84(80) \times 10^{-8}$ cm$^3$/g·s$^2$ |
| | $\quad\quad = 6.708\,37(80) \times 10^{-39}\,\hbar c\,(\text{GeV}/c^2)^{-2}$ |
| | $\quad\quad \approx 1.324 \times 10^{-55}$ MeV·cm/(MeV/$c^2$)$^2$ |
| Planck mass | $M_{Pl} = \sqrt{\hbar c / G_N} = 1.220\,93(7) \times 10^{19}$ GeV/$c^2$ |
| | $\quad\quad\quad\quad = 2.176\,51(13) \times 10^{-8}$ kg |
| 1 (tropical) year (yr) | $3.155\,692\,52 \times 10^7$ s |
| 1 fermi (fm) | $10^{-13}$ cm |
| 1 millibarn (mb) | $10^{-27}$ cm$^2 = 0.1$ fm$^2$ |

## ‖‖‖‖‖‖‖‖ FOR FURTHER READING ‖‖‖‖‖‖‖‖

**The physics of fundamental constants.** For a pedagogical introduction and extensive references, see the Resource Letter by
  P. J. Mohr and D. B. Newell, *Am. J. Phys*, **78**, 338 (2010).

**Constants and conversion factors.** The values tabulated here are drawn from
  J. Beringer et al. (Particle Data Group), *Phys. Rev. D* **86**, 010001 (2012),
    http://pdg.lbl.gov;
  2010 CODATA adjustment, http://j.mp/U5HBeM.

For a discussion of the 2010 adjustment, see

P. J. Mohr, B. N. Taylor, and D. B. Newell, *Rev. Mod. Phys.* **84**, 1527 (2012).

A brief overview of the 2010 adjustment may be found at http://j.mp/u2RYQ0.

**Système International d'Unités.** The modern metric system, or International System of Units, is known by its international abbreviation, SI. Extensive documentation is available at http://physics.nist.gov/cuu/Units.

# Author Index ||||||||||||||||||||||||||||||||||||||||||||||||||||||||||||||||||||||||||||||||||||||||||||

Numbers set in *italics* denote pages on which full citations appear. Numbers in parentheses denote reference numbers.

# Subject Index

Bold-face page numbers refer to principal treatments of a topic. Numbers set in *italics* denote pages on which supplementary readings appear.